ENVIRONMENTAL TAXATION LAW

To Angela, Caroline and Amelia

Environmental Taxation Law
Policy, Contexts and Practice

JOHN SNAPE
School of Law, University of Warwick, UK

and

JEREMY DE SOUZA
Consultant to White and Bowker, UK

Routledge
Taylor & Francis Group

LONDON AND NEW YORK

First published 2006 by Ashgate Publishing

2 Park Square, Milton Park, Abingdon, Oxfordshire OX14 4RN

52 Vanderbilt Avenue, New York, NY 10017

Routledge is an imprint of the Taylor & Francis Group, an informa business

First issued in paperback 2020

British Library Cataloguing in Publication Data
Snape, John
    Environmental taxation law : policy, contexts and practice
    1.Environmental impact changes - Law and legislation -
    Great Britain
    I.Title II.De Souza, Jeremy
    344.4'1046

Library of Congress Cataloging-in-Publication Data
Snape, John.
    Environmental taxation law : policy, contexts and practice / John Snape and
Jeremy de Souza.
        p. cm.
    Includes bibliographical references and index.
    ISBN 0-7546-2304-1
    1. Environmental impact charges--Law and legislation--Great Britain. 2.
Environmental impact charges--Law and legislation. I. De Souza, Jeremy. II.
Title.

    KD3382.E58S63 2005
    344.4104'6--dc22

                                                                2005048265

    ISBN 978-0-7546-2304-5 (hbk)
    ISBN 978-0-367-60419-6 (pbk)

Contents

About the Authors

Jeremy de Souza was educated at Charterhouse and New College, Oxford. Articled to Mr John Emrys Lloyd of Farrer & Co, he subsequently became an assistant solicitor and associate at that firm. Between 1976 and 1996, after which he became a consultant (enabling him to devote more time to writing), he was the firm's tax partner. During his 32 years with the firm, he contributed to the newsletters for the Agricultural Estates Group, the Employment and Pensions Group, the Charities Group and the *Briefing Service* for in-house lawyers.

In 1999, he became a part-time consultant to White and Bowker in Winchester, and has contributed to that firm's newsletters: *The Law of the Land, W&B Charity Law, Private Client Newsletter* and *Ins and Outs*.

He was a member of The Law Society Revenue Committee's Corporation Tax Sub-Committee between 1989 and 1992. Since 1995, he has been the Chairman of the Holborn and, from 2001, City of Westminster and Holborn, Law Society's Revenue Committee, writing a monthly article in the monthly journals, *Holborn Report* and *The Report*.

In 1969, he was a co-author of *Reform of Taxation and Investment Incentives*. He was a contributor to Volume 35 of *The Encyclopedia of Forms and Precedents (Sale of Land)* in 1989. He was co-author of *The Property Investor and VAT* in 1990 and of *The Conveyancer's Tax Primer* in 1999. Since 1991, he has been the General Editor of the Sweet and Maxwell looseleaf, *Land Taxation*. Since 2001, he has been a contributor to *The Lawyer's Factbook*.

He has contributed articles to *The Law Society's Gazette, STEP Journal, Private Client Business, The British Tax Review, Taxation, The Tax Journal, The Estates Gazette, Property Law Journal, Rural Practice, The Environmental Law Review, Trust Law & Practice, Trusts & Estates Tax Journal, Hampshire Chronicle, Negotiator, Butler & Co Farming News, Christie's Bulletin* and *The Investor's Chronicle*. He has also lectured on tax topics.

In addition to being a solicitor, he belongs to the Society of Trust and Estate Practitioners, the Stamp Taxes Practitioners Group and the Royal Institute of International Affairs.

John Snape was educated at St Mary's College, Blackburn, and St Edmund Hall, Oxford. He qualified as a solicitor in 1989 and subsequently worked in corporate and commercial legal practice. In 1993, he joined Nottingham Law School, the Nottingham Trent University, moving to the University of Leeds in 2002. In 2005, he took up a lectureship at the School of Law, the University of Warwick. He has written on a range of legal topics, including property law and tax law.

Authors' Preface

During the last seven years or so, the government of the United Kingdom has embarked on a range of far-reaching social and economic reforms. Among these, none is more striking, especially to those who have followed its development, than the enterprise of environmental taxation.

The use of a tax, for long regarded as a means of raising government revenue, redistributing wealth or of managing the economy, in order to discourage certain forms of undesirable industrial behaviour, had no systematic antecedent in UK law prior to 1997. Of course, there had for centuries been duties on cigarettes, petrol and alcohol, but no tax, not even landfill tax, introduced in the dying days of Prime Minister John Major's government, had been researched and designed specifically to steer the behaviour of those liable to pay it. True, landfill tax had attempted to put a price on the environmental costs of landfilling waste but, except to a slight extent, it had not been intended, when originally designed and implemented, to steer behaviour. By contrast, the two main taxes introduced after 1997, climate change levy, which is a tax on the industrial, commercial and agricultural consumption of non-environmentally-friendly forms of energy, and aggregates levy, a tax on the extraction of minerals, have been specially designed, via their structures and rates, to encourage the use of alternative resources. This is not the only way in which, if the reader will forgive the pun, these environmental taxes have been groundbreaking. The other way in which they have marked a departure, which the authors regard as highly characteristic, is their use in combination with other economic instruments, such as trading schemes, green certificates, road and cordon pricing and special reliefs within non-environmental tax codes.

All of the factors just rehearsed explain the book's title. We cannot, against the policy background outlined above, treat of environmental taxes in isolation from the other instruments which are designed to complement them. So the range of instruments involved is not the least of the contexts envisaged by the title. However, we have tried to look for other contexts, primarily legal ones, which will help us to explain the various aspects of environmental taxation more usefully. As an energy tax, climate change levy is imposed on gas and electricity supplies. Both the gas supply industry and the electricity supply industry are sectors of extreme technical complexity, as a result of their privatisation, in another characteristic adventure, this time of a decade or more ago and a different government. Grafting climate change levy onto these post-nationalisation structures, without a consistent energy policy, has contributed to producing a tax of such extraordinary complexity that it would be impossible to appreciate its subtleties without a knowledge of 'what lies beneath'. So discussions of the post-nationalisation energy sector structures and regulation have been included as well as environmental regulation and tax law. Again, since the regulation of air transport and road freight transport affects plans to introduce new airline taxes, emissions trading schemes and road pricing schemes, we have

Environmental Taxation Law

also included a brief discussion of how these sectors are structured and regulated. In every case, we seek to provide, not just a UK context for the material but an EU-wide and international one also. We hope, as a result, that this book will be no less useful to the overseas reader than to the UK one.

This brings us to another set of contexts: the European Union and international governance structures within which the UK's economic instruments for environmental protection have been designed and implemented.

The more obvious governance structure is the EU one, since the EU's institutions – notably the European Commission – provide both the impetus for action at national level, via the Sixth Environmental Action Programme, and the terms on which that action can be taken, via the European Treaty and the legislation made thereunder. Finding the right level of governance is an important question with economic instruments. In its enthusiasm to introduce emissions trading, partly as a sweetener to sectors already hit both by Integrated Pollution Prevention and Control and climate change levy, the government managed, with the UK Emissions Trading Scheme, to turn, in Sorrell's words, an 'early start into a false start'. This is because the period of the last 18 months to two years has also seen the creation of the EU Emissions Trading Scheme, under which carbon dioxide emissions will be capped, parcelled out and traded among all the Member States of the enlarged Union. The government claims that the UK scheme, set up in 2002, will be good practice for the EU scheme, due to commence in January 2005. Its detractors claim that the UK scheme was an irrelevance, nonsensically unilateral and designed simply to buy off opposition to the government's environmental policy from sectors most badly hit by climate change levy and IPPC. At the very least, there is a measure of over-regulation.

Although less obvious, governance is also felt at the international level, via the United Nations-sponsored 1992 Framework Convention on Climate Change, and its 1997 Kyoto Protocol, as well as the rules of the WTO/GATT 1994-based system of multilateral agreements on international trade. Both the EU itself and its Member States are parties to both sets of agreements. Although there are signs that the position may be changing, the values of the latter, which have governed international trade at least since the mid-twentieth century, may be difficult or impossible to reconcile with those of the former, whose values are, of course, much newer. The influence of Kyoto is to be felt, not only in the creation of the two emissions trading schemes referred to above, each of which take Kyoto as their inspiration, but also in the design of environmental taxes. Climate change levy, an energy tax, neatly sidesteps design problems created by GATT 1994's concept of the border tax adjustment and, in doing so, may sacrifice something of its environmental effectiveness.

The trade-off just referred to is, of course, a matter of political judgement. No less characteristic of the UK's environmental taxes and other economic instruments than their regulatory context and their place within European and international governance structures are the political choices they represent. These are shown both in present political compromises and in the political possibilities for the future. Lord Butler memorably described politics as 'the art of the possible'. Whether the structuring of climate change levy as a downstream energy tax owed more to fear of the political consequences of taxing domestic energy consumption and alienating what remained of the coal industry, than it did to economic and environmental principles, is a matter for historians to debate. The authors have some well-founded suspicions on these points. Nevertheless, it remains the case that, with these taxes in place, it would remain open to a future government, in a time of falling revenues, to supplement the Exchequer by increasing the rates of aggregates levy and climate change levy instead of turning to the much more politically sensitive expedient of increasing rates of income tax. The mechanisms are all now in place, thanks to the enterprise of environmental taxation.

The controversies on these points will surely continue. For our part, we have tried to offer a critical account of an area of UK policy, law and practice which has not, so far as we know, been systematically explored to date in this jurisdiction by any other lawyer authors. In an age when the expounding of a legal subject – however complex – does not perhaps enjoy the high reputation that once it did, we seek (as lawyers) to explain how the different instruments interact at a regulatory level. We do occasionally pass judgment but we do so, not on matters outside our expertise (neither of us being economists and, as yet, unwilling to succumb to the siren song of law and economics), but on the basis of the various accountability mechanisms in the UK's unwritten constitutional arrangements. In the last 12 to 18 months, the panoply of Select Committees at Westminster has begun to take a keen interest in the flutter of creative activity in environmental regulation in Whitehall. In fact, these accountability mechanisms and the network of departments, committees, advisory bodies and executive bodies that have been involved in the design and implementation of the measures under discussion, feature prominently in what follows. We have included the explanations of who they are and what they do because of the difficulties facing the non-specialist or non-UK lawyer in finding his or her way through the myriad of institutional actors.

What we offer, then, is a pragmatic, critical, account. We do not purport to offer any empirical conclusion but we think that we can at least provoke others to test our conclusions.

Each of the authors brings something different to the book. John Snape is an academic lawyer with an interest in international economic law and in the crossovers between environmental regulation, energy law and tax law. Jeremy de Souza is a senior practitioner who retired from full-time practice as a partner in Farrer & Co. in 1996. He has written extensively on tax law matters and is referred to in reference books as a leading specialist in the environmental taxation area. Thus do we seek to bring together our different but hopefully complementary skills. Although each author has read and commented in detail on the work of the other, we have each taken responsibility for writing different Chapters. Setting aside the very short Chapters 3, 9 and 10, which are very much joint efforts, John Snape was responsible for Chapters 1–12 and 28, while Jeremy de Souza was responsible for Chapters 13–27 and 29. In addition, Jeremy carried out the herculean task of compiling the tables and indices.

Warm thanks are due in several quarters and we would like to record them as follows: to Caroline de Souza and to Angela Kershaw, our respective wives, who, despite many commitments of their own, have taken time to bear with detailed verbal critiques of government policy; to former colleagues of John Snape at the University of Leeds, especially to Ann Blair, Michael Cardwell, Oliver Gerstenberg, Roger Halson and Anna Lawson, each for their wisdom and support; to colleagues of Jeremy de Souza at the City of Westminster and Holborn Law Society Revenue Committee and at White and Bowker, especially John Steel, Oliver Sowton and (in putting up with interruptions to her printing facilities), Abi Martin. John Snape began work on the book in 2000/2001 in a period of sabbatical leave from Nottingham Law School, the Nottingham Trent University. His thanks are due to Professors Michael Gunn and Peter Kunzlik, as well as to colleagues at Nottingham Trent who shouldered his teaching and administrative responsibilities in the period of his absence.

Over the period of writing the book, the subject matter has expanded almost daily. We must also record our thanks to John Irwin and Alison Kirk at Ashgate, for their patience and enthusiasm in awaiting a manuscript which therefore became rather later and rather larger than either of us authors could originally have envisaged.

In the footnotes, we have tended to confine case references to *Simon's Tax Cases*, the series now used by most UK tax academics and practitioners. References to other reports are to be found in the Tables.

We have attempted to reflect developments and to state the law, unless otherwise indicated, as at 2 December 2004 (the date of the Pre-Budget Report).

John Snape,
Jeremy de Souza,
20 December 2004.

Table of Cases

Table of Statutes

Table of Statutory Instruments

Table of Provisions of the European Treaty

Table of European Community Legislation

2   Regulations

3   **Council Decisions**

Table of Provisions of GATT 1994

Table of Other Treaty Provisions

Table of Abbreviations

AB	WTO Appellate Body
AC	Law Reports, *Appeal Cases*
ACBE	Advisory Committee on Business and the Environment
ACEA	*Association des Constructeurs Européens d'automobiles*
ACP	African-Caribbean-Pacific
AG	German form of public limited liability company
A-G	Attorney-General
AJIL	*American Journal of International Law*
All ER	*All England Law Reports*
ALSF	Aggregates Levy Sustainability Fund
APD	air passenger duty
Art.	Article
ASBL	Belgian form of limited liability company
BAT	best available techniques
BATNEEC	best available technology not entailing excessive cost
BETTA	British Electricity Trading and Transmission Arrangements
BISD	GATT *Basic Instruments and Selected Documents* Series
BMW	biodegradable municipal waste
BNFL	British Nuclear Fuels Ltd
BSE	bovine spongiform encephalopathy
BTA	border tax adjustment
BTR	*British Tax Review*
BV	Dutch form of private limited liability company
BYIL	*British Yearbook of International Law*
CA	Court of Appeal (England and Wales)
CAP	the Common Agricultural Policy
CBA	cost-benefit analysis
CBI	Confederation of British Industry
CCP	Common Commercial Policy
CCT	common customs tariff
C & E Commrs	HM Commissioners of Customs and Excise
CDM	Clean Development Mechanism
CEE	charge having an equivalent effect
CFCs	chlorofluorocarbons
CFI	Court of First Instance
CFIT	Commission for Integrated Transport
CGT	capital gains tax
Ch.	Law Reports, *Chancery Division*
ch.	chapter
CH_4	methane

CHP	combined heat and power
CHPQA	combined heat and power quality assurance
CLR	*Commonwealth Law Reports*
CLSA	*Current Law Statutes Annotated*
CMLR	*Common Market Law Reports*
Conv	*The Conveyancer and Property Lawyer*
COREPER	Committee of Permanent Representatives
CO_2	carbon dioxide
CTE	WTO Committee on Trade and the Environment
CTP	Common Transport Policy
Defra	Department of Environment, Food and Rural Affairs
DETR	Department of Environment, Transport and the Regions
DLR	*Dominion Law Reports*
DNO	distribution network operator
DSO	distribution system operator
DTI	Department of Trade and Industry
EAD	Electricity Acceleration Directive
ECCP	European Climate Change Programme
ECHR	European Convention on Human Rights
ECJ	Court of Justice of the European Communities
ECOSOC	Economic and Social Committee
ECR	*European Court Reports*
ECSC	European Coal and Steel Community
ECT	Energy Charter Treaty
EEA	European Environment Agency
EELR	*European Environmental Law Review*
EFTA	European Free Trade Area
EG	*Estates Gazette*
EIA	environmental impact assessment
ELM	*Environmental Law and Management*
ELR	*Environmental Law Review*
ERUs	emissions reduction units
ENDS Report	*Environmental Data Services Report*
Env LR	*Environmental Law Reports*
EPA	(US) Environmental Protection Agency
EPD	Energy Products Directive
ER	*English Reports*
ESC	Extra-Statutory Concession
ESH	*Electricity Supply Handbook*
EST	Energy Saving Trust
EU	European Union
EU ETS	European Union Emissions Trading Scheme
Euratom	European Atomic Energy Community
EWCA	Court of Appeal (of England and Wales)
EWHC	High Court of Justice (of England and Wales)
FFL	fossil fuel levy
FS	*Fiscal Studies*

FSA	Financial Services Authority
GAAP	Generally Accepted Accounting Practice
GAAR	General Anti-Avoidance Rule
GAD	Gas Acceleration Directive
GATS	General Agreement on Trade in Services
GATT	General Agreement on Tariffs and Trade
GDP	gross domestic product
GEMA	Gas and Electricity Markets Authority
GHG	greenhouse gas
GLA	Greater London Authority
GM	genetically modified
GmbH	German form of private limited liability company
HC	House of Commons
HCA	High Court of Australia
HFCs	hydroclorofluorocarbons
Hals	*Halsbury's Laws of England*
HL	House of Lords
HMSO	Her Majesty's Stationery Office (see TSO)
IAEA	International Atomic Energy Agency
ICAO	International Civil Aviation Organisation
ICCLR	*International Company and Commercial Law Review*
ICJ	International Court of Justice
ICJ Rep	*Reports of Judgments and Advisory Opinions of the International Court of Justice*
ICLQ	*International and Comparative Law Quarterly*
IEA	International Energy Agency
IELTR	*International Energy Law and Tax Review*
IGC	Inter-Governmental Conference
IHT	inheritance tax
ILM	*International Legal Materials*
IPC	Integrated Pollution Control
IPPC	Integrated Pollution Prevention and Control
IPT	Insurance Premium Tax
IRC	Commissioners of Inland Revenue
JAMA	Japanese Automobile Manufacturers Association
JEEPL	*Journal for European Environmental and Planning Law*
JEL	*Journal of Environmental Law*
JEPP	*Journal of European Public Policy*
JI	Joint Implementation
JLE	*Journal of Law and Economics*
JPL	*Journal of Planning Law*
JWT	*Journal of World Trade*
KAMA	Korean Automobile Manufacturers Association
KG	German limited partnership
LAAPC	Local Authority Air Pollution Control
LATS	Landfill Allowances Trading Scheme
LEC	levy exemption certificate

LIFO	last in, first out
LJQB	*Law Journal Reports (New Series) Queen's Bench (1831–1949)*
LNG	liquified natural gas
LPG	liquified petroleum gas
LS	*Legal Studies*
LT	*Law Times*
LTCS	landfill tax credit scheme
MEA	multilateral environmental agreement
MEP	Member of the European Parliament
MLR	*Modern Law Review*
MP	Member of the UK House of Commons
MSP	Member of the Scottish Parliament
MW	megawatt
NAFTA	North American Free Trade Area
NAP	national allocation plan
NDA	Nuclear Decommissioning Authority
NG	natural gas
NGG	National Grid Group
NICs	national insurance contributions
NIROC	Northern Ireland Renewable Obligation Certificate
NLJ	*New Law Journal*
NOA	National Audit Office
NOX	collective term for nitrogen oxides
N_2O	nitrous oxide
NV	Dutch form of public limited liability company
NWRU	National Waste Registration Unit
ODPM	Office of the Deputy Prime Minister
OECD	Organisation for Economic Co-operation and Development
OED	*Oxford English Dictionary*
Ofgem	Office of Gas and Electricity Markets
OHG	German partnership
OJ	*Official Journal of the European Communities*
PCBs	polychlorinated biphenyls
PCTs	polychlorinated terphenyls
PERN	package waste export recovery note
PFCs	perfluorocarbons
PL	*Public Law*
PLC	UK form of public limited liability company
Porteous	Andrew Porteous, *Dictionary of Environmental Science and Technology*, 3rd edn (Chichester: John Wiley, 2000)
PQ	Parliamentary Question
PRN	packaging waste recovery note
PSA	(departmental) public service agreement
PSO	public service obligation
QB	Law Reports, *Queen's Bench Division*
QPA	Quarry Products Association
RCEP	Royal Commission on Environmental Pollution

REC	regional electricity company
REGO	renewable energy guarantee of origin
RIA	regulatory impact assessment
RICS	Royal Institute of Chartered Surveyors
RIIA	Royal Institute of International Affairs
RO	Renewables Obligation
ROC	Renewables Obligation Certificate
ROO 2002	Renewables Obligation Order 2002
RRC	*Ryde's Rating Cases*
RSPB	Royal Society for the Protection of Birds
SA	public limited company in France, the Iberian Peninsular and Latin America
SARL	Form of limited liability company in some EU Member States
Sched.	Schedule
SCR	*Supreme Court Reports*
SDLT	stamp duty land tax
SEA	strategic environmental assessment
SEA 1986	Single European Act 1986
SEPA	Scottish Environment Protection Agency
SF_6	sulphur hexafluoride
SI	statutory instrument
SIP	single income (or farm) payment
SLT	*Scottish Law Times*
SO	Parliamentary Standing Order
SpA	Italian form of limited liability company
SpC	Special Commissioners' Decisions
SROC	Scottish Renewable Obligation Certificate
STC	*Simon's Tax Cases*
SWTI	*Simon's Weekly Tax Intelligence*
TAF	Trade Association Forum
TC	*Tax Cases*
TED	turtle excluder device
TFL	Transport for London
TJ	*Tax Journal*
TN	Treaty of Nice
TPA	third party access to gas and electricity transmission and distribution systems
TR	*Tax Reports*
TREC	tradeable renewable energy certificate
TSO	transmission system operator
TSO	The Stationery Office (see HMSO)
UK ETS	UK Emissions Trading Scheme
UKHL	United Kingdom House of Lords
ULR	*Utilities Law Review*
UNCLOS	United Nations Convention on the Law of the Sea
UNCSD	United Nations Commission on Sustainable Development
UNECE	United Nations Economic Commission for Europe

UNEP	UN Environment Programme
UNICE	Union of Industrial and Employers' Confederations of Europe
V&DR	*VAT and Duties Reports*
VAT	value added tax
VATTR	*Value Added Tax Tribunal Reports*
VED	vehicle excise duty
WA	written ministerial answer to a Parliamentary Question
WCED	World Commission on Environment and Development
WIP	Waste Implementation Programme
WCA	waste collection authority
WDA	waste disposal authority
WFD	Waste Framework Directive
WH	Westminster Hall debate of the House of Commons
WIP	Waste Implementation Programme
WLR	*Weekly Law Reports*
WN	*Weekly Notes*
WPRG	waste performance reward grant
WRAP	Waste and Resources Action Programme
WRA	waste regulation authority
WS	written Ministerial statement to Parliament
WTO	World Trade Organization
WTA	willingness to accept
WTP	willingness to pay

PART I
PROLOGUE

Chapter 1

Preliminaries

1.1 Introduction

The concept of the environmental tax is a new and controversial presence in the European regulatory landscape. The UK has been a pioneer in the transformation of this new type of tax from a twinkle in the eye of the environmental economist to a legal reality. This book seeks to give a comprehensive but critical account of the law relating to environmental taxes in the UK. It discusses the policies which have shaped those taxes, together with their relationship both with other forms of environmental regulation, and with other (non-environmental) taxes. It also offers insights into the application of environmental taxes in practice. Throughout the book, UK law and policy is placed in its wider international and European contexts.

In the absence of a general legislative definition,[1] the ways in which environmental taxes are to be identified and classified are, of course, key elements in this first chapter. The discussion begins,[2] however, by introducing the essence of some even more fundamental concepts, a number of which we shall encounter again and again: what is a tax? How does the concept differ from other levies, such as charges for government services? Within the concept of a tax, what distinction is to be drawn between direct and indirect taxes? What types of tax fall into each category? In each case, we explain the specific importance of each of these questions to the key term 'environmental taxes', both the nature and the classifications of which we discuss in some detail below.[3] Having explained the importance of each of these questions, as well as indicating something of the policies that inform them, we then go on to set out the approach which we have adopted to the material under consideration. The purpose of the study, as already stated, is to offer a critical account of the subject matter; para. 1.2.2 explains the viewpoint from which the criticism is made.

One of the all-embracing themes of the book is that of how – irrespective, almost, of the jurisdiction in which they are designed and implemented – the feasibility and design of environmental taxes (justified, as they often are, by reference to international environmental agreements), are subject to the law of international trade, as embodied in the 1994 revision of the General Agreement on Tariffs and Trade ('GATT 1994'), and its related agreements, and in the institutions of the World Trade Organization ('the WTO'). The UK is no exception to this general proposition and, furthermore, as a Member State of the EU, its own environmental levies are additionally subject to the provisions of the European Treaty. The nature and extent of both sets of treaty provisions, which coincide to a considerable extent, are considered in subsequent

---

[1]   None of the taxes with which this book is concerned are described as environmental taxes within the legislative codes that created them (see para. 1.4.2 below).
[2]   See para. 1.2.1 below.
[3]   See para. 1.2.1.5 below.

chapters.[4] In this Introduction, the explanation of the essential concepts and of our approach to the subject matter as a whole, is followed by a brief introduction to the international and European context of the material, as well as to the place, within that setting, of the UK and its constituent countries.[5] We seek to highlight throughout the book the governance questions to which these contexts give rise, since these help to explain, not only why European-wide action in the field of environmental taxation has proved so problematic, but also why the UK's environmental taxes are of UK-wide application, rather than being found (as, for example, in Spain)[6] in its constituent countries.

So far as the non-UK reader is concerned, the chief usefulness of the present volume, it is hoped, will be found in the detailed critical exposition, against the background discussed above, of the UK's environmental levies and their institutional framework. To this end, para. 1.4.2 below contains an introductory overview of the levies, most importantly (in the chronological order of their introduction), landfill tax, a levy on the landfilling of waste; climate change levy, a levy on the industrial, commercial and agricultural use of energy; and aggregates levy, which taxes the extraction of primary minerals. All of the levies have both a regulatory context[7] and a taxation context and paras 1.4.2 and 1.4.3 seek to give an overview of those contexts. Although, as will be seen, HM Customs and Excise ('Customs') is the government department mainly responsible for the administration and enforcement of each of these levies,[8] the non-UK reader (and possibly, also, even the UK reader new to the subject), will be struck by the myriad other bodies with some subsidiary role in at least the former process. Paragraph 1.5 briefly introduces these bodies, by way of a prelude to the detailed, as well as much more wide-ranging, discussion in Chapter 4 below. Together, the elements of the discussion in this chapter will equip the reader to appreciate the fiscal context of the levies to be discussed in the rest of the book.

These opening remarks having been made, we turn first to certain basic issues of terminology and, specifically, to the idea of environmental taxes and the implications of the constituent elements of the term.

---

4   See Chapters 8 and 12 below.
5   See para. 1.3 below.
6   See, for example, Pedro M. Herrera, 'Legal Limits on the Competence of Governments in Spain', in *Critical Issues in Environmental Taxation: International and Comparative Perspectives: Volume I*, ed. by Janet Milne, Kurt Deketelaere *et al.* (Richmond: Richmond Law and Tax, 2003), pp. 111–23. This chapter is a distilled version of Herrera's Spanish-language text, *Derecho tributario ambiental* (Madrid: Marcial Pons – Ministerio de Medio Ambiente, 2000). Neither of the present authors read Spanish.
7   Throughout the book, economic instruments (including regulatory taxes) are seen, not as an alternative to regulation, but as a form of regulation in themselves, albeit distinct from command and control regulation (see, for example, Robert Baldwin and Martin Cave, *Understanding Regulation – Theory, Strategy and Practice* (Oxford: Oxford University Press, 1999), pp. 1–2.
8   See para. 4.2.1.2(2) below.

1.2   Terminology, approach and sources

The use of the term 'environmental taxation' in popular discourse tends to elide the particular significances of each of its two constituent elements.

This para. 1.2 accordingly begins by setting out an initial definition of the concept of a tax and briefly explaining its relationship, first, with the distinct concept of a charge for government services and, secondly, with the concept of a fine. Both of these distinctions, important in the study as a whole, will be revisited in more detail in a later chapter.[9] The term 'levy',[10] also common in public finance literature, operates as a generic term for both taxes and charges: the allocation of a particular levy to one or other category, and the implications of that process, depend on isolating the characteristics of a tax, and distinguishing it from a charge (or fee) for government services.

Within the concept of a tax is entrenched a further distinction which, in the specific context of environmental taxes, is also of considerable significance: that between direct and indirect taxes. We examine the distinction in some detail here, since it will be of fundamental importance in explaining the international aspects of environmental taxation,[11] no less than its Community law implications.[12]

The preceding discussion forms the basis of an examination of the technical significance of the concept 'environmental taxation'. This in turn leads into an explanation of the approach taken in the book to the range of material under consideration,[13] as well as the sources to which we have referred.

1.2.1   Terminology

1.2.1.1   Taxes, fines and charges

The UK's constitutional arrangements, as will become apparent, have so far tended to discourage a narrow and prescriptive view of the tax as a legal concept.[14]

Fortunately, however, for constitutional reasons, there is in the Commonwealth jurisdictions a plethora of material on the legal nature of a tax and on its relationship with the concept of a charge or fee for government services.[15] A convenient (and enduring) definition of the former concept was applied in the Canadian case of *Re Eurig Estate*.[16] There, it was held that, assuming that it has legislative authority, a levy is a tax if it satisfies all of three conditions, that is, if it is:

1. legally enforceable;
2. levied by a public body; and
3. intended for a public purpose.

---

9   See Chapter 7 below.
10   See OED, 'levy' definition 2a.
11   See Chapter 8 below.
12   See Chapter 12 below.
13   See para. 1.2.2 below.
14   See para. 7.2.2.1 below.
15   See para. 7.2.1 below.
16   (1999) 165 DLR (4th) 1.

Each of these three characteristics, as well as the decision from which they are drawn, is analysed in more detail later in the study.[17] For present purposes, however, they provide us with a suitable working definition of a tax.

Whether and, if so, how, the tax concept is to be distinguished from a penalty or fine has been the subject of some debate. In asserting such a distinction, Hart, whilst acknowledging that it may often become blurred, expressed the distinction between fines and taxes in terms that the former involves '… an offence or breach of duty in the form of a violation of a rule set up to guide the conduct of ordinary citizens'.[18] Hart's expression of the distinction needs only to be stated for its problematic nature in relation to environmental taxes to be obvious; we shall return to it later in the book.[19]

By contrast with a tax or with a penalty, a 'charge' or 'fee' involves the provision of some service direct to a particular person, which is related to the amount charged but which does not preclude the possibility of a reasonable profit being made by the public body in question.[20] Spackman says that the essence of a charge is that it 'purchases a specific service and will generally vary with usage; the transaction should be a market exchange of a kind which the private sector could supply (as a principal rather than as an agent); and the charge *should* [emphasis added] merely cover the cost of providing the service',[21] in this last particular differing from the *legal* definition of a charge or fee.[22] Conversely, the essence of a *tax*, it might be said, is that it is a payment to general government which is compulsory and in relation to which the benefit to the taxpayer is usually not in proportion to the payment.[23] Obviously, certain payments will be difficult to allocate to one or other category.[24]

1.2.1.2   Direct and indirect taxes

Within the tax concept, both economists and lawyers traditionally make a further distinction, that between direct and indirect taxes. The classic articulation of the distinction comes from the *Principles of Political Economy* of John Stuart Mill (1806–1873) and is as follows:

> Taxes are either direct or indirect. A direct tax is one which is demanded from the very persons who, it is intended or desired, should pay it. Indirect taxes are those which are demanded from one person in the expectation and intention that he shall indemnify himself at the expense of another: such as the excise or customs. The producer or importer of a commodity is called upon to pay a tax on it, not with the intention to levy a peculiar

---

[17]   See para. 7.2.2.2 below.
[18]   H.L.A. Hart, *The Concept of Law* (Oxford: Clarendon Press, 1961), p. 39.
[19]   See para. 7.2.3 below.
[20]   See para. 7.2.1 below.
[21]   See Michael Spackman, 'Hypothecation: a view from the Treasury', in *Ecotaxation*, ed. by Timothy O'Riordan (London: Earthscan, 1997), pp. 45–51.
[22]   *Ibid.*
[23]   See Organisation for Economic Co-operation and Development, *Revenue Statistics 1965–2002* (Paris: OECD, 2003), p. 285.
[24]   See para. 7.3.1 below.

contribution upon him, but to tax through him the consumers of the commodity, from whom it is supposed that he will recover the amount by means of an advance in price.[25]

The distinction is sometimes expressed in terms of a dichotomy between those taxes imposed on *products* (that is, indirect taxes) and those imposed upon *producers* (that is, direct taxes).[26] Thus, the key examples of direct taxes are 'taxes on wages, profits, interests, rents, royalties, and all other forms of income, and taxes on the ownership of real property', whilst indirect taxes are exemplified by 'sales, excise, turnover, value added, franchise, stamp, transfer, inventory and equipment taxes, border taxes and all other taxes other than direct taxes and import charges'.[27] In *Re Eurig Estate*,[28] after referring to Mill's characterisation of the distinction between direct and indirect taxes, the majority of the Supreme Court of Canada held that probate fees of the Ontario Court, being chargeable on a deceased individual's executors in their representative capacity only, were direct taxes on the deceased's estate, rather than indirect taxes levied on the executors personally. For the majority, Major, J. said that:

> Applying Mill's definition, the tax would be indirect if the executor was personally liable for payment of probate fees, as the intention would clearly be that the executor would recover payment from the beneficiaries of the estate. However, the legislation does not make the executor personally liable for the fees. Payment is made by the executor only in his or her representative capacity … [A]s the amount is paid out of the estate by the executor in his or her representative capacity with the intention that the estate should bear the burden of the tax, the probate fees fall within Mill's definition of a direct tax.[29]

Despite the fact that it is not free from difficulty in economic terms,[30] the distinction between direct and indirect taxes has been highly influential, both in national legal orders (that is, in the laws of constitutions) and also in international and supranational legal orders (such as the EC and the GATT 1994-based system of multilateral trade agreements). In the context of constitutional law, the distinction between direct and indirect taxes has been used to delimit the respective taxing powers of federal and provincial legislatures. In *Re Eurig Estate*, for instance, the distinction was crucial

---

25 John Stuart Mill, *Principles of Political Economy with Some of their Applications to Social Philosophy* ('People's Edition') (London: Longmans, Green, 1892), Book 5, ch. 3, pp. 495–6. See also Sir William Blackstone (1723–1780), *Commentaries on the Laws of England*, Book I, 16th edn by C.J.T. Coleridge (London: Butterworths, 1825), pp. 316–317.

26 See WTO Secretariat, *Taxes and Charges for Environmental Purposes – Border Tax Adjustment* (WT/CTE/W/47), para. 31.

27 See 1994 WTO Agreement on Subsidies and Countervailing Measures (see para. 8.4.4 below), fn 58 thereto.

28 See para. 1.2.1.1 above.

29 (1999) 165 DLR (4th) 1, para. 26.

30 See John Tiley, *Revenue Law*, 4th edn (Oxford: Hart Publishing, 2000), p. 18. One distinction might be that indirect taxes are 'shifted into prices, not only more fully but more quickly' than are direct taxes (see Organisation for Economic Co-operation and Development, *Taxing Consumption* (Paris: OECD, 1988), para. 1.12). See also Paul Demaret and Raoul Stewardson, 'Border tax adjustments under GATT and EC law and general implications for environmental taxes' (1994) 4 JWT 5, 14–16.

to deciding whether the tax in question was within the competence of the provincial legislature. This was because, in Canada, direct taxes could be levied by both federal and provincial governments. Indirect taxes, by contrast, were leviable only by the federal government.[31]

In the world trade context, the distinction between direct and indirect taxes is crucial to the operation of two principles of international business taxation, the origin principle and the destination principle. The origin principle requires that products should be taxed in the country where they are made, regardless of where they are consumed.[32] The destination principle requires that, irrespective of where a commodity is produced, it should be taxed in the country where it is consumed.[33] Direct taxes reflect the origin principle, since they imply the taxation of the producer rather than of the product. Indirect taxes may be imposed in order to reflect either the origin principle, in which case products are taxed in the country of their production, or the destination principle, in which case they are taxed in the country of their consumption. In the context of international trade, the almost invariable rule is for indirect taxes to be applied in accordance with the destination principle. This, of course, is because the universal application of the origin principle in a world of divergent indirect tax rates would tend to encourage consumers to source products from countries with lower rates of indirect taxes.

The destination principle is put into effect through the rather misleadingly-named concept[34] of the 'border tax adjustment' (the 'BTA').[35] The BTA refers to the combined process of ensuring that products are exported from one country free of tax while being imported into another country subject to a tax designed to ensure that the products compete in tax terms on an equal footing with similar goods[36] produced domestically.[37] For present purposes, the crucial point is that, while the BTA is essential to giving effect to the destination principle, it has no application to the origin principle,[38] with the result that, traditionally, the BTA has had no application in relation to direct taxes.[39]

---

[31] See s.92, (Canadian) Constitution Act 1867. The rule might seem counter-intuitive to a European lawyer (but see Art 92, European Treaty (ex 98) (see para. 12.3.3.1(6) below)).

[32] OECD (1988), para. 7.4.

[33] OECD (1988), *op. cit.*, para. 7.3.

[34] Misleading, since, where exports are exempted from tax, or imports are taxed after importation, no adjustment is made at the border (see OECD (1988), *op. cit.*, p. 121). The term ('BTA') is nonetheless in universal use.

[35] See para. 8.4.3 below.

[36] See Won-Mog Choi, *Like Products in International Trade Law: Towards a Consistent GATT/WTO Jurisprudence* (Oxford: Oxford University Press, 2003).

[37] OECD (1988), *op. cit.*, para. 7.3. See also WTO Secretariat, *Taxes and Charges for Environmental Purposes – Border Tax Adjustment* (WT/CTE/W/47), para. 28. See para. 8.4.3 below.

[38] See the reports of the GATT panel of November 1976 in *Domestic International Sales Corporations ('DISCs')*, GATT BISD 23d Supp 98, 114, 127 and 137 (1977), which were adopted by the GATT Council in December 1981 (see Demaret and Stewardson, *op. cit.*, pp. 10–12).

[39] See Kalle Määttä, *Environmental Taxes – From an Economic Idea to a Legal Institution* (Helsinki: Finnish Lawyers' Publishing, 1997), p. 253, referring to Lans Bovenberg

Both the destination principle and the concomitant BTA have venerable theoretical pedigrees. They can each be traced back at least to David Ricardo (1772–1823), one of Mill's intellectual mentors,[40] who, taking the then topical example of corn, wrote in 1822 that:

> In the degree ... in which [domestic indirect] taxes raise the price of corn, a duty should be imposed on its importation ..., and a drawback of the same amount should be allowed on the exportation of corn. By means of this duty and this drawback, the trade would be placed on the same footing as if it had never been taxed, and we should be quite sure that capital would neither be injuriously for the interests of the country, attracted towards, nor repelled from it.[41]

Generally, although not invariably, environmental taxes fall into the category of indirect taxes. However, since discussions of environmental tax proposals have taken place both at the national and the international levels, some appreciation is necessary of the different types of direct tax, as well as of the various types of indirect tax. It is to these distinctions that we now turn.

1.2.1.3   Categories of direct taxes

The main examples of direct taxes are those on income, profits and capital gains. In the UK, the key direct taxes are income tax, capital gains tax ('CGT'), corporation tax[42] and national insurance contributions ('NICs').[43] Subject to the point made below about the classification of emissions and waste taxes,[44] the main significance of direct taxes to environmental taxation is twofold: (1) the extent to which the taxes just referred to contain environmentally-damaging 'tax subsidies';[45] and (2) whether and, if so then how, the statutory codes relating to them might be 'greened', that is, by incorporating reliefs, exemptions, etc., designed to encourage environmentally-friendly behaviour.[46] Additionally, however, there is the question of whether differential tax rates built into non-environmental tax codes (for example as in

---

[ ] and Jocelyn Horne, 'Taxes on Commodities: A Survey', in *Tax Harmonization in the European Community: Policy Issues and Analysis*, ed. by George Kopits (Washington DC: International Monetary Fund, 1992), pp. 22–51.

[40] See Samuel Hollander, *The Economics of John Stuart Mill*, vol. 1 (Toronto: University of Toronto Press, 1985), p. xii.

[41] See David Ricardo, 'On Protection to Agriculture', in *The Works and Correspondence of David Ricardo: Volume IV, Pamphlets and Papers 1815–1823*, ed. by Piero Sraffa with M.H. Dobb (Cambridge: Cambridge University Press, 1951), p. 218. See also Adam Smith (1723–1790), *The Wealth of Nations*, ed. by D.D. Raphael (London: David Campbell, 1991), pp. 440–45.

[42] All three of which fall within the OECD's classification category 1000 (see OECD (2003), pp. 190–92).

[43] See para. 1.4.3 below as to all of these. National insurance contributions fall within the OECD's classification category 2000 (see OECD (2003), *op. cit.*, p. 190).

[44] See para. 1.4.2.1(1) below.

[45] See para. 7.2.4 below.

[46] See Chapters 22–25 below.

relation to the income tax treatment of a company car as a benefit in kind)[47] and
intended to put a price on environmentally-undesirable behaviour might themselves
even be described as environmental taxes.[48]

1.2.1.4 Categories of indirect taxes

In the context of environmental taxes, the most relevant category of indirect taxes is
that of taxes on goods and services. This category, which appears as heading 5000
in the Organisation for Economic Co-operation and Development ('OECD')'s[49]
Classification of Taxes,[50] comprises sales taxes, excise, turnover, value added,
franchise, stamp, transfer, inventory and equipment taxes, border taxes and all other
taxes other than direct taxes and import charges. Taxes on goods and services are
in turn one of five categories of *consumption* taxes. The distinguished authors of
the OECD's 1988 report on consumption taxes[51] identified five ways of taxing
consumption:[52]

1   (as just mentioned) by imposing 'taxes on goods and services themselves';
2   through the use of 'fiscal monopolies and public utilities';
3   by making levies for the grant of licences;
4   by making other levies 'that may contain an element of consumption taxation';
    and
5   by a personal expenditure tax.

Of these, 5 is not relevant in the present context, since it refers to an as yet theoretical
alternative to existing consumption taxes. The most important to the present study
of the classifications in the list are 1 and 3. Of the rest, 2 refers to the possibility that
governments may produce or distribute certain goods and sell them at a higher price
than if they had been sold untaxed in a competitive market.[53] At the time that the 1988
report was written, the most important such goods would have been electricity and
gas but, following the 'unbundling' of public utilities across Europe in the 1990s,[54]
this category is now so unimportant as not to be worth further consideration here. The
residual category in 4 refers to the hypothesis that elements of consumption taxation
may be discovered in taxes which are not specifically imposed on consumption.[55]
This leaves us with categories 1 and 3.

Within the category of taxes on goods and services themselves, the authors of the
OECD's 1988 report drew a distinction between selective consumption taxes on

---

[47]   See para. 23.2 below.
[48]   See para. 1.2.1.5(2) below.
[49]   See para. 4.4.2 below.
[50]   OECD (2003), *op. cit.*, pp. 283–300.
[51]   OECD (1988), *op. cit.* The report's authorship was particularly, perhaps, distinguished,
       including (as it did) Professors Cedric Sandford and Sijbren Cnossen.
[52]   OECD (1988), *op. cit.*, paras 1.7–1.13.
[53]   *Ibid.*, *op. cit.*, para. 1.9.
[54]   See paras 2.4, 6.4.3 and 12.2.6.3 below.
[55]   OECD (1988), *op. cit.*, para. 1.11.

particular products or services, on the one hand, and general consumption taxes on the other.[56] Examples of the former are excise duties on hydrocarbon oils. The latter, general consumption taxes, are further subdivided, however, into single stage taxes and multi-stage taxes.[57] Single-stage, or 'sticking', taxes, may be levied at any one of the stages of supply,[58] that is, from manufacturers to wholesalers, from wholesalers to retailers or from retailers to consumers. As will be apparent later on in this chapter, the UK's indirect environmental taxes are of this single-stage type. Multi-stage taxes are, however, taxes levied each time goods or their components are sold.[59] Unless credit is given for tax paid at each of the earlier stages,[60] such taxes generate multiple taxation (hence their label as 'cascade' taxes) and are therefore generally regarded as bad. The common system of value added tax ('VAT'), which does provide for such a system of credits, is not a cascade tax. Article 33 of the Sixth Council Directive on the Harmonisation of the Laws of the Member States relating to Turnover Taxes and a Common System of Value Added Tax ('the Sixth VAT Directive')[61] bans multi-stage taxes similar to VAT. However:

> Without prejudice to other Community provisions ... this Directive shall not prevent a Member State from maintaining or introducing taxes on insurance contracts, taxes on betting and gambling, excise duties, stamp duties and, more generally, any taxes, duties or charges which cannot be characterised as turnover taxes, provided however that those taxes, duties or charges do not, in trade between Member States, give rise to formalities connected with the crossing of frontiers.

Article 33 operates against the general background of the obligation on EU Member States, which dates from 1967, to replace their turnover taxes with VAT.[62] In a series of decisions, the ECJ has construed Art 33 as banning the adoption by a Member State of any wide-based turnover tax that might rival VAT.[63] In, for instance, *Dansk Denkavit ApS v. Skatteministeriet*,[64] which concerned a now-repealed levy imposed in Denmark, the ECJ stated that:

> [F]or a tax to be characterized as a turnover tax, it is not necessary for it to resemble VAT in every respect; it is sufficient for it to exhibit the essential characteristics of VAT.

---

[56]   *Ibid.*, *op. cit.*, para. 1.8.
[57]   *Ibid.*, *op. cit.*, para. 1.8(b).
[58]   Note the potential significance of a 'prior-stage' indirect tax, that is, an indirect tax levied on goods or services used directly or indirectly in the manufacture of the product in question (see 1994 WTO Agreement on Subsidies and Countervailing Measures (see para. 8.4.4 below), fn 58 thereto).
[59]   OECD (1988), *op. cit.*, para. 1.8(b).
[60]   See 1994 WTO Agreement on Subsidies and Countervailing Measures (see para. 8.4.4 below), fn 58 thereto.
[61]   Council Directive 77/388/EEC, (1977) OJ L145 1.
[62]   Council Directive on the Harmonisation of Legislation of Member States Concerning Turnover Taxes, Council Directive 67/227/EEC/EEC, (1967) OJ Sp Edn 14, Art 1.
[63]   See David Williams, *EC Tax Law* (London: Longman, 1998), p. 91. See para. 12.3.2 below.
[64]   C–200/90, [1992] ECR I–2217.

In the present case, the differences which have been mentioned [that is, by the Danish Government] do not affect the nature of a levy such as the Danish levy, which resembled VAT in all essential respects.[65]

Thus, within the Member States of the EU, it is in principle permissible for a single-stage consumption tax to be imposed, provided that it is not one that can be taken to mimic VAT.[66] An obvious example of such a tax is that imposed on plastic bags in Eire.[67] This provision gives rise to the possibility of indirect taxes on consumption with an environmental purpose and it is one which has been embraced by a number of Member States, including the UK.

Category 3. in the list given above creates a conundrum for anyone involved in the study of taxes as regulatory instruments.[68] As already stated, levies may be divided into taxes and charges, the latter having a quality of reciprocity which is not shared by the former. As suggested in the 1988 OECD report, however, there may be an intermediate category of levies made for services which are provided direct to the individual but which lack this element of reciprocity. *Re Eurig Estate*[69] suggests that such levies are taxes rather than charges.[70]

The relevance of the category of indirect taxes to environmental taxes is that most, although not all, environmental taxes, are taxes on the consumption of goods and services (that is, 'product taxes') and are therefore indirect taxes. Subject to the points made below,[71] the UK's aggregates levy and landfill tax are probably both to be viewed in this way, whilst climate change levy and excise duties (excise duties being environmental taxes in a broad sense and only if there is some environmental differential) are certainly to be regarded as product taxes.

---

[65]   [1992] ECR I–2217 (para. 14). From its earlier jurisprudence, the ECJ concluded that the essential characteristics of VAT were that it 'applies generally to transactions relating to goods or services; it is proportional to the price of those goods or services; it is charged at each stage of the production and distribution process; and finally it is imposed on the added value of goods and services, since the tax payable on a transaction is calculated after deducting the tax paid on the previous transaction' ([1992] ECR I–2217 (para. 11)).

[66]   See, for example, the consideration of, respectively, national meat and dairy products marketing levies in *Fazenda Pública v. Fricarnes SA*, C–28/96, [1997] STC 1348, and *Fazenda Pública v. União das Cooperativas Abastacedoras de Leite de Lisboa*, C–347/95, [1997] STC 1337, stamp duty on construction contracts in *Fazenda Pública v. Solisnor-Estaleiros Navais SA*, C–130/96, [1998] STC 191, and a local tax upon the gross receipts from entertainment performances in *NV Giant v. Commune of Overijse*, C–109/90, [1993] STC 651. See also a case involving municipal parking charges, *Fazenda Pública v. Câmera Municipal do Porto (Ministério Público, third party)*, C–446/98, [2001] STC 560.

[67]   See para. 12.3.2 below.

[68]   Most environmental taxes being of this kind (see Chapter 5 below).

[69]   See paras 1.2.1.1 and 1.2.1.2 above.

[70]   See OECD (2003), *op. cit.*, p. 286.

[71]   See para. 1.4 below.

1.2.1.5   *Environmental taxes*

The preceding discussion enables us to reflect on the nature and classification of environmental taxes generally.[72] This should serve as a useful introduction to setting out, in para. 1.2.2 below, the approach which we have taken to the material in the rest of the study, as well as the sources on which the authors have placed reliance in its preparation. Prior to that, however, it is worth noting the significance of the key word 'environmental', as it relates to the expression 'environmental taxes'.

Environmental taxes, as mentioned above, are one type of economic instrument for environmental protection, the other main examples being transferable permit systems, deposit-refund systems and financial assistance from government.[73] Within the last category, that is, financial assistance from government, we include measures for 'greening' a tax system, so the concept of economic instruments is taken to include 'green' exemptions and reliefs which have been built into taxes such as income tax and corporation tax.[74] Emissions trading systems, whilst schematically distinct from taxes, operate, in one or another form, in conjunction with two of the UK's environmental taxes, that is, climate change levy and landfill tax. We shall therefore make extensive reference to them throughout the rest of the study.

(1) *The environmental aspect of environmental taxes*
The term 'environmental', as it is used in the expression 'environmental taxation', suggests a wide area of operation for the instruments under consideration in this book.[75] This is consistent with the use of the word 'environmental' in relation to 'environmental law' and 'environmental regulation' generally. Krämer, for instance, in writing about the objectives, principles and conditions of European environmental law, looks (naturally enough) to the European Treaty for guidance on the scope of the notion of 'the environment'. He says:

> It follows from Articles [174(1) (ex 130r(1)] and [175(2) (ex 130s(2))][76] that the environment includes humans, town and country planning, land use, waste and water management and use of natural resources, in particular of energy. This list, which was not meant to be exhaustive, includes practically all facets of the environment, in particular fauna and flora which are part of the natural resources, as well as climate. The inclusion of town and country planning makes it clear that 'environment' is not limited to the natural environment.[77]

---

72 No distinction is drawn in this book between the expressions 'environmental taxes', 'ecological taxes', 'ecotaxes' and 'green taxes'.

73 Organisation for Economic Co-operation and Development, 1991 *Recommendation of the Council on the Use of Economic Instruments in Environmental Policy*, C(90)177/Final, January 31, 1991 (available from www.oecd.org).

74 See Part III, Section B below.

75 See Sean Coyle and Karen Morrow, *The Philosophical Foundations of Environmental Law* (Oxford: Hart Publishing, 2004).

76 See para. 12.2.1 below.

77 See Ludwig Krämer, *EC Treaty and Environmental Law*, 2nd edn (London: Sweet and Maxwell, 1995), p. 41.

Bell and McGillivray, quoting Einstein as having been reported to remark that '[t]he environment is everything that isn't me', adopt a wide definition of the terms for the general purposes of their textbook on UK environmental law. In discussing the expression 'environmental law', and therefore the scope of their book, they say:

> We intend to concentrate on those laws and practices which relate primarily to *the protection of the whole or part of the general surroundings*, as opposed to those where the true objective is the protection of public health, or individual people such as workers or consumers. [Emphasis added.][78]

Finally, Winter defines the term 'environmental law' as follows:

> ... the law regulating the relationship of us to nature, understood both as the world around us *and* as the nature we carry within ourselves.[79]

From all of the above, a number of areas of application for environmental taxes may be envisaged. Obvious applications of environmental taxes are in the waste management and water protection areas, as well as in the control of air and atmospheric pollution. Less obvious, perhaps, are resource use, town and country planning and (not specifically referred to in the above quotations) noise abatement. Although the UK has not yet applied an environmental tax in relation to water pollution[80] or noise, landfill tax operates in the area of waste management, now being intended (at least in part) to assist in reducing the amount of waste being sent to landfill,[81] while climate change levy is a key instrument in the control of air and atmospheric pollution.[82] Landfill is unusual in that it can perhaps be identified quite clearly with a single environmental problem. The same may not, however, be true of some of the less obvious applications. In relation to resource use, it might be a matter for debate whether the predominant justification for the tax in question should be the conservation of resources or the control of pollution. This is the case, for example, with aggregates levy, which, besides being justified by reference to a need to conserve the supply of virgin aggregate, is also designed to address the problem of the noise and visual intrusion caused by quarrying.[83] Similarly, with projects for road charging or cordon pricing (such as London's congestion charge),[84] it might be a matter for debate as to whether the predominant

---

[78] Stuart Bell and Donald McGillivray, *Ball and Bell on Environmental Law: The Law and Policy relating to the Protection of the Environment*, 5th edn (London: Blackstone Press, 2000), p. 4. They also adopt the definition of the environment in the Environmental Protection Act 1990, s.1, that is that the environment consists of 'all, or any, of he following media, namely, the air, water and land' (*ibid.*).

[79] Gerd Winter, 'Perspectives for Environmental Law – Entering the Fourth Phase' (1989) 1 JEL 39, quoted in John F. McEldowney and Sharron McEldowney, *Environmental Law and Regulation* (London: Blackstone, 2001), p. 78.

[80] But see para. 29.1 below.

[81] See para. 6.3 below.

[82] See para. 6.4 below.

[83] See para. 11.3.2 below.

[84] See para. 18.2 below.

environmental concern should be traffic congestion or the pollution caused by motor vehicles. Even in the case of air and atmospheric pollution, there is the additional concern of the conservation of energy, something which (as we shall see) is a key element in climate change levy. All these points made, it should be stressed that environmental taxation, at least in the UK, tends not to operate specifically in relation to plants, animals and habitats.[85]

Finally, it should be noted that, in referring to the use of natural resources, in particular of energy, Krämer signposts the difficulties which have arisen from attempts, both at Community and Member State level, to 'graft' environmental policies onto the process of 'unbundling' Europe's energy markets. This is a recurrent theme throughout the rest of the study, since the design of UK's tax on the industrial, commercial and agricultural consumption of non-renewable energy, that is, climate change levy, evidently seeks to marry up apparently contradictory objectives within Community and UK energy policy itself.[86]

(2) The taxation aspect of environmental taxes
It has already been stated that none of the UK's environmental taxes are specifically referred to as such in the legislation which creates them.[87] This may be problematic, since, as already mentioned, environmental taxes may be susceptible to different evaluative criteria from non-environmental taxes. The OECD has espoused different definitions at various times but the definition reproduced in its 1997 report, *Environmental Taxes and Green Tax Reform*[88] has gained a certain currency. This states that:

> A tax falls into the category environmental if the tax base is a physical unit (or a proxy for it) of something that has a proven specific negative impact on the environment, when used or released.[89]

The intellectual appeal of this definition is that, in concentrating on the tax base, it emphasises the way in which a particular tax works, rather than whatever justification

---

[85]   It seems that the abolition in 1988 of the income tax charge on commercial woodlands was motivated by a desire to protect the income tax base by removing a tax shelter, rather than by environmental considerations. The same point applies to the (heavily circumscribed) inheritance tax ('IHT') relief in respect of growing trees and underwood forming part of an individual's estate on death (see Inheritance Tax Act 1984, ss.125–30).

[86]   See para. 21.5.2 *et seq.*, below. Helm likens New Labour's capacity to formulate energy policy in this way to George Orwell's 'doublethink', that is, 'the power of holding two contradictory beliefs in one's mind simultaneously, and accepting both of them' (see Dieter Helm, *Energy, the State, and the Market: British Energy Policy since 1979*, revised edn (Oxford: Oxford University Press, 2004), p. 294).

[87]   Unlike, exceptionally, s.369 of the Ecotax Law in Belgium, in which an environmental tax, ['*écotaxe*'] is defined as a '*taxe assimilée aux accises, frappant un produit mis à la consomation en raison des nuisances écologiques qu'il est reputé générer*' (quoted in Määttä, *op. cit.*, p. 41).

[88]   Paris: OECD, 1997.

[89]   OECD (1997), p. 18.

may be given for the introduction of the tax.[90] The definition is certainly capable of capturing energy, landfill and severance taxes.[91]

Faced with the problem of categorisation (which may, or may not be significant),[92] Määttä makes a useful distinction[93] between environmental taxes in the strict sense and environmental taxes in the broad sense.[94] Taxes which do not fall into either category (for example income tax and corporation tax) are, naturally enough, 'non-environmental taxes'[95] and this is so even if they embody environmentally-motivated design features.[96] One tax which would appear to be a non-environmental tax, whatever the government's protestations to the contrary, is air passenger duty ('APD'), a flat-rate indirect tax which, since 1994, has been chargeable on passenger flights from the UK.[97] Tiley draws our attention to the fact that APD merely imposes an excise duty on an item which is zero-rated for VAT purposes,[98] thus reducing the disparity between items which come within and those which are outside the VAT net.[99]

Environmental taxes in the strict sense are 'taxes about which a long-established unanimity exists that they can be classified as environmental taxes'.[100] The examples that Määttä instances are effluent taxes, taxes on beverage containers and incentivising tax differentiations between leaded and unleaded petrol. What each of these have in common, whether it is to reduce the flow of effluents into water, to reduce litter or to encourage consumers to buy unleaded petrol, is an intention to influence behaviour.[101] Such environmental taxes are *incentive* environmental taxes, a category which includes, not only taxes whose revenues are less important than their capacity to steer behaviour, but also redistributive environmental taxes, some or all of the proceeds of which are used to incentivise taxpayers to purchase, for example, environmental protection equipment, such as energy-saving plant and machinery.[102]

---

[90]   *Ibid.*, p.18. General definitions of environmental taxes which rely on their proffered justifications are, as Määttä reminds us, problematic (see Määttä, *op. cit.*, p. 47).

[91]   Nevertheless, in its 2001 report, the OECD preferred the concept of the 'environmentally related tax' (see Organisation for Economic Co-operation and Development, *Environmentally Related Taxes in OECD Countries: Issues and Strategies* (Paris: OECD, 2001)). Severance taxes are taxes levied on the extraction of minerals; they are common in the US.

[92]   See Määttä, *op. cit.*, p. 45.

[93]   *Ibid.*, pp. 40–45.

[94]   Määttä actually uses the Latin tags of environmental taxes *sensu stricto* and those *sensu largo* but we have eschewed the use of the Latin here.

[95]   Määttä, *op. cit.*, pp. 40–53.

[96]   *Ibid.*, p. 41.

[97]   See Finance Act 1994, ss.28–44 and Sched. 6. It is an excise duty (*ibid.*, s.40(1)) and, as such, is under the care and management of HM Customs and Excise. The rates are £10 per flight, for flights within Europe, and £20 for flights to destinations elsewhere (see Finance Act 1994, s.30(4)).

[98]   See Value Added Tax Act 1994, Sched. 8, Group 8, item 4(a), (c).

[99]   Tiley, *op. cit.*, p. 11. See para. 27.5 below (proposed taxes on aviation).

[100]   See Määttä, *op. cit.*, p. 41.

[101]   Anthony Ogus, 'Corrective Taxes and Financial Impositions as Regulatory Instruments' (1998) 61 MLR 767; Anthony Ogus, 'Nudging and Rectifying: The Use of Fiscal Instruments for Regulatory Purposes' (1999) 19 LS 245–66.

[102]   *Ibid.*, p. 64.

The question raised earlier,[103] of whether differential tax rates which are built into non-environmental tax codes and are intended to put a price on environmentally-undesirable behaviour, might themselves be described as environmental taxes, is thus answered in the affirmative, since the differential will itself constitute an environmental tax.[104] A more difficult question arises, however, when taxes whose taxable characteristics are not, *per se*, the harmful characteristics of the product (for example a tax on the carbon content of fuels) but the product itself (for example a tax on plastic bags).[105] Whether or not the revenue raised by a tax is earmarked (or 'hypothecated') for an environmental purpose might also determine whether or not a given tax is to be regarded as environmental in the strict sense.[106] In connection with incentive environmental taxes, it should also be stressed that a tax which *intentionally* steers behaviour (that is, distorts the market) does not breach the neutrality principle. As Tiley says: '[t]he principle of neutrality simply asserts that all distortions should be conscious and so subject to justification through the political process'.[107]

By contrast with environmental taxes in the strict sense, environmental taxes in the broad sense are taxes as to 'which no unanimity exists about the nature of the tax in the sense in question'.[108] They are characterised by the fact that, whilst they are chiefly concerned with raising revenue, they also have considerable effects on the environment.[109] Such taxes include, says Määttä, taxes on energy and taxes on motor transport.[110] Taxes falling into one or other of the above categories can be divided into *emissions taxes* and *product taxes* in accordance with the official classifications of the OECD and the European Commission.[111] The OECD's classification is contained in the 1991 *Recommendation of the Council on the Use of Economic Instruments in Environmental Policy.*[112] The key part of the 1991 Recommendation reads as follows:

2. Emission charges or taxes are payments on the emission of pollutants into air or water or onto or into soil and on the generation of noise. Emission charges or taxes are calculated on the basis of the quantity and type of pollutant discharged.
...[113]
4. Product charges or taxes are levied on products that are harmful to the environment when used in production processes, consumed or disposed of. Product charges or taxes can act as a substitute for emission charges or taxes when charging directly for emissions is not feasible. They may be applied to raw materials, intermediate or final (consumer) products. Product tax differentiation may be designed for the same purpose.[114]

---

103 See para. 1.2.1.3 above.
104 *Ibid.*, p. 48.
105 *Ibid.*
106 *Ibid.*, pp. 49–50. See para. 11.2.2 below (hypothecation).
107 See Tiley, *op. cit.*, p. 11.
108 See Määttä, *op. cit.*, p. 41.
109 *Ibid.*, pp. 50–51.
110 *Ibid.*, p. 41.
111 See para. 4.3.2 below.
112 C(90)177/Final, 31 January 1991. See para. 8.2.6n below.
113 Paragraph 3 of the 1991 Recommendation refers to 'User charges or taxes, [which are] payments for the costs of collective treatment of effluent or waste'.
114 C(90)177/Final, 31 January 1991.

The Commission's classification is contained in its 1997 Communication on the use of environmental taxes and charges.[115] This is generally to similar effect and is worth setting out in full:

> Emission levies involve payments that are directly related to the real or estimated pollution caused, whether emitted into air, water or on the soil, or due to the generation of noise. Existing examples are charges on emissions of NOx from large combustion plants, and charges on pollution to water from waste water treatment plants. Levies on the emission of noise exist in the field of aviation. In so far as these levies are applied to stationary sources (such as industrial plants) they will to a large extent fall outside the scope of this communication, as the cost of paying them falls only on domestic producers.

> Product levies are applied to raw materials and intermediate inputs such as fertilizers, pesticides, natural gravel, and ground water, and on final consumer products such as batteries, one way packaging, car tyres and plastic bags. Some product levies that have existed for many years, mainly in the field of energy, are increasingly regarded as contributing to the integration between environmental and energy policies. Typical examples are taxes on gasoline, diesel and heating oils and electricity.[116]

On the basis of the distinctions set out above, it is clear that environmental taxes imposed on *products* are *indirect* taxes; that they are subject to the *destination* principle;[117] and that, as such, they are eligible for BTA.[118] Equally clear is that environmental taxes imposed on *producers* are *direct taxes*; that they are subject to the *origin* principle;[119] and that, being subject to the origin principle, they do not have the benefit of BTA. Effluent or emissions taxes (including carbon taxes)[120] are, in principle, direct taxes; by the same token, energy taxes are, in principle, indirect taxes (that is, as product taxes). Problems of classification come with waste taxes, depending on the nature of the taxable event in relation to the tax in question.[121]

1.2.2   Approach and sources

The title of the present volume reflects its emphasis on the legal aspects of the UK's environmental levies. As was mentioned at the beginning, our overriding aim has been to give a comprehensive, contextual and critical account of the law relating to environmental taxation in the UK. Such an aim, although relatively modest, seems to us to be amply justified by the absence of any comparable account of the area.

Our approach has been to centre the work on the legal analysis of the measures under consideration. Whilst this analysis has broadly been informed by scholarship on

---

[115]   Commission Communication, *Environmental Taxes and Charges in the Single Market*, COM (97) 9 final, (1997) OJ C224 6 (see para. 12.3.1 below).

[116]   COM(97) 9 final, para. 12.

[117]   See para. 1.2.1.2 above.

[118]   *Ibid.*

[119]   *Ibid.*

[120]   See Thomas C. Schelling, 'Prices as Regulatory Instruments', in *Incentives for Environmental Protection*, ed. by Thomas C. Schelling (London: MIT Press, 1983), pp. 1–40, esp. p. 15.

[121]   See para. 1.4.2(1) below (landfill tax).

environmental regulation, it has not been possible, in what is already rather a large book, to take a unifying theoretical standpoint, based, for example on the economic analysis of law.[122] Such a theoretical approach will have to await another opportunity.[123]

In providing a comprehensive account of the material, we have referred to a range of sources, both theoretical and practical. The detailed reports on both environmental and energy law and policy appearing in the *Financial Times*[124] have proved particularly valuable; the reader will find many references to them in the footnotes. So fast-moving are the areas of law and policy discussed in the book that it is difficult to imagine keeping abreast of regulatory developments without the data provided by that newspaper. We have also made much use of the Environmental Data Services Report ('the ENDS Report')[125] which, containing, as it does, briefings on the latest developments in environmental regulation, is also invaluable. Again, there are many references throughout the footnotes to reports from this source. Finally, we have referred quite extensively to reports and other documentation published by Chatham House (the premises of the Royal Institute of International Affairs ('the RIIA') which operates from them).[126]

Website addresses were correct on 30 September 2004.

1.3   International, European, national and regional contexts

The task of providing a clear account of the UK's environmental taxes is complicated by a number of factors, at least one of which is not present in other areas of tax law. This is the interaction between environmental law and policy on the one hand, and tax law and policy on the other. We shall return to this interaction in paragraph 1.4 below. Another complicating factor is the place of UK environmental regulation, including its environmental taxes, within the international and European Community's legal order. Before going any further, it would perhaps be helpful to give a brief indication of the issues raised by the place of the UK and its constituent countries in that legal order.

The UK is almost, although not quite, a federal state.[127] The detailed implications of this statement will be followed up at a later stage of the discussion.[128] It should, however, be noted at the outset that, except in environmentally-related taxation matters,

---

[122] This is the approach taken by Määttä, *op. cit.* See Richard Posner, *Economic Analysis of Law*, 6th edn (New York: Aspen, 2003). See also the discussion of Posner's work in David Campbell and Sol Picciotto, 'Explaining the interaction between law and economics: the limits of formalism' (1998) 18 LS 249–78.

[123] On this footing, we are content to accept the theoretical economic arguments and to analyse the findings of the bodies (mainly Parliamentary Select Committees – see para. 4.2.1.1(3) below) whose duty it is to report on how these theories have worked in practice. This is consistent with the approach taken in, for example, Baldwin and Cave, *op. cit.*, throughout.

[124] See www.ft.com.

[125] See www.endsreport.com.

[126] See www.chathamhouse.org.uk. Chatham House is the origin of the so-called 'Chatham House Rule', first promulgated in 1927.

[127] See Chris Hilson, *Regulating Pollution – A UK and EC Perspective* (Oxford: Hart Publishing, 2000), p. 47.

[128] See para. 4.2.2 below.

the Scottish Parliament and the (currently dissolved) Northern Ireland Assembly have full legislative powers in relation to the environment.[129] Although the UK Parliament retains competence to pass primary legislation in relation to both England and Wales, the Welsh Assembly has competence as regards Wales in relation to secondary legislation. Since this includes the environment but, again, not environmentally-related taxation, and, since most environmental legislation is secondary legislation, the Welsh Assembly possesses considerable power in relation to the environment. Thus, except in relation to taxes, as Hilson points out, the result of this structure is that, '[t]he UK Parliament does not appear to have maintained any power to legislate on the environment for the UK as a whole'.[130] One practical consequence of this situation is that environmentally-related statutory instruments covering the same subject-matter either exist in two forms, one relating to England and Wales and the other to Scotland,[131] or, possibly, in three forms, relating to each of England, Wales and Scotland.[132] A second, more important, point is that none of Wales, Scotland or Northern Ireland has any power to create its own environmental tax.[133]

If it is correct that the UK Parliament has not retained power to legislate on the environment for the UK as a whole, one explanation may be that the legislative activity of the UK Parliament in relation to environmental matters consists in transposing the provisions of European Directives. This is so, even though competence to legislate in the environmental sphere is shared between the Community and the Member States.[134] Shared, too, is competence in relation to taxation (including environmental taxation), although, here, any decision must be unanimous[135] as between the Member States of the EU.

Finally, since the UK is a Member State of the EU, and since both the UK and the EU are members of the WTO, any taxes designed and implemented in the UK must comply with GATT 1994/WTO rules.[136] These are very similar to Community rules on taxation and free movement of goods[137] but, despite the existence of an 'environmental' exception,[138] do not easily lend themselves to environmental measures. Thus, the

---

[129] *Ibid.*

[130] See *op. cit.*, p. 47.

[131] Where this situation obtains, we refer only to the SI for England and Wales, in the absence of some special consideration. This is simply for reasons of space and we hope not thereby to offend readers from UK countries other than England.

[132] Or possibly three (four) forms, where there is a separate instrument for Northern Ireland under the current direct rule arrangements (see para. 4.2.2 below).

[133] Or any other tax, for that matter. Part IV of the Scotland Act 1998 empowers the Scottish Parliament to pass a resolution with the effect of varying the basic rate of income tax by up to three percentage points, upwards or downwards, for a particular tax year (Scotland Act 1998, s.73(1)). If resolved upon, such a variation affects 'Scottish taxpayers', that is, individuals who are treated as resident in the UK and whose closest connection is with Scotland (Scotland Act 1998, s.75). The Act grants no other tax-raising power to the Scottish Parliament, however. 'Scotland' is defined in the Scotland Act 1998, s.126(1), (2).

[134] See para. 12.2.1 below.

[135] See paras 4.3.1, 12.2.1 and 12.3.2 below.

[136] See para. 8.4.1 below.

[137] See paras 12.3 and 12.4 below.

[138] See para. 8.4.2 below.

rule that a direct tax (for example an emissions tax) cannot benefit from BTA has controversially contributed to the inhibition of taxes on carbon emissions.[139]

1.4 Regulatory and taxation contexts

1.4.1 Introduction

One of the principal contextual aims of the book is to integrate our discussion of environmental taxes into a discussion of environmental regulation more generally. To this end, in subsequent chapters, we consider the UK's environmental regulation, and the place of environmental levies and other economic instruments within it, prior to analysing the taxation context of those levies.[140]

The discussion of the regulatory context of the levies in Chapter 6 below is divided into the main areas of application of environmental taxes, that is, waste management and control of air and atmospheric pollution, as well as air passenger and road freight transport and mineral extraction. The regulatory background to each of these areas is discussed in some detail in Chapter 6. Meanwhile, it may assist the reader to have a brief regulatory and taxation overview, with emphasis, in relation to the former, on the economic instruments currently employed in the particular area of application.

It is worth noting that, whilst landfill tax and climate change levy form part of a package of regulatory instruments in their respective areas of application, aggregates levy forms the sole, or at least the main,[141] basis of environmental policy in the area to which it applies. Määttä would therefore classify aggregates levy as an *independent* environmental tax, climate change levy and landfill tax being *complementary* environmental taxes.[142]

1.4.2 Regulatory context of the UK's environmental taxes

1.4.2.1 Waste management regulation

The command and control framework of waste management law forms an intricate regulatory pattern.[143] The main elements are: (1) the Integrated Pollution Prevention and Control ('IPPC') permit system, which covers, *inter alia*, the disposal of waste by

---

139   There are, of course, other formidable obstacles to such a tax!
140   See Chapters 6 and 7 below.
141   Town and country planning law and Environmental Impact Assessment also having a part to play (see para. 6.6 below).
142   See Määttä, *op. cit.*, p. 70.
143   The expression 'command and control' is customarily used to refer to regulation involving the bringing of influence to bear on the regulated by requiring standards to be met, on pain of criminal sanction (for example via the imposition of fines). Since command and control typically relies on licensing processes involving payments to government or government agencies, it is often difficult for the layperson to distinguish such payments from taxation (see category 3 in para. 1.2.1.4 above). This, together with the fact that economic instruments (see para. 1.2.1.5 above) cannot operate satisfactorily without being backed up by criminal sanctions, makes the distinction between each type of regulation a rather slippery one (see Baldwin and Cave, *op. cit.*, pp. 35–9).

incineration or by landfill, as well as waste recovery and fuel production from waste; (2) a waste management licensing system and a statutory duty of care in relation to the handling of waste (both under Part II of the Environmental Protection Act 1990); (3) the banning of the co-disposal of hazardous and non-hazardous wastes, of tyres[144] and of liquid, clinical and hazardous wastes at landfill sites (pursuant to legislation implementing the Landfill Directive);[145] (4) special rules for the disposal of special and hazardous wastes; and (5) rules on waste imports and exports.[146]

In addition to the above, however, the following economic instruments have a prominent part to play,[147] most importantly, landfill tax.

(1) Landfill tax

Landfill tax was introduced in the Finance Act 1996.[148] It is chargeable on taxable disposals of material as waste, by way of landfill, at landfill sites.[149] The person liable to pay the tax charged on a taxable disposal is the landfill site operator.[150] It is generally payable at a fixed amount for each whole tonne of material disposed of, plus a proportionately reduced sum for any additional part of a tonne, or, where less than a tonne is disposed of, just a proportionately reduced amount.[151]

Landfill tax is discussed in detail in Chapter 15 below but several key features of the levy will be apparent from the foregoing. First, there is no doubt that landfill tax is indeed a tax within the definition set out above.[152] Secondly, it is also clearly an environmental tax, since its tax base is a physical unit of something that has 'negative impact' on the environment when released.[153] Whether it is treated as a direct tax on emissions (as theoretically it might be) or an indirect tax on the consumption of landfill services (as in practice it is) is probably unimportant.[154]

(2) Packaging waste recovery notes

Packaging waste recovery notes ('PRNs') are a market-creating mechanism for the recovery and recycling of packaging waste. They operated originally[155] on a non-

---

144 That is, whole tyres. Landfilling shredded tyres will be banned in 2006 (see para. 6.3.2(4) below).
145 See para. 12.2.5.1(2)(b) below.
146 See para. 6.3 below (in relation to all of these).
147 Note also the projected waste performance reward grant ('WPRG'), currently (December 2004) under consultation (see para. 6.3.3(5) below).
148 Finance Act 1996, ss.39–71, Sched. 5; Finance Act 1997, ss.50–53 and 113, Sched. 5; Finance Act 1999, s.124; Finance Act 2000, ss.140–42, Sched. 37. There are also at least ten statutory instruments specifically relating to landfill tax: these are referred to where appropriate in the text below.
149 Finance Act 1996, s.40.
150 *Ibid.*, s.41(1).
151 *Ibid.*, s.42(1).
152 See paras 1.2.1.1 above and 8.3.1 below.
153 See para. 1.2.1.5(2) above.
154 See Määttä, *op. cit.*, p. 254.
155 Since 1 January 2004, both PRNs and the newly invented PERNs have operated under the auspices of Producer Responsibility Obligations (Packaging Waste) (England) (Amendment) Regulations 2003, S.I. 2003 No. 3294: see para. 19.8 below.

statutory basis and as an adjunct to a statutory command and control regime creating producer responsibility for packaging waste and imposing penalties for a range of offences. That regime obliges businesses of a certain size who deal in virtually any way with the packaging of goods to register with the relevant government agency;[156] to take reasonable steps to recover and recycle particular percentages of packaging handled in the preceding year; and to submit a certificate of compliance to the agency at the end of the current year. Obligated business may meet their recovery obligation, either individually, or by joining a registered compliance scheme, which will fulfil the obligation on their behalf. The submission of PRNs is the main way in which obligated businesses, including compliance schemes, demonstrate their compliance. They acquire the PRNs either on their issue by accredited packaging waste reprocessors or by purchase on the open market from other organisations.

Salmons epitomises PRNs by saying that, '[a]lthough not conceived as such when they were introduced in 1998, [PRNs] ... have rapidly evolved into a functioning tradeable compliance credit system'.[157] Packaging waste recovery notes are discussed in detail in Chapter 19 below.

(3) The Waste Recycling Credits Scheme

A waste recycling credits scheme ('WRCS') operates, which is designed to ensure that neither waste disposal authorities ('WDAs') nor waste collection authorities ('WCAs')[158] are penalised for the costs of recycling. It is not considered further in this study.[159]

(4) The Landfill Allowances Trading Scheme ('the LATS')

The framework for the landfill allowances trading scheme ('the LATS'), which is due to come into operation in England and Wales in 2005, is contained in the Waste and Emissions Trading Act 2003. The relevant minister[160] is empowered by the 2003 Act to allocate to WDAs, in each of the four countries of the UK, allowances which authorise the sending to landfill[161] of specified amounts of biodegradable municipal waste ('BMW'), for each year between 2004 and 2020.[162] The LATS, a 'cap and trade scheme',[163] is discussed in detail in para. 20.7 below.

---

156 That is, the Environment Agency (see para. 4.2.1.3 below).
157 See Roger Salmons, 'A New Area for Application of Tradeable Permits: Solid Waste Management', in Organisation for Economic Co-operation and Development, *Implementing Domestic Tradeable Permits: Recent Developments and Future Challenges* (Paris: OECD, 2002), pp. 187–226, esp. pp. 199–211.
158 See paras 2.3 and 6.3.2(1) below.
159 It is currently (December 2004) the subject of government consultation (see www.defra. gov.uk).
160 The identity of whom varies for England, Scotland, Wales and Northern Ireland (see Waste and Emissions Trading Act 2003, s.24(1)).
161 Defined in Waste and Emissions Trading Act 2003, s.22.
162 Waste and Emissions Trading Act 2003, s.4(1).
163 See Salmons, *op. cit.*, esp. pp. 211–17.

1.4.2.2 Control of air and atmospheric pollution

This area of environmental regulation is dominated by the UK's commitment under the Kyoto Protocol to reduce its 1990 levels of all greenhouse gases ('GHGs') by 12.5 per cent by 2010.[164] Although there are command and control regimes relating to emissions and air quality, the chief regulatory instruments at work in the UK's climate change programme are economic ones. They are as follows:

(1) Climate change levy
Climate change levy, which was enacted in the Finance Act 2000,[165] has been imposed on supplies for industrial, commercial and agricultural purposes[166] of electricity, gas, liquid natural gas, coal, lignite and coke,[167] made after 31 March 2001.[168] Excluded from the list of taxable commodities are hydrocarbon oil, road fuel gas and waste (within the scope of the Environmental Protection Act 1990).[169] There is a series of environmentally-inspired exemptions, *inter alia*, for renewable source electricity and electricity generated in combined heat and power ('CHP') stations.[170]

Normally, the person liable to account for the levy is the person making the supply.[171] However, where a taxable supply is made by a non-UK resident non-utility, the person liable to account for the levy charged on the supply is the person to whom the supply is made.[172] Exports of taxable commodities are not within the scope of the levy, an exemption that mirrors the controversial aggregates levy relief for the supply of aggregate to a destination outside the UK.[173] The levy is charged at different poundages, depending on the type of energy supplied, reduced rates being applicable to supplies to horticultural producers[174] and to supplies made by certain facilities, where the operator has entered into a so-called 'climate change agreement'.[175] Climate change levy is discussed in detail in Chapter 14. Introduced

---

[164] See para. 6.4.2 below.

[165] Finance Act 2000, s.30, Scheds 6 and 7. See Stephen Smith, 'Environmental and Public Finance Aspects of the Taxation of Energy', in *Environmental Policy: Objectives, Instruments and Implementation* (Oxford: Oxford University Press, 2000), pp. 172–202. There are in addition around at least nine statutory instruments relating specifically to climate change levy, and these are referred to as relevant in the text below.

[166] Finance Act 2000, Sched. 6, para. 8 (this effect is achieved by excluding supplies for domestic or charity use from the scope of the levy).

[167] Finance Act 2000, Sched. 6, paras 2 and 3(1). Coke is the 'solid porous fuel that remains after gases have been driven from coal by heating'; lignite is 'a brownish black coal that is harder than peat but usually retains the texture of the original wood' (see the *Longman Concise English Dictionary* (London: Longman Group, 1985)).

[168] Finance Act 2000, Sched. 6, para. 10.

[169] *Ibid.*, para. 3.

[170] See paras 14.3 and 14.4 below.

[171] Finance Act 2000, Sched. 6, para. 40(1).

[172] *Ibid.*, Sched. 6, para. 40(2).

[173] Finance Act 2000, Sched. 6, para. 11 (see para. 8.4.5.1 below).

[174] *Ibid.*, Sched. 6, para. 43.

[175] *Ibid.*, Sched. 6, para. 44. Gas for burning in Northern Ireland is exempted from the levy on a temporary basis (*ibid.*, Sched. 6, para. 11A).

in Chapter 14, as a prelude to the discussion in Chapter 20, is the UK Emissions Trading Scheme ('the UK ETS'), the UK ETS having been designed to bolster the network of climate change agreements.

Despite its rather coy name, climate change levy is, like landfill tax, a true tax.[176] It is also, on the basis of the 1997 OECD definition given above,[177] an environmental tax, which takes energy consumption as a proxy for the carbon emissions released into the atmosphere when the various taxable commodities to which it applies are used or consumed. Since it is an energy tax rather than a carbon tax, as well as a tax on energy products rather than a tax on emissions, it is an indirect tax. It therefore makes use of BTA in taxing imports of taxable commodities, whilst exempting their exportation.

(2) The UK Emissions Trading Scheme ('the UK ETS')

The UK ETS was created in March 2002, under powers conferred on the Secretary of State for Environment, Food and Rural Affairs,[178] by ss.3(5)(a) and 153, Environmental Protection Act 1990. Initial participation in the scheme was on a voluntary basis.[179]

There are two main types of participant in the UK ETS. First, there are 'direct participants', who undertook emissions reduction targets in return for incentive payments from a specially-designated government fund of £215 million. Emission reduction targets could either relate to carbon dioxide alone or to all six Kyoto GHGs.[180] Secondly, there are 'agreement participants', whose emissions reduction target has been fixed, not in return for an incentive payment, but in return for a one-fifth rate of climate change levy as parties to climate change agreements.

For each year of the UK ETS, direct participants receive allowances corresponding to their emissions target for that year. Each direct participant must then possess enough allowances at the end of the year to cover its emissions during that year. If it does not have enough allowances, then the direct participant must buy more allowances from other participants, to cover its excess emissions. Failure to have enough allowances disentitles the direct participant to a slice of incentive payment and results in the tightening of the following year's target. If, however, the direct participant has more than enough allowances, it may sell them to other participants.

Likewise, where an agreement participant overachieves (that is, makes fewer emissions than its climate change agreement requires), then it is issued with allowances to the extent of the overachievement. It can then trade these with participants lacking sufficient allowances. Equally, where an agreement participant underachieves, it may either purchase allowances from other participants or pay the full rate of the levy.

The detail of the UK ETS is discussed in Chapter 20 below. The scheme is an entirely novel concept in UK law and has been highly controversial, especially with

---

176 See para. 1.2.1.1 above and para. 7.3.1 below.
177 See para. 1.2.1.5 above.
178 See para. 4.2.1.2(1) below.
179 For an early legal assessment of the UK ETS, see Anthony Hobley, 'The UK Emissions Trading System: Some Legal Issues Explored', in *Economics, Ethics and the Environment*, ed. by Julian Boswall and Robert Lee (London: Cavendish Publishing, 2002), pp. 61–79.
180 See para. 8.3.1.4 below.

regard to the targets actually set. It is in part a 'cap and trade scheme' and in part a 'baseline and credit scheme'.[181]

(3) The EU Emissions Trading Scheme ('the EU ETS')
The EU ETS is due to start in January 2005.[182] The Directive which creates the scheme[183] is based on a provision of the European Treaty which requires the EU Council to decide on the Community action necessary to achieve the Treaty's environmental objectives.[184] It is thus a key example of Community level governance in an area where competences are shared, which, given the transboundary nature of the environmental problem that it is seeking to address, seems entirely appropriate.[185]

Participants in the EU ETS are the operators of specified industrial installations, each of which must have a permit in order to emit carbon dioxide. The initial limitation of the scheme to emissions of carbon dioxide is one point of difference with the UK ETS. Another difference is the compulsory nature of participation, initial participation in the UK ETS having been voluntary only. A third difference is that the installations covered include electricity generators, who were excluded from eligibility to participate in the UK ETS.

Under the EU ETS, an operator of a specified installation is allocated with allowances for each year, which correspond to the relevant installation's target for the year. This is a fourth point of difference with the UK ETS, the target being imposed rather than being agreed in return for a share in an incentivisation fund. Allocation of allowances is in accordance with a Community-wide emissions cap on the industries covered, the cap being divided between the Member States in accordance with so-called 'national allocation plans' ('NAPs').

At the end of each year, the operator of the relevant installation is required, under the terms of its emission permit, to surrender allowances equal to the carbon dioxide that it has actually emitted in that year. If the operator has insufficient allowances to cover its emissions, then it must either purchase extra allowances from other operators or pay a penalty. However, if the operator has more than enough allowances to cover its emissions, then it may sell them to other operators.

The detail of the EU ETS, which is basically a 'cap and trade' scheme similar to earlier US examples designed to combat acid rain,[186] is discussed in Chapter 28 below. Whilst unquestionably groundbreaking, as the first trans-national emissions trading scheme, the design of the EU ETS has been described as 'a pragmatic

---

[181]   See Kumi Kitamori, 'Domestic GHG Emissions Trading Schemes: Recent Developments and Current Status in Selected OECD Countries', in OECD (2002), pp. 69–103.
[182]   See the overview of the EU ETS in Fiona Mullins and Jacqueline Karas, *EU Emissions Trading: Challenges and Implications of National Implementation* (London: Royal Institute of International Affairs, November 2003), pp. 12–14.
[183]   See para. 28.1 below.
[184]   See para. 28.2.1 below.
[185]   See para. 1.3 above.
[186]   See Steve Sorrell, 'Turning an Early Start into a False Start: Implications of the EU Emissions Trading Directive for the UK Climate Change Levy and Climate Change Agreements', in Organisation for Economic Co-operation and Development, *Greenhouse Gas Emissions Trading and Project-based Mechanisms* (Paris: OECD, 2004), pp. 129–151, p. 130.

compromise between economic efficiency and political acceptability'.[187] Even that pragmatic compromise may come in time to be severely tested.

(4) The Renewables Obligation ('the RO')
The Renewables Obligation was introduced in April 2002. It was imposed under powers granted to the relevant authorities by the Electricity Act 1989, as amended by the Utilities Act 2000. As its name suggests, unlike the UK ETS, there was nothing optional about it.

The RO requires an electricity supplier to prove to the regulator, the Office of Gas and Electricity Markets ('Ofgem'),[188] that, either alone or in combination with other suppliers, it has supplied specified quantities of renewable source electricity[189] to customers in Great Britain within a specified period. The supplier must satisfy the RO by producing so-called 'Green Certificates' or 'ROCs' to the regulator. Green Certificates are traded, since the RO may be satisfied by the production of certificates which were originally issued to a different supplier. Under the rules of the UK ETS, the holder of Green Certificates may convert them into UK ETS allowances; the converse is not, however, possible.[190]

The concept of Green Certificates is well-documented in the literature; they exist in various forms throughout OECD countries.[191]

1.4.2.3 Regulation of mineral extraction

The key regulatory instrument here is, as mentioned above, aggregates levy.

The levy was introduced in the 2001 Finance Act.[192] It has been chargeable, since 1 April 2002, whenever quantities of taxable aggregate have been subjected to commercial exploitation.[193] Aggregate is rock, gravel or sand, together with whatever substances are incorporated in the rock, gravel or sand or naturally occur mixed with it.[194] Aggregate is not taxable, *inter alia*, if it has previously been used for construction purposes,[195] a point which underlines the fact that the levy is aimed at reducing the environmental impact of commercial quarrying. The width of the 'commercial exploitation' notion means that it is envisaged that imported aggregate should be treated in the same way as aggregate originating in the UK. In line with

---

[187] See Sorrell, *op. cit.*, p.130.
[188] See para. 4.2.1.3 below.
[189] See para. 6.4.3.1(2)(a) below.
[190] See paras 21.5 and 20.6 below.
[191] The concept of Green Certificates is referred to in the literature as a 'tradeable renewable energy certificate' (a 'TREC'). See the survey of TRECs in Richard Baron and Ysé Serret, 'Renewable Energy Certificates: Trading Instruments for the Promotion of Renewable Energy', in Organisation for Economic Co-operation and Development, *Implementing Domestic Tradeable Permits: Recent Developments and Future Challenges* (Paris: OECD, 2002), pp. 105–40.
[192] Finance Act 2001, ss.16–49 and 109–11, Scheds 4–10.
[193] *Ibid.*, ss.16(1),(2) and 17(2).
[194] *Ibid.*, s.17(1).
[195] *Ibid.*, s.17(2)(b).

the points discussed above, the supply of aggregate to a destination outside the UK is relieved from the levy.[196] Anyone who is responsible for subjecting aggregate to commercial exploitation is liable to pay the levy.[197] It is payable at a fixed amount for each tonne of aggregate subjected to commercial exploitation, the amount of levy payable on a part of a tonne being the proportionately reduced amount.[198]

Aggregates levy is discussed in detail in Chapter 13 below. Like climate change levy and, despite an early governmental equivocation,[199] its status as a tax is not in doubt.[200] Unlike climate change levy, it is by comparison a relatively 'small' environmental tax, lacking a panoply of associated instruments. Secondly, it is clearly an environmental tax, the subjection to commercial exploitation of the material in question being a proxy for the environmental effects of quarrying. Although the argument that aggregates levy is a direct tax might be a good one, there is judicial authority to the effect that the levy is an indirect tax.[201] On this basis, the BTA for which the tax provides is a lawful one.

1.4.2.4   Transport regulation

Two economic instruments are currently in operation (setting aside, for example, excise duty differentials):[202]

1   Powers to introduce workplace parking levies were introduced by the legislation instituting the new Mayor and Assembly for Greater London.[203] Similar powers were subsequently granted to other local authorities in England and Wales,[204] which legislation also amended the earlier measures relating to London.[205] Workplace parking levies are discussed in detail in Chapter 17 below. To the authors' knowledge, no local authority has so far[206] exercised these powers.

2   The legislation setting up the Mayor and Assembly for Greater London also introduced the concept of road user charging.[207] Again, similar powers were subsequently granted to other local authorities in England and Wales,[208] and again this legislation also amended the earlier provisions relating to London.[209]

---

[196]   See para. 8.4.5.1 below.
[197]   Finance Act 2001, s.16(3).
[198]   *Ibid.*, s.16(4).
[199]   See para. 13.1 below.
[200]   See para. 7.3.1 below.
[201]   See *R (on the application of British Aggregates Association and others) v. C & E Commrs*, [2002] EWHC 926 (Admin), [2002] 2 CMLR 51, paras 68–78 (Moses, J.).
[202]   See para. 22.2 below.
[203]   Greater London Authority Act 1999, s.296 and Sched. 24.
[204]   Transport Act 2000, ss.178–190.
[205]   *Ibid.*, Sched. 13, paras 19–43.
[206]   That is, as at December 2004.
[207]   Greater London Authority Act 1999, s.295 and Sched. 23. 'Road user charging' is used here in the sense of 'cordon pricing' (see Stephen Ison, *Road User Charging: Issues and Policies* (Aldershot: Ashgate, 2004), p. 14).
[208]   Transport Act 2000, ss.163–77.
[209]   *Ibid.*, Sched. 13, paras 1–18.

In 2002, the government also took powers to tax road user by heavy lorries.[210] Road user charging schemes are discussed in Chapter 18 below.

To date,[211] only Durham, in the northeast of England, has introduced a road user charging scheme outside London.

1.4.2.5 *Geographical and jurisdictional considerations*

Each of landfill tax, climate change levy and aggregates levy apply throughout the UK. For most tax purposes, the UK consists of England, Wales, Scotland and Northern Ireland, plus, dating from the 1973 Finance Act, the territorial sea and continental shelf. Statutory provisions governing the territorial effect of a tax are, of course, of the first importance on traditional tax law principles.[212]

Although there is no special definition of the UK in Finance Act 1996, the nature of landfill tax means that such a definition may, in any event, be unnecessary. This is because Finance Act 1996, s.66, provides, *inter alia*, that land is a landfill site at a given time if at that time a site licence for the purposes of Part II of the Environmental Protection Act 1990 or an IPPC permit is in force in relation to the land, in each case authorising disposals in or on the land.[213] In an electronic age, the attractions of such a 'land-bound' tax to the resourceful legislator are obvious, and this is again a point to which we shall return.[214]

In relation to aggregates levy, whose status as a property tax is rather more doubtful, the legislator has sought to achieve a comparable effect by casting very wide the net of accountability and registrability.[215] This is a particularly worrisome feature of the levy, and it is with some relief that we find that the interpretation provisions of Finance Act 2000 (the principal statute relating to climate change levy) do define the UK, in the way familiar from non-environmental taxes as including the territorial waters adjacent to any part of the UK.[216]

It was mentioned above that all of the devolved administrations have extensive powers in relation to the environment.[217] The non-tax measures referred to above illustrate some of the consequences of the proposition. A good example is the LATS. In England, the allocation of allowances under the LATS is the responsibility of the Secretary of State for Environment, Food and Rural Affairs;[218] in Scotland, it is the responsibility of the Scottish Ministers;[219] in Wales, it is the National Assembly

---

[210] Finance Act 2002, s.137.

[211] That is, as at December 2004.

[212] Two fundamental jurisdictional principles, each of which reflects an axiom of international practice, apply with full force in UK tax law. See *Government of India v. Taylor*, [1955] AC 491 (in the absence of express agreement – which there is within the EU – one state will not enforce the revenue law of another state) and *Clark v. Oceanic Contractors Inc.*, [1983] STC 35 (UK tax laws apply only in the UK).

[213] See the discussion in para. 15.2 below.

[214] See Chapter 26 below.

[215] See para. 1.4.2.3 above and para. 13.3 below.

[216] See Finance Act 2000, Sched. 6, para. 147.

[217] See para. 1.3 above.

[218] See Waste and Emissions Trading Act 2003, s.24(1)(a). See para. 4.2.1.2(1) below.

[219] *Ibid.*, s.24(1)(b). See para. 4.2.2 below.

for Wales;[220] and in Northern Ireland, it is the Department of the Environment.[221] Different commencement dates are proposed, for example, 1 October 2004 (Wales) and 1 April 2005 (Scotland) and there will be regional variations in the operation of the scheme. For example, in Wales, allowances will not be tradeable, while in Northern Ireland, although they will not be capable of being traded, they will be capable of being banked and borrowed.[222]

1.4.3   Taxation context of the UK's environmental taxes

1.4.3.1   Relationship between environmental and non-environmental taxes

There are seven main non-environmental taxes in the UK: income tax, VAT, national insurance contributions ('NICs'),[223] corporation tax, capital gains tax ('CGT'), stamp duties (which are being supplemented, and largely replaced, by stamp duty reserve tax and stamp duty land tax) and inheritance tax ('IHT').

Each one except IHT[224] will be mentioned with sufficient frequency in the rest of the study for a brief note of its essential features to be justifiable at this stage,[225] together with a brief note of its relevance to the environmental regulatory instruments mentioned above.[226]

1   Income tax is chargeable on the income of individuals, trusts and estates, for so-called 'tax years'.[227] In essence – although this is probably an over-simplification – the amount charged for each tax year is based on the statutory income of the person(s) in question for the tax year. The person's statutory income is then reduced by any available personal allowances, with the remainder being taxed at the starting, basic or higher rates.[228]

---

[220]   *Ibid.*, s.24(1)(c). See para. 4.2.2 below also.
[221]   *Ibid.*, s.24(1)(d), reflecting the dissolution of the Northern Ireland Assembly and the re-imposition of direct rule.
[222]   See 353 ENDS Report (2004) 47–48.
[223]   These ought to be treated as taxes, rather than charges, since payments are 'graduated in a way which does not relate directly to the graduation in benefit' (see Tiley, *op. cit.*, p. 6, and *Metal Industries (Salvage) Ltd v. ST Harle (Owners)*, [1962] SLT 114 (employers' NICs are taxes)). As such, NICs are *hypothecated taxes* (see para. 11.2 below).
[224]   Inheritance tax is a tax which has but a small part to play in what follows. In essence, however, it is a tax on non-commercial transfers of capital, whether during an individual's lifetime or on death. Lifetime transfers, where they are taxed, are taxed more favourably than transfers on death, the latter being taxed at double the rate of the former.
[225]   General guides to UK tax law which can be especially recommended are: *Davies: Principles of Tax Law*, ed. by Geoffrey Morse and David Williams, 4th edn (London: Sweet and Maxwell, 2000); Tiley, *op. cit.*; and Lesley Browning *et al.*, *Revenue Law – Principles and Practice*, 22nd edn (London: LexisNexis Tolley, 2004).
[226]   These points are developed in much greater detail in the rest of the book, especially in Chapters 23 and 24 below.
[227]   6 April to the following 5 April.
[228]   Or the special rate on dividend income.

2   Capital gains tax is chargeable for tax years on the realised chargeable gains, less realised allowable losses, of individuals, trusts and estates. It is charged for individuals at the same rate as their marginal income tax rate, and for personal representatives and trustees at a single rate.

3   Corporation tax is chargeable at starting, small company and main rates, on the income and net chargeable gains of companies' accounting periods.

4   National insurance contributions are chargeable on the earnings of employed earners, and payable by both the employee ('primary contributions') and by the employer ('secondary contributions'). Primary and secondary contributions alike are calculated as percentages of the employee's earnings for so-called 'earnings periods', from lower earning thresholds, up to an upper earnings limit.

Certain general points can be made about the four taxes summarised above: (1) not all income is taxable (in the case of income tax and corporation tax), and not all payments are earnings (in the case of NICs); (2) in the case of income tax and corporation tax on income alike, the question of what is income and what is capital is, in general, ascertained by reference to the same tedious and rather arbitrary rules;[229] (3) for both income tax and corporation tax on income, income must fall within one of the so-called 'Schedules' (A, D (Cases I and II) and F)[230] to be taxable, except where (in the case of an individual) it is employment income, in which case it is taxed under the Income Tax (Earnings and Pensions) Act 2003; and (4) in the case of both CGT and corporation tax on chargeable gains, the realisation of gains on the disposal of business assets may generally be postponed when the assets in question are being replaced.

    An obvious way of 'greening' each of the above taxes (but especially income tax and corporation tax) is to create exemptions, deductions, reliefs or tax credits to reflect various 'environmentally-friendly' forms of behaviour.[231]

5   Value added tax is chargeable on domestic supplies of goods and services, on cross-border acquisitions from other Member States of the EU and upon imports from countries outside the EU. Depending on the circumstances, when it involves a cross-border supply, it may be charged either on the origin principle or on the destination principle.[232] It differs from all of the taxes discussed above in a number of major respects: (1) it is an indirect, as opposed to a direct, tax;[233] (2) rather than merely having a counterpart in other EU Member States, it is a European tax in the sense of being created pursuant to the terms of the European Treaty; and (3) it features a much more complex rate structure, a structure which in the UK context has been described as '... one of the most complex rate structures for VAT in the developed world'.[234]

---

[229]   See, for example, *British Insulated and Helsby Cables Ltd v. Atherton*, (1925) 10 TC 177; *IRC v. Church Commissioners for England*, [1976] STC 339.

[230]   Individuals, trusts and estates only, in the case of Schedule F.

[231]   See Chapters 23–25 below. The Scedules will be discontinued in 2005 for income tax.

[232]   See para. 1.2.1.2 above.

[233]   *Ibid.*

[234]   See Davies, *op. cit.*, p. 385.

6   Stamp duties have recently attained a significance in the UK tax system which
     only a few years ago would have seemed barely credible. This is in some measure
     due to recent legislative steps taken to 'green' the UK tax system[235] but it is
     also symptomatic of a current tendency of governments to favour property and
     product taxes in an attempt to combat erosions in the tax base.[236] Stamp duties
     are taxes on particular types of document, at rates that are either *ad valorem* or
     fixed.

Provided that the strict requirements of s.74(1), Income and Corporation Taxes
Act 1988, are met, payments in respect of environmental taxes for which no credit
has been obtained[237] should be deductible in calculating trading profits[238] for the
purposes of income tax and corporation tax. The same point should in principle apply
for payments made under the PRN scheme, the UK ETS, the EU ETS and the RO.[239]
Penalties under, for example, the UK ETS or EU ETS, are presumably not deductible
in calculating trading profits, however.[240] Clearly not deductible for these purposes,
furthermore, are civil penalties imposed under the three main environmental tax
codes.[241] Value added tax is, in principle, chargeable on the environmental tax-
inclusive amount of the price charged for, for example, landfill services supplied
(landfill tax) or electricity supplied (climate change levy).[242]

1.4.3.2   Geographical and jurisdictional considerations

In the cases of income tax and corporation tax on income, the general rule is that,
where a person is resident in the UK, he is liable to tax on his worldwide income;[243]
where a person is not resident in the UK, then he is liable to tax only on income
arising to him in the UK. These principles reflect a series of agreed assumptions
about international taxation applicable to all OECD members, and which differ from
state to state only in matters of detail.[244] They are consistent with the idea, already
referred to, that direct taxes are applied consistently with the origin principle.[245] The

---

[235]   See Part II, Section B below.
[236]   See Chapter 26 below.
[237]   See para. 16.9 below.
[238]   Note the peculiar point on the deductibility of 10 per cent of contributions to environmental
        trusts (see paras 21.3.2 and 24.8 below).
[239]   See para. 24.7 below for a detailed discussion of these points.
[240]   See *McKnight v. Sheppard*, [1999] 3 All ER 491.
[241]   *Ibid.* See para. 16.14 below.
[242]   See the Sixth VAT Council Directive 77/388/EEC, Art 11A(2)(a), which states that, for
        the purposes of VAT, 'the taxable amount shall include taxes, duties, levies and charges,
        excluding the value-added tax itself'.
[243]   This is subject to the qualification that, where an individual is resident in the UK but
        domiciled elsewhere, he will generally be taxed on his foreign income (and capital gains)
        'only if it is remitted to the UK' (see summary in Browning *et al.*, *op. cit.*, para. [13.1]).
[244]   See also Williams, *op. cit.*, p. 15, and, generally, Brian Arnold and Michael McIntyre,
        International Tax Primer, 2nd edn (The Hague: Kluwer, 2002) and Roy Rohatgi, *Basic
        International Taxation* (London: Kluwer, 2002), p. 12.
[245]   See para. 1.2.1.2 above.

same basic principles, with appropriate modifications, apply in relation to CGT and to corporation tax on chargeable gains.[246] Because of double taxation in the case of a person resident in one jurisdiction who has a source of income in another, credits or exemptions are available either unilaterally, or under the provisions of the relevant double tax convention.[247]

Liability to each of NICs and VAT depend on a range of other rules, the effect of which may perhaps be summarised as follows. In relation to NICs, liability depends upon residence or presence in the UK.[248] No doubt because of its nature, the position in relation to VAT is somewhat different. Liability to VAT depends on the State in which the supply, acquisition or importation[249] has taken place, as to the determination of which there are detailed rules,[250] reflecting the origin and destination principles discussed above.[251]

In relation to income tax, corporation tax and CGT, the UK is defined as Great Britain and Northern Ireland.[252] However, it is also deemed to include the territorial sea of the UK and every designated area designated under the Continental Shelf Act 1964, s.1(7).[253]

As regards NICs, the Social Security Contributions and Benefits Act 1992 states that any reference to Great Britain in the Act 'includes a reference to the territorial waters of the UK adjacent to Great Britain, [and] ... any reference to the UK includes a reference to the territorial waters of the UK'.[254] The theme is continued in the VAT legislation, the UK including, for the purpose of that tax, the territorial sea of the UK.[255] The somewhat anomalous status of the Channel Islands and the Isle of Man should be noted in this context. They are not included within the general definition of the UK in the Interpretation Act 1978, Sched. 1. However, for VAT purposes only, the Isle of Man, although not the Channel Islands, and the UK are treated as a single area.[256]

---

[246] It should not be assumed from this, however, that the position is straightforward, since there are detailed exceptions (see Whitehouse, *op. cit.*, ch. 20 (CGT) and paras [32.121]–[32.140]).

[247] See the discussions in Williams, *op. cit.*, pp. 12–17, and Davies, *op. cit.*, paras 26–07–26–09.

[248] Social Security Contributions and Benefits Act 1992, s.1(6) and Social Security (Contributions) Regulations 2001, S.I. 2001 No. 1004, reg. 145.

[249] That is, from countries outside the EU.

[250] Value Added Tax Act 1994, ss.7 (place of supply), 13 (place of acquisition) and 15(1) (place of importation).

[251] See para. 1.2.1.2 above.

[252] Interpretation Act 1978, Sched. 1.

[253] See Income and Corporation Taxes Act 1988, s.830 (inserted following the discovery of oil in the North Sea (see para. 2.4.3 below)).

[254] See Social Security Contributions and Benefits Act 1992, s.172.

[255] See Value Added Tax Act 1994, s.96(11).

[256] See the Isle of Man Act 1979, s.6.

1.5   Institutional framework

It has already been mentioned that Customs is the government department mainly responsible for administering each of landfill tax, climate change levy and aggregates levy.[257] The detail of the department will be discussed in para. 4.2.1.2(2) below. For the moment, it should simply be noted that Customs is the department traditionally responsible for the administration of the UK's indirect taxes, plus customs duties and excise duties. The overall position is not as straightforward, however, as this may suggest, since various ancillary aspects of the taxes are administered by other Departments and bodies. For example, with climate change levy, in addition to Customs, the Department for Environment, Food and Rural Affairs ('Defra'), the Department of Trade and Industry ('the DTI') and Ofgem also have responsibility for related aspects of the levy. The relevant responsibilities of each of these are also examined in detail in Chapter 4 below.

Direct taxes (that is, in this case, non-environmental taxes) are traditionally the responsibility of the Board of Inland Revenue (generally referred to simply as 'the Revenue'). Again, the institutional detail of the Inland Revenue will be examined below.[258] There are long-term plans to amalgamate the Revenue and Customs Departments, following the controversy surrounding a number of recent high-profile cases. The Revenue's main interest in the subject matter of this book has been in devising the environmentally-inspired tax subsidies discussed in Chapters 23 and 24 below, with a view to 'greening' the UK's direct tax system.

It should be stressed that the institutional framework of the study as a whole is much wider than simply the departments and the executive body mentioned above. For the moment, however, the key point is that Customs has been the tax authority more closely involved with the UK's environmental levies. The Department for Environment, Food and Rural Affairs, arguably, has become more involved in the economic instruments enterprise in the last seven or so years than either of them.

1.6   Scheme of the book

All of the foregoing will have assisted the reader in gaining an idea of the basic concepts at work in the areas covered by the book, as well as the contexts in which those concepts operate. Chapter 2 looks at the industrial sectors whose activities are sought to be regulated by the instruments under discussion. This should assist in an appreciation of the design issues involved in each of the instruments under discussion.

From Chapter 2 onwards, the discussion progresses in three main Parts. In Part II, we examine the institutional framework referred to in the previous paragraph in some detail. This is intended to give the reader some idea of the provenance of the various measures under discussion. The discussion is followed, in Chapter 5, by an analysis of the government's proffered technical justifications for the environmental taxes referred to above.

---

[257]   See para. 1.1 above.
[258]   See para. 4.2.1.2(2) below.

The detail of the regulatory and taxation contexts referred to earlier in this introductory chapter forms the matter of Chapters 6 and 7. Chapter 8 then widens the scope of the discussion to look at the international context of the instruments under discussion. This is all the more necessary given the international context of the climate change problem which at least one of the taxes is professedly seeking to address. This discussion is intended to provide a context for the European aspects discussed in Chapter 12.

Part III is mainly concerned with the fine detail of the measures outlined in the present chapter. It should thus be of interest, not only to practitioners, but also to policy-makers, reflecting, as it does, insights into both the practical operation of the measures under discussion and also the criticisms which can be made of them. As to the latter point, the key chapter is Chapter 21, which, rather ambitiously, seeks to draw together the various policy and industrial aspects of the measures under discussion. Practitioners will no doubt be very interested in the synthesis of the administrative provisions applicable to the three main environmental taxes, which are discussed in Chapter 16.

Part IV concentrates on new directions in the great experiment with which the book is concerned. Chapter 28 is devoted to the EU ETS. Taxes and other measures canvassed but as yet not implemented are discussed in Chapter 27 of the book. Under discussion in Chapter 27, therefore, is a combination of measures, some already implemented, and some as yet only projected (for example an incineration tax and a pesticides tax).

Finally, in Part V, we attempt to gather together the threads of the discussion, with a view to offering some general conclusions about the UK's experience of environmental taxes thus far. This is the fruit of an immensely detailed review of the published sources but it is hoped that it will inspire readers to test through 'empirical' study some of the conclusions to which it points.

The material under discussion is characterised by its diversity. To assist the reader in navigating it, we have included brief 'orientation' chapters at the beginning and end of Part II and at the beginning of Part III. Although they are very short and do not, in general, contain new material, they will help the reader to anticipate the main themes of each of the chief components of the book.

Chapter 2

Regulated Sectors

2.1　Introduction

Having described the scope of the study, we next present a sketch of the structures of the main sectors whose activities are subject to the environmental levies, subsidies and other economic instruments introduced in the previous chapter. Such a brief description is justifiable by the need to refer, in general terms, to the environmental externalities created by particular sectors of industry. However, it is also desirable because, unless the reader has at least an outline knowledge of how the regulated sectors actually operate, together with some awareness of the commercial issues facing those sectors, it will be difficult to gain a full appreciation of the significance of the instruments under discussion.

In this chapter, therefore, the reader will find brief descriptions of the sectors affected by the three environmental taxes and other instruments introduced in Chapter 1. The sectors in question are waste management, energy and mineral extraction. Of the three, the first and third are relatively straightforward and their discussion is therefore commensurately brief. With the energy sector, however, the position is somewhat different. With the obvious exception of oil, the relevant industries are all former state monopolies and the relevant post-nationalisation structures, especially those relating to electricity and gas, are extremely complex ones.[1] The issues arising from these ex-state industries are not merely ones of technical complexity, however. The post-nationalisation structures have also proved more than a little difficult to reconcile with key objectives of environmental policy. Nowhere is this more apparent than in the problematic relationship between the UK's energy policy and the design of climate change levy, together with its associated economic instruments. This is a theme to which we shall return at various points in the book.

Landfill tax, climate change levy and aggregates levy are all single-stage indirect taxes.[2] This being so, and although each impose considerable compliance burdens on the industries referred to above, the three taxes rely for their incentivising effect on the fact that, when chargeable, they are passed on to their customers. Of equal importance, therefore, to the structures of the three industries referred to above are those of the customers themselves. In outlining the waste management and mineral extraction industries, therefore, we have included some brief comments on the businesses and other bodies that they serve and on the impact on the latter of the instruments concerned. In the case of the energy sector, we have devoted a separate paragraph to enumerating briefly those industries which, as intensive users of energy, are particularly relevant to climate change levy and its associated

---

[1]　The regulatory provisions are discussed in para. 6.4 below. The role of the gas and electricity markets' regulator, Ofgem, is discussed in para. 4.2.1.3 below.

[2]　See paras 1.4.2.1(1), 1.4.2.2(1) and 1.4.2.3 above.

instruments.[3] Besides being liable to pay climate change levy on their energy use, they are shortly also to become participants in the EU Emissions Trading Scheme ('the EU ETS') which is scheduled for introduction in January 2005.[4] The relationship between the levy and the EU ETS is an intricately-woven thread in the discussion both of the levy and of the EU ETS. It is further complicated, moreover, by the interaction of each one with the pre-existing UK Emissions Trading Scheme ('the UK ETS').[5]

Neither of the authors is, nor professes to be, a specialist in industrial economics or sectoral analysis. As discussed in para. 1.2.2 above, the principal aim of the book is to offer a critical account of the environmental taxes and other instruments making up its subject matter. The details which we have attempted to capture in the present chapter are therefore intended to be sufficient only to render more intelligible some of the technical issues involved in the larger discussion. It is for this reason that we have included at the end of the chapter a brief paragraph on the issues facing the air transport and road freight transport industries. Although the civil aviation industry falls outside the scope of existing economic instruments for environmental protection,[6] its prospective inclusion in the EU ETS, together with the possibility of the 'greening' of airport charges,[7] makes a quick sketch of the industry useful here. Likewise with the road freight transport industry, especially given that one of its trade associations has – surprisingly perhaps – given cautious support to the proposed scheme for charging heavy lorries by reference to road usage,[8] which is now due to be introduced in 2007.[9]

Since the larger discussion involves an examination, not only of the institutional aspects of the instruments under consideration, but also of the processes by which they have passed into law, reference is often made in what follows to the responses of trade associations and environmental pressure groups to the various policy initiatives under discussion. We have also taken the opportunity to introduce, where appropriate, the main trade associations and environmental pressure groups that continue to participate in those processes.

2.2   Trade associations, policy-makers and pressure groups

The commercial concerns of each of the sectors described in the present chapter are represented by a number of trade associations. Equally, the environmental externalities

---

[3]   Together with the landfilling of waste, these are activities covered by the Integrated Pollution Prevention and Control ('IPPC') regime (see para. 6.2.3 below).

[4]   See para. 28.3 below.

[5]   See Chapter 20 below. Unlike the UK ETS, the EU ETS includes power stations (see para. 28.3 below). However, the beautiful simplicity of this statement is disfigured by the fact that an electricity generator may convert Renewable Obligation Certificates into UK ETS allowances (see paras 6.4.3.1(2), 20.6 and 21.5 below).

[6]   That is, on the basis that air passenger duty is not an environmental tax (see para. 1.2.1.5(2) above).

[7]   See para. 2.6 below.

[8]   See para. 27.3 below.

[9]   See Wright, *Financial Times*, 6 February 2004, p. 5.

attributable to each of those sectors have long been the preoccupation of a number of high profile environmental pressure groups. The development of the present government's policy on environmental levies, subsidies and other instruments has attempted conspicuously to build a consensus around their differing viewpoints.[10]

There is an extensive literature, not only on the relative influence of the interests of economic groups and ideas on policy-making generally, but also on their influence specifically on environmental policy-making.[11] It is beyond the scope of this book either to attempt to add to that literature or to do more than to speculate on the relative weight of ideas and interest groups in shaping the various instruments under consideration. However, in describing the various industrial sectors affected by the levies, subsidies and other instruments under discussion, it is not inappropriate to begin by drawing attention to two organisations with high profiles in debates on economic instruments in environmental protection: the Confederation of British Industry ('the CBI')[12] and the Friends of the Earth.[13] The former body is generally recognised as being representative of the views of British business; the latter is one of the UK's leading environmental pressure groups. The CBI's activities are financed by industry and commerce; Friends of the Earth is largely financed by individual donations. Representatives of both groups are strident voices in the British media.[14] The efforts of both bodies are highly coordinated and both have some greater or lesser international presence.

Despite the undoubtedly high profiles of the organisations just referred to, it would be too simplistic to view the trade associations, the policy-makers and the environmentalists as representing three mutually antagonistic corners of a noisy triangular debate. Setting aside both the ritual complaints of regulated industries about the effects of regulation on competitiveness and the dissatisfactions of environmental campaigners on the lack of progress in attaining environmental goals, arguments about economic instruments in environmental protection tend to operate within a broad consensus of the need for some regulatory response to commonly recognised problems. Thus, even allowing for occasional exceptions,[15] the contributions of each to the public debate on economic instruments for environmental protection tends to be a nuanced and balanced one. The CBI's consistent viewpoint on economic instruments is well demonstrated from its comment in a policy brief of April 2002 that, although '... the use of environmental economic instruments can be justified,

---

[10] This approach has often been attributed to Anthony Giddens' argument for the renewal of social democracy in *The Third Way: The Renewal of Social Democracy* (Oxford: Polity Press, 1998) and *The Third Way and its Critics* (Oxford: Polity Press, 2000). Both books draw attention to the influence of ecological movements and the latter endorses the technical justifications for green taxes discussed in Chapter 5 below (see *The Third Way and its Critics*, pp. 100–101).

[11] See, for example, Anthony Giddens, *The Third Way: The Renewal of Social Democracy*, pp. 54–64.

[12] See www.cbi.org.uk. The current Director-General of the CBI is Sir Digby Jones, formerly senior partner of a large firm of Birmingham solicitors.

[13] See www.foe.co.uk.

[14] For example, Sir Digby Jones's eloquent objections to the EU ETS on BBC's *Today* programme on 19 January 2004.

[15] See, for example, 348 ENDS Report (2004) 18–22, 20.

... the theory does not always translate well into practical design. This leads to sub-optimal results for both business competitiveness and the environment, which we believe the government should put right'.[16] For its part, Friends of the Earth, commenting on the 2003 Pre-Budget Report,[17] said that '[t]he Chancellor [that is, of the Exchequer][18] has accepted the concept of sustainability as a basis for the UK's future economy. However recently steps taken have been increasingly cautious. The clearest indicator is the considerable fall in environmental taxation in years, for 9.7 per cent of total taxes in 1999 to 8.8 per cent in 2002'.[19] Both assessments are, of course, highly problematic,[20] although both accept the *principle* of using economic instruments in environmental protection.

If, as is widely, although not universally, held to be the case, the government's approach to the use of economic instruments has become unnecessarily cautious, this may itself be indicative of its preoccupation with creating an atmosphere of consensus around its policies on environmental economic instruments. Whether this preoccupation, with its seemingly endless streams of consultative documents and policy justifications, masks a rather less palatable reality, is something to which we shall presently return. Suffice it to say for the present that it is noteworthy that all of the trade associations referred to in the subsequent paras of this chapter have been involved, to a greater or lesser extent, and with varying degrees of success, in the consultations leading to the introduction of the instruments discussed in this book. The Department of Trade and Industry (the 'DTI'), whose role in environmental regulation is discussed in a subsequent chapter,[21] even supports[22] a body called the Trade Association Forum ('the TAF'),[23] whose membership includes a number of the trade associations referred to in succeeding paras.

2.3   Waste management industry

According to the trade association which represents firms providing waste management and associated services,[24] almost 430 million tonnes of waste are generated in the UK

---

[16]   See Confederation of British Industry Business Environment Brief, *Green Taxes: Rhetoric and Reality* (London: April 2002), p. 1. This viewpoint is closely consistent with the view of the theory and reality of environmental taxes taken by a former Director-General of the CBI (see Adair Turner, *Just Capital: The Liberal Economy* (London: Pan Books, 2002), pp. 310–15).

[17]   See para. 21.3.2 below.

[18]   See para. 4.2.1.2(2) below.

[19]   See Friends of the Earth Briefing, *Time for a Sustainable Economy?* (London: Friends of the Earth, November 2003), p. 1.

[20]   The CBI's in that it is always possibly to lament 'the shadow between the idea and the reality', especially from a viewpoint of some self-interest; the Friends of the Earth's because a drop in revenue from environmental taxes might indicate that they are actually having some environmental effect!

[21]   See para. 4.2.1.2 below.

[22]   Together with the CBI and the TAF's own membership.

[23]   See www.taforum.org.

[24]   The Environmental Services Association Ltd. See para. 2.2.3 below.

every year.[25] The association divides this into four main categories, that is, waste arisings from agriculture, from mining and quarrying, from construction and demolition and, finally, from other industries, commerce and households. The waste arisings in the fourth category are estimated at 110 million tonnes per annum, over 45 per cent coming from the relevant industrial sectors, with a little over 27 per cent arising in each of the commercial and household sectors. Almost 60 per cent of all of the waste arising in the UK each year is, according to the same source, sent to landfill.

The structure of the industry whose business it is to sort out, or at least to bury or burn, the mess referred to above has recently and usefully been outlined by Barrow.[26] Responsibilities relating to the collection and disposal of waste are imposed by legislation on local authorities,[27] although the services relating thereto are supplied under contract between those authorities and private sector firms which provide the collection and disposal services. Waste disposal is the responsibility of the waste disposal authority ('WDA'), which is usually the county council,[28] while waste collection is the responsibility of the waste collection authority ('WCA'), usually the relevant district council.[29] Where, instead of county and district councils, there is a unitary authority, the authority in question combines the collection and disposal functions. The relevant WCA contracts either with its own refuse collection service or a private sector firm for the collection of waste from commercial premises and households. Then, under the direction of the WDA, the firm takes the waste to a landfill site, incinerator or other disposal site. Landfill sites (usually disused quarries or mine workings) are generally owned and operated by private sector firms with whom the WDA has negotiated contracts. By Finance Act 1996, s.41, the landfill site operator is, of course, the person liable to pay the landfill tax charged on a taxable disposal.[30] The contracts, made between WDAs and landfill site operators, specify minimum and maximum annual quantities of waste, are usually concluded for relatively long periods of time and contain provisions for renegotiation in appropriate circumstances. The gate price per tonne of waste, exclusive of landfill tax, is usually related to a price index for the contract term. The costs of collecting the waste and transporting it to the landfill site are borne by the WCA, while the WDA pays the gate price to the landfill site operator. It will thus be appreciated that the landfill tax collected by the site operator has itself been paid by a public authority out of public funds.[31]

The environmental effects of landfilling waste[32] and the understandable public enthusiasm for recycling are well known. Landfill produces emissions to air, water and soil and causes disamenities such as visual intrusion, noise, odour, vermin and

---

[25]   See www.esauk.org/waste, from which some of the information in this paragraph is drawn.
[26]   Michael Barrow, 'An Economic Analysis of the UK Landfill Permits Scheme', (2003) 24 *FS* 361–81, 363. This paragraph is much indebted to the portrait of the industry structure contained in that paper.
[27]   See para. 6.3.2.1 below.
[28]   See para. 4.2.3 below.
[29]   *Ibid.*
[30]   See paras 1.4.2.1(1) above and 15.2 below.
[31]   See para. 21.4.1 below.
[32]   See, for example, Cambridge Econometrics, *A Study to Estimate the Disamenity Costs of Landfill in Great Britain* (London: Department for Environment, Food and Rural Affairs, 2003), section 1.

litter.[33] Despite suggestions that recycling may be more equivocal a good than it may at first appear,[34] enthusiasm for household recycling is undiminished.[35]

The waste management industry is characterised by a close engagement with policy development, both at the sectoral level, via the Environmental Services Association ('the ESA'), and at the level of individual firms. The ESA, which is a member of the TAF,[36] proclaims a vision of 'an economically and environmentally sustainable waste management industry for the United Kingdom' and, to this end, vows to assist, not only its members, but all levels of government in the UK, to achieve the vision.[37] Among particular firms, especially prominent perhaps is Biffa Waste Services Ltd,[38] which has issued a number of publications dealing with policy issues in waste management[39] and has also funded mass-balance research[40] through the landfill tax credit scheme ('the LTCS').[41]

2.4   Energy industries and consumers

2.4.1   Electricity

The denationalised electricity[42] industry in England and Wales[43] comprises four main activities:[44] generation,[45] transmission, distribution and supply. Of these,

---

[33]   Cambridge Econometrics, *op. cit.*, para. 1.5.

[34]   See, for example, the works referenced in Bjørn Lomborg's *The Skeptical Environmentalist: Measuring the Real State of the World* (Cambridge: Cambridge University Press, 2001), p. 209. On the current status of Lomborg's controversial work, see Houlder and MacCarthy, *Financial Times*, 18 December 2003, p. 15. See also Richard D. North, *Life on a Modern Planet* (Manchester: Manchester University Press, 2001).

[35]   This has recently been illustrated by the easy passage through Parliament of the symbolically important Household Waste Recycling Act 2003, originally a Private Member's Bill introduced by a Labour MP, Mrs Joan Ruddock (see 346 ENDS Report (2003)).

[36]   See para. 2.2 above.

[37]   See www.esauk.org.

[38]   See www.biffa.co.uk. Interestingly, Biffa is a subsidiary of Severn Trent plc (www. severn-trent.com*)*, a water utility. The latter acquired Biffa to diversify out of regulated water activities, which are subject to severe 'capping'.

[39]   See www.biffa.co.uk.

[40]   See www.massbalance.org.

[41]   See para. 4.2.1.2(2) below. Biffa's sponsorship of mass-balance research seems to have been hit by the government's decision in November 2002 to withdraw two-thirds of the landfill tax credit scheme ('the LTCS') (see para. 21.3.2 below and 336 ENDS Report (2003)).

[42]   See Walt Patterson, *Transforming Electricity: the Coming Generation of Change* (London: Earthscan, 1999), pp. 3–5. Since electricity cannot be stored (except indirectly in pumped storage schemes), demand must exactly match supply. At every moment of every day, therefore, the electricity being fed into the system must match the electricity being used from the system.

[43]   Different arrangements apply for Scotland and Northern Ireland.

[44]   See Utility Week, *The Electricity Supply Handbook 2004*, 57th edn (Sutton: Reed Business Information, 2004), p. 13 ('ESH'). See also www.energynetworks.org below.

[45]   With which 'production' is synonymous.

generation and supply are subject to competition, whilst transmission and distribution are monopolies.[46] The two monopoly elements have been separated by legislation[47] from the competitive elements,[48] making the industry among the most liberalised electricity industries in the world.

The first competitive activity, generation, refers to the conversion of primary and renewable energy sources into electricity.[49] Most electricity is generated at power stations fired by gas or coal, or at nuclear power stations,[50] with renewables as yet accounting for only a small proportion of the total electricity generated in the UK in each year.[51] There are around 40 major electricity generators, including Powergen (UK) plc,[52] RWE Innogy Holdings plc,[53] Drax Power Ltd[54] and British Energy plc.[55]

The second competitive element, supply, is the business of buying electricity in bulk from generators and selling it on to domestic, commercial and industrial consumers.[56] The electricity industry therefore has a wholesale stage (generators and suppliers) and a retail stage (suppliers and customers). Under Finance Act 2000, sch. 6, paras 5(1) and 40(1), it is the supplier who is generally the person liable to account for climate change levy,[57] which makes it a downstream, rather than an upstream, energy tax.[58] The wholesale trade in electricity has taken place since March 2001 under the New Electricity Trading Arrangements ('NETA').[59] A replacement for NETA, the British Electricity Trading and Transmission Arrangements ('BETTA'), to include Scotland

---

46 See Callum McCarthy, 'Ofgem: Characteristics and Issues of the British Electricity Market' 12 [2001/2002] 6 ULR 170–72.

47 See para. 6.4.3.1 below.

48 The 'unbundling' of European energy markets has been a major theme of European Energy policy for decades (see Carlos Ocana *et al.*, *Competition in Electricity Markets* (OECD/IEA, 2001) and para. 12.2.6.3 below). The unbundled energy markets of Europe, in which Enron was a major player, fortunately survived Enron's collapse (see 'Power play', *Economist*, 24 July 2003).

49 ESH, p. 13.

50 See DTI/National Audit Office, *Digest of United Kingdom Energy Statistics 2004* (London: HMSO, 2004). This is an annual publication.

51 *Ibid.*

52 See www.powergenplc.com.

53 See www.rweinnogy.com.

54 See www.draxpower.co.uk. See para. 21.4.3 below.

55 See www.british-energy.com. British Energy plc, whose core business is nuclear generation, is currently the subject of a proposed £5 billion government-backed rescue, the background to which has been heavily criticised by the House of Commons Public Accounts Committee (see Taylor, *Financial Times*, 12 February 2004, p. 2). See paras 6.4.3.1(3) and 21.4.3 below.

56 ESH, p. 13.

57 Being the holder of a supply licence under Electricity Act 1989, s.6(1)(d) (see Finance Act 2000, Sched. 6, para. 150(2)(a) and paras 7.4 and 15.1 below).

58 See para. 11.3.1 below.

59 The New Electricity Trading Arrangements replaced the 'pool' originally created on the privatisation of the electricity market in 1990. On the introduction of NETA, see, for example, 'Beyond the pool', *Economist*, 1 March 2001. For the background to NETA, see Dieter Helm, *Energy: The State and the Market British Energy Policy since 1979*, revised edn (Oxford: Oxford University Press, 2004), ch. 17.

as well as England and Wales, is expected to be introduced in 2005.[60] The main electricity suppliers, among many, include npower Ltd[61] and SEEBOARD Energy Ltd.[62] The retail trade utilises published tariffs and involves the reading of meters, the issuing of bills and the processing of payments.

The monopoly elements of the electricity industry, as mentioned above, are transmission and distribution. Transmission refers to the transport of electricity in bulk, on the national grid,[63] from the generators' power stations to the companies responsible for distributing the electricity to consumers.[64] The sole owner and operator of the national grid is the National Grid Group ('NGG'),[65] which also operates the national transmission system for gas.[66] Besides being interconnected to the transmission systems of Scotland and Northern Ireland,[67] the national grid is connected to the French national grid.[68]

Distribution, the second monopoly activity, is the process of delivering electricity from the national grid, via regional distribution networks, to consumers.[69] Each regional distribution network is owned and operated by a Distribution Network Operator ('DNO'). There are presently 14 DNOs,[70] including Aquila Networks plc[71] (the West Midlands), Northern Electric Distribution Ltd[72] (the North East) and Southern Electric Power Distribution plc[73] and EDF Energy Networks (SPN) plc[74] (London and the South East). Suppliers pay for electricity to be transmitted across the national grid and distributed to their customers.[75]

The probable effects on climate of burning the primary fuels used in electricity generation are both notorious and well-documented.[76] The possible effects of nuclear power generation on the environment also remain controversial, possibly even more so.[77] In these circumstances and, given the various legal and policy

---

[60]   See para. 6.4.3.1(4) below.

[61]   See www.npower.com. Npower is RWE Innogy's retail business (see ESH, p. 94).

[62]   See www.seeboardenergy.com. SEEBOARD Energy is the retail business of EDF Energy (see ESH, p. 73).

[63]   Also referred to as the transmission grid. The most obvious manifestations of the national grid are the overhead pylons which carry the system of high voltage transmission lines. High voltage lines may travel underground instead, although underground cables are of course much more expensive to install and maintain than overhead ones.

[64]   ESH, p. 13.

[65]   See www.ngtgroup.com.

[66]   See para. 4.2.1.3 below.

[67]   See this para. below.

[68]   Thus facilitating the importation of 'renewable electricity' from France and Belgium (see 331 ENDS Report (2002)).

[69]   See ESH, p. 13.

[70]   *Ibid.*, p. 36.

[71]   See www.aquila-networks.co.uk.

[72]   See www.ce-electricuk.com.

[73]   See www.scottish-southern.co.uk.

[74]   See www.edfenergy.com.

[75]   ESH, p. 13.

[76]   See, for example, the works referenced in Lomborg, *op. cit.*, ch. 24.

[77]   See, for example, 348 ENDS Report (2004) 13 on emissions of radioactive gas from British Energy's Hartlepool power station.

commitments,[78] it is unsurprising that the renewable energy sector, especially in the shape of wind farms, has been growing,[79] although it should be noted that NETA, among other factors, has not been favourable to renewables generators.[80] Moreover, environmental benefits are promised by combined heat and power ('CHP') electricity generation, which enables the simultaneous generation of electricity and heat at the point of use.[81]

The electricity industry has been characterised since 1990 by considerable merger and acquisitions activity,[82] which has produced corporate structures of extreme complexity.[83] For over a decade, the industry's trade association was the Electricity Association,[84] but this has been replaced by three new associations, from the beginning of October 2003. The three new associations reflect the differing interests of the participators in the market: the Association of Electricity Producers;[85] the Energy Networks Association;[86] and the Energy Retail Association.[87] There is also the British Wind Energy Association[88] and the Combined Heat and Power Association.[89]

2.4.2   Gas

Although, historically, gas has been produced from both coal and oil,[90] the gas used in Britain today is natural gas, sourced from around 100 offshore gas fields around the UK.[91]

---

78   See para. 6.4 below.
79   See, for example, 348 ENDS Report (2004) 13. A number of suppliers offer 'green electricity' tariffs (see 348 ENDS Report (2004) 29–30).
80   See Ross Fairley and Karina Ng, 'Green Energy in the NETA World' 12[2001/2002] 3 ULR 57–60. See also para. 21.4.2 below.
81   See www.chpa.co.uk. See in this para. below. The term 'combined heat and power' is synonymous with 'cogeneration' and 'total energy'. CHP/cogeneration/total energy refers to the simultaneous generation of useful thermal energy (such as heat or steam) and electricity in a single process (see Finance Act 2000, s.148, for the definition applicable to climate change levy).
82   See Electricity Association, *Electricity Companies in the United Kingdom – a Brief Chronology*, June 30, 2003 (still available from www.electricity.org.uk).
83   See Electricity Association, *Who Owns Whom in the UK Electricity Industry*, June 30, 2003 (still available from *www.electricity.org.uk*).
84   Whose website (www.electricity.org.uk) is/was a mine of useful information.
85   See www.aepuk.com.
86   See www.energynetworks.org.
87   See www.energy-retail.org.uk.
88   See www.bwea.com.
89   See www.chpa.co.uk.
90   This was so-called 'town gas', which, prior to the first commercial offshore gas finds in British Waters in November 1965, supplied the relatively small amounts of gas required in the UK. Town gas was smelly; natural gas, by contrast, is odourless and is artificially 'stenched', to enable leaks to be detected (see David Upton, *Waves of Fortune: the Past, Present and Future of the United Kingdom Offshore Oil and Gas Industries* (Chichester: John Wiley, 1996), pp. 20–22).
91   There is also an interconnector gas pipeline, permitting gas imports, which runs between Bacton, in East Anglia, and Zeebrugge in Belgium. Plans are currently under way for

The gas industry[92] comprises six main activities: production, storage, shipment, transmission, distribution and supply. As with the electricity industry, legislation[93] has separated transmission and distribution (the monopoly components) from production, shipment and storage and supply, all four of which are subject to competition.

Gas production, the first competitive element, comprises the offshore extraction of gas, its delivery onshore under contracts concluded with gas shippers and its subsequent treatment at seven beach terminals located around the UK. There are around 30 offshore gas producers, including BP plc,[94] Shell UK Exploration and Production plc [95] and ExxonMobil International Ltd.[96]

The second and third competitive elements, gas shipment and storage, refer to the activities of the gas shippers once the gas has been brought ashore. Shipment describes the process by which shippers arrange with the transmission system operator ('the TSO') for gas to be transported on, and taken out of, the national transmission system ('NTS'). This is the wholesale stage of the gas industry, shippers also being responsible for the large-scale storage of gas in massive submarine gas storage facilities. The wholesale trade has taken place since March 1996 under the Network Code. There are around 90 shippers, including PowerGen Gas Ltd[97] and Shell Gas Direct Ltd.[98]

Gas supply, the fourth competitive element, refers to the purchase of gas by suppliers from shippers and its onward sale to customers. Gas supply thus generally corresponds to the retail part of the industry, major suppliers including npower[99] and British Gas.[100] Under Finance Act 2000, Sched 6, paras 6(1) and 40(1), it is the supplier[101] who is generally the person liable to account for climate change levy; this replicates the position for electricity and, as mentioned above, makes the tax a 'downstream' tax.

The monopoly components of the gas industry, transmission and distribution, are the responsibility of the TSO, that is, NGG[102] ('Transco').[103] Transmission refers to

---

upgrading the interconnector and for the construction of two further interconnectors (see Ofgem Factsheet 37, *Securing Britain's Gas Supply* (London: Ofgem, 2003)).

[92]   It is, of course, somewhat artificial to split out the gas industry from the oil industry (see para. 2.4.3 below). However, we have embraced this artificiality, given that the sketch of the industry included here is intended chiefly to elucidate elements of climate change levy, which is inapplicable to supplies of hydrocarbon oil (see paras 1.4.2.2(1) above and 14.1 below).

[93]   See para. 6.4 below.

[94]   See www.bp.com.

[95]   See www.shell.co.uk.

[96]   See www.exxonmobil.com.

[97]   See www.powergen-power.co.uk.

[98]   See www.shellgasdirect.co.uk.

[99]   See www.npower.com.

[100]   See www.gas.co.uk.

[101]   Being the holder of a supply licence under Gas Act 1986, s.7A(1) (see Finance Act 2000, Sched. 6, para. 150(3)(a) and paras 6.4 and 14.1 below).

[102]   See para. 2.4.1 above.

[103]   See www.transco.uk.com.

the high pressure transport of gas in bulk on the NTS, from the beach terminals to 40 power stations,[104] to a small number of large industrial consumers and to the 12 Local Distribution Zones ('LDZs'). Distribution refers to the low pressure transport of gas on the LDZs, from the NTS to most business and to domestic consumers. As TSO, Transco is responsible for ensuring that the transmission and distribution systems remain in balance, being empowered to buy and sell gas to ensure that supply matches demand.

Like the electricity industry, the gas industry has been for long characterised by considerable merger and acquisitions activity, many gas suppliers supplying electricity too. In 2003, the UK was only one of two G7 countries that at that time were self-sufficient in gas, although this was liable to change.[105] Gas consumption, which is forecast to rise between 14 per cent between 2002 and 2011, presently stands at 113bn cubic meters per annum and has grown by 66 per cent since 1992.[106] A significant proportion of this growth is explicable by the increased use of gas in electricity generation, 29.7 per cent of all gas consumption being accounted for in this way in 2002.[107] As with the electricity industry's trade associations, different associations represent the interests of different participators. The Energy Networks Association[108] and the Energy Retail Association[109] are relevant here just as in relation to electricity and a list of trade associations for gas (and also oil) appears on the DTI website.[110] There is also the UK Offshore Operators Association,[111] which represents the offshore oil industry as well as the offshore gas industry.

One of the three fossil fuels,[112] natural gas is used as a catch-all term for natural hydrocarbon gases associated with oil production, of which the main ones are methane (CH_4) and some ethane (C_2H_6).[113] Although the burning of natural gas produces minimal sulphur dioxide emissions, methane is one of the six GHGs listed in Annex A to the Kyoto Protocol.[114] Scientific opinion is divided as to the sustainability of natural gas reserves.[115]

---

[104] See para. 2.4.1 above.
[105] Ofgem, *op. cit.*
[106] Ofgem, *op. cit.*
[107] Ofgem, *op. cit.* There is an exemption from climate change levy for taxable commodities used in the generation of electricity (see Finance Act 2000, Sched. 6, para. 14).
[108] See para. 2.4.1 above.
[109] *Ibid.*
[110] See www.dti.gov.uk/sectors.
[111] See www.ukooa.co.uk.
[112] That is, coal, oil and natural gas, all of which are 'derived from organic matter deposited over geological time-scales' (Porteous).
[113] Porteous.
[114] See para. 8.3.1.4 below.
[115] See Lomborg, *op. cit.*, p. 126, for the (highly controversial!) view that natural gas becomes more abundant over time.

2.4.3 Oil

The UK has been a major producer of crude oil ever since the Argyll field in the North Sea became the first offshore oil field to become operational in June 1975.[116] Because the offshore oil production industry has never been a state-owned industry,[117] and since hydrocarbon oils are outside the scope of the climate change levy,[118] the structure of the oil industry is not of central relevance to the present study. However, since oil refineries fall within the scope of the EU ETS,[119] and since we devote a chapter below to a discussion of fuel excise duties,[120] it may be useful simply to draw attention to the fact that the oil industry is usually seen as having 'upstream' and 'downstream' components. The upstream component includes the activities of exploration, production and transportation (via oil tankers and pipelines), while the downstream component involves the refining, distribution and supply of oil products, such as transport fuels.

The main oil producers are the same group of around 30 companies who are also gas producers.[121] As for gas, the interests of the main UK oil producers are represented by the UK Offshore Operators Association,[122] while those of the downstream industry are represented by the UK Petroleum Industry Association ('UKPIA').[123]

Since the Second World War, oil has achieved a hegemony over coal as the main fossil fuel. The use of oil and its distillates such as petrol and DERV for transport purposes is a major source of carbon dioxide emissions.[124] Carbon dioxide is, of course, one of the GHGs listed in Annex A to the Kyoto Protocol.[125] It also does terrible damage when leaked or spilt from tankers, of course.[126]

The significance of oil should not simply be seen in terms of its use as a fuel. Oil is also used in the chemical industry for the production of ethylene, which is in turn used in the production of a range of end products, such as the higher glycols, acetic acid and acetic fibre and butadiene.

---

[116] See Upton, *op. cit.*, p. 56. Upton relates that oil was originally discovered in commercially-worthwhile quantities on the British mainland at Eakring, near Sherwood Forest, in 1938, the discovery then being kept secret. The Petroleum (Production) Act 1934 had vested onshore petroleum reserves in the Crown (see Upton, *op. cit.*, p. 20).

[117] See Martha M. Roggenkamp *et al.*, *Energy Law in Europe: National, EU and International Law and Institutions* (Oxford: Oxford University Press, 2001), paras 13.26 and 13.259. With the exception of the long-defunct British National Oil Company, the UK Government 'has been content to licence the industry and collect its share of the rent' (*ibid.*, para. 13.259).

[118] Finance Act 2000, s.3(2)(a).

[119] See para. 28.3 below.

[120] See paras 22.2 and 22.3 below.

[121] See para. 2.4.2 above.

[122] See www.ukooa.co.uk above.

[123] See www.ukpia.com.

[124] See, for example, Department of Environment, Transport and the Regions, *A New Deal for Transport: Better for Everyone*, 1998 (Cm 3950, 1998), paras 1.6–1.11.

[125] See para. 8.3.1.4 below.

[126] One thinks of the cases of *Atlantic Empress* in 1979 and *Exxon Valdez* in 1989.

2.4.4 Coal

Following the privatisation of the coal industry in 1994, UK coal production has been entirely in the hands of private firms, the largest of which is UK Coal plc.[127] Despite large helpings of state aid,[128] and the fact that the privatisation arrangements meant that the coal industry's biggest customers would be the electricity generating companies, the industry is now a shadow of its former self.[129] In November 2002, there were 15 deep mines in the UK, producing about 18 million tonnes of coal per annum;[130] in 1975, there were around 200 collieries, producing around 130 million tonnes of coal.[131] The coal industry's trade association is the Confederation of UK Coal Producers ('Coalpro').[132]

Climate change levy applies to supplies of coal and lignite, as well as to coke and semi-coke of coal and lignite.[133] Tax is chargeable on any supply of these commodities where the supply is made in the course or furtherance of a business.[134] However, coal supplies to electricity generators are generally exempt from the levy![135]

The partial displacement of coal by oil as the most important fossil fuel has not eroded the status of coal as 'Environmental Enemy No. 1'.[136] Although it is the dirtiest of the fossil fuels, it is possible to reduce by various technological means the sulphur dioxide and nitrogen oxide emissions caused by the burning of coal.[137] In addition to coal's environmental costs, there are, of course, the social costs of what is still, even in 2004, an industry fraught with physical danger.

2.4.5 Energy-intensive sectors

Certain sectors of industry are extremely high consumers of the three fossil fuels. Apart from oil refining and electricity generation, the main energy-intensive sectors are the steel industry, the chemical industry, the paper industry, the glass industry, the ceramics industry, the gypsum[138] industry, the china clay industry, the cement and concrete industries and the aluminium industry. There is an umbrella trade organisation, the Energy Intensive Users Group,[139] but also individual trade associations for each of the individual sectors themselves.[140] The significance of

---

[127] See www.ukcoal.com.
[128] See para. 12.2.6.3 below.
[129] See 'Bottomless pits', *Economist*, 18 April 2002.
[130] See www.ukcoal.com.
[131] See 'Bottomless pits', *Economist*, 18 April 2002.
[132] See www.coalpro.co.uk.
[133] Finance Act 2000, Sched. 6, paras 3(1)(d) and 3(1)(e).
[134] *Ibid.*, Sched. 6, para. 7(2).
[135] *Ibid.*, Sched. 6, para. 14.
[136] See 'Environmental enemy No 1', *Economist*, 6 July 2002, p. 11.
[137] See Lomborg, *op. cit.*, p. 127, citing the work in Danish of Jesper Jesperson and Stefan Brendstrup.
[138] Gypsum is hydrate calcium sulphate ($CaSO_4 \cdot 2H_2O$), which is used in the manufacture of plasterboard (Porteous).
[139] See www.eiug.org.uk. There is a separate trade association for the cement and concrete industries, that is, the British Cement Association (see www.bca.org.uk).
[140] See the links at www.euig.org.uk.

the energy-intensive sectors in the rest of the study is that they fall within the scope of the command and control Integrated Pollution Prevention and Control ('IPPC') regime.[141] This, in turn, makes sector participators eligible to enter into climate change agreements ('CCAs') and thereby to obtain an 80 per cent reduction in climate change levy.[142] As parties to CCAs, participators in each sector are likely also be Agreement Participants in the UK ETS.[143] Not voluntary, however, will be the EU ETS which, from January 2005, will apply to all the energy-intensive sectors covered by the earlier instruments, as well as to electricity generation and oil refining.[144]

Before leaving energy and its heavy consumers, it is appropriate to mention the highly influential UK Emissions Trading Group ('the ETG'), originally formed by the CBI and the Advisory Committee on Business and the Environment ('ACBE')[145] 'to represent the UK business interest in greenhouse gas emissions trading'.[146] Although the ETG has been involved in the implementation of the EU ETS in the UK, it was the ETG that originally convinced the government of the need for the cash subsidy to get the UK ETS going,[147] using the argument that the development of a market in GHG emissions would make the City of London a world leader in trading emissions.[148]

2.5   Mineral extraction industry

There are, according to one of the quarry operators' trade associations, around 1,300 quarries in the UK, producing £3bn worth of quarry products each year.[149] Quarried, as opposed to recycled, aggregates, are generally referred to either as 'virgin' or 'primary' aggregates. For the purposes of this study, two points are perhaps significant. First, that the consumer of a considerable proportion of the products quarried in any year is the public sector, aggregates being of course essential to public sector construction projects (for example, school and university improvements, road maintenance and hospital building). Secondly, that the use of recycled aggregates in such projects seems to be increasing.

For the purposes of aggregates levy, the chargeable person is the person responsible for subjecting aggregate to commercial exploitation,[150] a concept which includes the removal of the aggregate from the quarry in question,[151] its sale,[152] its use for

---

[141]   See paras 6.2.3 and 12.2.2 below.
[142]   See para. 14.6 below.
[143]   See para. 1.4.2.2(2) above and para. 20.4 below.
[144]   See para. 1.4.2.2(3) above and para. 28.3 below.
[145]   See para. 4.2.1.4 below.
[146]   See www.uketg.com.
[147]   See para. 20.2 below.
[148]   See Helm, *op. cit.*, pp. 358–9. This was at the time that there seemed to be a possibility that, with the possible imposition of withholding tax on interest payments on eurobonds, London would lose its pre-eminence in that particular field.
[149]   That is, the Quarry Products Association. See www.qpa.org.
[150]   Finance Act 2001, s.16(3).
[151]   *Ibid.*, s.19(1)(a) and 19(2)(a).
[152]   *Ibid.*, s.19(1)(b). See para. 13.2 below.

construction purposes[153] or its mixing, other than in permitted circumstances, with any substance other than water.[154] This means that the person primarily liable for aggregates levy will be the quarry operator.

The quarrying industry has two main trade associations, the Quarry Products Association ('the QPA')[155] and the British Aggregates Association.[156] The latter was involved in an unsuccessful challenge to the legality of the aggregates levy in April 2002.[157]

The environmental costs of quarrying are controversial. Certainly, there has been by no means the same level of concern, at least at a European level, in relation to its effects as in relation, say, to emissions of GHGs and energy consumption. A later chapter[158] will show how the government has characterised the environmental costs of quarrying in terms of noise, dust, visual intrusion and biodiversity loss.[159] However, there remains the concern that the level of opposition to the tax, as compared to that mounted in relation to, say, climate change levy, is a sign that the government may not have succeeded in demonstrating the environmental case for the tax.

2.6   Air passenger and road freight transport sectors

Two parts of the transport sector figure prominently in the present study: the road freight transport and the air transport industries. Although neither is currently subject to any form of environmental levy,[160] yet, as mentioned above, there are currently plans for the introduction of a nationwide road-user charging ('road pricing') scheme for heavy lorries,[161] as well as continuing concern about the international exemption from fuel duties applicable to aircraft fuel.[162] In addition, the government has decided to work to bring the aviation industry within the EU ETS and has promised to prioritise the matter during the UK's 2005 EU Presidency.[163] The issues facing road freight transport and air transport therefore merit at least brief mention in the present context.

The Freight Transport Association,[164] which, together with the Road Haulage Association,[165] is one of the two main trade associations for the road freight

---

153   *Ibid.*, s.19(1)(c).

154   *Ibid.*, s.19(1)(d).

155   See www.qpa.org.

156   See www.british-aggregates.com.

157   See *R (on the application of British Aggregates Association and others) v. C & E Commrs*, [2002] EWHC 926 (Admin), [2002] 2 CMLR 51.

158   See para. 11.3.2 below.

159   *Ibid.*

160   Air passenger duty is not an environmental tax (see para. 1.2.1.5(2) above).

161   See para. 27.3 below.

162   See paras 8.5 and 27.5 below. See also Friends of the Earth, *op. cit.*, p. 10 and Chris Nash, 'Transport and the Environment', in *Environmental Policy: Objectives, Instruments, and Implementation*, ed. by Dieter Helm (Oxford: Oxford University Press, 2000), pp. 241–59.

163   See Department for Transport, *The Future of Air Transport: Summary* (London: Department for Transport, 2003), p. 8.

164   See www.fta.co.uk.

165   See www.rha.net.

transport industry in the UK, estimates that road freight transport constitutes 64 per cent of domestic transport as a whole.[166] The same source also figures that a total of around 65,000 companies operate the UK's commercial vehicles fleet,[167] which is a 31 per cent reduction on the figures for 1996.[168] The reduction has mainly been among operators with less than half-a-dozen vehicles, with the number of businesses operating in excess of 50 vehicles actually increasing.[169] Although heavy lorries are legendarily associated with graphic levels of noise, air pollution and congestion, it is clear that the sector as a whole is itself contending increasingly with congestion, whilst complaining vociferously about what it regards as disproportionately high levels of fuel duties and road taxes in comparison with other EU Member States.[170] The proposed introduction of the road-user charging scheme for heavy lorries will be made sectorally possible only on the basis of reductions in fuel excise duties.[171]

The aviation industry in the UK, unlike a number of other EU Member States, has been entirely in private hands since British Airways was privatised in 1987.[172] The main UK airlines include BMI British Midland, Britannia Airways, British Airways and Virgin Atlantic Airways, all of which are members of the industry's trade association, the British Air Transport Association.[173] Something in excess of 105 million passengers were carried on UK airlines in 2000, over 33 million on charter flights and nearly 72 million on scheduled flights.[174] The airlines pay airport (or 'landing') charges to the owners of the industry's infrastructure, the airports. Seven of the UK's major airports are owned by BAA plc[175] (that is, Aberdeen, Edinburgh, Gatwick, Glasgow, Heathrow, Southampton and Stansted), whilst Manchester Airport, in common with a number of others, is owned by the relevant local authority. The government's White Paper on the future of air transport, of December 2003, identified aircraft noise and the property price blight caused by proposals for airport development as two areas of environmental concern.[176] Besides envisaging schemes requiring airport operators to tackle the latter concern, the White Paper promised legislation on the air pollution side ('when Parliamentary time permits') to allow airport charges to have an emissions-related element.[177] In addition, as mentioned above, and as an acknowledgement of the growing contribution of civil aviation to

---

[166] See Freight Transport Association, *Freight Transport: Delivering for a Successful Economy* (Tunbridge Wells: Freight Transport Association, 2002), p. 2. The Road Haulage Association puts the figure rather higher, at 80 per cent!
[167] Freight Transport Association, *op. cit.*, p. 11.
[168] *Ibid.*
[169] *Ibid.*
[170] Freight Transport Association, *op. cit.*, p. 14.
[171] Which is why the trade associations are broadly in favour of the scheme (see Wright, *Financial Times*, 6 February 2004, p. 5).
[172] There is an excellent overview of the UK aviation sector in *Whitaker's Almanack 2004* (London: A. & C. Black, 2004), pp. 453–4, to which this paragraph is greatly indebted.
[173] See www.bata.uk.com.
[174] See *Whitaker's Almanack*, above, p. 454.
[175] See www.baa.plc.
[176] See Department for Transport, *The Future of Air Transport: Summary* (London: Department for Transport, 2003), p. 3.
[177] *Ibid.*, p. 8.

carbon dioxide emissions, the government has committed itself to working to bring the aviation industry within the scope of the EU ETS.[178]

2.7   Concluding remarks

The sketches of the regulated sectors given in this chapter are intended only to clarify some of the technical issues addressed in our wider critical account of the taxes and other instruments forming the subject matter of the study. It is appropriate that these sketches should detail the trade associations – both general and sectoral – representing the industries in question, since, consistently with the government's attempts at consensus-building, they have engaged closely in the debates leading to the implementation of the instruments in question. This engagement is well, though not uniquely, illustrated by the involvement of the waste industry in governmental policy debates, both at a sectoral level and at the level of particular firms. An even more striking illustration is the success of the ETG in apparently influencing government policy on emissions trading so dramatically.

However beautiful the empathy may – or may not – be as between government and industry, such a phenomenon is not more important in the context of the present study than the structures of the industries in question. It is not insignificant, for instance, that the main bodies intended to be incentivised by landfill tax are themselves tax-raising, albeit local tax-raising, bodies. Equally, as regards climate change levy, it is not insignificant that the person liable to account for the tax, in the cases of gas and electricity, is not the generator – for such a thing would make the levy an 'upstream tax' – but the holder of the gas or electricity supply licence, which, in combination with other features of the tax, makes it a downstream levy on the industrial, commercial and agricultural consumption of non-renewable energy. If, as is widely believed, this structure severely compromises the environmental effectiveness of the levy, it is no less important than the effect of the structure of the electricity industry on the potential for greater levels of renewables generation.[179] Accordingly, even if the industry sketches given above may appear somewhat cursory, they should at least enable a closer understanding of the evaluative issues to be referred to later in the study.

Whatever the technical strengths and weaknesses of climate change levy may be, there is at least some measure of agreement of the environmental problem that it is seeking to address. This, no doubt, goes a considerable way to legitimising the tax in the eyes of those in industry who are most closely affected by it. The same cannot, alas, be said for aggregates levy; the highly problematic environmental basis of this tax, as well as the difficulties of making it acceptable to those affected by it, will concern us in later chapters. The difficulties of justifying aggregates levy, an existing tax, contrast rather vividly with the ease with which the application of economic instruments might be justified, at least in environmental terms, to correct the externalities created by the civil aviation industry.

---

[178] *Ibid.*
[179] See para. 21.6.1 below.

Although, in this chapter, we have referred to the environmental externalities created by each of the sectors under discussion, we have sought mainly to highlight the structures of those industries. This is not because we are not, in Joanne Scott's words, 'on the side of the angels',[180] or that we do not care for environmental issues. It is simply that such issues are covered in other texts by specialists in the fields. We have sought merely to illuminate, if only briefly, the often obscure factual background to the regulatory and fiscal issues to be discussed in the rest of the study.

---

[180]   See Joanne Scott, *EC Environmental Law* (London: Longman, 1998), p. 1.

PART II
POLICY AND CONTEXTS

Chapter 3

Introduction to Part II

Although we have already referred to the fact that various aspects of the UK's environmental economic instruments, including the three main environmental taxes, are administered by several different government departments, we have not so far elaborated on that statement. This part of the book accordingly continues our account with an analytical discussion of the roles of those departments. Such is the intricacy of the web of departmental involvement that this discussion is unfortunately rather long and detailed. However, as the material much later in the book unfolds, and the references to the different departments and government agencies accumulate, we think that the discussion will have proved indispensable, especially for the reader whose 'home' jurisdiction is one other than the UK. This is not simply because it may be convenient to gather all this information into one place. It is also because standard reference works on environmental, energy and transport regulation or, for that matter, tax law have not, historically, been written specifically with environmental economic instruments in mind.

The discussion of government departments in Chapter 4 forms part of an overview of the institutions of UK government divided into the central (that is, the national), the regional and the local. Our discussion of central government begins, however, not with government departments, but with a discussion of the committee structure of the UK parliament. The reader may wonder why we have deemed such a discussion of the nature and role of Parliamentary Committees to be relevant at this stage. The reason, quite simply, is that, in Part III of the book, we refer rather extensively to reports of Parliamentary Select Committees in attempting to map how the various instruments under consideration have operated in practice. It is our contention that, for the lawyer or political scientist, there is no more valuable source for discovering what effect the various instruments are actually having (as it were) on the ground. However, though valuable, Select Committee reports are the products of a somewhat idiosyncratic system and, rather than accepting them at face value, it is well to be aware of the weaknesses, as well as the strengths, of the committee system as it exists at Westminster. The early sections of Chapter 4 accordingly attempt to convey some sense of these points.

Besides Select Committees, whose role is to hold the government to account after the event, there are also the Standing Committees, whose duty it is to scrutinise legislation and thereby to ensure a measure of accountability before the event. We have included a brief discussion of the composition of these Committees, and the conditions under which they work, especially with regard to the length of time allowed for debate, because pressures on parliamentary time seem too often to have prevented the proper scrutiny of much of the legislation discussed in this book. This is a matter to which we return in Part III below.

Whilst the accountability of the relevant government departments to parliament is an important theme of this Part II, so also is the accountability of the various bodies which have been delegated to administer different aspects of the measures under consideration. These are bodies such as the Environment Agency, which itself has responsibility for the unit tasked with administering the system of PRNs, and Ofgem, which, among other things, is responsible for monitoring the operation of the RO and for issuing Green Certificates. The National Audit Office ('the NAO'), which is discussed in the same part of Chapter 4 (as an executive non-departmental body), has, however, the distinctive statutory task of providing financial advice and assurance on the application of public funds. In later chapters, we shall see how its qualified approval of the UK ETS, in a special report on the scheme, has disappointed many who regard the incentive payments made under the UK ETS to have been a serious misapplication of public funds. The discussion of these bodies is then followed by a brief outline of various advisory bodies, such as the Commission for Integrated Transport ('CFIT'), whose activities feature prominently in the public debate on environmental matters in the UK, and whose nature and role must be understood in order to assess the nature and value of their various contributions.

All of the foregoing relates to central government but, as mentioned in para. 1.3 above, the UK has for some years past had a quasi-federal structure. The almost non-existent tax-raising powers of each of the devolved governments, together with the national application of the UK's environmental levies, means that this is not as important a point, at present at least, as it may at first appear. However, it will also be recalled that the devolved governments do have responsibility for various matters other than taxation, such as environment and energy, which means that, at various points throughout the book, we shall be drawing attention to regional differences in the material under discussion. Also, as is perhaps implied by the foregoing, some emphasis should be placed on the phrase 'at present', since the absence of tax-raising powers vested in the devolved governments will almost certainly lead in time to discussions at governmental level, either as to whether these restrictions should be removed, or as to whether it is nonetheless possible, even as matters currently stand, for the devolved governments to create economic instruments which nonetheless fall short of taxes properly so called. A significant part of the discussion in Chapter 4 is therefore devoted to the detail of the devolution arrangements.

It will be appreciated that, just as the relevant issues differ somewhat as between central and regional government, so also do they vary at the local level. Suffice it to say, at this stage, that local government in the UK has even more limited taxation powers than its regional counterpart. For instance, although they are empowered to collect and retain local taxes, local authorities have no control over the structure of those taxes. Subject to two exceptions, there is therefore no question of unilaterally imposed environmental levies at local level. The two exceptions, of course, are the statutorily-conferred powers to impose workplace parking levies and road user charging schemes which, to date (December 2004), remain overwhelmingly and conspicuously unimplemented. That said, as will be seen, local authorities do have extensive powers in relation to environmental command and control regulation and the nature and extent of those powers are also briefly described in Chapter 4.

The near-federal nature of the UK, and the issues relating to environmental regulation, whether by economic instruments or by command and control, are mirrored in the near-federal nature of the EU. Of all global regional organisations, the EU is the most developed and, as we shall see in Part III, it shares competences in both the environmental and taxation spheres with its Member States. In such circumstances and, given the international context to be described in a moment, it is not surprising that there have been a number of key Community-level initiatives over the last decade or so to tackle air and atmospheric pollution, the quintessentially transnational environmental issue. Early attempts at a carbon/energy tax failed but – against the expectations of many – the EU-wide emissions trading system, the 'EU ETS', did not fail and it is due to become a reality in January 2005. Although the detail of these measures is for consideration in Part III, this Part of the book underlines the importance of these issues by describing the relative nature and roles of the EU institutions which have had some part in the design of the EU ETS and may yet need to adjudicate on disputes arising from its implementation. Most importantly, of course, there are the European Commission, the Council of the European Union and the European Court of Justice ('the ECJ'), each of which, together with other Community institutions, have played a part in bringing the EU ETS to fruition.

Just as the 1990s saw an attempt to create an EU-wide carbon/energy tax, so also did they witness academic suggestions for a global excise on carbon. Interesting as this notion has been, it is not however the reason for the inclusion in the latter part of Chapter 4 of a discussion of international organisations. Even a cursory glance through this volume will indicate the considerable influence on the development of environmental taxes and other economic instruments of the Paris-based Organisation for Economic Co-operation and Development ('the OECD'). The reader is accordingly introduced to the OECD as a way of illustrating the provenance of many of the economic ideas which recur throughout the study. Equally important in this process, however (especially as regards the development of emissions trading) is the United Nations ('UN'), whose 1992 Framework Convention on Climate Change, with its 1997 Kyoto Protocol, has provided the international consensus which, with the notable exceptions of the US, China and (until September 2004) Russia, has enabled concepts such as multinational emissions trading schemes even to be conceivable. The UN's benevolent role in these matters is frequently contrasted with that of the World Trade Organization ('the WTO'), whose GATT 1994-based system of multilateral agreements on international trade is still seen as a major obstacle to the development on a global scale of economic instruments for environmental protection.

If it were necessary to illustrate the pervasive influence of the OECD in the realm of environmental taxes and their associated instruments, it would be necessary to look no further than the technical justifications offered by the UK government for the package of economic instruments introduced in the UK since 1997. Chapter 5 analyses an HM Treasury policy document, which sets out these technical justifications in some detail. The discussion covers four main areas, of which the second, third and fourth, dealing respectively with the taxation of environmental 'bads', the correction of market failures resulting from pollution and the price level at which intervention in the market should be determined, follow closely the OECD's thinking on these matters, as expressed in various policy documents and recommendations. The discussion of these justifications is offered in this part as a prelude to the account in Part III of the processes by which the UK's main environmental levies passed into law.

We asserted at the beginning of the book that the distinctive characteristic of the UK's environmental taxation enterprise was the use of environmental taxes in combination with other economic instruments and with traditional command and control instruments.[1] The UK's environmental taxes therefore have both a taxation context and a regulatory context, involving environmental, market and transport regulation. Given the regulatory nature of the taxes themselves, it is perhaps useful to consider regulatory context first. Chapters 6 and 7 seek to explore the two contexts in detail, bearing in mind that readers whose 'home' discipline is tax law will be unfamiliar with the regulatory background, just as those whose 'home' discipline is environmental or energy law and regulation will be unfamiliar with the taxation context. Chapter 6 therefore attempts the ambitious task of describing within a single chapter the regulatory landscape of which environmental taxes are now such a prominent feature. The discussion is largely concerned to map out the environmental regulation of waste, of air and atmospheric pollution, of aircraft emissions and road congestion, as well as the loss of amenity caused by mineral extraction. However, it also contains a detailed discussion of the economic regulation of the 'unbundled' energy markets, as well as an examination of the economic regulation of air passenger and road freight transport. The reader might wonder why we have cast the net so wide. So far as the energy markets are concerned, the authors' contention is that it is not possible to appreciate the technical legal problems associated especially with climate change levy, nor yet the renewables generation policies that its exemptions and exclusions are designed to complement, without understanding the economic regulation of the energy sector. Chief among these problems is the status, already discussed, of climate change levy as a downstream energy tax.[2] Parallel comments apply to the economic regulation of air transport, given the continued interest of both the European Commission and the UK government either in bringing air transport within the EU ETS or (an interest evinced so far by the Commission only) in imposing an environmental tax on airliners' profligate consumption of kerosene.[3]

While Chapter 6 considers the status of economic instruments as environmental regulation, Chapter 7 discusses the tax context of those instruments. This part of the discussion, which is more analytical than the rather descriptive material in Chapter 6, has two main aspects. First, we are concerned to expand the definition of a tax offered in Chapter 1 and to investigate whether and, if so, which payments made by those regulated by the UK's economic instruments are payments of taxes. Secondly, we seek initially to identify the provisions by which it has been sought to 'green' the UK's non-environmental tax codes. These latter are indeed economic instruments and are perhaps best seen as a form of financial assistance from government.[4] Since, irrespective of whether they occur in the codes on environmental or non-environmental taxes, they involve an element of revenue foregone by government, such incentivising 'green' measures can be seen as tax expenditures or tax subsidies. This is borne out, for example, by the state aid implications of the climate change levy exemptions and reliefs which are discussed in Part III below.

Chapter 9 seeks to draw together the strands of the discussion in Part II but, before that, Chapter 8 seeks to place the economic instruments under discussion in the

---

1   See Authors' Preface above.
2   See para. 2.4.1 above.
3   See Done, *Financial Times*, 4 October 2004, p. 8.
4   See para. 1.2.1.5 above.

context of the UK's international obligations. Whilst it may well be the case that, at least in the environmental and world trade spheres, the UK's role on the international stage is as a Member State of the EU, such a statement obscures two little-known but important analytical points. First, that, although the UK's Kyoto obligations are reallocated with other EU Member States under the so-called 'burden-sharing agreement', the UK has, in its own right, both signed and ratified the Kyoto Protocol. Secondly, that, whilst the ECJ subsequently held that it should not have been, the UK was itself a signatory, along with 14 other Member States, to the GATT 1994 Uruguay Round Treaties signed at Marrakesh in April 1994. These points apart, it is nonetheless the case that, both in its own right and as an EU Member State, any economic instrument designed and implemented by the UK government must satisfy, not only the rules of Community law, to be discussed in Part III below, but also those of GATT 1994. The inability of the Kyoto mechanisms to 'mesh' satisfactorily with the rules of GATT 1994/WTO is a key theme in Chapter 8.

Chapter 4

Institutional Framework

4.1   Introduction

It is not easy to locate the UK's environmental taxes and other economic instruments into the framework of institutions responsible for their design, implementation and administration. This is in part a function of the wide range of governmental, Community and international institutions which have had at least some involvement in each of these processes. It is also, however, an aspect of the authors' dilemma in deciding what, in an already large book, should be regarded as significant in the involvements of these various institutional actors.

With regard to the former point, we have divided the discussion into three main parts. First, a description of the institutions of UK central, regional and local government with an interest, or potential interest, in the design, implementation and operation of the taxes and other instruments under consideration. Next, the same, although for the relevant EU institutions, and finally the same again for the relevant international institutions. Far from being counter-intuitive, this working outwards from the UK's domestic institutions emphasises that, unlike other forms of environmental regulation, environmental taxes at least have implications for national sovereignty. In each case, we emphasise the question of who is, or was, responsible for what, since this remains one of the most opaque aspects of the combination of economic instruments currently being operated in the UK.

In relation to the other point made at the beginning of this chapter, we have contented ourselves with merely drawing attention to the presence or absence of accountability mechanisms in relation each of the institutions under discussion. It is not possible, obviously, in a book of this nature, to embark upon a legitimacy-oriented examination of the taxes and other instruments under consideration.

4.2   Institutions of central, regional and local government

The initial stage of the discussion is intended to reflect the three main levels of UK government: the central, the regional and the local. Within the first of these, we make a fourfold division into the respective roles of the UK Parliament; of the departments of government; of non-departmental agencies and public bodies; of bodies with an advisory role; and of bodies with a judicial role, namely the courts. The first of these involves a discussion of the Committee system, the key accountability mechanism in the UK's parliamentary system. The discussion of the second, that is, of government departments, is further subdivided into a consideration of those involved in environmental, energy and transport regulation and those involved in taxation. Discussion of the third and fourth (that is, the advisory bodies and the courts) is, for reasons of space, relatively short. When it comes to regional and local government, a detailed discussion is necessary, given the former's role in developing environmental policy, as well as the role of the latter in tackling traffic congestion.

4.2.1   Central government

Following devolution and the other constitutional reforms introduced after 1997,[1] the expression 'central government' has come to be used in order to distinguish the institutions of government of the UK as a whole, not only from local government, but also from the devolved governments of Wales, Scotland and Northern Ireland.[2]

Central government in the UK has never been characterised by a strict separation of powers. The principal executive body under the UK's unwritten constitution is the Cabinet, headed by the Prime Minister,[3] but all of its 20 or so members are also members of one or other House of the UK Parliament. Equally, the UK Parliament is not simply a legislature, since it sets up Select Committees of Inquiry and Committees tasked with scrutinising the administration.[4] The structures of central government, as they have developed since 1997, are such as to make the scrutinising role of the UK Parliament no less important, although often considerably less effective, than once it was. This is a theme to which we shall return, not only in the present chapter, but throughout the book.

4.2.1.1   The UK Parliament

(1) Generally
Subject to the provisions of Community law, the UK Parliament sitting at Westminster has responsibility both for the general shape of environmental, energy and transport regulation in England and Wales[5] and for the taxation law of the UK as a whole.

---

1   The date is of some significance to the present study, since the General Election in May of that year was when the current Labour Government was returned for the first time. It was returned again in 2001, on the back of a reduced voter turnout. As at 8 September 2004, the government had 407 seats (a majority of 159), the Conservatives (the main Opposition party) had 163 seats and the Liberal Democrats had 55 (see www.parliament. uk/directories/hcio/stateparties.cfm). With the exception of landfill tax, the economic instruments discussed in this book are from the period 1997 onwards (see para. 11.4 below).

2   See paras 1.3 above and 4.2.2 below.

3   Currently (December 2004) The Rt Hon. Tony Blair, MP.

4   See O. Hood Phillips and Paul Jackson, *Constitutional and Administrative Law*, ed. by Paul Jackson and Patricia Leopold, 8th edn (London: Sweet and Maxwell, 2001), paras. 2–020.

5   See www.parliament.uk. The leading reference works on the law and customs of the UK Parliament are *Sir Thomas Erskine May's Treatise on the Law, Privileges, Proceedings and Usages of Parliament*, ed. by Donald Limon, W.R. McKay, *et al.*, 22nd edn (London: Butterworths, 1997) and *Griffiths and Ryle on Parliament: Functions, Practice and Procedures*, ed. by Robert Blackburn and Andrew Kennan, 2nd edn (London: Sweet and Maxwell, 2003). For a comprehensive, but more succinct account, see Hood Phillips and Jackson, *op. cit.*, Part II.

Being the UK's highest legislative authority, Parliament may not only pass laws for the UK as a whole but also for any part of it individually.[6] Made up of the Queen in Parliament, the House of Lords and the House of Commons, the House of Lords consists of life peers, hereditary peers,[7] the Law Lords[8] and 26 Church of England diocesan bishops, while the House of Commons comprises a membership ('MPs') elected by universal adult suffrage.[9] The maximum length of a Parliament is five years, elections to the Commons being by secret ballot, with (unlike in other countries) voting in parliamentary elections being non-compulsory. Proposed Acts of Parliament, in draft form, are referred to as Bills, and must normally be passed by both Houses before becoming law.[10] Of the various stages through which legislation passes before becoming law,[11] of greatest interest for the purposes of the present study is the Committee Stage, since this is the point at which Bills have traditionally been considered clause by clause and amendments made by a Standing Committee of the House of Commons.[12]

With the exceptions of the UK ETS, which was created under pre-existing statutory powers,[13] and PRNs, which seem to have developed almost by accident, all of the economic instruments discussed in this book have a statutory basis. Whether it can be inferred from this that they have all been fully debated is a matter which we leave until Chapter 11.[14]

(2) Standing Committees
The principal purpose of Standing Committees is to consider and to amend Bills that have been through their Second Reading and stand committed to one of the

---

6   By way of illustration, Hood Phillips and Jackson, *op. cit.*, record that, 'in the first two years of devolution [see para. 1.3 above], 14 Bills that fell within the legislative powers of the Scottish Parliament [see para. 4.2.2 below] were passed by Westminster'. They note, however, that the consent of the Scottish Parliament was sought and given in each case.

7   Albeit now elected by their peers, except in the case of a few office holders.

8   See para. 4.2.1.5 below.

9   See above, n, for the state of the parties as at 8 September 2004. The total number of seats is 659, consisting of 529 for England, 40 for Wales, 72 for Scotland and 18 for Northern Ireland.

10   Bills may be public, private or hybrid, although only public ones are referred to in this book (see Hood Phillips and Jackson, *op. cit.*, for further information on the consequences of the distinction). Most Private Members' Bills are public Bills.

11   It is not appropriate to detail the procedure here, although it should be noted that, of the stages in the passing of a Bill, legally the most informative tends to be the 'Committee Stage', since it involves a detailed technical debate on the text of the Bill, rather than a broader argument over political priorities. Nonetheless, as the Committee Stage of the parts of Finance Bill 2000 relating to climate change levy demonstrated, even apparently technical debate can uncover important information about the nature of a tax.

12   See Kenneth Bradshaw and David Pring, *Parliament and Congress* (London: Constable, 1972), ch. 5 (esp. pp. 258–62) for a comparison of the UK system with that of the committee system of the US Congress.

13   See para. 1.4.2.2 above.

14   See para. 11.3 below.

Committees.[15] The Standing Orders[16] of the House of Commons allow the appointment of as many Standing Committees, consisting of anything between 16 and 50 members, as may be necessary for the consideration of Bills in this way. Members of the Standing Committees are nominated by the Committee of Selection, a body which is required to have regard both to the qualifications of members and to the composition of the House in making such nominations. Standing Committee debates are often extremely searching and, although they have little or no interpretative status, the minutes of their debates[17] provide fascinating clues, not only as to the strengths and weaknesses of the clauses of a Bill, but also (via, mainly, the contributions of MPs from the parties other than that of government) as to the possible governmental motivation in shaping them as they are.[18] This is well illustrated by the Standing Committee debates on climate change levy.[19]

Whilst Standing Committee debates are an important factor in the accountability of the parliamentary process, it is important to note that the government of the day has drastic powers to curtail debate on particular Bills. Of these, the most controversial is probably what is referred to as 'the Guillotine', a procedure the increased use of which in the current parliament is a matter of mounting concern to a number of commentators. The procedure, the more formal title of which is an Allocation of Time Order, allows a government minister to move either that: (1) particular dates and days be allocated to the various stages of a particular Bill, or (2) that the relevant Committee must report the Bill to the Commons by a certain date, the detail being left to the Business Committee of the House or to a business subcommittee of the Committee.[20] Guillotine motions are debated in each case for a maximum of three hours and are almost invariably subject to a division.[21] The procedure inevitably means that large parts of Bills, including ones of great importance, might receive very little parliamentary scrutiny. A particularly striking example of the consequences of this situation is the curtailment of debate on the provisions of the 2003 Finance Bill.[22]

(3) Select Committees
Before moving on to consider the government departments, it is important to comment on another aspect of the Committee system of the UK Parliament. This is the operation of House of Commons Select Committees,[23] especially the departmental ones. Select

---

[15]   That is, public Bills.
[16]   As to which, see Hood Phillips and Jackson, *op. cit.*, para. 11–004.
[17]   Available in Hansard on the web at www.parliament.uk. Finance Bill 2003, for instance, was considered by Standing Committee B (see below, *passim*).
[18]   See Robert Baldwin and Martin Cave, *Understanding Regulation – Theory, Strategy and Practice* (Oxford: Oxford University Press, 1999), ch. 3, for the significance of this point.
[19]   HC, Standing Committee H, 7th and 8th Sittings, 16–18 May 2000.
[20]   See Hood Phillips and Jackson, *op. cit.*, para. 11–014.
[21]   As to which, see Hood Phillips and Jackson, *op. cit.*, para. 11–010.
[22]   HC, 1 July 2003, cols 178–180 (Report).
[23]   These are most relevant to the present study, although it should be noted that the House of Lords also has a number of Select Committees, for example, on the EU, on Science and Technology and on Economic Affairs (see www.parliament.uk).

Committees are made up of MPs chosen by the Selection Committee, such MPs being either regularly re-appointed or appointed from time to time.[24] The powers and authority of Select Committees, which include the power to send for persons, papers and records, are delegated to them by the House.[25] The eight Committees of particular importance in the present context are the Environmental Audit Committee; the Environment, Food and Rural Affairs Committee; the Transport Committee; the Treasury Committee; the Trade and Industry Select Committee; the Public Accounts Committee; the Science and Technology Committee; and the Public Administration Committee. Some comments on some of these might be appropriate.

The Trade and Industry Select Committee has the task of examining, on behalf of the House of Commons, the expenditure, administration and policy of the Department of Trade and Industry ('the DTI').[26] Likewise, the remit of the Treasury Committee is to examine the same three areas in relation to HM Treasury, the Board of Inland Revenue, the Board of HM Customs and Excise[27] and certain related public bodies. The Transport Committee, which dates from July 2002, was one of two committees appointed to replace the former Transport, Local Government and the Regions Committee.[28] It examines the expenditure, administration and policy of the Department for Transport[29] and the public bodies associated with it.

All three of the Treasury, the Trade and Industry and the Transport Committees are, as their names suggest, departmental Select Committees. Of the remaining three Select Committees referred to above, the Environment, Food and Rural Affairs Committee is a Departmental Committee, whilst the Environmental Audit Committee and the Committee of Public Accounts ('the Public Accounts Committee') are not. The first of these oversees the work of Defra,[30] while the second, which dates from 1997, is intended to monitor '... the contribution made by Government departments and agencies to environmental protection and sustainable development and ... [to audit] ... progress against targets'.[31] One of the predecessor Committees to the Environment, Food and Rural Affairs Committee, the Environment, Transport and Regional Affairs Committee, produced an early report on the operation of landfill tax in 1999.[32]

---

[24] In their present form, Select Committees are a development of a change in the system made when Mr Richard Crossman was Leader of the House. Before that, scrutiny was through the Estimates Committee, before the event, the Public Accounts Committee (which obviously still exists) after the event, and with other Committees scrutinising Statutory Instruments (a vital task still performed by that Committee) and (while they still existed) enquiring into the operation of nationalised industries.

[25] Standing Orders of the House of Commons – Public Business 2002(2), SO 152 (see www.publications.parliament.uk).

[26] See www.parliament.uk.

[27] See below in this para.

[28] See below in this para. for the process of reorganisation that has resulted in the formation of Defra.

[29] See below in this para.

[30] *Ibid.*

[31] See www.parliament.uk.

[32] That is, House of Commons Environment, Transport and Regional Affairs Committee, *The Operation of the Landfill Tax*, HC Papers, Session 1998–1999, HC 150–I (London: Stationery Office, 1999).

The sixth of the Select Committees listed above, that is, the Public Accounts Committee, is also rather important in the present context. Consisting of not more than 16 members, and chaired by a member of the Opposition,[33] the Public Accounts Committee's task is to examine how funds granted by Parliament to meet public expenditure have been spent. The Committee reports to Parliament, one day in each session being devoted to debating its reports. The government undertakes to make a reply to the debate, which is an indication of the importance attached to the Committee's work. The Public Accounts Committee is the only committee that has, in the shape of the NAO,[34] a strong bureaucracy.[35]

In recent years, both the Science and Technology Committee and the Environmental Audit Committee have produced reports highly critical of the government's climate change policy and of the three economic instruments designed to tackle air and atmospheric pollution (that is, climate change levy, the UK ETS and the RO). The reports in question are analysed later in the book.[36]

The strength of the system of Select Committees is said to be that the Committees are allowed to choose the questions that they examine. However, whatever accountability this might be thought to lend to the parliamentary process may be compromised by the way in which their membership is selected. Although, as mentioned above, there is a Selection Committee tasked with appointing MPs to the Select Committees, the House of Commons Liaison Committee's 2000 report found that nominations were in fact controlled by the party whips.[37] Despite the potentially negative impact on public confidence of such a system, the government rejected reform to the selection process on the basis that it would result in less time being spent on substantive questions than upon questions of Committee membership.[38] Other accountability concerns arise out of the use of the Guillotine. The procedure was used before the Second Reading of the Finance Bill 2003, an extremely complex and controversial piece of legislation, even by the standards of Finance Bills.[39] Indeed, the increasing size of Finance Bills since the mid-1990s has brought the Committee system very close to collapse.

---

[33]   Since 2001, the Conservative, Mr Edward Leigh, MP.

[34]   See para. 4.2.1.3 below.

[35]   See Hood Phillips and Jackson, *op. cit.*, para. 12–020n.

[36]   See para. 21.6.2 below.

[37]   See House of Commons Liaison Committee, *Shifting the Balance: Select Committees and the Executive*, HC 1999–2000 (London: Stationery Office, 2000).

[38]   HM Government, *The Government's Response to the First Report from the Liaison Committee on Shifting the Balance*, 2000 (Cm 4737, 2000).

[39]   An example is provided by cll 18 and 19, Finance Bill 2003, which introduced a new s.77A and an amended Sched. 11, para. 4(2), to Value Added Tax Act 1994. These provisions give draconic powers to HM Customs and Excise to collect tax owing by an independent third party from a bona fide trader who they think might have been negligent in not noticing that he was dealing with a defaulter. On the Second Reading, The Rt Hon. Dawn Primarolo, MP, the Paymaster-General (see para. 4.2.1.2 below) referred to certain 'in-built safeguards' [HC, 6 May 2003, col. 629] but in Standing Committee B, Mr John Healey, MP, the Economic Secretary to the Treasury (see para. 4.2.1.2 below, also) ran into a barrage of opposition, including some from Labour backbenchers, and gave an undertaking that Customs would only exercise the powers after having given a written warning [*Hansard* HC, 15 May 2003, col. 066].

4.2.1.2 Government departments

Government departments, or ministries, originated from three sources. First, there were the holders of the Great Offices of State, insofar as they were political, rather than ceremonial – only one of these, that of Lord High Chancellor, still exists and is about to be abolished, but a second, Lord High Treasurer, is technically in commission; second, the position of Secretary of State, which is technically a single office held jointly by a number of people, specialist responsibilities only being allocated since 1782;[40] and, thirdly, what were originally committees of the Privy Council,[41] called 'boards',[42] being absorbed into ministries following the First World War,[43] they being, in turn, amalgamated into departments from the 1970s onwards.

The head of each department is now a Secretary of State, who will also be a member of the Cabinet.[44] The Secretary of State will be assisted by a number of junior ministers of two ranks – Minister of State and Parliamentary Under-Secretary of State – and will divide the responsibilities of his department between them.

Whilst questions of policy are always reserved to a minister for decision, the detailed administrative and research work involved in policy development has traditionally been carried out by non-political civil servants appointed during good behaviour.[45] During the spring and summer of 2003, however, there was increasing concern as to the position of special advisers, not least the authority of the Prime Minister's Chief of Staff's authority to give instructions to civil servants.[46]

Statutory authority is usually required both for the creation of new departments and for the transfer of functions between them. The latter may, under the Ministers of the Crown Act 1975, be effected by Order in Council. Environmental regulation has been an area of government policy beset by departmental reorganisations in recent years.

---

40   See David Kynaston, *The Secretary of State* (Lavenham: Terence Dalton, 1978).
41   The Privy Council is a body that used to advise the Crown on government policy, a role which is now taken by the Cabinet. One of its roles is to make secondary legislation in cases where a statute has delegated legislative powers to her Majesty in Council.
42   The last surviving such Board being the Board of Trade, the post of President of the Board of Trade now being held by the Secretary of State for Trade and Industry (see below). The title was last used by Mr Michael Heseltine, the deputy Prime Minister in the Major Government.
43   The first 'minister' was Mr David Lloyd George, when asked to take responsibility for Munitions in 1915. He had been a senior minister and it would not have been appropriate to rate him as President of a Board. The last full department minister was Mr Nick Brown, whose Ministry of Agriculture, Fisheries and Food was absorbed into Defra after the food and mouth disease outbreak of February 2001. Departmental ministers were sometimes not in the Cabinet. The rank of Minister of State, which is now the senior grade of non-departmental minister, was first created for Mr Richard Law (the son of the former Prime Minister Andrew Bonar Law) in 1943.
44   Since the June 2003 reshuffle, this principle is almost honoured in the breach, with the Scotland and Wales Offices being entrusted to ministers with primary responsibilities elsewhere, that is, Transport and Leadership of the House of Commons respectively.
45   See Hood Phillips and Jackson, *op. cit.*, paras 18–021–18–031.
46   Timmins, *Financial Times*, 12 September 2003, p. 6.

Most government departments are subject to the supervision of the Parliamentary Commissioner for Administration ('the Ombudsman') in cases of maladministration.[47]

(1) Environmental, energy and transport regulation
At least four departments of central government contribute to the design, implementation and enforcement of various aspects of environmental regulation, that is, the Cabinet Office;[48] the Department for Environment, Food and Rural Affairs ('Defra');[49] the Department of Trade and Industry ('the DTI');[50] and the Department for Transport ('the DFT').[51] In addition to these four, the Foreign and Commonwealth Office[52] and the Department for International Development[53] each have a role to play in international environmental policy issues. A further department, the Office of the Deputy Prime Minister ('the ODPM'),[54] might also be added to the list, since this has responsibility for a wide range of policy issues touching on environmental matters, including housing, planning, devolution and regional and local government. The ODPM and the DFT were together, prior to May 2002, parts of the Department of Transport, Local Government and the Regions ('the DTLR'), a Ministry that is now itself defunct.

The government describes the role of the Cabinet Office as being 'to support the Government's delivery and reform programme'[55] by helping other departments to strike a balance between under- and over-regulating. Whilst not having a direct involvement in the evolution of the UK's environmental taxes and other economic instruments, one of the Units within the Cabinet Office, that is, the Regulatory Impact Unit, seems to have been highly influential in translating regulatory theory into practice. Part of the Unit's role is to work with other government departments, agents and regulators to help ensure that regulations are fair and effective.[56] However, it has also been responsible for promoting a 'business friendly' approach to the enforcement of regulation[57] and for supporting the work of the Better Regulation

---

47 See the Parliamentary Commissioner Act 1967, Scheds 2 and 3. See also Sir Cecil Clothier, 'The Value of an Ombudsman' [1986] PL 204 and Mary Seneviratne, *Ombudsmen: Public Services and Administrative Justice* (London: Butterworths, 2002).
48 See www.cabinet-office.gov.uk.
49 See www.defra.gov.uk.
50 See www.dti.gov.uk.
51 See www.dft.gov.uk.
52 See www.fco.gov.uk.
53 See www.dfid.gov.uk.
54 See www.odpm.gov.uk. Although the position of Deputy Prime Minister is not one recognised by the constitution, the title has been awarded to a senior minister during Wartime Coalitions and at various times since 1962. The Rt Hon. John Prescott, MP has held this title since May 1997.
55 See www.cabinet-office.gov.uk/min-org/organisation/index.asp.
56 See www.cabinet-office.gov.uk/regulation/Role/Index.htm.
57 See also the Regulatory Reform Act 2001, which gives ministers extensive powers to use secondary legislation to reform primary legislation (subject to certain safeguards) as well as a reserve power to set out a code of good enforcement practice.

Task Force.[58] The Prime Minister's Strategy Unit,[59] another of the Units within the Cabinet Office[60], is tasked with addressing long-term cross-sectoral problems. One such problem is that of how to reduce dependence on landfill as the primary method of disposing of waste, to which the Unit has responded by developing the government's 'Waste Not, Want Not' policy,[61] to which reference will be made as relevant in the rest of the study.[62]

The remit of Defra is a wide one, including pollution issues (for example, policies on waste and recycling), climate change policies and international negotiations on sustainable development.[63] The Department is headed by the Secretary of State for Environment, Food and Rural Affairs, who is a Cabinet member.[64] The self-proclaimed aim of Defra, shortly after its formation in 2001, was to:

> ... enhance the quality of life through promoting: a better environment; thriving rural economies and communities; diversity and abundance of wildlife resources, a countryside for all to enjoy, sustainable and diverse farming and good industries that work together to meet the needs of consumers.[65]

The Department is the culmination of a complex process of organisation and reorganisation after the General Elections of 1997 and 2001. In 1997, the then Department of Transport was merged with the then Department of the Environment, to form the Department of Environment, Transport and the Regions ('the DETR').[66]

---

58 Established in 1997 to promote the idea of better, as opposed to, de- regulation (see Cabinet Office Press Release CAB 85/97, *New Better Regulation Task Force Launched*, 17 September 1997).
59 See www.strategy.gov.uk.
60 Its work has since been taken over by the Performance and Innovation Unit.
61 See Cabinet Office Strategy Unit, *Waste Not, Want Not: A Strategy for Tackling the Waste Problem in England* (London: Cabinet Office Strategy Unit, 2002).
62 See paras 6.3.2 and 21.4.1 below.
63 As to sustainability, Defra is responsible for the £58.6m Aggregates Levy Sustainability Fund ('the ALSF') (see para. 21.3.1 below). Financed, as its name suggests, by receipts from aggregates levy, the ALSF makes grants for: 'improving areas where aggregates extraction has taken place; helping to reduce demand for primary materials by research into alternatives; encouraging the recycling and re-use of aggregates and promoting new methods for extracting and moving aggregates to reduce environmental damage' (see Defra News Release, *Cleaner Quarries and Fewer Lorries*, 11 April 2002). Funds are distributed through various agencies, including the Countryside Agency (see www.countryside.gov. uk), English Heritage (see www.english-heritage.org.uk), English Nature (see www. english-nature.org.uk) and the Waste and Resources Action Programme ('WRAP') (see www.wrap.org.uk and www.aggregain.org.uk – this latter is an information database funded through the ALSF).
64 Currently (December 2004), the original holder, The Rt Hon. Mrs Margaret Beckett, MP.
65 *Department for Environment, Food and Rural Affairs: Aims and Key Tasks*, Defra, 2001, quoted in David Hughes *et al.*, *Environmental Law*, 4th edn (London: Butterworths LexisNexis, 2002), p. 33.
66 Secretary of State for the Environment, Transport and the Regions Order 1997, S.I.1997 No. 2971.

The thinking had been to integrate transport and environment policies,[67] so it was surprising when, following the 2001 General Election, the two policy areas were again uncoupled, transport being moved to the newly-formed Department for Transport, Local Government and the Regions ('the DTLR'), with environment being moved to another newly-formed ministry, Defra.[68] The main other component in Defra, besides the 'environmental' part of the DETR, was the larger part of the Ministry of Agriculture, Fisheries and Food, whose responsibilities for the promotion of public health and for the safeguarding of agriculture had long been regarded as being in mutual conflict.[69]

Below the Secretary of State for Environment, Food and Rural Affairs, who has overall responsibility for all departmental issues, there are two Ministers of State, respectively for Environment and Agri-Environment[70] and for Rural Affairs and Urban Life.[71] There are then two Parliamentary Under Secretaries, to whom the responsibilities of food, farming and sustainable energy,[72] and Nature Conservancy and fisheries[73] have been allocated.

Directly responsible to the Secretary of State is the Permanent Secretary, a civil servant, to whom a number of Directorates and Directorates General report directly. Most importantly, for present purposes, these include the Legal Services Directorate General, different Directorates of which deal with state aid issues,[74] investigations and prosecutions on behalf of the Department, waste regulation and climate change; and the Environmental Protection Directorate General, whose responsibilities include waste management policy and policy on climate, energy and environmental risk. Within Defra is also the 'shadow' Emissions Trading Authority, which is responsible for the administration of the UK ETS,[75] and the Combined Heat and Power Quality Authority.

In sum, Defra has been responsible for the design, implementation and operation of PRNs,[76] the LATS (in England),[77] climate change agreements under climate change levy[78] and the UK ETS.[79] As if all this were not enough, it has also been tasked with the preparation of the UK's national allocation plan ('NAP') under the EU ETS.[80]

---

[67]   Department of the Environment Press Release 216, 11 June 1997.
[68]   See the Secretaries of State for Transport, Local Government and the Regions and for Environment, Food and Rural Affairs Order 2001, S.I. 2001 No. 2568.
[69]   Most notably, of course, in relation to the BSE crisis of the mid-1990s and the foot and mouth crisis of 2001.
[70]   Currently (December 2004) Mr Elliot Morley, MP (who replaced The Rt Hon. Michael Meacher, MP in June 2003).
[71]   Currently (December 2004) The Rt Hon. Alun Michael, MP.
[72]   Currently (December 2004) Lord Whitty, the former General Secretary of the Labour Party, who also deals with all the Department's business in the House of Lords.
[73]   Currently (December 2004) Mr Ben Bradshaw, MP.
[74]   See para. 12.2.7 below.
[75]   The government had signalled its intention to establish this as an independent statutory body when Parliamentary time allows.
[76]   See para. 1.4.2.1(2) above.
[77]   See para. 1.4.2.1(4) above.
[78]   See para. 1.4.2.2(1) above.
[79]   See para. 1.4.2.2(2) above.
[80]   See para. 1.4.2.2(3) above and para. 28.5 below.

The DTI, which dates from 1983,[81] is responsible for trade policy, not only within the UK, but also as regards the UK's relationship with the EU and its Member States, as well as third countries. At the head of the DTI is the Secretary of State for Trade and Industry, who, like the Secretary of State for Environment, Food and Rural Affairs, is a member of the Cabinet.[82] Directorates within the Department deal with, *inter alia*, company law, competitiveness, international trade policy[83] and matters relating to the production, generation and supply of energy.[84] As regards the last of these:

> The ... [Energy Group] is committed to working with others to ensure competitive energy markets while achieving safe, secure and sustainable energy supplies. Its role is to set out a fair and effective framework in which competition can flourish for the benefit of customers, the industry and suppliers, and which will contribute to the achievement of the UK's environmental and social objectives. These include the alleviation of fuel poverty, and maintaining the security and diversity of the UK energy sources.[85]

Thus, within the Group are units dealing with energy strategy, British Energy plc,[86] nuclear and coal liabilities, energy markets, energy innovation and business and licensing and consents. Certain aspects of emissions trading are the responsibility of the strategy unit; European Commission approval for state aids are dealt with by the British Energy unit;[87] international treaties and negotiation, as well as electricity trading reform and regulatory issues are covered by the energy markets unit;[88] renewables and coal state aid and industry sponsorship are dealt with by the energy innovation and business unit; and oil and gas environmental matters are dealt with by the licensing and consents unit. The DTI, in conjunction with Defra, was responsible in February 2003 for the Energy White Paper, *Our Energy Future – Creating a Low Carbon Economy*.[89]

In connection with the promotion of a low carbon economy, the DTI has a further significance, as being one of four sponsors[90] of a body called the Carbon Trust.[91] An independent company limited by guarantee, the Carbon Trust was set up by

---

81 That is, following the amalgamation of two other ministries. Its origins can be traced much further back, however.

82 Currently (December 2004), The Rt Hon. Mrs Patricia Hewitt, MP.

83 Including relations with the WTO and the OECD (see paras 4.4.1 and 4.4.2 below).

84 See www.dti.gov.uk/energy.

85 See www.dti.gov.uk/energy/about.shtml.

86 See para. 21.4.4 below.

87 *Ibid.*

88 See paras 2.4 above and 6.4.3 below.

89 DTI/Defra, *Our Energy Future: Creating a Low Carbon Economy*, 2003 (Cm 5761, 2003). See further Chapters 20 and 21 below.

90 The others are the Department for Environment, Food and Rural Affairs (see para. 4.2.1.2(1) below); the Scottish Executive (see para. 4.2.2 below); from the National Assembly for Wales and from the Northern Ireland Assembly (see para. 4.2.2 below for both). The Carbon Trust, whose total funding is £50 million per annum, also receives part of the receipts from climate change levy (see para. 21.3.1 below).

91 See www.thecarbontrust.co.uk. This is a singularly uninformative website, high on political rhetoric and low on technical information.

the government 'in partnership with business', to invest in the development and deployment of low carbon technologies.[92] Since July 2002, it has had responsibility for the everyday administration of the enhanced capital allowances scheme for expenditure on specified environmentally-friendly energy equipment.[93] The members of the board of directors of the Carbon Trust are primarily, though not exclusively, from the private sector. The Carbon Trust has a sibling called the Energy Saving Trust ('the EST'),[94] set up by the pre-1997 Conservative Government after the United Nations' Conference on Environment and Development at Rio de Janeiro in 1992.[95] The EST runs two schemes aimed at stimulating the market: the Community Energy programme and the Photovoltaic Demonstration programme.[96] It is difficult to see the justification for either body, or why the functions of either body could not be transferred to central government.[97]

The DFT is the result of a further stage in the reorganisation process outlined above in relation to Defra. In 2002, the DTLR was split yet again, certain of its transport functions being moved to this new 'Department for Transport'.[98] The idea was for the DFT to provide a stronger focus on delivering the government's transport strategy.[99] Its self-proclaimed aim is 'transport that works for everyone' and, to that end, it works in partnership with others to 'tackle congestion; improve accessibility; reduce casualties; respect the environment; and support the economy'.[100] It describes itself as working with Defra on air-quality, climate change and sustainability and with HM Treasury,[101] Defra and with the DTI on the government's Powering Future Vehicles strategy. The DFT has four groups, that is, the Railways, Aviation, Logistics, Maritime and Security Group; the Strategy, Finance and Delivery Group; the Roads, Regional and Local Transport Group; and the Driver and Vehicle Operator Group.[102]

---

[92]   See House of Commons Science and Technology Committee, 4th Report. *Towards a Non-Carbon Fuel Economy: Research, Development and Demonstration* (HC Papers, Session 2002–2003, HC 55–I) (London: Stationery Office, 2003), para. 38. See further para. 21.3.1 below.

[93]   See para. 21.3.1 below.

[94]   See www.est.org.uk.

[95]   See paras 8.2.3 and 8.3.1.3 below.

[96]   The Community Energy programme is funded by Defra (see para. 4.2.1.2(1) below) and managed by the EST and the Carbon Trust and the Photovoltaic Demonstration programme is funded by the DTI. In 2002–2003, the EST's budget was £90 million, made up of funds from central and regional government (see *Towards a Non-Carbon Fuel Economy: Research, Development and Demonstration*, above, para. 43).

[97]   See *Towards a Non-Carbon Fuel Economy: Research, Development and Demonstration*, above, para. 44.

[98]   See the Transfer of Functions (Transport, Local Government and the Regions) Order 2002, S.I. 2002 No. 2626. Note the difference in nomenclature between the post-2002 'Department for Transport' and the pre-1997 'Department *of* Transport'.

[99]   See www.dft.gov.uk.

[100] *Ibid.*

[101] See para. 4.2.1.2(2) below.

[102] The DVLA is also part of the DFT.

(2) Taxes and tax policy

Governmental responsibilities in relation to taxation are located in three departments: HM Treasury;[103] the Board of Inland Revenue;[104] and the Commissioners of Customs and Excise.[105] The latter two are to be amalgamated under the name 'HM Revenue and Customs', a process outlined briefly below.

Her Majesty's Treasury has ultimate responsibility for the other two departments in the list, each being one of the UK's two tax authorities.[106] However, the Treasury's responsibilities extend far beyond tax administration, covering 'the supervision and control of national finance',[107] as well as the development and coordination of economic policy. The Department's responsibility for economic policy is of relatively recent origin.[108] That it is today one of the most important of the Department's functions is, however, plain from the Treasury's website,[109] where the Department's aim and objectives are set out. The former includes a specific commitment to the environment.[110]

The Treasury Board[111] comprises the First Lord of the Treasury, that is, the Prime Minister,[112] the Chancellor of the Exchequer[113] and five Junior Lords.[114] (Although called 'Lords', these posts are held by members of the House of Commons.)

The Chancellor is the key Finance Minister and under him are a number of Junior Ministers, that is, the Chief Secretary to the Treasury,[115] the Paymaster-

---

103　See www.hm-treasury.gov.uk. The Treasury has, over the years, been the subject of several interesting studies, for example, Joel Barnett, *Inside the Treasury* (London: André Deutsch, 1982) and Colin Thain and Maurice Wright, *The Treasury and Whitehall* (Oxford: Oxford University Press, 1995).

104　See www.inlandrevenue.gov.uk.

105　See www.hmce.gov.uk.

106　An introductory outline of the operational aspects of the two authorities has already been given (see para. 1.5 above).

107　See Hood Phillips and Jackson, *op. cit.*, para. 18–007.

108　See Edmund Dell, *The Chancellors* (London: HarperCollins, 1996), p. 110.

109　See www.hm-treasury.gov.uk.

110　See www.hm-treasury.gov.uk/about/about_aimsobject.cfm.

111　Which constitutes, technically and historically, the Commissioners running the Treasury while the office of Lord High Treasurer is vacant (as indeed it has been since the Duke of Shrewsbury was appointed by Queen Anne on her deathbed in 1714).

112　As First Lord of the Treasury. The post of Prime Minister was not created until 1905, although the person so regarded usually held that position since the era of Sir Robert Walpole (1721–1742). Since the early 1970s, the Prime Minister has also been Minister for the Civil Service.

113　Currently (December 2004), The Rt Hon. Gordon Brown, MP, New Labour's original appointment in May 1997 and the Chancellor with the longest tenure since Nicholas Vansittart (1812–1823).

114　Although they are used to sign statutory instruments, in practice the junior Lords of the Treasury are medium ranking House of Commons whips who rank considerably below Parliamentary Under-Secretary in the political hierarchy. The Chief Whip holds the (nominal) position of Parliamentary Secretary to the Treasury and the next three in seniority the posts of Treasurer, Vice-Chamberlain and Comptroller of HM Household (see, further, Hood Phillips and Jackson, *op. cit.*, para. 18–007).

115　Who is the Cabinet Minister responsible for expenditure, currently (December 2004) The Rt Hon. Paul Boateng, MP.

General,[116] the Financial Secretary to the Treasury[117] and the Economic Secretary to the Treasury.[118] Of these, the Chief Secretary to the Treasury is also, like the Chancellor, a Cabinet Minister. The Paymaster-General used – before the proposed departmental amalgamation – to be responsible for the Inland Revenue and has overall responsibility for each Finance Bill. Responsibility for Customs and Excise used to lie with the Financial Secretary to the Treasury. The fact that the Treasury has two Cabinet Ministers tends to underline the popular impression that the Treasury is the senior department of government. The Treasury is made up of seven main Directorates, each of which consists of a number of Directorate Standing Teams. Environmental taxes are dealt with by the Budget and Public Finances Directorate and, within that, by the Environment and Transport Taxes and Saving Incentives Team. The Treasury's key contribution to the evolution of policy relating to environmental taxes and other economic instruments is reflected in its policy document of 2002, *Tax and the environment: using economic instruments*.[119]

The Board of Inland Revenue is responsible for the administration of the UK's direct taxes.[120] By the Inland Revenue Regulation Act 1890, s.2, the Board of Inland Revenue is constituted by a quorum of Commissioners,[121] who 'may do and order and direct and permit to be done throughout the United Kingdom or in any part thereof all acts, matters, and things related to inland revenue'.[122] Currently, four of the nine Commissioners have specific areas of responsibility, which the Revenue's website lists as Strategic Service Delivery, General Policy and Technical, Corporate Services and the Valuation Office Agency.[123] Below the level of the Board, the structure of the Department is a complex one,[124] made the more so by the Department's ongoing programme of reorganisation.[125] The three aspects that are worthy of note in the

---

[116]   Currently (December 2004) The Rt Hon. Dawn Primarolo, MP, who has been the principal Treasury Minister in charge of the Inland Revenue since May 1997.

[117]   In September 2004, Miss Ruth Kelly, MP was replaced by Mr Stephen Timms, MP. He had been her predecessor in the post.

[118]   Currently (December 2004), Mr John Healey, MP, whose primary responsibility is for HM Customs and Excise.

[119]   See Chapter 5 below.

[120]   See para. 1.5 above. The Board of Inland Revenue was originally constituted under the Inland Revenue Board Act 1849.

[121]   Usually two. The current membership of the Board, the last permanent Chairman of which was Sir Nicholas Montagu, KCB, until his retirement in March 2004, can be viewed at www. inlandrevenue.gov.uk/board. Mrs Ann Chant is the current acting Chairman. In Budget 2004, delivered on 17 March 2004, proposals were announced for the amalgamation of the Inland Revenue with Customs. Shortly thereafter the government started advertising for an individual of major corporate experience to manage the merger, in the course of which it was anticipated that a significant reduction in personnel would be achieved.

[122]   'Inland revenue' is defined, in Inland Revenue Regulation Act 1890, s.39 as '... the revenue of the United Kingdom collected or imposed as stamp duties, taxes, and placed under the care and management of the Commissioners, and any part thereof'.

[123]   See www.inlandrevenue.gov.uk/board.

[124]   The organisation of the Department is envisaged as being along the lines of a Next Steps Executive Agency (see para. 4.2.1.3 below).

[125]   See, for example, Inland Revenue, *Annual Report 2003*, pp. 16–17.

present context are: first, the Revenue Policy Division, at least three units of which have an input into the environmentally-inspired provisions of the direct tax codes;[126] secondly, the Strategic Service Delivery Division, which contains the units concerned with enforcement; and, thirdly, the Solicitor's Office, which conducts the Board's civil litigation in England, Wales and Scotland, as well as its prosecutions in England and Wales. Within the second of these, the UK is divided up into a large number of Tax Districts in a regional structure. Inspectors and Collectors of Taxes, who are to act under the Board's direction, are appointed for each of these Districts and are assisted by the specialist units in the same Division.[127] Given its direct tax-orientation, the Revenue has not had a close involvement with environmental taxes; it has, however, contributed to the 'greening' of the UK's direct tax system, through the measures discussed in Chapters 23 and 24 below and has been working on the income tax and corporation tax treatment of the UK ETS, as well as of the EU ETS.[128]

Whereas the Board of Inland Revenue is responsible for the administration of the UK's direct taxes, the Commissioners of Customs and Excise are tasked with administering its indirect taxes. Indirect taxes include, of course, not only VAT and duties of customs and excise, but also landfill tax, climate change levy and aggregates levy. By the Customs and Excise Management Act 1979, s.6(1):

Her Majesty may from time to time, under the Great Seal of the United Kingdom,[129] appoint persons to be Commissioners of Customs and Excise, and any person so appointed shall hold office during Her Majesty's pleasure and may be paid such remuneration and allowances as the Treasury may determine.

Subsection (2) of s.6 then goes on to provide that, in addition to specific duties imposed by statute, and subject to the general control of the Treasury, the Commissioners '… shall be charged with the duty of collecting and accounting for, and otherwise managing, the revenues of customs and excise'. This involves the care and management of the taxes and duties laid on them, enforcing restrictions and prohibitions, dealing with individual taxpayer's affairs and dealing with criminal investigations.[130] Specifically entrusted to the Commissioners are the care and management of VAT;[131] the care and management of landfill tax;[132] the care and management of climate change levy;[133] and the care and management of aggregates levy.[134] However, the Department has also been responsible for the management of the consultations relating to the design, implementation and ongoing review of each of these taxes.[135]

---

[126] For example, the Benefits and Expenses Team, which deals with transport-related benefits, company cars and vans and green transport. See further Chapter 23 below.

[127] For example, the National Insurance Contributions Office, the Tax Credit Office, etc.

[128] See para. 24.7 below.

[129] See Hood Phillips and Jackson, *op. cit.*, paras 18–001–18–003, as to the Great Seal.

[130] See, for example, Customs and Excise, *Annual Report 2001–2002*, p. 113.

[131] Value Added Tax Act 1994, Sched. 11, para. 1(1).

[132] Finance Act 1996, s.39(2).

[133] Finance Act 2000, Sched. 6, para. 1(2).

[134] Finance Act 2001, s.16(5).

[135] See paras 11.3 and 11.4 below.

The Department was reorganised from April 2001 to focus all of Customs' operational activities into two 'core' functions: first, the Business Services and Taxes Directorate; and, secondly, the Law Enforcement Directorate.[136] As part of this reorganisation, an attempt was made to simplify the Department's management structure, as well as to separate support services from so-called 'front-line functions'. The two 'core' functions, though largely self-explanatory, perhaps merit some brief elaboration. Landfill tax, climate change levy and aggregates levy, as well as VAT and excise duties, are collected and managed by Business Services and Taxes. Law Enforcement's functions are primarily in relation to indirect tax fraud, it having responsibility for investigation, detection and the recovery of criminal assets. Within Business Services and Taxes, there is a team that deals exclusively with environmental taxation issues, including not only aggregates levy, climate change levy and landfill tax, but also the environmental aspects of excise duties and air passenger duty.[137] Other teams, whilst not being solely dedicated to environmental taxes, deal with the environmental aspects of the VAT code (for example, reduced rates of tax for domestic installation of energy-saving materials).[138]

An important function of Customs in the context of the environmental taxes is the Commissioners' oversight of Entrust, the regulatory body for the landfill tax credit scheme ('the LTCS').[139] Section 53 of the Finance Act 1996 gave power to the Commissioners to make regulations for securing that a landfill site operator would be entitled to a credit against landfill tax if he paid a sum to an environmental body, that is, one whose objects are or include the protection of the environment and in relation to which certain other prescribed conditions are satisfied.[140] The same section also gave power to Customs to make regulations for requiring environmental bodies to be approved by a regulatory body.[141] The Landfill Tax Regulations 1996, as amended,[142] accordingly contain detailed provisions both as to approved bodies and as to the appointment of the regulatory body, that is, Entrust.[143]

Customs have also been given control of the LRUC Management Authority, which has been set up to run the lorry road user charge.[144]

Both Customs and the Revenue are subject to the supervision, not only of the Ombudsman,[145] but also of the Revenue Adjudicator, whose function is to examine complaints from taxpayers about the manner in which the Department in question

---

[136]   As with the Revenue, this has been inspired by the Next Steps Executive Agencies (see para. 4.2.1.3 below).

[137]   See para. 1.2.1.5(2) above.

[138]   See para. 21.3.1 below.

[139]   See www.entrust.org.uk. See also paras 16.9 and 21.3.2 below. For details of the LTCS, see www.ltcs.org.uk.

[140]   See Finance Act 1996, s.53(1).

[141]   See Finance Act 1996, s.53(2)(b).

[142]   See Landfill Tax Regulations 1996, S.I. 1996 No. 1527, as subsequently amended, most recently by Landfill Tax (Amendment) Regulations 2003, S.I. 2003 No. 605. See further para. 21.3.2 below.

[143]   Entrust's website states that the body is funded by an enrolment fee from approved bodies and an administrative fee of 2 per cent of landfill operators' donations.

[144]   See para. 27.3 below.

[145]   See the beginning of para. 4.2.1.2(1) above.

conducts the taxpayer's affairs.[146] The office is rendered all the more important because of both Departments' enforcement powers, particularised as necessary throughout the rest of the study.[147] Because of its 'policing' role,[148] Customs has the widest prosecution and enforcement powers of any government department.

Controversy over their handling of a number of high-profile cases,[149] no less than the involvement of the Boards of both the Inland Revenue and Customs and Excise in a PFI scheme involving an offshore intermediate lessor with existing tax losses,[150] prompted the Treasury to announce a major review of the way in which the two departments operate.[151] It was announced on 17 March 2004, that the amalgamation of the two bodies is to result from this.[152] In May 2004, the chairman[153] and deputy chairman[154] of the amalgamated Department, to be called 'HM Revenue and Customs',[155] were named.

4.2.1.3   Non-departmental public bodies and agencies

Of primary relevance to the environmental regulation strand in this account are a number of non-departmental public bodies and agencies. There are essentially four separate bodies: the Environment Agency;[156] the Gas and Electricity Markets Authority; the Office of Gas and Electricity Markets;[157] and the National Audit Office.[158] All four are executive non-departmental bodies (or 'quangos'),[159] rather

---

146   See, for example, the useful summary of the Revenue Adjudicator's functions in Lesley Browning *et al.*, *Revenue Law – Principles and Practice*, 22nd edn (London: Lexis Nexis Tolley, 2004), para. 2.52.

147   See, esp. para. 16.12 below.

148   For example, in relation to smuggling, etc.

149   Resulting in Customs' prosecuting role being brought under the direct supervision of the Attorney-General.

150   See House of Commons Treasury Select Committee, Tenth Report of Session 2002–03, HC 834, Inland Revenue Matters (published 23 July 2003).

151   See HM Treasury Press Release 78/03, *Chancellor Announces Major Review of Inland Revenue and HM Customs and Excise*, 2 July 2003. The issues are of some standing. Tiley quotes Denis Healey (Lord Healey), the Labour Chancellor from 1974 to 1979, as saying in his memoirs (characteristically) that the Inland Revenue, like Customs and Excise, 'considered itself to be at least as independent of the Treasury as the three armed services were of the Ministry of Defence!' (see John Tiley, *Revenue Law*, 4th edn (Oxford: Hart Publishing, 2000), p. 58n, quoting Denis Healey's *The Time of My Life* (London: Michael Joseph, 1989), p. 373).

152   The Inland Revenue has recently absorbed the Contributions Agency, which was responsible to the Department of Social Security for the collection of NICs.

153   Mr David Varney, from a leading listed company.

154   Mr Paul Gray, a Second Permanent Secretary from a Ministry.

155   See, for example, Anon., *Financial Times*, 13 July 2004, p. 8 (discussing job losses as a result of the projected merger of the two departments).

156   See www.environment-agency.gov.uk.

157   See www.ofgem.gov.uk.

158   See www.nao.gov.uk.

159   That is, quasi-autonomous non-governmental organisation.

than Next Steps Executive Agencies.[160] To the list might be added at least two other bodies, who figure to a greater or lesser extent in what follows: the Civil Aviation Authority,[161] which deals with the economic and safety regulation of UK aviation and airlines, and the Coal Authority,[162] which is responsible for licensing coal mining operations and ensuring the safety of abandoned mine workings.[163]

Set up under the Environment Act 1995, the Environment Agency took over a range of functions and powers, including those of her Majesty's Inspectorate of Pollution, of the National Rivers Authority and, in relation to waste-management, of local authorities.[164] The Agency is funded partly by Defra and the National Assembly for Wales and partly by its own activities (for example, levies on local authorities and licence fees).[165] It is organised on a regional basis, with the water management boundaries being based on the National Rivers Authority's eight river catchment areas, and the pollution control functions being organised largely on the same areas, but with modifications to match the local authority boundary closest to the water management boundary.[166]

By s.4(1), Environment Act 1995, the principal aim of the Agency is to discharge its functions so to protect or enhance the environment, taken as a whole, as to contribute to the objective of achieving sustainable development.[167] The principal aim is qualified in two ways: (a) it is subject to, and the discharge of the Agency's functions must be in accordance with, other statutory provisions, including those of the Act itself; and (b) the Agency must take into account any likely costs. Whether or not the wording of the principal aim creates a legally enforceable duty, in view of its wide scope, is uncertain. As to (a), the principal aim is overridden by other statutory provisions in cases where such

---

[160] Environment Act 1995, s.1(1), establishes the Environment Agency as an independent body corporate. As to Next Steps Executive Agencies and the constitutional issues they raise, see Hood Phillips and Jackson, *op. cit.*, para. 18–023.

[161] See www.caa.co.uk.

[162] See www.coal.gov.uk.

[163] For completeness, we should also mention the roles of the Office of Fair Trading (www. oft.gov.uk) and of the Competition Commission (www.competition-commission.org.uk) in relation to the privatised electricity and gas markets discussed in Chapter 2 above and Chapter 6 below.

[164] The history of the Environment Agency is usefully outlined in Bell and McGillivray, *op. cit.*, pp. 162–163. The Agency has a Scottish equivalent in the Scottish Environment Protection Agency ('SEPA'), see www.sepa.org.uk.

[165] See, for example, Environment Agency, *Annual Report and Accounts 2001–2002*.

[166] See Susan Wolf and Neil Stanley, *Wolf and Stanley on Environmental Law*, 4th edn (London: Cavendish, 2003), p. 38.

[167] The Rio Declaration (see paras 8.2.3 and 8.2.6 below) contains 27 Principles of international environmental law, which together constitute the main outlines of the concept of sustainable development. There is no definition of the term in the Environment Act 1995, the Agency's own guidance using the Brundtland Report's definition, that is: '[D]evelopment that meets the needs of the present without compromising the ability of future generations to meet their own needs'. The Brundtland Report (World Commission on Environment and Development, *Our Common Future* (Oxford: Oxford University Press, 1987) was the work of the World Commission on Environment and Development ('WCED') and its definition of sustainable development has been described as 'somewhat Delphic' (see Patricia Birnie and Alan Boyle, *International Law and the Environment*, 2nd edn (Oxford: Oxford University Press, 2002)), p. 41. See, further, Bell and McGillivray, *op. cit.*, pp. 39–46.

provisions place the Agency under a duty to have regard to particular considerations or to carry out particular acts. The latter qualification, that is, in (b), includes the costs to any person and to the environment.[168] The Secretary of State must give guidance from time to time to the Agency with respect to the objectives that he considers it appropriate for the Agency to pursue in the discharge of its functions.[169] Specifically, such guidance must include advice on how, having regard to the Agency's responsibilities and resources, it is to achieve the objective of sustainable development.[170] Prior to issuing such guidance, the Secretary of State is obliged to consult the Environment Agency and any other appropriate bodies or persons.[171] One of the key functions of the Agency, in the context of the present study, is its administrative and enforcement role, via its National Waste Registration Unit ('NWRU') of the system of producer responsibility for packaging waste and the non-statutory PRN regime.[172]

Other duties are imposed upon the Agency by ss.5 and 6, Environment Act 1995. Section 5(1) provides that the Agency's pollution control powers are to be exercisable for the purpose of preventing or minimising, or remedying or mitigating the effects of, pollution of the environment. A very general duty is likewise imposed on the Agency by s.6(1), which provides that the Agency must, 'to such extent as it considers desirable, … generally promote the conservation and enhancement of the natural beauty and amenity of inland and coastal waters' (plus associated land), 'the conservation of flora and fauna that are dependent on an aquatic environment' and 'the use of such waters and land for recreational purposes'.

Section 39(1) of the Environment Act 1995 imposes a duty on the Agency to 'take into account the likely costs and benefits of the exercise or non-exercise of a power' conferred on it, by or under an enactment, or its exercise in the manner in question. However, the subsection continues, the duty does not apply if it is considered unreasonable to embark upon this calculation by reference to the nature or purpose of the power in question or in the circumstances of the particular case. The Agency's own guidance suggests that one consequence of this wording is that there is no duty to conduct a cost/benefit assessment in relation to an enforcement decision.[173] The Agency's enforcement powers, including those of inspection and entry, are wide, and are set out in Environment Act 1995, s.108. Subsection (1) of s.108 provides that the powers are exercisable for the purpose of: determining whether any pollution control legislation is being complied with; exercising or performing any of the Agency's pollution control functions; and determining 'whether and, if so, how, such a function should be exercised or performed'. Subsection (4) of s.108 empowers an officer appointed by the Agency, *inter alia*, to enter premises; to be accompanied by a police officer in certain circumstances; and make any necessary investigation. The Agency has published its policy in relation to the prosecution of environmental offences.[174]

---

[168] Environment Act 1995, s.56(1).
[169] *Ibid.*, s.4(2).
[170] *Ibid.*, s.4(3).
[171] *Ibid.*, s.4(5).
[172] See para. 1.4.2.1(2) above and Chapter 19 below.
[173] Environment Agency Sustainable Development Series SD3, *Taking Account of Costs and Benefits*, para. 11.1.
[174] See www.environment-agency.gov.uk.

The Office of Gas and Electricity Markets ('Ofgem') regulates the UK's gas and electricity industries.[175] It carries out its work subject to the direction of the first of the bodies in this list, the Gas and Electricity Markets Authority ('GEMA'), a body corporate with the primary duty of carrying out its functions in the way best calculated to protect the interests of gas and electricity consumers, wherever appropriate by promoting effective competition. The Gas and Electricity Markets Authority was established by the Utilities Act 2000 to carry out the functions formerly carried out by the Director General of Gas Supply and the Director General of Electricity Supply, as well as the other functions assigned to it under that Act.[176] By s.1(2), Utilities Act 2000, GEMA's functions are performed by it on behalf of the Crown. It must make arrangements with the Gas and Electricity Consumer Council for cooperating and exchanging information. As to Ofgem, according to its website, Ofgem's work concentrates on the following areas: making gas and electricity markets work effectively; regulating monopoly businesses intelligently; securing Britain's gas and electricity supplies; and meeting its increased social and environmental responsibilities.[177] The Office has enforcement powers under the Competition Act 1998.

Finally, there is the National Audit Office ('the NAO'), established by the National Audit Act 1983. The NAO is tasked with providing independent information, advice and assurance, both to the public at large and to Parliament, about the financial operations of government departments and other bodies that receive public funds. The NAO is made up of the Comptroller and Auditor-General and the staff appointed by him.[178] The NAO's expenses are defrayed out of money provided by Parliament,[179] neither the Comptroller and Auditor-General nor any of his staff being regarded as holding office under her Majesty or as discharging any functions on behalf of the Crown.[180]

4.2.1.4   Advisory bodies

Before going on to consider regional and local government, reference should briefly be made at this point to four advisory bodies: the Royal Commission on Environmental Pollution ('RCEP');[181] the Advisory Committee on Business and the Environment ('ACBE');[182] the Commission for Integrated Transport ('CFIT');[183] and the Commission on Sustainable Development.[184]

---

[175]   Under the Gas Act 1986 and the Electricity Act 1989 (as amended by the Utilities Act 2000).
[176]   Utilities Act 2000, s.1(1).
[177]   See www.ofgem.gov.uk.
[178]   National Audit Act 1983, s.3(1).
[179]   *Ibid.*, s.4(1).
[180]   *Ibid.*, s.3(5).
[181]   See www.rcep.org.uk.
[182]   See www.defra.gov.uk/environment/acbe/default.htm.
[183]   See www.cfit.gov.uk.
[184]   See www.sd-commission.gov.uk; see also the Defra sustainability website: www.defra.gov.uk/environment/sustainable/index.htm. The Sustainability Commission's Chairman is Sir Jonathon Porritt, CBE.

The Royal Commission on Environmental Pollution is an independent standing body established in 1970, to advise the Queen, the government, parliament and the public on environmental issues. As a standing Royal Commission, with its own secretariat, it is, as Bell and McGillivray say, 'a rather rare beast'.[185] Besides its 20-odd reports,[186] the most visible aspect of its work, it also responds to consultations and submits memoranda to Committees of the UK Parliament.[187] Despite being described as 'constitutionally independent' from government departments, its funding is provided by Defra, which requires the Commission to supply it with evidence that the funding has been put to good use and that RCEP is providing value for money.[188]

The other three committees mentioned above can be dealt with fairly briefly. The Advisory Committee on Business and the Environment ('ACBE') 'provides for dialogue between government and business on environmental issues and aims to help mobilise the business community in demonstrating good environmental practice and management'.[189] Hughes *et al.* pinpointed the problem with ACBE as being that, given that there was no attempt to combine the interests of the industrial sector and pro-environment groups in ACBE's membership, there was a danger that it would become, or would be seen as becoming, 'an institutionalised form of lobbying for industry'.[190] These words proved prophetic in relation to the UK Emissions Trading Group,[191] the body formed by ACBE and the CBI[192] jointly, in July 1999, to represent the business interest in emissions trading.[193]

The role of CFIT was identified in the government's 1998 White Paper[194] as being '... to provide independent advice to government on the implementation of integrated transport policy, to monitor developments across transport, environment, health and other sectors and to review progress towards meeting ... [the government's objectives under its 10-year transport plan]'.[195] Since then, CFIT has championed a number of

---

185 See Bell and McGillivray, *op. cit.*, p. 157.
186 Some of the most important of these have been its Fifth Report, *Air Pollution Control: An Integrated Approach*, 1976 (Cmnd 6371, 1976); its Tenth Report, *Tackling Pollution – Experiences and Prospects*, 1984 (Cmnd 9149, 1984); its Eleventh Report, *Managing Waste: The Duty of Care*, 1985 (Cmnd 9675, 1985); its Twelfth Report, *Best Practicable Environmental Option*, 1988 (Cmnd 310, 1988); its Twentieth Report, *Transport and the Environment – Developments since 1994*, 1997 (Cm 3759, 1997); its Twenty-first Report, *Setting Environmental Standards*, 1998 (Cm 4053, 1998); and its Twenty-second Report, *Energy – The Changing Climate*, 2000 (Cm 4749, 2000).
187 See para. 4.2.1.1 above.
188 See Hughes *et al.*, *op. cit.*, p. 54.
189 See www.defra.gov.uk/environment/acbe/default.htm.
190 See Hughes *et al.*, *op. cit.*, p. 55.
191 See www.uketg.com.
192 See para. 2.2 above.
193 As the incentive payments made under the UK ETS have graphically demonstrated (see paras 12.2.7.2, 20.1 and 20.4 below).
194 Department of the Environment Transport and the Regions, *A New Deal for Transport: Better for Everyone*, 1998 (Cm 3950, 1998). For the DETR, and its subsequent fate, see para. 4.2.1.2(1) above.
195 See *A New Deal for Transport*, para. 4.4.

causes, including congestion charging, in relation to which it provides excellent up-to-date information via its website.[196]

Finally, the Commission on Sustainable Development is the successor to the UK Round Table on Sustainable Development and to the British Government Panel on Sustainable Development.[197] The Commission was originally set up in October 2000, following up a commitment to do so in the 1999 report *A Better Quality of Life*.[198] Its task is to monitor government progress in relation to sustainable development across all relevant fields and specifically to identify policies currently undermining such progress, with a view to their remediation. Recently, it has produced a detailed response to the joint Treasury and DFT consultation paper on the use of economic instruments in regulating the environmental effects of aviation.[199]

4.2.1.5   The court system

The quasi-federal structure of the UK has a single tax system and three different legal systems.[200] The legal system of England and Wales has a different system of appeals from those of Scotland and Northern Ireland. The factor common to all three systems is the ultimate appeal to the House of Lords and to the European Court of Justice ('the ECJ'). In this para, we outline briefly the legal system of England and Wales, as it relates to tax and environmental matters.[201]

Taxpayers may appeal against assessments to income tax, CGT and corporation tax under certain conditions.[202] They may choose whether to appeal to the General Commissioners or to the Special Commissioners. The General Commissioners are part-time and unpaid, often lay people rather than experts, and assisted by a clerk, who is usually a solicitor.[203] The Special Commissioners, by contrast, are barristers, advocates or solicitors of at least ten years' standing.[204] The Commissioners, when they hear a case, have the right to look anew at any matter of law or fact referred to them, and they can dismiss or alter an assessment in any way they consider to be justified.[205] By Value Added Tax Act 1994, ss.83 and 84, VAT appeals are made to the VAT and Duties Tribunals,[206] which consist of a chairman sitting either with two

---

[196]   See www.cfit.gov.uk/congestioncharging/index.htm.
[197]   As Hughes *et al.* point out, this body is not to be confused with the United Nations Commission on Sustainable Development (see para. 4.4.3.1 below).
[198]   Cm 4345, 1999. See para. 5.2 below.
[199]   See www.sd-commission.gov.uk/pubs/aviation/index.htm. For a discussion of the joint report itself, see para. 27.5 below.
[200]   See *Davies: Principles of Tax Law*, ed. by Geoffrey Morse and David Williams, 4th edn (London: Sweet and Maxwell, 2000), para. 2–06.
[201]   See, for example, *Smith, Bailey and Gunn on the Modern English Legal System*, ed. by S.H. Bailey, J.P.L. Ching, M.J. Gunn and D.C. Ormerod, 4th edn (London: Sweet and Maxwell, 2002).
[202]   Taxes Management Act 1970, s.31.
[203]   *Ibid.*, ss.2, 3.
[204]   *Ibid.*, s.4.
[205]   Davies, *op. cit.*, para. 1–12.
[206]   Value Added Tax Act 1994, Sched. 12.

other members or with one other member or alone.[207] Tribunal chairmen must have held a right of audience before the tribunal for at least seven years.[208] By Finance Act 1996, s.54(3), a person affected by a decision of Customs on landfill tax may by notice in writing require them to review the decision. Thereafter, appeal lies to a VAT and Duties Tribunal under Finance Act 1996, s.55. Virtually identical procedures apply for the purposes of climate change levy[209] and for aggregates levy.[210] Such, in essence, is the relevant system of tax appeals.

If the involvement of the UK courts in tax matters is as a result of appeals against assessments, in environmental, energy and transport matters it tends to occur as a result of applications for judicial review of administrative decisions,[211] for example, refusals to grant licences under the various command and control mechanisms[212] or refusals of, or the imposition of conditions on, planning permissions.[213] Review may be sought in the High Court on a number of grounds, including procedural flaws, procedural unfairness and irrationality. It is seen as providing a valuable accountability mechanism after the event.[214]

The High Court of Justice is divided into three administrative Divisions: the Queen's Bench, Chancery and Family Divisions. It is served by over 100 High Court judges. The Chancery Division hears appeals from the General and Special Commissioners on points of law only.[215] However, appeals from VAT and Duties Tribunals lie to the Administrative Court of the Queen's Bench Division, since they are treated as administrative law appeals.[216] Applications for judicial review are generally made to the Queen's Bench Division Administrative Court.

Appeals from both the Chancery Division and the Queen's Bench Division go to the Court of Appeal (Civil Division). Around 35 Lords Justices of Appeal serve the Court of Appeal, but Law Lords may also occasionally sit. There is usually a panel of three judges on the hearing of appeals, decisions being taken by a majority of them. The House of Lords, sitting in its judicial capacity, is the final court of appeal within the UK legal systems. Appeals are usually heard by five Law Lords, their decision being by a simple majority. Appeals may reach the Lords either from the Court of Appeal or, under the so-called 'leapfrog procedure', direct from the High Court.[217]

---

207   *Ibid.*, Sched. 12, para. 5(1).
208   *Ibid.*, Sched. 12, para. 7(4).
209   Finance Act 2000, Sched. 6, paras 121 and 122.
210   Finance Act 2001, ss.40 and 41.
211   See, generally, de Smith, Woolf and Jewell, *Judicial Review of Administrative Action*, 5th edn (London: Sweet and Maxwell, 1995). See also Chris Hilson, *Regulating Pollution – A UK and EC Perspective* (Oxford: Hart Publishing, 2000), pp. 52–53 and 61–64. There is also a useful overview at Wolf and Stanley, *op. cit.*, pp. 64–66.
212   See Chapter 6 below.
213   See para. 6.4.3.1(3) below.
214   But see Hilson, *passim*, on the value of this.
215   Taxes Management Act 1970, s.56(6) and 56A(1). For the distinction between points of law and points of fact, see, for example, *Edwards v. Bairstow and Harrison*, [1956] AC 14.
216   See Civil Procedure Rules 1998, S.I. 1998 No. 3132, Sched. 1, para. 2; Davies, *op. cit.*, para. 1–12.
217   Administration of Justice Act 1969, ss.12–15.

4.2.2   Devolved regional government

The UK is now close to being, as already mentioned, a federal state.[218] This has had a number of paradoxical consequences. England is now the only part of the UK that does not have its 'own particular institutions',[219] and MPs in the UK Parliament who represent Scottish seats may vote in the UK Parliament on matters that do not affect Scotland.[220]

There are three devolution schemes, that is, those for each of Wales, Scotland and Northern Ireland. They have certain similarities:[221]

1   unlike in the UK Parliament, there is a single legislative chamber;
2   elections are based on proportional representation;
3   powers are delegated by central government to the regions 'without relinquishment of sovereignty';
4   there is the same scheme for referring devolution issues to higher courts;
5   each of the devolved bodies is subject to the 1950 European Convention on Human Rights and the Human Rights Act 1998;
6   the Executive in each province is constituted from elected members, although the functions and powers of each Executive differs considerably;
7   whereas Wales and Northern Ireland are entirely dependent for revenue on Westminster, the Scottish Parliament may increase or decrease the basic rate of income tax fixed by the UK Parliament by a maximum of 3 per cent;[222] and
8   whereas the Scottish Parliament has primary legislative power in relation to pollution for Scotland, and the Northern Ireland Assembly has that power in relation to Northern Ireland, the UK Parliament retains this power in relation to England and Wales.

Setting aside such similarities, the devolution scheme for each province is rather different. At the time of writing, the Northern Ireland Assembly has been dissolved, following a period of suspension.[223]

The Government of Wales Act 1998 provides for the election of a National Assembly for Wales[224] for a term of four years, no provision existing for early

---

[218]   See para. 1.3 above. See also Hilson, *op. cit.*, p. 47, referring to for the distinction between devolution and federalism, B. Burrows and G. Denton, *Devolution or Federalism? Options for a United Kingdom* (London: Macmillan, 1980).

[219]   See Hood Phillips and Jackson, *op. cit.*, para. 5–042.

[220]   This is the so-called 'West Lothian Question', so-called after the Constituency of the MP who reputedly first raised the possibility in the 1970s (Mr Tam Dalyell, MP): see Hood Phillips and Jackson, *op. cit.*, para. 5–045.

[221]   See Hood Phillips and Jackson, *op. cit.*, paras 5–013–5–019, to which the following material in para. 4.2.2 is considerably indebted.

[222]   See para. 1.3 above.

[223]   It was suspended from midnight on 14 October 2002 and dissolved on 28 April 2003. Elections to the new Assembly were held, but the majority parties on both sides of the political divide changed and, as at December 2004, they had been unable to reach an accommodation.

[224]   See www.wales.gov.uk/index.htm.

dissolution.[225] The same period is provided for by the Scotland Act 1998, although the Scottish Parliament[226] may force an election 'before the end of its four year term if such a move has the support of two-thirds of its members, or if a vacancy arises in the office of First Minister and no member is able to win sufficient support to form a new government within 28 days'.[227] The Northern Ireland Act 1998 also provides for the election of the Northern Ireland Assembly for four years.[228]

The Welsh Assembly, a body corporate,[229] has no power to enact primary legislation, although it does have the power to enact secondary legislation (called 'Assembly Orders'). The Secretary of State for Wales[230] is required to 'carry out such consultation [that is, with the Welsh Assembly] about the government's legislative programme for the session as appears to him to be appropriate'.[231] By s.33 of the Government of Wales Act 1998, the Welsh Assembly may consider and make 'representations [to the UK government] about any matter affecting Wales'. Besides enacting Assembly Orders, the Welsh Assembly may issue policy statements in the form of circulars and give guidance on the exercise of statutory powers. Within this pattern of possibilities and constraints, the powers of the Welsh Assembly are broad, covering agriculture, forestry, fisheries and food, culture, economic development, education and training, the environment, health and the health services, highways, housing, industry, local government, social services, sport and recreation, tourism, town and country planning, transport, water and flood defence and the Welsh language.[232] Limitations on the Assembly's powers are contained in ss.106, 107 and 108, Government of Wales Act 1998 (respectively EU obligations, human rights and international obligations). By s.121, Government of Wales Act 1998, the Assembly must set a new economic agenda for Wales as well as promote sustainable development.

Like the Welsh Assembly, the Scottish Parliament only has functions falling within its devolved competencies.[233] Unlike the Welsh Assembly, however, the Scottish Parliament has a general power to make laws[234] that fall within its legislative competence.[235] Thus, whilst being unable to modify provisions of, for example, the Human Rights Act 1998 and parts of the European Communities Act 1972, and whilst being unable to legislate in certain areas,[236] the Scottish Parliament may legislate in a wide range of areas, including local government (and even 'local taxes to fund local

---

[225] Government of Wales Act 1998, s.3(2). A provision for early dissolution was deemed unnecessary, since the Welsh Assembly has no power to refuse to pass government Bills.

[226] www.scottish.parliament.uk.

[227] See Hood Phillips and Jackson, *op. cit.*, para. 5–028 (Scotland Act 1998, ss.3 and 46).

[228] Northern Ireland Act 1998, s.31(1).

[229] Government of Wales Act 1998, s.1(2).

[230] A Cabinet Minister (see para. 4.2.1 above). The office is currently held by The Rt Hon. Peter Hain, MP (December 2004), whose main function is Leader of the House of Commons.

[231] Government of Wales Act 1998, s.31.

[232] *Ibid.*, Sched. 2.

[233] See Colin Reid and Gerardo Ruiz-Rico Ruiz, 'Scotland and Spain: The Division of Environmental Competencies' (2003) 52 ICLQ 209–25.

[234] Scotland Act 1998, s.28.

[235] *Ibid.*, s.29.

[236] *Ibid.*, Sched. 5.

authority expenditure'), fishing, forestry and economic development. In particular, as mentioned above, the Scottish Parliament may vary the basic rate of income tax up to three percentage points up or down but (not mentioned above), in the event that it reduces the basic rate, it must make a payment to the Board of Inland Revenue to make up the resulting shortfall, out of the Scottish Consolidated Fund.[237]

The (dissolved)[238] Northern Ireland Assembly's powers fall somewhat between those of the Welsh Assembly and of the Scottish Parliament.[239] Like the Scottish Parliament, it has a general legislative power but, unlike the Scottish Parliament, the power is somewhat closely restricted. Crown matters, defence, elections, Parliament, international relations, treason and national security are all so-called 'excepted matters',[240] on which the Northern Ireland Assembly has no power to legislate. Even where it does have power to legislate, however, the Assembly must act within its legislative competence and not, for example, pass an Act that is incompatible with human rights or European law.

The Executive takes different forms, and has different powers, in each of the three provinces. With Scotland, there is a Scottish Executive ('the Scottish Ministers'), consisting of a First Minister, other ministers and the Scottish Law Officers,[241] the Executive exercising devolved executive powers on behalf of the Crown.[242] In Northern Ireland, a complicated structure (designed to move Northern Ireland towards power-sharing) provides for the election of a First Minister and Deputy First Minister by the Assembly. Their fellow ministers on the Northern Ireland Executive Committee are elected by the Assembly, by a system of proportional representation designed to ensure that parties have ministerial posts in proportion to their strength in the Assembly. The Queen continues to exercise executive power in Northern Ireland, except with respect to 'transferred matters', where ministers on the Executive Committee and their departments exercise it on her behalf.[243] With Wales, since the Assembly is the executive, there is a Welsh Assembly Cabinet, made up of the Assembly First Secretary, the Chairman of the Executive Committee and the Assembly Secretaries.[244] The First Secretary is known as the First Minister, and the other members of the Executive Committee as ministers. None of these have powers by virtue of their office. Unlike under either of the Scotland or Northern Ireland settlements, the Welsh Cabinet exercises its functions on behalf of, and in cooperation with, the Assembly.

The devolved administrations in Wales and Scotland have been responsible for allocating allowances under the LATS, each choosing different commencement

---

[237] *Ibid.*, s.78.
[238] Following the 2003 Election in which the Democratic Unionist Party became the majority party on the Unionist side and Sinn Fein on the Nationalist side.
[239] See www.ni-assembly.gov.uk.
[240] Northern Ireland Act 1998, Sched. 2.
[241] Scotland Act 1998, s.44. See www.scotland.gov.uk.
[242] *Ibid.*, s.53(2).
[243] Northern Ireland Act 1998, s.23.
[244] Government of Wales Act 1998, s.56. See the website: www.wales.gov.uk. The Executive Committee is the formal name of the Welsh Assembly Cabinet, the Welsh Assembly having exercised its power to designate it as such.

dates.[245] In Northern Ireland, following the re-imposition of direct rule, allocation has been the responsibility of the Department of the Environment for Northern Ireland.[246]

Under all three sets of arrangements, Committees have an important part to play. Section 54 of the Government of Wales Act 1998 requires the Assembly to create, not only the Executive Committee[247] but also subject committees, a subordinate legislation scrutiny committee, an audit committee and regional committees, these being elected by the Assembly.[248] Besides being able to hold the administration to account, the subject committees have a role in scrutinising legislation. In Northern Ireland, s.29 of the Northern Ireland Act 1998 provides for the establishment of statutory committees 'to advise and assist each minister in the formulation of policy'. These committees have both scrutinising functions in relation to legislation and the power to hold the Executive Committee to account. Finally, the Scottish Parliament has mandatory committees (for example, standards, finance, audit, equal opportunities, European, etc) but it has also established special subject committees, one of which is that for Transport and the Environment. Unlike with the Government of Wales and Northern Ireland Acts, the Scotland Act 1998 does not contain detailed provisions on the establishment of committees.

The relationship between the UK and the devolved institutions is dealt with in a 1999 Memorandum of Understanding,[249] together with a number of supplementary agreements. There is also a Joint Ministerial Committee to supply a central coordination of the relationships between the devolved institutions and the UK.[250]

4.2.3 Local government

Local government in England and Wales 'consists of the administration by locally elected bodies of powers conferred and duties imposed by Parliament'.[251] England, except London and the Isles of Scilly, is divided into counties and, within those counties, into districts.[252] Districts and counties may each be either metropolitan or non-metropolitan.[253] A non-metropolitan county, a district or a London Borough is a principal area.[254] Although each principal area must have a principal council,[255]

---

245   See para. 1.4.2.5 above.
246   See www.doeni.gov.uk. This is currently (December 2004) the responsibility of the Secretary of State for Northern Ireland and the Northern Ireland Office.
247   See above in this para.
248   Government of Wales Act 1998, s.57.
249   Scottish Executive, *Memorandum of understanding and supplementary agreements between the United Kingdom Government, Scottish Ministers and the National Assembly for Wales,* 1999 (Cm 4444, 1999). (Subsequently revised in 2001 as Cm 5240.)
250   On both the Memorandum of Understanding and the Joint Ministerial Committee, see Hood Phillips and Jackson, *op. cit.*, paras 5–0475–049.
251   See 29(1) *Hals*, para. 1 (to which this para. 4.2.3 is indebted).
252   Local Government Act 1972, s.1(1).
253   *Ibid.*, ss.1(2)–1(4) (but note that metropolitan county *councils* were abolished by the Local Government Act 1985).
254   Local Government Act 1972, s.270(1).
255   *Ibid.*, ss.2 and 270.

there is an exception to this rule where, instead of both a district council[256] and a county council, there is a unitary authority.[257] All three of counties, districts and London Boroughs are local authorities within the legislative terminology.[258]

Constituted as corporations, local authorities are persons distinct in law from the residents of the areas that they govern. To the extent that Parliament, whether by primary or secondary legislation, has provided for ministers of central government to direct, control or supervise the exercise of their powers and duties, local authorities are subordinate to the direction, control and supervision of central government.[259] Councillors for each principal area are elected by local government electors[260] by universal adult suffrage. Councillors, who must not be in paid employment with the local authority in question, are elected for four years and must have a sufficient local connection.[261]

Local authorities have for long collected and retained local taxes,[262] in relation to which they have wide powers of enforcement, but they have no control over the structure of those taxes nor of the rules that provide for the imposition of liability or the grant of relief.[263] Under the terms of the Transport Act 2000, however, they have acquired powers to impose workplace parking levies[264] and to create road user charging schemes.[265] The exercise of these powers is discussed elsewhere in the book.[266] Waste disposal authorities, usually county councils,[267] are subject to the LATS,[268] designed to restrict the landfilling of biodegradable municipal waste, while both WDAs and WCAs[269] have the benefit of the much older WRCS, which is designed to ensure that WCAs are not penalised for the costs of recycling.[270]

---

[256] A district council may petition for borough status (see Local Government Act 1972, s.245).

[257] Sometimes referred to as 'single tier' authorities. The government has taken powers in Regional Assemblies (Preparations) Act 2003 to initiate referenda in the English regions with a view to setting up Regional Authorities in place of one of the tiers of government. Where set up, such authorities are likely to have environmental responsibilities.

[258] Local Government Act 1972, s.270(1). In England, there are at present 34 non-metropolitan counties, which are all divided into non-metropolitan districts, 45 unitary authorities and 238 non-metropolitan districts.

[259] See 29(1) *Hals*, para. 1.

[260] Local Government Act 1972, s.270(1).

[261] *Ibid.*, s.79.

[262] That is, council tax (see Local Government Finance Act 1992) and the uniform business rate (see Local Government Finance Act 1988, Part III). Council tax is a property tax payable in respect of domestic property to contribute to the cost of local government. The uniform business rate, which is payable on non-residential property, is also a property tax. The uniform business rate is fixed by central government but council tax rates are determined locally.

[263] Brief accounts of both council tax and the uniform business rate can be found in CCH Editions, *CCH Tax Handbook 2003–04* (Banbury: CCH, 2003), para. 10050.

[264] See Chapter 17 below.

[265] See Chapter 18 below.

[266] *Ibid.*

[267] See para. 2.3 above.

[268] See para. 1.4.2.1(4) above.

[269] See para. 2.3 above.

[270] See para. 1.4.2.1(3) above.

By contrast with their still somewhat limited tax-raising powers, the continuing involvement of local authorities with environmental regulation has been only slightly diminished with the transfer of their waste management functions to the Environment Agency in the mid-1990s.[271] Their involvement covers town and country planning under the planning legislation; the investigation and abatement of statutory nuisances under the Environmental Protection Act 1990; the control of smoke emissions, under the Clean Air Act 1993; the authorisation of certain atmospheric emissions under the Environmental Protection Act 1990, Part I, and under the Pollution Prevention and Control Act 1999; the identification of areas of contaminated land under the Environmental Protection Act 1990, Part IIA; and, in the case of County Councils, London Borough Councils and district councils, responsibilities as so-called 'hazardous substance authorities', under the Planning (Hazardous Substances) Act 1990.[272] Local authorities have extensive powers of enforcement in relation to these matters.

London is still a special case in the complex of local government in the UK. In addition to the 32 London Borough Councils is the Greater London Authority ('GLA'), which was created by the Greater London Authority Act 1999.[273] It consists of the Mayor of London and the London Assembly.[274] The GLA has eight main areas of responsibility: transport, planning, economic development and regeneration, the environment, police, fire and emergency planning, culture and health. In relation to transport, s.154 of the Greater London Authority Act 1999 provides for the creation of Transport for London ('TFL'). Transport for London has a number of different functions, including the provision of public passenger transport services to, from or within Greater London.[275] The Mayor of London is elected and holds office for four years;[276] he has the duty of setting the annual budgets for the GLA[277] and its functional bodies,[278] including TFL. The Mayor appoints the members of TFL and the other functional bodies.[279] On the environmental front, he is responsible for the London biodiversity action plan, the municipal waste management strategy, the London Air Quality Strategy and the London Ambient Noise Strategy. The London

---

271 That is, under the Environment Act 1995 (see para. 4.2.1.3 above and para. 6.3.2.1(1) below).

272 This list is based on the more detailed one in Wolf and Stanley, *op. cit.*, pp. 54–55.

273 See www.london.gov.uk.

274 Greater London Authority Act 1999, ss.1, 2 and Sched. 1.

275 *Ibid.*, s.173.

276 The first Mayor, Mr Ken Livingstone (who had been the last Leader of the old Greater London Council abolished in 1985), was elected, as an independent, on 4 May 2000 (see www.london.gov.uk). He was readmitted to the Labour Party before his five year expulsion period had expired in order to stand as the official Labour candidate in the 2004 election. He won that election also. It should be noted that the office of Mayor of London is entirely separate and distinct from the ancient office of the Lord Mayor of London. The latter's authority is confined to the City, that is, 'the square mile' financial district.

277 Greater London Authority Act 1999, s.122(1).

278 The four functional bodies are each separate bodies corporate: TFL, the London Development Agency, the Metropolitan Police Authority and the London Fire and Emergency Planning Authority (see Greater London Authority Act 1999, s.424(1)).

279 Greater London Authority Act 1999, s.154, Sched. 10, paras 2 and 3.

Assembly, which consists of 25 elected members,[280] has 14 constituency members, that is, one constituency member for each Assembly constituency, and 11 London members, that is, members for the whole of Greater London.[281]

4.3   European Union institutions

Some tension between environmental regulation, on the one hand, and taxation, on the other, is apparent from an examination of the place of each of these areas in the institutional framework of the EU.[282] Whilst there is broad consensus among Member States as to the role of environmental concerns in EU policy-making,[283] the scope, and even the meaning, of tax harmonisation, is very much more controversial.[284]

Not discussed below, although, given the importance of energy policy in the study, nonetheless relevant to the subsequent discussion in the book, is the European Atomic Community ('Euratom'). Eurotom was created in 1957, to support the Member States' non-military nuclear industries, by sponsoring research, laying down safety standards, overseeing implementation issues and monitoring the distribution of fissionable material.[285]

4.3.1   The Council of the European Union[286]

By Art. 203, European Treaty (ex 146), '[t]he Council shall consist of a representative of each Member State at ministerial level, authorised to commit the government of that Member State'.[287] From this wording, it follows that Council members are therefore politicians rather than civil servants.[288] Also by Art. 203, European Treaty

---

280 Defined *ibid.*, s.424(1).
281 *Ibid.*, s.2.
282 For the institutional framework of the EU, see: Stephen Weatherill and Paul Beaumont, *EU Law*, 3rd edn (London: Penguin, 1999), chs 2–6 and Paul Craig and Gráinne de Búrca, *EU Law: Text, Cases and Materials*, 3rd edn (Oxford: Oxford University Press, 2003), ch. 2. See also Sir Leon Brittan, 'Institutional Development of the European Community' [1992] PL 567; J. Lewis, 'The Methods of Community in EU Decision-Making and Administrative Rivalry in the Council's Infrastructure' (2000) 7 *JEPP* 67; A. Stevens, with H. Stevens, *Brussels Bureaucrats? The Administration of the European Union* (Basingstoke: Palgrave, 2001); and Neill Nugent, *The Government and Politics of the European Union*, 5th edn (London: Palgrave Macmillan, 2003).
283 See Chapter 12 below.
284 *Ibid.*
285 Euratom Treaty, Arts 1 and 2. See *Energy Law in Europe: National, EU and International Law and Institutions*, ed. by Martha M. Roggenkamp *et al.* (Oxford: Oxford University Press, 2001), paras 3.120–3.123.
286 The name is the result of Council Decision 591/93/EC, (1993) OJ L281 18. The Council of the European Union must not be confused with the European Council (that is, the Heads of State or government of the Member States, plus the President of the Commission (see Craig and de Búrca, *op. cit.*, pp. 71–75)).
287 http://ue.eu.int/en/main.htm.
288 Craig and de Búrca, *op. cit.*, p. 65.

(ex 146), it is provided that the office of President is to be held in turn by each Member State in the Council for a term of six months in the order decided by the Council acting unanimously.

Council meetings, of which there are approximately 80–100 per year, are generally arranged according to the subject matter under discussion, different ministers attending from the Member States according to the subject matter under consideration. Article 204, European Treaty (ex 147) provides that the Council shall meet 'when convened by its President, on his own initiative or at the request of one of its members or of the Commission'.

By Art. 202, European Treaty (ex 145), the Council must ensure that the objectives set out in that Treaty are attained. To this end, it must: '... ensure co-ordination of the general economic policies of the Member States'. However, neither this wording nor, indeed, the wording of the rest of Art. 202, conveys a sense of the true role of the Council as a legislative body. Such a sense is apparent only from a consideration of the European Treaty as a whole:

1   The Council's approval is required for legislative proposals by the European Commission. The Council must generally act by a majority of its members, whether simple or qualified,[289] the nature of the majority depending on the Treaty Art. under which a particular measure is enacted.[290] However, in certain crucially important areas, it is provided that Council approval must be unanimous.[291]

Under Title XIX, European Treaty (ex XVI), on the Environment, the areas in which the Council's decisions must be unanimous are specified as follows:

a.   'provisions primarily of a fiscal nature';
b.   measures affecting town and country planning;
c.   quantitative management of water resources or affecting, directly or indirectly, the availability of those resources;
d.   land use; and
e.   measures 'significantly affecting a Member State's choice between different energy sources and the general structure of its energy supply'.

Excepted from d. are waste management and measures of a general nature.[292]

---

[289]   Qualified majority voting, which is a system of weighted voting, is provided for by Art. 205(2) (ex 148). The Treaty of Nice, together with its Protocol on the Enlargement of the EU and Declaration 20 of the Declarations adopted by the Nice IGC, has effected a number of amendments, both current and prospective, to Art. 205(2), European Treaty (see Craig and de Búrca, *op. cit.*, pp. 155–6). The Treaty of Nice came into force on 1 February 2003 (see Art. 12(2), TN), the Republic of Ireland having deposited its instruments of ratification in December 2002. For all this, see http://europa.eu.int.

[290]   See Art. 205(1), European Treaty (ex 148).

[291]   Abstentions cannot block a measure that requires unanimity (see Art. 205(3) (ex 148(3)). For the cases where unanimity is required, see Weatherill and Beaumont, *op. cit.*, pp. 89–91.

[292]   See Art. 175, European Treaty (ex 130s), as amended by TN.

Environmental issues not mentioned in this list must normally be dealt with by qualified majority voting.[293]

In relation to taxation, two Arts are significant: Art. 93, European Treaty (ex 99), and Art. 95(2), European Treaty (ex 100a). Both of these appear in Title VI, European Treaty (ex V) on Common Rules on Competition, Taxation and Approximation of Laws. Article 93, European Treaty, provided for the unanimous harmonisation of legislation concerning turnover taxes, excise duties and other forms of indirect taxation to the extent that such harmonisation was necessary in the run up to the completion of the internal market on 31 December 1992.[294] Article 95(2), European Treaty, reaffirms the need for unanimity when the Council is considering a proposal of the European Commission relating to fiscal provisions, provisions relating to the free movement of persons and provisions relating to the rights and interests of employed persons.

2   By Art. 208, European Treaty (ex 152) the Council may request the European Commission to undertake '... any studies the Council considers desirable for the attainment of the common objectives, and to submit to it any appropriate proposals'. This is a powerful weapon in the hands of the Council, since it can be used to require the Commission to consider specific legislative proposals,[295] which can themselves be evolved through COREPER[296] and its working parties.

3. Since the Council can delegate powers to the Commission, under which the latter may make detailed regulations in a particular area, and since such regulations can be closely scrutinised by the Council, the Council can exercise close control over delegated legislation.[297]

4.3.2   The European Commission

The College of Commissioners consisted until 1 May 2004, of 20 members[298] and, since enlargement, consists of 25. They are chosen on the grounds of their general competence and on the basis that their independence is beyond doubt.[299] Individual Commissioners are appointed for a renewable period of five years.[300] Under the former wording of Art. 214(2), European Treaty (ex 158), the Commissioners and the Commission President were appointed by common accord of the Member States,

---

[293]   See para. 12.2.1 below.
[294]   See also Art. 14 (ex 7a). As to whether there may still be life in Art. 93, European Treaty, see David Williams, *EC Tax Law* (Longman: London, 1998), p. 34.
[295]   See Brittan, *op. cit.*, pp. 568–9.
[296]   Article 207 (ex 151) provides that a committee consisting of the Permanent Representatives of the Member States ('COREPER') is to be responsible for preparing the work of the Council and for carrying out the tasks assigned to it by the Council. See Lewis, *op. cit.*, generally.
[297]   See Council Decision 468/99/EC, (1999) OJ L184 23.
[298]   www.europa.eu.int/comm/index_en.htm. See Neill Nugent, *The European Commission* (Basingstoke: Palgrave, 2001).
[299]   Art. 213 (ex 157).
[300]   Art. 214(1) (ex 158).

subject to a vote of approval by the European Parliament. Following the entry into force of the Treaty of Nice, these provisions have been modified, as a result of which the European Parliament should in future have a greater say in the appointment process.[301] By Art. 213, European Treaty, (ex 157) the Commission had to include at least one, and no more than two, Commissioners from each Member State.[302] Each Commissioner has a specific portfolio.[303] From November 2004, each Member State of the enlarged EU is to have only one Commissioner.

If and when the European Constitutional Treaty, which was agreed in principle in June 2004 but has not yet been signed or ratified, comes into force, the procedure for selecting the Commission will change. The Council will have a full-time president. It will also, by weighted majority voting, propose a candidate for the approval of the Parliament for the post of President of the Commission. Once confirmation has been received, the Commission will be selected by its President and the Council taking into account suggestions made by Member States. One member of the Commission will be the Union Minister for Foreign Affairs, a position which is seen as forming part of a geographically spread troika with the two presidents. From 2014, the size of the Commission is to be reduced from one per Member State to two-thirds of the Member States, with an equal rotation of representation. Thus each Member State will be represented in the Commission for only two five yearly periods out of three.

Under the present arrangements, particular legislative proposals usually begin in one or more[304] of the Commission's Directorates General ('DGs'). The DGs of greatest relevance to the present study are: Energy and Transport,[305] Environment,[306] Taxation and the Customs Union[307] and Trade.[308] From the relevant DG, the draft of the proposal is then sent to the personal staffs (or *cabinets*) of the relevant Commissioners

---

[301] See Art. 214(2), as amended by Art. 1(21), TN. The Council, meeting in the composition of Heads of State or government and acting by qualified majority, nominates the person it intends to appoint as Commission President. This nomination must be approved by the European Parliament. The Council, acting by qualified majority and by common accord with the nominee for President, adopts a list of the other proposed Commissioners, drawn up in accordance with the proposals made by each Member State. The President and other Commissioners are then subject to a vote of approval by the European Parliament. Following this, the Council, by qualified majority, appoints the President and the other members of the Commission.

[302] This arrangement changed in November 2004, following enlargement (see Craig and de Búrca, *op. cit.*, pp. 55–6).

[303] So far as relevant to the present study, they had been, prior to mid-November 2004 (when the Prodi Commission's extended mandate came to an end): Mr Frits Bolkestein (Internal Market, Taxation and Customs Union); Mrs Loyola de Palacio (Energy and Transport); Mrs Margot Wallström (Environment) and Mr Pascal Lamy (Trade). Only Mrs Wallström is a member of the new (Barroso) Commission, but she has not retained her previous environment portfolio (see, for example, Avril, *Figaro*, 28/29 August 2004, p. 4).

[304] See Laura Cram, 'The European Commission as a Multi-Organisation: Social Policy and IT Policy in the EU' (1994) 1 JEPP 194.

[305] www.europa.eu.int/comm/dgs/energy_transport?index_en.html.

[306] www.europa.eu.int/comm/dgs/environment/index_en.htm.

[307] www.europa.eu.int/comm/dgs/taxation_customs/index_en.htm.

[308] http://europa.eu.int/comm/trade/index_en.htm.

and thence to the weekly meeting of the *chefs de cabinet*. Having been considered in that forum, the proposal then goes to the College of Commissioners, 'which may accept it, reject it or suggest amendments'.[309]

The Commission has three main functions, that is: as the initiator of legislation, as guardian of the Treaties and as the Community's executive. It has the right of initiation in relation to the Treaties because, under their common format, when making legislation, the Council and the European Parliament act on a proposal from the Commission.[310] This right of initiation is a particularly powerful weapon in the Commission's hands, since it enables it to develop policies via its work programme for a particular year. Increasingly, for example, in relation to energy, the Commission is making use of industry for its policy development.[311] Undoubtedly, the most striking achievement of the Prodi Commission (1999-2004), in the environmental sphere, has been the design and implementation of the EU ETS.[312]

The Commission's role as guardian of the Treaties arises from Art. 211, European Treaty (ex 155), first indent:

> [The Commission shall] ... ensure that the provisions of this Treaty and the measures taken by the institutions pursuant thereto are applied.

This wording enables the Commission to bring two types of proceedings:

1   actions under Art. 226, European Treaty (ex 169) against Member States who are in breach of Community law; and
2   investigations and adjudications on Treaty violations (for example, in relation to unlawful state aids under Art. 88, European Treaty (ex 93)).[313]

Craig and de Búrca describe the Commission's powers under (1) and (2) as 'judicial powers'. State aids have a particular importance in the context of the law and policies under discussion in this book and are discussed in detail in Chapter 12 below.[314]

Finally, some of the most important of the Commission's executive powers, so far as environmental and tax policy is concerned, are in the field of external relations. Three of these can be enumerated as follows:

1   to determine and conduct the EU's external trade relations;
2   to negotiate and manage the EU's various external agreements with third countries and groups thereof; and
3   to represent the EU at a range of international organisations.[315]

---

[309]   See Craig and de Búrca, *op. cit.*, p. 58.
[310]   *Ibid.*, pp. 59–60.
[311]   See Peter Cameron, *Competition in Energy Markets* (Oxford: Oxford University Press, 2002), p. 284.
[312]   See para. 28.2 below.
[313]   See para. 12.2.7 below.
[314]   *Ibid.*
[315]   See Nugent, *Government and Politics, op. cit.*, p. 145.

The most important of the negotiations referred to in 1 are the tariff negotiations conducted under the auspices of the WTO.[316] As to 3, Art. 302, European Treaty (ex 229) stipulates that it is for the Commission to ensure the maintenance of all appropriate relations with the organs of the UN and of its specialised agencies; Art. 303, European Treaty (ex 230) stipulates that the Community must establish all appropriate forms of communication with the Council of Europe; and Art. 304, European Treaty (ex 231) provides that the Community shall '... establish close cooperation with the Organisation for Economic Co-operation and Development, the details of which shall be determined by common accord'.

4.3.3   The European Parliament

The European Parliament consists 'of representatives of the peoples of the States brought together in the Community'.[317] Until June 2004 it had 626 Members; the ceiling on membership of 700 in the European Treaty, as amended by the Treaty of Amsterdam, has been increased to 732 by the Treaty of Nice.[318] Supporting the Parliament in its work is a secretariat numbering thousands.

The Bureau of the Parliament is the body responsible for organisational, staff and administrative matters, as well as the Parliament's budget. The Bureau consists of the President,[319] plus the 14 Vice-Presidents, each of whom hold office for two-and-a-half years. Both President and Vice-Presidents are elected by the Parliament as a whole. The Bureau is assisted by five Quaestors, who also oversee financial and administrative matters concerning MEPs. By Art. 190(5), European Treaty, (ex 138(5)) the Parliament must, after seeking an opinion from the Commission and with the approval of the Council acting unanimously, lay down the regulations and general conditions governing the performance of the duties of MEPs. Also, by Art. 199, European Treaty (ex 142), the Parliament must adopt its own procedural rules.[320]

Committees play an important part in the Parliament's work. It has 17 standing committees, plus a range of subcommittees, with provision also being made for committees of inquiry and temporary committees. Of the 17 standing committees, particularly relevant in the present context are: the Committee on Industry, External Trade, Research and Energy;[321] the Committee on the Environment, Public Health and Consumer Policy;[322] and the Committee on Regional Policy, Transport and Tourism.[323] The Committee on the Environment, Public Health and Consumer Policy unsuccessfully floated 73 amendments to the proposal for the EU ETS on its first reading in October 2002.[324]

---

[316]   See para. 4.4.1 below and para. 8.4 below.
[317]   Art. 189, European Treaty (ex 137). See www.europarl.eu.int.
[318]   See Art. 1(17), TN.
[319]   That is, of the Parliament.
[320]   (1999) OJ L202 1.
[321]   See www.europarl.eu.int/meetdocs/committees/itre.
[322]   See www.europarl.eu.int/meetdocs/committees/envi.
[323]   See www.europarl.eu.int/meetdocs/committees/rett.
[324]   See para. 28.2.3 below.

Of the eight political groupings in the Parliament elected in June 2004,[325] the largest is the European People's Party, with 276 seats;[326] next is the Party of European Socialists, which has 202; and third is the enlarged Group of the European Liberal, Democrat and Reform Party, with 84 seats. There are then several smaller groupings, including the Greens and (for the first time) Eurosceptics, plus various MEPs who are unaligned.[327]

Members of the European Parliament ('MEPs') are elected by direct universal suffrage,[328] although the representative legitimacy of the institution is undermined by the fact that the number of MEPs for each Member State is far from proportionate to population size, with the result that smaller countries are disproportionately over-represented.[329] Unhelpful too is the fact that, despite litigation and amendment to the European Treaty by the Treaty of Amsterdam, the uniform electoral procedure provided for by Art. 190, European Treaty (ex 138) has never come into existence.[330] Finally, the turnout in Member States for European Parliament elections has been worryingly low.[331]

Substantial powers are exercised by the Parliament in three areas: legislative, budgetary and supervisory. The third of these involves the monitoring of the activities of the other institutions by questions and committees of inquiry. The legislative powers of the Parliament have been greatly enhanced by successive Treaty amendments, the general effect of which has been to extend the sphere of operation of the co-decision procedure.[332] The procedure, which applies wherever the European Treaty refers to it for the adoption of an act, accords equal status to the Council and to the Parliament in the adoption of Community legislation. It prevents the adoption of a measure without the approval, not only of the Council, but also of the Parliament.[333] In the process, the procedure places emphasis on the legislative text in question being approved by both bodies.

In relation to the Parliament's legislative role, a number of points are important in the present context. First, although the Parliament has no right to initiate legislation, it does have the power, by a majority of MEPs, to request the Commission to submit an appropriate proposal on matters 'on which it considers that a Community act is

---

[325] See Art. 191, European Treaty (ex 138a).
[326] It consists of Christian Democrats and European Democrats and is a party of the centre-right.
[327] See www.europarl.eu.int/presentation/default_en.htm.
[328] See Art. 190(1), European Treaty (ex 138).
[329] See Craig and de Búrca, p. 76. See also, *ibid.*, p. 77, for a breakdown of the composition of the Parliament in terms of seats numbers to Member States, both before and after enlargement.
[330] See Craig and de Búrca, p. 76.
[331] See, for example, *The 1999 Elections to the European Parliament*, ed. by Juliet Lodge (Basingstoke: Palgrave, 2001). Craig and de Búrca (*op. cit.*, p. 78n, cite the fact that turnout in the EU as a whole apparently dropped from 56.5 per cent in the 1994 elections to 49.7 per cent in 1999, the turnout in the first elections held in 1979 having been 63 per cent.
[332] The procedure is contained in Art. 251, European Treaty (ex 189b) and is summarised in Craig and de Búrca, *op. cit.*, pp. 144–6.
[333] See Craig and de Búrca, *op. cit.*, p. 144.

required for the purpose of implementing [the] Treaty'.[334] Secondly, not merely does the Parliament have the power to censure the Commission,[335] ever since the Treaty of Amsterdam, its approval must also be sought in relation to the appointment of the President of the Commission.[336] Taken together with the changes made by the Treaty of Nice to the appointment procedure of the President,[337] this marks a significant change in the political nature of the Commission, as well as in the procedure by which the President of the Commission is appointed.[338]

Despite all of these legitimising factors, there are still a number of areas in which the Parliament has no role. In the present context, a surprising one perhaps is the Parliament's meaningful exclusion from matters involving international agreements.[339] Article 133(3), European Treaty (ex 113(3)) does not accord to the Parliament any role in the negotiation of such agreements and Art. 300(2), European Treaty (ex 228(2)) merely gives the Parliament the right to be informed about matters such as their suspension.

4.3.4   The Economic and Social Committee

The Economic and Social Committee ('ECOSOC') is an advisory body representing the various socio-economic organisations in the Member States.[340] Article 7(2), European Treaty (ex 4(2)) provides that the Council and Commission are to be assisted by ECOSOC and also by the Committee of the Regions, acting in an advisory capacity. Although the Commission or Council may, without being obliged to do so, consult ECOSOC on certain matters, in other cases, particular Treaty Arts require that ECOSOC must be consulted. The Committee, which has 224 members appointed for four years by the Council, acting unanimously,[341] has an increasing role to play in the legitimisation of the EU institutions.[342] This is underlined by the fact that the TN has altered Art. 257 (ex 193), to provide that ECOSOC must consist of 'representatives of the various categories of economic and social activity', especially 'representatives of producers, farmers, carriers, workers, dealers, craftsmen, professional occupations and representatives of the general public'.

---

334   Art. 192, European Treaty (ex 138b).
335   Art. 201, European Treaty (ex 144).
336   Art. 214(2), European Treaty (ex 158(2)).
337   See para. 4.3.2 above.
338   See Craig and de Búrca, *op. cit.*, p. 82, referring to Simon Hix, 'Executive Selection in the European Union: Does the Commission President Investiture Procedure Reduce the Democratic Deficit?', in *European Integration after Amsterdam*, ed. by K. Neunreither and A. Weiner (Oxford: Oxford University Press, 2000), pp. 95–111.
339   See para. 8.4.1 below.
340   See www.esc.eu.int/pages/en/home.asp.
341   See Art. 258, European Treaty (ex 194).
342   See ECOSOC, *The ESC: A Bridge between Europe and Civil Society* (Brussels: European Economic and Social Committee, 2001).

4.3.5   The European Environment Agency

An agency rather than an EU institution, the European Environment Agency ('EEA')[343] gathers and supplies information to assist in the implementation of Community policy on environmental protection and improvement. Established in 1993,[344] it must publish a report on the state of the environment every three years[345] but it has no enforcement or policing powers. It has also published material on the implementation and environmental effectiveness of environmental taxes.[346]

4.3.6   The Committee of the Regions

Tasked with providing opinions on matters of particular regional concern, the membership of the Committee of the Regions is drawn from local and regional bodies, rather than from governments.

4.3.7   Court of Justice of the European Communities and Court of First Instance

Finally, the ECJ sits in Luxembourg, and its task is to ensure that Community law is observed in the interpretation and application of the European Treaty.[347] *Inter alia*, the ECJ has jurisdiction to give preliminary rulings concerning the interpretation of the European Treaty and the validity and interpretation of acts of the Community institutions.[348] Where such an issue is raised before any court or tribunal of a Member State, the court or tribunal may, if it considers that a decision on the issue is necessary to enable it to give judgment, request the ECJ to give a ruling thereon.[349] The Court of Appeal of England and Wales, in *Bulmer v. Bollinger*,[350] laid down guidelines for determining whether a reference was necessary for these purposes.[351] 'Where any such question is raised in a case pending before a court or tribunal of a Member State against whose decisions there is no judicial remedy under national law, [reads the relevant Treaty Art.] that court or tribunal must bring the matter before [the ECJ]'.[352] The Court of First Instance, designed to relieve the pressure on the ECJ, has, since 1993, heard judicial review cases, as well as actions for damages against

---

343   See www.eea.eu.int.

344   See Council Regulation EEC/1210/90, (1990) OJ L120 1.

345   See European Environment Agency, *Europe's Environment: The Third Assessment* (Luxembourg: Office for Official Publications of the European Communities, 2003).

346   See European Environment Agency, *Environmental Taxes – Implementation and Environmental Effectiveness* (Copenhagen: European Environment Agency, 1999).

347   Art. 220, European Treaty (ex 164).

348   Art. 234, European Treaty (ex 177).

349   Art. 234, first indent, European Treaty (ex 177).

350   [1974] Ch. 401.

351   See also, for example, *Naturally Yours Cosmetics Ltd v. C & E Commrs*, C–230/87, [1988] STC 879. This was a case of a VAT and Duties Tribunal obtaining a ruling direct from the ECJ.

352   Art. 234 (ex 177), European Treaty (second indent). See, for example, the reference from the House of Lords in *C & E Commrs v. Sinclair Collis Ltd*, [2001] STC 989.

Community institutions, although not proceedings brought by the Member States or the Community Institutions.[353]

It should be noted that the ECJ is a completely distinct entity from the European Court of Human Rights, the latter having been established by the European Convention on Human Rights of 1950.

4.4 International institutions[354]

4.4.1 World Trade Organization

The World Trade Organization is the pre-eminent international trade association.[355] It is intended to 'provide the common institutional framework for the conduct of trade relations among its members'.[356] It is the product of the most recent (1986–1993) round of trade negotiations under the General Agreement on Tariffs and Trade ('GATT 1994'), a process which has become known as the Uruguay Round.

The four principal instruments that establish the WTO and create the multilateral trading system of which it is the principal institution are as follows:

1   the Final Act embodying the results of the Uruguay Round of Multilateral Trade Negotiations;[357]
2   the Agreement establishing the World Trade Organization;[358]
3   the General Agreement on Tariffs and Trade in Goods, 1994 Revision ('GATT 1994');[359] and
4   the General Agreement on Trade in Services ('GATS').[360]

The supreme decision-making body of the WTO, which meets at least once every two years, is the Ministerial Conference. After that, the next level down is the General Council, which meets on a number of occasions every year and is made up mainly of heads of delegations and ambassadors in Geneva. The General Council, which is responsible for issues of governance not dealt with by the Ministerial Conference, also meets under the guise of the Trade Policy Review Body and the Dispute Settlement Body.

One of the roles assumed by the Dispute Settlement Body is the appointment of 'Panels' to investigate and report on complaints. Another is the adoption of reports

---

353   See Art. 224, European Treaty (ex 168).
354   In addition to the website details given below, see *Bowett's Law of International Institutions*, 5th edn, by Philippe Sands and Pierre Klein (London: Sweet and Maxwell, 2001).
355   See www.wto.org. Most of the WTO-related legal texts and other documents mentioned below, together with summaries thereof, can be downloaded from this site.
356   See Art. 2(1), Agreement Establishing the World Trade Organization.
357   (1994) 33 ILM 1.
358   (1994) 33 ILM 13.
359   (1994) 33 ILM 28.
360   (1994) 33 ILM 46.

of the Appellate Body (the 'AB'). The AB was established during the Uruguay Round by the Dispute Settlement Understanding,[361] with the power to hear appeals on questions of law from GATT panels. Consisting of seven members, the AB is a standing body, unlike the GATT panels. The members of the AB are individuals of recognised authority with expertise in law, international trade and the GATT/WTO agreements. They are elected by the Dispute Settlement Body for a four-year period that can be renewed once.[362]

The various organs of the WTO obviously require massive technical support, such support being the responsibility of the WTO Secretariat, based in Geneva, and headed by the WTO's Director-General.[363] The Director-General of the WTO is appointed by the Ministerial Conference and his responsibilities are 'exclusively international in character', since he is not to seek nor to accept 'instructions from any government or any other authority external to the WTO'.[364]

The WTO system requires compliance with the decisions both of the Panels of the AB. The enforcement procedure operates by removing the right of a WTO member to suspend WTO concessions or to take other measures against another member. In the words of Sands and Klein, 'the entire scheme militates strongly against unilateral determinations and establishes a central role for the [Dispute Settlement Body] ...'.[365]

The importance of the WTO agreements to environmental levies and subsidies in general, whether existing or projected, is profound and far-reaching.[366] In recognition of this fact, the WTO founded the Committee on Trade and Environment. Unfortunately, despite the fact that it is tasked with investigating 'the relationship between the provisions of the multilateral trading system and charges and taxes for environmental purposes', it has so far achieved relatively little.[367] The rules contained

---

[361] That is, the Understanding on Rules and Procedures Governing the Settlement of Disputes (1994) 33 ILM 112.

[362] Art. 17(3), Dispute Settlement Understanding, reads: 'The Appellate Body shall comprise persons of recognised authority, with demonstrated expertise in law, international trade and the subject matter of the covered agreements generally. They shall be unaffiliated with any government. The Appellate Body membership shall be broadly representative of membership in the WTO. All persons serving on the Appellate Body shall be available at all times and on short notice, and shall stay abreast of dispute settlement activities and other relevant activities of the WTO. They shall not participate in the consideration of any disputes that would create a direct or indirect conflict of interest.'

[363] The details of the post are contained in the Agreement Establishing the World Trade Organization, Art. VI. The current holder is HE Dr Supachai Panitchpakdi of Thailand, whose three-year term began in September 2002. His predecessor was The Rt Hon. Michael Moore, a former Prime Minister of New Zealand (see para. 8.2.1n below), who held the position from September 1999.

[364] Art. VI(4), Agreement Establishing the World Trade Organization.

[365] See Bowett, *op. cit.*, para. 12–073.

[366] See Zen Makuch, 'The World Trade Organization and the General Agreement on Tariffs and Trade', in *Greening International Institutions*, ed. by Jacob Werksman (London: Earthscan, 1996), pp. 94–115.

[367] See Ole Kristian Fauchald, *Environmental Taxes and Trade Discrimination* (London: Kluwer Law International, 1998), p. 4.

in GATT 1994 create an international legal structure that raises considerable design problems for them.[368]

4.4.2   *Organisation for Economic Co-operation and Development*

The Organisation for Economic Co-operation and Development ('the OECD') is the successor to an organisation called the Organisation for European Economic Co-operation.[369] It has been very influential, at least since the early 1970s, in developing the economic arguments in favour of environmental levies and subsidies.[370]

The organs of the OECD are the Council, the Committees and the Secretariat. Of these, the principal one is the Council, which is 'the body from which all acts of the Organisation derive'.[371] The Council, which consists of Ministers or Permanent Representatives of the member countries, develops priorities for the OECD's activities and produces directives on future work. The Organisation has a rule that decisions and recommendations of the Council must be passed unanimously.[372] There is one vote per member country. The unanimity rule has never become a veto, however, because of the provision that abstentions do not invalidate decisions nor prevent them from becoming binding on other members.[373] Thus, decisions bind only those member countries that have voted for them and once their own constitutional procedures have been complied with.[374]

The Committees, which number over 200, consist of experts and officials from member countries. The OECD's most recent work on environmentally-related taxes has been produced by the Joint Meeting of Experts on Tax and Environment, which is convened under the joint auspices of the Committee on Fiscal Affairs and the Environment Policy Committee.[375] Individuals who are members of the Committees are thus in a position to feed OECD thinking into the development of national policies. The Secretariat, headed by a Secretary-General,[376] includes individuals seconded from academia, commerce and industry, who compile and analyse economic data.[377] Indeed, the OECD's work on environmental levies and subsidies has been of an economic, rather than of a legal, nature.[378]

---

[368] See paras 1.2.1.4 above and 8.4.5 below.
[369] The older organisation ('the OEEC') had been established in 1948 to administer American aid to Europe under the Marshall Plan (see Bowett, *op. cit.*, para. 6–021). The OECD was established by the Convention signed in Paris on December 14, 1960 and which came into force on 30 September 1961.
[370] See Chapter 5 below.
[371] 1960 Paris Convention, Art. 7.
[372] *Ibid.*, Art. 6.
[373] *Ibid.*
[374] *Ibid.*
[375] See Organisation for Economic Co-operation and Development, *Environmentally Related Taxes in OECD Countries* (Paris: OECD, 2001), p. 3.
[376] Currently (December 2004), Mr Donald J. Johnston, since June 1996.
[377] See the Organisation's database of environmentally-related taxes in OECD countries, at: www.oecd.org.
[378] See the Bibliography of Organisation for Economic Co-operation and Development, *Environmentally Related Taxes in OECD Countries* (Paris: OECD, 2001), pp. 137–42.

Article 1 of the 1960 Paris Convention states the purposes of the OECD as being:

- to achieve the highest sustainable economic growth and employment and a rising
  standard of living in member countries, while maintaining financial stability, and thus to
  contribute to the development of the world economy;
- to contribute to sound economic expansion in member as well as non-member countries
  in the process of economic development; and
- to contribute to the expansion of world trade on a multilateral, non-discriminatory basis
  in accordance with international obligations.

Thus, by Art. 2 of the 1960 Paris Convention, member countries assume duties regarding the efficient use of economic resources, regarding research and regarding the general economic development of both member and non-member countries.[379] Possibly because of the highly technical nature of the OECD's work, possibly because of its status as a debating forum and possibly because of the unanimity rule referred to above, the institutional accountability or otherwise of the OECD has not been the focus of such attention as that accorded to the WTO.

In 1974, in response to the oil crisis, the OECD established an autonomous agency, called the International Energy Agency,[380] funded by member countries via the OECD itself, primarily to ensure energy security. Although its remit is fairly wide, one of its key areas of activity is the relationship between energy and the environment.[381]

4.4.3   United Nations subsidiary bodies

The United Nations' General Assembly has established two subsidiary bodies to deal with environmentally-related matters: the UN Environment Programme ('UNEP')[382] and the UN Commission on Sustainable Development ('the UNCSD').[383] The former dates from 1972, following the Stockholm Conference on the Human Environment;[384] the latter was established in 1992, on the basis of the mandate given by the UN Conference on Environment and Development.[385] The United Nations Economic Commission for Europe ('the UNECE')[386] has also had an important role to play in promoting international environmental agreements.[387]

---

[379] 1960 Paris Convention, Art. 2. See also Organisation for Economic Co-operation and Development, *Policy Brief. Environmentally related taxes: Issues and strategies* (OECD; Paris, 2001). Organisation for Economic Co-operation and Development, *OECD Environment Programme 2003–2004* (Paris: OECD, 2003).

[380] See www.iea.org.

[381] See Roggenkamp, *op. cit.*, paras 3.133–3.179. The IEA's constituent instrument is the International Energy Programme, a treaty document created following the suggestion for greater solidarity in energy matters among OECD member countries, originally made by US Secretary of State, Henry Kissinger, in 1973 (*ibid.*, paras 3.133–3.134).

[382] See www.unep.org.

[383] See www.un.org/esa/sustdev/csd/csd12/csd12.htm.

[384] See para. 8.2.3n below.

[385] See para. 8.2.3 below.

[386] See www.unece.org and www.unece.org/env/welcome.html.

[387] For example, in the promotion of the 1979 Geneva Convention for the Control of Long-Range Transboundary Air Pollution (see Chapter 8 below); the 1991 Espoo Convention

The United Nations Commission on Sustainable Development, which meets in New York or Geneva, and is supported by a secretariat in the former city, is the main UN body for sustainable development issues, and has the task of monitoring progress and making recommendations on the implementation of Agenda 21.[388] The United Nations Environment Programme, which is based in Nairobi, is the sole UN body exclusively dedicated to international environmental matters, with a range of tasks, including those of promoting international environmental cooperation, providing policy guidance on environmental programmes within the UN and promoting scientific knowledge and information. The United Nations Environment Programme is made up of 58 members who are elected by the General Assembly and has an Environmental Secretariat, chaired by an Executive Director. Although it has been responsible for the promotion of a number of international agreements, not least the 1985 Vienna Convention for the Protection of the Ozone Layer[389] and for the 1992 Convention on Biological Diversity,[390] the underfunding of its operations, no less than its relative lack of status within the UN structure, means that its performance continues to disappoint many people.[391]

The UN's closest environmental association is the 1992 UN Framework Convention on Climate Change and its 1997 Kyoto Protocol. Each of these raise certain interesting issues at an institutional level. As will be seen in Chapter 8 below,[392] and as is well-known, these international agreements commit the developed parties thereto (referred to as 'Annex I Parties')[393] to targets for the reduction of their GHG emissions.[394] One of the economic instruments for achieving these targets is the Clean Development Mechanism ('the CDM'), which 'marks an interesting and innovative new structure in international institutional arrangements, including a formal role for non-state actors'.[395]

Given that the present book is much concerned with the energy markets and with the relationship between energy and the environment, as a background to energy taxation, it is perhaps useful at this juncture to mention the existence of the International Atomic

---

on Environmental Impact Assessment in a Transboundary Context (1991) 30 ILM 802; and (see Chapter 8 below) the 1998 Aarhus Convention on Access to Information, Public Participation in Decision-making and Access to Justice in Environmental Matters (1999) 38 ILM 517 (in force 30 October 2001).

[388] See Chapter 8 below.

[389] See para. 8.3.1.1 below.

[390] See para. 8.3.1.3 below.

[391] Proposals for a new UN specialised agency are discussed by Daniel Esty in *Greening the GATT: Trade, Environment and the Future* (Washington DC: Institute for International Economics, 1994). See also Birnie and Boyle, *op. cit.*, pp. 54–7.

[392] See paras 8.3.1.3 and 8.3.1.4 below.

[393] The Annex I Parties, not all of whom have ratified the Protocol (see para. 8.2.2 below), are: Australia, Austria, Belarus, Belgium, Bulgaria, Canada, Czechoslovakia, Denmark, European Community, Estonia, Finland, France, Germany, Greece, Hungary, Iceland, Ireland, Italy, Japan, Latvia, Lithuania, Luxembourg, Netherlands, New Zealand, Norway, Poland, Portugal, Romania, Russian Federation, Spain, Sweden, Switzerland, Turkey, Ukraine, United Kingdom and the US.

[394] See para. 8.3.1.4 below.

[395] Bowett, *op. cit.*, para. 4–024. See para. 8.3.1.4 below.

Energy Agency ('the IAEA').[396] Despite its name, the IAEA, which works closely with Euratom, is an independent intergovernmental agency founded in 1957, which is obliged to report to the UN in certain situations.[397] Also relevant to proposals to introduce emissions trading schemes for airlines is the International Civil Aviation Organisation ('the ICAO'),[398] a specialised agency of the UN, which is the treaty organisation of the 1944 Chicago Convention on International Civil Aviation.

4.4.4   *Various international treaty organisations*

Before closing this part of the chapter, it is useful briefly to refer for completeness to some of the myriad treaty organisations covering environmental matters. It might also help to refer to one of the most recent of international treaty organisations, the 1994 Energy Charter Treaty.

No doubt the significance of such organisations has been enhanced somewhat by the relative unimportance of the UNEP. For instance, the 1973 Washington Convention on International Trade in Endangered Species of Wild Fauna and Flora[399] has a Conference of the Parties, which meets at least every two years to consider amendments to the provisions limiting trade in listed species.[400] Similar Conferences exist for the 1992 Convention on Biological Diversity referred to above;[401] for the 1985 Vienna Convention for the Protection of the Ozone Layer, also mentioned above, as well as its 1987 Montreal Protocol;[402] for the 1992 UN Framework Convention on Climate Change;[403] for the 1997 Kyoto Protocol;[404] for the 1972 London Dumping Convention;[405] and for the 1989 Basel Convention.[406]

The Energy Charter Treaty which, as its name suggests, is limited in scope to the energy sector, nonetheless has an obvious environmental significance. It is independent of the EU and has its own Secretariat and Conference, both of which are based in Brussels.[407]

4.5   Concluding comments

Finally, it falls to us to gather together the main strands of the discussion. We began this chapter by considering the organs of central government: Parliament, government departments, public bodies and agencies and advisory bodies. Thence,

---

[396] See www.iaea.org/worldatom.
[397] See Roggenkamp, *op. cit.*, paras 3.49–3.57.
[398] See www.icao.int.
[399] See para. 8.4.2n below.
[400] See www.wcmc.org.uk/CITES/eng/index.shtml.
[401] See www.biodiv.org.
[402] See para. 8.3.1.5 below.
[403] See para. 8.3.1.3 below.
[404] See para. 8.3.1.4 below.
[405] See para. 8.3.2 below.
[406] *Ibid.*
[407] See www.encharter.org. See para. 8.6 below.

we analysed the contribution to the design, implementation and enforcement of taxes and other economic instruments, of what is a considerable range of government departments, non-departmental public bodies, agencies and advisory bodies. The chief characteristic of this institutional pattern was one of almost overwhelming complexity but, so far as central government was concerned, it indicated that, in the UK, chief responsibility for the design and implementation of economic instruments lies, not with any one body, but mainly with Customs, with Defra and, so far as they relate to electricity, with Ofgem.

Next, we considered the devolution schemes applicable to Wales, Scotland and Northern Ireland. The present significance of these to the subject matter is constrained by the fact that, although all of them have environmental competencies, importantly in relation to the LATS, only Scotland has any tax raising power, and that is not relevant to environmental taxation. It seems inevitable, at this juncture, that pressure for additional tax-raising powers will increase, pre-eminently, perhaps, in the environmental taxation field.

With regard to the wider picture, we have seen that the ability of the EU institutions to create environmental taxes is circumscribed by the fiscal veto provisions of the European Treaty provisions. The successful creation of the EU ETS marks a neat sidestepping of the problems which had been associated with the 'shelved' EU carbon/energy tax.[408]

The close relationship of the EU institutions with the OECD, with its commitment to the development of policy in the environmental taxation field, will ensure however that its work on environmental taxes and other economic instruments retain considerable influence globally. Meanwhile, the impact on the possibilities for certain kinds of environmental taxes (especially carbon taxes), of the rules of the multilateral trading system, in GATT 1994, continue to be controversial and we return to them in Chapter 8 below.

---

[408] Note, however, the Energy Products Directive of December 2003 (see para. 12.3.4 below). For the carbon/energy tax proposal, see para. 28.1n below.

Chapter 5

Technical Justifications

5.1   Introduction

This chapter analyses the reasons that the UK government has itself provided, via
HM Treasury, for regulating by means of environmental taxes and other economic
instruments. These reasons should be seen in the broader context of the government's
policy on sustainable development.[1]

The chapter takes the form of a commentary on HM Treasury's 2002 paper
containing the government's most recent detailed justifications for its policy on the
use of economic instruments in pollution regulation.[2] Our discussion assumes that,
in its legislative programme on the use of economic instruments for this purpose,
the government is acting in pursuit of the public interest.[3] This is in keeping with
the disciplinary approach explained in Chapter 1 above.[4] The assumption as to
public interest objectives serves to underline the important distinction between the
technical justification for introducing a particular measure and a theory about why
the measure has in fact been introduced. It does not necessarily imply that, in a
particular instance, there will be found to be a disparity been the two, since a theory
that seeks to explain why a measure has been introduced may reaffirm the public
interest reason. In other words, the proffered reason, which is a good reason, may
also be the real reason.[5] We have sought specifically to emphasise in this chapter the
profferred reasons, rather than the theoretical arguments for green levies generally.
Although there is some discussion of the latter, this has been minimised for reasons
of space.[6]

The 2002 Treasury paper covers aggregates levy, climate change levy and landfill
tax, as well as explaining the rationale for the UK ETS[7] and the RO.[8] It does not deal

---

1   See para. 4.2.1.3n above and para. 5.2 below.
2   HM Treasury, *Tax and the Environment: Using Economic Instruments* (London: HMSO,
     2002), available from the Treasury's website, at www.hm-treasury.gov.uk. It is referred
     to alternately below as 'the 2002 paper', 'the 2002 Treasury paper', 'the paper' or 'the
     Treasury paper'.
3   See Robert Baldwin and Martin Cave, *Understanding Regulation – Theory, Strategy
     and Practice* (Oxford: Oxford University Press, 1999), p. 9n, for a survey of the relevant
     literature on why and how the *motivation or explanation* for regulation may differ from
     the regulation's technical justification.
4   See para. 1.2.2 above.
5   The allusion is to a statement attributed to an eminent American financier (see Jean
     Strouse's *Morgan: American Financier* (London: Harvill Press, 1999), p. xiii).
6   See, generally, for example, *Environmental Policy: Objectives, Instruments, and
     Implementation*, ed. by Dieter Helm (Oxford: Oxford University Press, 2000), esp. Part
     One. The literature is vase (see nn in this book, *passim*).
7   See para. 1.4.2.2(2) above.
8   See para. 1.4.2.2(4) above.

in any detail with the nascent LATS,[9] possibly since the enabling legislation was introduced in the House of Lords at or around the time that the paper was published.[10] Also not dealt with in much detail in the 2002 paper is the diverse package of measures introduced in 1999 and 2000 which were designed to assist in dealing with traffic congestion and road traffic pollution.[11] These are the income tax reliefs introduced in Finance Act 1999, to encourage employees to choose environmentally friendly ways of commuting,[12] and the provisions allowing local authorities to implement road user charging and workplace parking levy schemes,[13] brought in by the Transport Act 2000.

The present chapter is divided into four parts:[14] an overview of the government's policy with regard to sustainable development; a discussion of the government's taxation objectives in the light of that policy; a review of the economic arguments for using fiscal measures to contribute to sustainable development by correcting market failures resulting from pollution; and, finally, an examination of how the decision to use fiscal measures for this purpose is taken and the price level at which intervention in the market is determined. Our discussion of how economic policy on environmental taxes and other economic instruments has been translated into reality appears in Chapter 11 below.[15]

5.2   Sustainable development

In a rather surprising ordering of the material, the third chapter of the 2002 Treasury paper is devoted to the government's sustainable development targets and indicators.[16] It draws on the former DETR's[17] Sustainable Development Strategy White Paper, in which the government's aim is stated as being 'to achieve a better quality of life for everyone, now and for future generations'.[18] The White Paper reflects the government's approach to the commitments made at the 1992 Earth Summit in Rio; this is somewhat different from that taken by the previous – Conservative

---

9   See para. 1.4.2.1(4) above.
10   The Waste and Emissions Trading Bill was only introduced in the House of Lords on November 14, 2002.
11   Although certain of these are used as brief examples in paras 6.18–6.23 of the 2002 paper.
12   See para. 23.3 below.
13   See Chapters 17 and 18 below.
14   Corresponding to the subject matter of four of the chapters of the 2002 paper.
15   See para. 11.3 below.
16   See Giles Atkinson, 'Sustainable Development and Policy', in Helm, *op. cit.*, pp. 29–47, and Victoria Jenkins, 'Placing Sustainable Development at the Heart of Government in the UK' (2002) 22 LS 578–601.
17   See para. 4.2.1.2(1) above.
18   See Department of Environment, Transport and the Regions, *A Better Quality of Life: A Strategy for Sustainable Development in the United Kingdom*, 1999 (Cm 4345, 1999). This was a revision of the previous Conservative Government's strategy, *Sustainable Development, the UK Strategy*, 1994 (Cm 2426, 1994).

– administration prior to May 1997. The Rio Declaration, it will be recalled, contains the 27 Principles of international environmental law that together constitute the main outlines of the sustainable development concept.[19]

The 2002 paper reflects the 15 headline indicators of the Sustainable Development Strategy White Paper, grouped into four 'dimensions'. Only two of these four dimensions specifically contain indicators relating to environmental matters and they read as follows:

Effective protection of the environment:
– emissions of greenhouse gases;
– days when air pollution is moderate or high [*sic*];
– road traffic;
– rivers of good or fair quality;
– populations of wild birds; and
– new homes built on previously-developed land.
Prudent use of natural resources:
– waste arisings and management.[20]

These indicators, which 'should move in the right direction over time',[21] are designed to show, in an accessible way, the progress that society as a whole is making towards sustainable development.[22] The government acknowledges in the paper that, helpful as these indicators are in providing a snapshot of society's progress, they do not measure impacts directly, since, for example, road traffic levels are not in themselves measures of emissions affecting air quality, GHG emissions, congestion or noise implications of road use.[23] Nonetheless, in reaching its aim of achieving a better quality of life, now and for future generations,[24] the government clearly sees them as being very useful.[25]

The government's comments on performance to date on the environmental dimensions of sustainable development have an air of disarming candour about them.[26] As to

---

19   *Ibid.* and para. 4.2.1.3n above.
20   The other two groups of headline indicators are: '(1) maintaining high and stable levels of economic growth and employment, which group comprises: total output of the economy ('GDP'), investment in public, business and private assets and proportion of people of working age who are in work; and (2) social progress which recognises the needs of everyone, and which comprises: poverty and social exclusion (fuel poverty etc.); qualifications at age 19; expected years of healthy life; homes judged unfit to live in; and level of crime' (see 2002 paper, Box 3.1, p. 9). The social dimension had been missing from the 1994 Conservative strategy.
21   See 2002 paper, para. 3.1.
22   *Ibid.*, para. 3.3.
23   *Ibid.*, para. 3.4.
24   The reference to future generations echoes the definition of sustainable development used in the Brundtland Report of 1987 (see para. 4.2.1.3n above).
25   *Ibid.*, para. 3.1.
26   That is, as distinct from economic growth and social progress. Progress against the environment-related indicators contained in the White Paper is set out in Annex A to the 2002 paper. The material in Annex A is derived from Department for Environment, Food and Rural Affairs, *Achieving a Better Quality of Life – Review of Progress towards*

GHGs, although on current trends, the UK is one of the few EU Member States likely to meet its Kyoto target, after 2012 emissions targets 'are likely to become tighter, and the UK needs to be ready to take further action to continue to reduce emissions in the longer term'.[27] Likewise, with the amount of biodegradable municipal waste sent to landfill, despite the fact that the Landfill Directive[28] fixes a reviewable target of a 35 per cent reduction on 1995 levels by 2020, 'volumes of waste sent to landfill have continued to rise over recent years'.[29] Most worrying of all for the government, perhaps, is the situation with regard to the environmental effects of transport, since '[a]lthough the fuel efficiency of new vehicles is improving, transport continues to be a major user of energy and source of carbon dioxide emissions'.[30]

Lastly, in this third chapter of the 2002 paper, the Treasury seeks to illustrate its own commitment to making progress on the headline indicators by drawing attention to the various Departmental Public Service Agreements ('PSAs'). The PSAs, which closely reflect the headline indicators referred to above, set out each Department's priorities and give targets against which the Department in question's progress can be monitored.[31] For instance, the Treasury has agreed to: 'Protect and improve the environment by using instruments that will deliver efficient and sustainable outcomes through evidence-based policies'; Defra[32] has promised to: '[p]romote sustainable development across government and the country as a whole as measured by achieving positive trends in the government's headline indicators of sustainable development'; and the DFT[33] has promised to: 'Reduce congestion on the inter-urban trunk road network and in large urban areas in England below 2000 levels by 2010'.[34] It is interesting that the DFT seems to view the chief environmental problem of too much road traffic as primarily one of congestion, rather than of emissions. This is fully in tune with modern economic thinking on the subject.

5.3   Taxation and sustainable development

Against the background of government policy on sustainable development, the 2002 Treasury paper begins[35] by recalling the government's Statement of Intent

---

    Sustainable Development in 2001 (London: Defra, 2002), available from www.sustainable-development.gov.uk.

27   See 2002 paper, para. 3.6.
28   Council Directive 99/31/EC, (1999) OJ L182 1 (see para. 12.2.5.1(2) below). See further para. 6.3.2.4 below.
29   See 2002 paper, para. 3.7.
30   *Ibid.*, para. 3.8.
31   *Ibid.*, para. 3.11.
32   See para. 4.2.1.2(1) above.
33   *Ibid.*
34   See 2002 paper, Annex B (pp. 51–2). This is not a complete list of the PSAs, which are quite extensive, and, specifically, does not include the promise of the ODPM (see para. 4.2.1.2(1) above), also extracted in Annex B to the 2002 paper.
35   *Ibid.*, ch. 2.

on environmental taxation dated 2 July 1997.[36] An important passage in the 1997 document is reproduced in the 2002 paper and reads as follows:

> How and what governments tax sends clear signals about the economic activities they believe should be encouraged or discouraged, and the values they wish to entrench in society. Just as work should be encouraged through the tax system, environmental pollution should be discouraged.[37]

The Statement of Intent goes on to say that, accordingly, where environmental taxes 'met the general tests of good taxation', then the government would use them to achieve its economic objectives, which were expressed as follows:

> The Government's central economic objectives are the promotion of high and sustainable levels of growth and high levels of employment. By that we mean that growth must be both stable and environmentally sustainable. Quality of growth matters; not just quantity ... [T]he Government will explore the scope for using the tax system to deliver environmental objectives – as one instrument, in combination with others like regulation and voluntary action.[38] Over time, the Government will aim to reform the tax system to increase incentives to reduce environmental damage. That will shift the burden of tax from 'goods' to 'bads';[39] encourage innovation in meeting higher environmental standards; and deliver a more dynamic economy and a cleaner environment, to the benefit of everyone.[40]

The translation of this statement of policy into a legal reality is discussed below.[41]
The Treasury seeks to draw a distinction between '[t]axes on broad aspects of economic activity such as energy, waste and transport'[42] and 'much smaller taxes which target specific environmental impacts'.[43] The former, presumably meaning climate change levy, landfill tax and environmentally-friendly excise duty differentials, the 2002 paper states, raise significant levels of revenue which can be used to 'offset' other taxes. The latter, by contrast, 'would be unlikely to raise very much revenue and therefore would not have any significant impact on the overall tax

---

36 That is, Budget day. This was the Rt Hon. Gordon Brown, MP's first Budget as Chancellor (see HM Treasury, *Tax Measures to Help the Environment*, News Release, 2 July 1997, available from HM Treasury's website at www.hm-treasury.gov.uk).

37 See *Tax Measures to Help the Environment*, above.

38 Probably for presentational reasons, the government contrasts the use of economic instruments, such as environmental levies and subsidies, with regulation. In reality, the idea of regulation is wide enough to encompass the economic instrument concept, and it is in this wider sense that the term is used in this study (see Baldwin and Cave, *op. cit.*, pp. 1–2, and Chapter 1 above). See also Chris Hilson, *Regulating Pollution – A UK and EC Perspective* (Oxford: Hart Publishing, 2000), p. 103.

39 See Department of the Environment, *First Report of the British Government Panel on Sustainable Development* (Sir Crispin Tickell, Chairman) (London: 1995), p. 12. See also, more recently and (possibly) more influentially, Anthony Giddens, *The Third Way and its Critics* (Cambridge: Polity Press, 2000), pp. 100–101.

40 See *Tax Measures to Help the Environment*, above.

41 See Ch. 11 below.

42 See 2002 paper, para. 2.10.

43 *Ibid.*, para. 2.11.

base'.[44] There is no clue from the brief discussion of the 'much smaller taxes' whether those alluded to are actual or as yet only proposed, although the context seems to suggest that the latter meaning is intended.[45] If this conclusion is correct, then these paragraphs of the 2002 paper underline the point made elsewhere in the present book that the hypothecation of tax revenues, even in the context of environmental taxes, is something that the Treasury still strongly resists.[46] The government instead limits itself to saying that the significant levels of revenue raised by green taxes 'can be used to offset other taxes'.[47] The first part of its subsequent claim that:

> The Government has used revenue from taxes such as the climate change levy and aggregates levy to reduce employers' national insurance contributions and has also introduced enhanced capital allowances to reduce costs of investments in environmentally-friendly technologies ...[48]

is examined below, as also is the rather laconic statement which follows the one just quoted, that is, that: 'Some of the revenue [that is, from taxes such as the climate change levy and aggregates levy] has also been used to support related spending programmes'.[49] From a technical point of view, it should be noted that the reference to the use of environmental tax revenues to reduce labour taxes is commonly known as 'the employment double dividend' and is much contested by economists.[50]

The Treasury specifically relates each of the levies and subsidies discussed in this study to the government's sustainability policy. Thus, the goal of the totality of

---

44   *Ibid.*

45   Unless this is a reference to road user charging schemes.

46   See para. 11.2 below. As discussed there, the landfill tax code contains some mechanisms by which an effect akin to hypothecation is achieved (see para. 11.2 below). The Treasury's resistance to hypothecation is highlighted by the fact that, in para. 2.11 of the 2002 paper, it is stated that, with the much smaller taxes targeting specific environmental impacts, '... there may be a stronger case for using most or all of the revenue to encourage a response to the tax'.

47   See 2002 paper, para. 2.10.

48   *Ibid.*

49   *Ibid.* See para. 21.3 below.

50   A subject on which there is a considerable literature: see, for example, Lawrence Goulder, *Environmental Taxation and the 'Double Dividend': A Reader's Guide* (Cambridge, MA: National Bureau of Economic Research, 1994); Stephen Smith, *'Green' Taxes and Charges: Policy and Practice in Britain and Germany* (London: Institute for Fiscal Studies, 1995), p. 14; Kalle Määttä, *Environmental Taxes – From an Economic Idea to a Legal Institution* (Helsinki: Finnish Lawyers' Publishing, 1997), pp. 157–8; ed. by Timothy O'Riordan, *Ecotaxation* (London: Earthscan, 1997), Part II; C.J. Heady *et al.*, *Study on the Relationship Between Environmental/Energy Taxation and Employment Creation* (Bath: University of Bath, 2000); Organisation for Economic Co-operation and Development, *Environmentally-Related Taxes in OECD Countries: Issues and Strategies* (Paris: OECD, 2001), paras 1.6.2 and 2.3; Adair Turner, *Just Capital: The Liberal Economy* (London: Macmillan, 2002), pp. 310–11; and Kurt Kratena, *Environmental Tax Reform and the Labour Market: The Double Dividend in Different Labour Market Regimes* (Cheltenham: Edward Elgar, 2002). See also the 2002 paper, paras 7.9–7.13, esp. para. 7.11.

climate change levy, of the UK ETS,[51] of the RO[52] and of tax reliefs such as capital allowances for expenditure on energy-saving plant and machinery,[53] is to assist the UK in meeting its carbon emissions reduction targets under the post-Kyoto EU burden-sharing agreement.[54] The objectives of the other levies, given their nature and scope, is closer to home. Aggregates levy is designed to 'tackle environmental costs of aggregate extraction including noise, dust visual intrusion [and], biodiversity loss', whilst the goal of landfill tax is '[t]o internalise [the] environmental costs of landfill e.g. methane emissions, nuisance, groundwater pollution; to give better price signals for alternatives to landfill; and to assist in meeting waste targets in [the] most efficient way'.[55] Landfill tax is to be supported by the LATS,[56] whose aim is to restrict the landfilling of biodegradable municipal waste.[57]

5.4 Economic instruments and market failures

Government policy with regard to tax and the environment is thus to use the tax system to ensure the attainment of environmental objectives and to reform it to increase incentives to reduce environmental damage.[58]

The basis on which the policy is justified is that of welfare economics, a discipline which has achieved a 'near hegemonic status' as a theoretical basis for pollution control.[59] Under the heading 'Why not leave it to the market?', the 2002 paper justifies the use of economic instruments such as environmental taxes by reference to market failures and their distributional effects.[60] To the lawyer, the discussion reads almost as a digest of the economic learning in this area.[61]

Both a vibrant and influential academic literature[62] and the 'soft law'[63] and other policy documentation produced by the OECD[64] make a persuasive welfare economics case for the use of economic instruments in environmental regulation, the broad outlines of which might be summarised as follows. The economic rationale for

---

[51]   See para. 1.4.2.2(2) above.
[52]   See para. 1.4.2.2(4) above.
[53]   See 2002 paper, paras 6.39–6.47. See also Chapter 24 below.
[54]   See paras 8.2.2n and 8.3.1.4 below.
[55]   See 2002 paper, Table 7.1 (p.42).
[56]   See para. 1.4.2.1(4) above.
[57]   *Ibid.*
[58]   See *Tax Measures to Help the Environment*, above.
[59]   See Hilson, *op. cit.*, pp. 6–8.
[60]   See 2002 paper, ch. 4.
[61]   As to which, see, for example, Tom Tietenberg, *Environmental and Natural Resource Economics*, 3rd edn (New York: Harper Collins, 1992), to take one distinguished text from among many.
[62]   See Benjamin J. Richardson and Kiri L. Chanwai in 'Taxing and Trading in Corporate Energy Activities: Pioneering UK Reforms to Address Climate Change' (2003) 14(1) ICCLR 18–27, 19.
[63]   See para. 8.2.6 below.
[64]   See para. 4.4.2 above.

environmental regulation via economic instruments[65] is the elimination of economic waste in situations where 'the unregulated price of a good does not reflect the true cost to society of producing that good'.[66] Baldwin and Cave, in a review of the relevant arguments, provide the following example:

> [A] ... manufacturer of car tyres might keep costs to consumers down by dumping pollutants arising from the manufacturing process into a river. The price of the tyres will not represent the true costs that production imposes on society if clean-up costs are left out of account. The resultant process is wasteful because too many resources are attracted into polluting activities (too many tyres are made and sold) and too few resources are devoted by the manufacturer to pollution avoidance or adopting pollution-free production methods.[67]

The disparities between 'true social costs' and 'unregulated price' are 'spillover' costs, which environmental economists generally refer to as 'externalities'[68] or 'diseconomies'.[69] Some form of regulation – of state intervention – is necessary to compel the internalisation of such spillover costs, on the polluter pays principle.[70] There has been a market failure, that is, a specific instance of where 'the competitive outcome of markets is not efficient from the point of view of the economy as a whole'.[71] The correction of market failures by regulation might take the form of traditional command and control mechanisms, voluntary mechanisms or economic instruments, including environmental taxes or charges.[72] The inspiration for the last of these possibilities is usually attributed to the distinguished early-twentieth century economist, Arthur Pigou.[73] Pigou's writing is sometimes portrayed as being much

---

65   See Chapter 6 below.

66   Stephen Breyer, 'Typical Justifications for Regulation', in *A Reader on Regulation*, ed. by Robert Baldwin, Colin Scott and Christopher Hood (Oxford: Oxford University Press, 1998), pp. 59–92, p. 68.

67   See Baldwin and Cave, *op. cit.*, pp. 11–12.

68   See Breyer, *op. cit.*, p. 68. Spillover costs are the norm, but Breyer acknowledges also the possibility of a 'spillover benefit' in a different situation, for example where 'honeybees fertilise nearby apple orchards, the beekeepers provide a spillover benefit to the orchard owners, so long as the latter do not pay the former for their service' (*ibid.*).

69   Or 'external diseconomies': see, for example, Määttä, *op. cit.*, p. 7.

70   See Baldwin and Cave, *op. cit.*, p. 12.

71   See 2002 paper, para. 4.2. Factors other than externalities may cause market failure, for example, information failures, the absence of perfect competition and even government intervention itself (see the 2002 paper, para. 4.9, and Baldwin and Cave, ch. 2).

72   A voluntary approach, rather than a tax, has been adopted, for the time being, in relation to pesticide use (see para. 21.9.5 below). A bid for a voluntary arrangement was tried, and failed, in relation to aggregates levy (see para. 11.3.2 below). The role of voluntary measures is discussed in para. 7.8 of the 2002 paper.

73   Hence the expression 'Pigouvian taxes', which is sometimes used as a synonym for pollution (that is, environmental) taxes (see para. 1.2.1.5 above). Arthur Cecil Pigou (1877–1959) held the Chair of Political Economy at the University of Cambridge from 1908 until 1944. Memorably, Pigou saw the inspiration of economic science as being the 'sordidness of mean streets and the joylessness of withered lives'. His ideas for pollution taxes were expounded in *The Economics of Welfare*, 4th edn (London: Macmillan, 1952), Part II, chs 2 and 3, and in *A Study in Public Finance*, 3rd edn (London: Macmillan, 1947),

more straightforward than it is; in Andersen's words, 'Pigou is much more cautious about his pollution taxation scheme than certain of his followers'.[74]

The academic rationale for correcting market failure is mirrored in the OECD's policy documentation:

> The basic theoretical premise behind the introduction of environmental instruments, including environmentally related taxation, to correct for [*sic*] environmental damage is the existence of negative environmental externalities in unregulated economies. A negative externality is a cost that one economic agent imposes on another but does not take into account when making production or consumption decisions. When the costs of pollution or resource use are not reflected in prices, market inefficiencies result with excessive production or consumption of products and activities that impose social costs. Externalities exist because of the public goods nature of the environment. In the absence of property rights for clean air, clean water, etc. economic agents use these services without regard for the impact their decisions have on other economic agents, including future generations. Even where charges or taxes are raised on a polluting activity, for example on municipal waste disposal, often they do not fully internalise the cost of the externality. Where environmental costs are fully internalised into the price of a product or activity a reallocation of resources in the economy occurs according to fair and efficient prices.[75]

Besides the internalisation of externalities, the OECD material also claims for economic instruments in general opportunities to realise static and dynamic efficiencies.[76] Static efficiency implies that marginal abatement costs are equalised between polluters, without the need for regulators to seek out information about the abatement costs of particular firms.[77] Dynamic efficiency implies that firms have an 'ongoing incentive to reduce pollution abatement costs, rather than simply to meet specific standards, which require constant review'.[78]

Both the academic rationale and the OECD policy statements are closely reflected in Chapter 4 of the 2002 paper.[79] The paper acknowledges the possibility, not only of negative externalities, but also of positive externalities.[80] An example of a positive externality would be a developer cleaning up contaminated land, since this would encourage the regeneration of the area surrounding that land.[81] Negative externalities, on the other hand, might include, for example, 'the visual and noise impacts of quarries', says the Treasury.[82] It supplies a further example of negative

---

Part II, ch. 8, both of which are far more difficult reading than the oft-seen 'summaries' of his work may appear to suggest.

[74] See Mikael Skou Andersen, *Governance by green taxes: making pollution prevention pay* (Manchester: Manchester University Press, 1994), p. 5.

[75] See OECD, *op. cit.*, para. 1.1.

[76] *Ibid.*, paras 1.3 and 1.4.

[77] *Ibid.*, para. 1.3.

[78] *Ibid.*, para. 1.4.

[79] Both are referred to in the Bibliography thereto (see the 2002 paper, pp. 57–58).

[80] A possibility contemplated by Pigou himself: see the example of the lighthouse in *A Study in Public Finance*, above, p. 94, and Breyer's example of the honeybees in n above.

[81] See 2002 paper, para. 4.5.

[82] *Ibid.*, para. 4.5.

externalities which develops, though not expressly, the one given by Baldwin and Cave:

> In a simple example, two firms, a factory and a fishery, use the same river as an input good. By using the river to dispose of waste, the factory imposes costs on the fishery and reduces its productive capacity; but the market does not reflect this cost in prices because there are no property rights for the pollution.[83] There would be an overall gain for the economy as a whole if the amount of pollution was set where the marginal benefit accruing to the factory from each additional unit of waste disposal was equal to the marginal cost to the fishery of each additional unit of river pollution ... The efficient outcome cannot occur while the factory does not face the full costs of its activities – its own private costs and the wider social costs.[84]

Moreover, the Treasury embraces the concepts of static and dynamic efficiency, referred to above. It emphasises that a 'well-designed economic instrument can equalise the marginal abatement costs between polluters', because industries that face lower abatement costs will cut back on pollution relatively more.[85] Also, since polluters are required to pay tax for 'residual emissions', as well as for abatement costs, they will have a strong financial incentive to invest in technological developments providing greater environmental protection.[86]

Having identified environmental market failures, Chapter 4 of the 2002 paper then goes on to look at their distributional implications. Environmental externalities can lead to differential impacts on different sectors of the economy and on different income groups; they also impose costs on those not responsible for the externalities. Traffic congestion, according to the Treasury, impacts on business in a different way from that in which it impacts on other sectors.[87] Air pollution not only impacts disproportionately on low income groups in the inner cities but, on a transboundary scale, it impacts more severely on less developed than on more developed countries. Again, adverse effects on air quality, river pollution, sulphur emissions and climate change, can have international and even global implications, not just for this generation but for the ones to come.[88] To the extent that addressing these distributional impacts requires value judgments to be made, these should, says the Treasury, 'be underpinned by economic analysis so that the debate is well-informed'.[89]

It should be said that a number of objections could be made to the classical rationale described above, although they do not seem currently to garner much favour and are not reflected in government policy. It has, for instance, been argued that 'spillover costs do not call for government intervention but, rather, for a rearrangement of

---

[83] This seems to be intended to refer to a regulation-free situation.
[84] See 2002 paper, Box 4.1, p. 13.
[85] See 2002 paper, para. 6.9.
[86] *Ibid.*, para. 6.10.
[87] This part of the 2002 paper is a little difficult to follow. If business is defined as economic activity, it is difficult to exclude anyone from its scope.
[88] *Ibid.*, para. 4.14.
[89] *Ibid.*, para. 4.17.

private property rights'.[90] Ronald Coase, writing in 1960, famously took this position, arguing that where there are 'well-defined property rights, and where the costs of bargaining are small enough, the affected parties can bargain with one another and agree on an efficient allocation of resources'.[91] Thus, in the examples given above, those suffering from the pollution would get together and offer to pay the polluter either to clean up or to adopt a pollution-free production method.[92] Such an argument might be significant in the context of small-group externalities, where there are only a few affected parties whose rights have clearly been defined[93] but the unreality of the argument in relation to large-group externalities is apparent as soon as it is stated. Even more unreal would be this argument in relation to mass externalities, such as emissions causing climate change. The costs of bargaining become ever more significant with the increase in the number of people affected, and a 'clear consensus is harder to obtain'.[94] In this situation, there is a continuance of spillover costs, because of the bargaining costs to those affected in banding together.[95]

Especially in the light of the subsequent development of environmental regulation, it now seems almost inconceivable that arguments based upon a rearrangement of private property rights could ever reassert themselves at a political level, at least in relation to mass externalities. As the Treasury itself says, 'if the market does not reflect costs properly, there will implicitly be subsidies within the economy to those causing pollution ...'.[96] Be that as it may, governments should clearly beware of justifying every environmental intervention in terms of spillover cost. Breyer suggests that, if the spillover rationale is to be 'intellectually useful, [it] should be confined to instances where the spillover is large, fairly concrete, and roughly monetizable'.[97] One can always find some – broadly defined – spillover cost rationale for regulation, since 'there is always some possible beneficial effect in reversing a market-made decision'.[98] In fairness to the authors of the 2002 Treasury paper, this is a problem of which they are clearly aware, since they refer to market failures caused by information failures, the absence of competition and government intervention.[99]

Besides providing general justifications for the creation of environmental levies and subsidies, the Treasury specifically states the justification for all except one of the measures that have actually been taken as being the correction of negative

---

90 See Breyer, *op. cit.*, p. 69. This paragraph is indebted to Breyer's analysis of the objections to the rationale rehearsed in the 2002 paper.

91 See Ronald Coase, 'The Problem of Social Cost' (1960) 3 JLE 1–44 (in Kalle Määttä's succinct summary (see Määttä, *op. cit.*, pp. 7–8)).

92 See Breyer, *op. cit.*, p. 59.

93 See Määttä, *op. cit.*, p. 8.

94 See Breyer, *op. cit.*, p. 69, and the works cited therein: J Buchanan, 'An Economic Theory of Clubs' (1965) 32 *Economica* 1; Mancur Olson, *The Logic of Collective Action: Public Goods and the Theory of Groups* (Cambridge, MA, London: Harvard University Press, 1965).

95 See Breyer, *op. cit.*, p. 69.

96 See 2002 paper, para. 4.6.

97 See Breyer, *op. cit.*, p. 72.

98 *Ibid.*

99 See 2002 paper, para. 4.9.

externalities.[100] The exception is the positive externality which is intended to be corrected by corporation tax relief at a notional 150 per cent of expenditure for the costs of remedial work on contaminated land.[101]

5.5   The efficient level of fiscal intervention

Having elucidated the concept of market failures caused by externalities, the Treasury then poses, in Chapter 5 of the paper, three further issues for policy consideration.[102] These are the questions of:

1   how the government decides whether intervention in order to address the externality is worthwhile; if so,
2   what the most efficient extent of any intervention would be; and
3   whether it would be most appropriate to intervene at the local, at the national or at the international level.

The Treasury reports that the government recognises that, given that decisions should be made on the basis of good scientific evidence, the precautionary principle[103] should be invoked where the scientific case is uncertain and, in accordance with the principles of good regulation, the intervention should be proportionate and consistent.

    The question of whether intervention would be worthwhile is answered using a cost-benefit analysis ('CBA'), part of the regulatory impact assessment ('RIA') procedure discussed below.[104] 'Cost-benefit analysis' is not here being used in a lay sense but is 'a highly specialised decision-making tool developed by economists, which provides a formal, systematic assessment of the costs and benefits of a proposed course of action'.[105] The aim of CBA is to find the point at which the total benefits of control outweigh the total costs by the greatest margin, since this will be the optimal level of pollution control. This is unlikely to be the level eventually adopted, however, since other factors have a part to play in the process of decision-making.[106]

    The CBA involves a two-stage process, whereby an environmental good is first valued, with the cost of taking action then being subjected to economic appraisal. Valuing environmental goods involves either inferring the 'price' of those goods or the cost of remediating damage caused by pollution. The price of environmental goods

---

100   *Ibid.*, Tables 6.1 (p. 23) and 7.1 (p. 42).
101   See Finance Act 2001, Sched 22. See paras 6.7, 7.3.4 and 24.5.1 below.
102   See 2002 paper, ch. 5.
103   See para. 8.2.6 below.
104   See 2002 paper, para. 5.2. Somewhat later in ch. 5, and somewhat inconsequentially, given what has gone before, the 2002 paper emphasises the importance of cost-benefit analysis in high-level target setting (see 2002 paper, para. 5.20). See also para. 11.2 below and 'Cost-Benefit Analysis and Environmental Policy', in Helm, *op. cit.*, pp. 48–74.
105   See Hilson, *op. cit.*, p. 73. He gives a full discussion of the process at Hilson, *op. cit.*, pp. 73–78, as to which see Chapter 11 below also.
106   See Hilson, *op. cit.*, pp. 73–74.

may be inferred either from consumers' behaviour or from their stated valuation of a good. Customers' behaviour may be tested, as appropriate, by using hedonic pricing techniques, by travel cost models or by random utility models.[107] Stated valuations involve eliciting estimates of consumers' willingness to pay ('WTP') or willingness to accept ('WTA') a particular outcome, through the use of specially-designed questionnaires. The cost of remediating damage, by contrast, is measured simply by estimating the costs involved in removing the source of the pollution. Inferring the price of an environmental good places emphasis on the value of that good to society, whereas measuring the cost of remediating the damage places emphasis on the resource implications of the environmental damage.[108] The latter, says the Treasury, is useful for testing data obtained by the former (price inference) method.

The implications of valuing the environmental good having been described, the Treasury then goes on to consider how the government subjects the cost of taking action to economic appraisal. Economic appraisal also helps to determine the most efficient extent of any intervention. There is a standard appraisal method, which is contained in what is colloquially known as the Treasury Green Book,[109] although the government acknowledges that, particularly where significant changes are required, the costs of taking action may be difficult to substantiate. The Treasury uses as an example the process through which the former DETR went in deciding whether to impose what became the aggregates levy and, in the event that it was imposed, what would be the appropriate rate at which it should be charged.[110] Essentially, the latter question involved estimating how much people valued avoiding the effect of quarrying for rock, sand or gravel both in their locality and in landscapes of national importance.[111]

In the aggregates levy consultation, 10,000 people, residing in the vicinity of 21 aggregates production sites, including quarries,[112] were asked how much they would

---

[107] For the distinctions between these, see Hilson, p. 74, and the 2002 paper, para. 5.10.

[108] Pigou had not been particularly troubled about fixing the appropriate rate of tax, although he did refer to an inquiry in Manchester showing that an annual loss of £290,000 resulted from 'the extra laundry costs, artificial light and damage to buildings as a result of heavy air pollution' (quoted in David Gee, 'Economic Tax Reform in Europe: Opportunities and Obstacles', in *Ecotaxation*, ed. by Timothy O'Riordan (London: Earthscan, 1997), pp. 81–105, p. 87).

[109] That is, HM Treasury, *The Green Book: Appraisal and Evaluation in Central Government: Treasury Guidance* (London: TSO, 2003). The methodology described therein involves, for example, discounting future costs and benefits to reflect the value that society places on the consumption of goods and services now, as compared with future consumption.

[110] See para. 11.3.2 below.

[111] See 2002 paper, Box 5.1, p. 19. See London Economics, *The Environmental Costs and Benefits of the Supply of Aggregates* (DETR: London, 1998); Susana Mourato and David Pearce, *Environmental Costs and Benefits of the Supply of Aggregates: A Review of the London Economics Report* (DETR: London, 1998); and London Economics, *The Environmental Costs and Benefits of the Supply of Aggregates Phase 2* (DETR: London, 1999) (this last is referred to below as 'Phase Two').

[112] The list of 21 was made up of eight sites carrying out sand and gravel operations; eight sites with hard rock quarries; three sites carrying out recycling operations and two marine aggregate wharves (see Phase Two, above, Annex 7). Phase Two included Swinden Quarry, a hard rock quarry near Skipton, North Yorkshire, in the Yorkshire Dales National Park (see Phase Two, photographs at Annex 6 thereto).

pay in the form of taxes, over a five-year period, for the quarry to be closed.[113] This was a way of attributing a value to the environmental damage by those directly affected, so these were referred to as the 'local surveys'. Following this, another 1,000 people, living in 21 postcode areas not located near quarries, were asked what they would pay to close a quarry in a National Park. This was a way of attributing a value to the damage by those who were only indirectly affected, so these were referred to as the 'national survey'. The results of both local and national surveys having been processed, national estimates were calculated for the average amount that people were willing to pay for the environmental benefit of shutting down a quarry. The national average, though weighted according to the type of output, was calculated to be £1.80 per tonne.[114]

Finally, as to the question of whether, with regard to an externality, it is most appropriate to intervene at a local, national or international level, the Treasury makes some interesting distinctions. Road traffic congestion is characterised as a localised problem, best tackled using a road user charging ('congestion charging' or 'cordon pricing') scheme,[115] at a local level. Problems of a national nature only (not particularised in detail) are best dealt with at the UK level,[116] whilst transboundary problems (such as acid rain) are best dealt with at a European level and global problems (such as climate change) are best dealt with at an international level.[117]

5.6   Concluding remarks

Following our discussion in Chapter 4 of the institutional structures within which environmental levies and subsidies are developed, implemented and enforced, we have sought to elucidate in the current chapter the technical justifications for these instruments. Thus, the Treasury, and therefore, UK government, justifies its green levies and subsidies primarily by reference to its policy on sustainable development, which is itself designed to implement the environmental commitments to which the UK is bound by various international treaties.[118] We have seen how the economic arguments for economic instruments as a means of carrying through these commitments, have achieved an almost unchallengeable status, such that they may be

---

[113]   See Phase Two, para. 2.2. The specially-designed questionnaires used a specially-evolved (stated) contingent valuation method, based on WTP. In relation to the early closure of the quarry in question, people were generally asked to assume that the site was restored in keeping with the surrounding landscape, and that the workers found new employment (see 2002 paper, Box 5.1, p. 19, second para). See, generally, *Valuing Environmental Preferences – Theory and Practice of the Contingent Valuation in the US, EU and Developing Countries*, ed. by Ian J. Bateman and Ken G. Willis (Oxford: Clarendon Press, 1999).

[114]   When the aggregates levy was eventually introduced in April 2002, the rate was put at the more conservative £1.60 per tonne (see para. 13.2 below).

[115]   See 2002 paper, paras 5.26 and 5.32.

[116]   *Ibid.*, paras 5.26 and 5.30.

[117]   *Ibid.*, paras 5.26 and 5.28–5.29.

[118]   See Chapter 8 below.

seen to prevail even where the evaluation methods they employ are, to say the least, controversial. That these arguments also have a part to play in the ongoing review of taxes dating back before 1997 is illustrated by the ongoing review of landfill tax, which will be considered in a later chapter.[119]

The 2002 Treasury paper is a profoundly interesting document, not least because, despite being separated from the government's original Statement of Intent on environmental taxation by almost five years, it is remarkably consistent with that original Statement. In para. 8.6 of the paper, the Treasury states its aims of continuing to explore the use of economic instruments to achieve environmental and sustainable development objectives; of continuing to keep under review the impact of environmental policy on innovation; and of continuing to engage with stakeholders[120] on the use and design of economic instruments.[121] It also states its intentions of taking the lessons of the UK experience to European and international discussions on the issues surrounding the use of economic instruments in meeting environmental challenges.[122]

What is particularly interesting about the final chapter of the 2002 paper, however, is that, at the very end, it clearly acknowledges the limitations of the economic discipline as a basis for policy making, whilst at the same time emphasising the arguments that the discipline has to offer. In para. 8.4, it is said that '[e]conomics provides a useful framework for assessing the extent and nature of government action to deal with environmental issues, helping to inform judgments on how to balance environmental, economic and social impacts. So far as possible [it continues], the actions that the government takes and the targets that it sets or agrees to need to reflect the costs and benefits of those actions'.

---

[119] See para. 11.4 below.
[120] See above, n.
[121] See 2002 paper, p. 45.
[122] *Ibid.*

Chapter 6

Regulatory Context

6.1   Introduction

The main purpose of the present chapter is to locate the UK's environmental levies and their associated economic instruments within the broader context of UK environmental and market regulation.[1] It would be possible to evolve a number of different structures, more or less theoretical, for such an investigation. Pragmatically, the one that we have evolved is to identify the range of regulatory instruments used to address particular environmental concerns, their selection being suggested by the spheres of operation of the various environmental levies and their associated economic instruments.

Before turning, in the next chapter, to particular examples of the instruments just referred to, it will be necessary to introduce a pervasive feature of the regulatory scene, not only in the UK, but in the rest of the EU also: Integrated Pollution Prevention and Control ('IPPC'). Our discussion of this concept, together with that of its older national forebear, Integrated Pollution Control ('IPC'), is designed to form a backdrop to the subsequent examination of the specific regimes relating to the regulation of waste management,[2] the control of air and atmospheric pollution;[3] the regulation of air passenger and road freight transport,[4] the regulation of mineral extraction[5] and (briefly) the regulation of contaminated land.[6]

The list just recited of the various regulatory spheres as they have developed in Western Europe draws attention both to possible overlaps between the categories (for example, as between air and atmospheric pollution and transport regulation) and to the need to separate environmental regulation out from market regulation. Both are, in a sense, forms of economic regulation, depending on how widely the boundaries of economic science are set. To that extent, to draw a line of demarcation between the 'environmental' and 'the market' is artificial.[7] Nonetheless, that the dichotomy

---

[1]   See, *inter alia*: Stuart Bell and Donald McGillivray, *Environmental Law*, 5th edn (London: Blackstone, 2000); Richard Burnett-Hall, *Environmental Law* (London: Sweet and Maxwell, 1995); David Hughes *et al.*, *Environmental Law*, 4th edn (London: Butterworths LexisNexis, 2002); John F. McEldowney and Sharron McEldowney, *Environmental Law and Regulation* (London: Blackstone, 2001); Susan Wolf and Neil Stanley, *Wolf and Stanley on Environmental Law*, 4th edn (London: Cavendish, 2003); Justine Thornton and Silas Beckwith, *Environmental Law*, 2nd edn (London: Sweet and Maxwell, 2004); and Maurice Sunkin, David Ong and Robert Wight, *Sourcebook on Environmental Law*, 2nd edn (London: Cavendish, 2002).

[2]   See para. 6.3 below.

[3]   See para. 6.4 below.

[4]   See para. 6.5 below.

[5]   See para. 6.6 below.

[6]   See para. 6.7 below.

[7]   See Authors' Preface above.

is to be found in the legislative materials under consideration appears to us to be incontrovertible. This is especially true, for instance, of energy regulation,[8] where, as stated elsewhere, the 'unbundling' of formerly nationalised industries, such as gas and electricity supply, has been accompanied by the parallel but distinct development of environmental law and policy.[9] More recently still, there have been attempts to graft onto the unbundling process a series of measures to safeguard environmental interests.[10] The problems raised by this afterthought are, of course, a central preoccupation of the book.

In what follows, we have tended to give the greatest weight to the explanation of regulatory issues which are either not covered elsewhere in the book or of which some preliminary understanding is necessary in order to be able to gain some insight into the strengths and weaknesses of the various levies and subsidies under consideration later on. In indicating in the present chapter the place of environmental levies and subsidies in the overall regulatory jigsaw, we have tended simply to cross-refer to the more detailed explanations of them which appear elsewhere in the study. In order to highlight the economic instruments, such as the environmental taxes, we have also adopted the traditional classification of command and control instruments, on the one hand, and economic instruments on the other. We are, of course, aware of the fact that, whilst superficially appealing, such a dichotomy may, on close examination, cease to exist. For instance, waste management licensing is seen as a command and control instrument, yet the licence fee may be viewed as an economic one, a pure case of the polluter having to pay. Likewise, an emissions trading system may seem to typify the concept of the economic instrument, yet no such system will succeed unless there is built into it a system of sanctions for non-compliance.[11]

We begin, as mentioned above, with cross-sectoral regulation. After considering IPC and IPPC, we go on briefly to consider the Environmental Impact Assessment ('EIA') regime and the newer regime for Strategic Environmental Assessment ('SEA'). Like much of what follows, the policy inspiration here is a Community-wide one and, to the extent that these two concepts shift attention back to the process of planning operations with potentially very significant environmental impacts, they are of the greatest significance to an important element in the rest of the study. Ironically, as we shall see, EIA and SEA may actually prove of particular significance in the

---

8　　See *Energy Law in Europe: National, EU and International Law and Institutions*, ed. by Martha M. Roggenkamp *et al.* (Oxford: Oxford University Press, 2001). See also *United Kingdom Oil and Gas Law*, ed. by Terence Daintith and Geoffrey Willoughby (London: Sweet and Maxwell, 1984); Patricia D. Park, *Energy Law and the Environment* (London: Taylor and Francis, 2002); and *Energy Law and Regulation in the European Union*, ed. by Robert H. Tudway *et al.* (London: Sweet and Maxwell, 1999).

9　　It seems appropriate to deal with UK energy regulation first, since, during the 1990s, the 'British model of privatisation' has been taken up by the European Commission (see Dieter Helm, *Energy, the State and the Market: British Energy Policy since 1979*, revised edn (Oxford: Oxford University Press, 2004), pp. 372–3).

10　　See, for example, s.4AB, Gas Act 1986, and s.3B, Electricity Act 1989 (Guidance on social and environmental matters) and s.33BC, Gas Act 1986, and 41A, Electricity Act 1989 (Promotion of the efficient use by consumers of gas/electricity).

11　　See para. 1.4.2.1 above.

electricity sector, where there is currently some evidence that they are hindering the development of renewables generation.[12]

6.2 Integrated Pollution Control, Integrated Pollution Prevention and Control and Environmental Assessment

6.2.1 Introduction

United Kingdom environmental regulation takes effect within the framework created by two overarching regulatory regimes: Integrated Pollution Control ('IPC') and Integrated Pollution Prevention and Control ('IPPC').

The basis of the IPC regime is Part I of the Environmental Protection Act 1990, whilst that of IPPC is the Integrated Pollution Prevention and Control Directive,[13] as transcribed into the law of England and Wales by the Pollution Prevention and Control Act 1999 and the Pollution Prevention and Control (England and Wales) Regulations 2000.[14]

The IPC regime represents an early attempt to coordinate the control of pollution by specific regulation of all of the emissions from certain prescribed industrial processes.[15] The objective of IPPC, by contrast, is much wider, the general approach of the IPPC legislation being to 'prevent, reduce and (if possible) to eliminate pollution and environmental impact as a whole'.[16] Integrated Pollution Prevention and Control has enhanced the role of local authorities[17] in the prevention and control of pollution;[18] it should have replaced IPC by 2007.

6.2.2 Integrated Pollution Control

Part I of the Environmental Protection Act 1990 creates two systems of pollution regulation: IPC, now under the control of the Environment Agency,[19] and Local Authority Air Pollution Control ('LAAPC'). The former was intended to regulate the most seriously polluting processes, whilst the latter was intended for those processes which are clearly in need of regulation but which are not so grave as to require centralised control.

---

[12]   See para. 6.2.4 below.
[13]   Council Directive 96/61/EEC, (1996) OJ L257 26 (see para. 12.2.2 below).
[14]   S.I. 2000 No. 1973. According to Bell and McGillivray, *op. cit.*, the IPPC Directive was one of the first items of Community environmental legislation to be transposed and implemented in the devolved UK (*ibid.*, p. 383). See paras 1.3 and 4.2.2 above.
[15]   For the background, see Royal Commission on Environmental Pollution, *Fifth Report on Air Pollution Control: An Integrated Approach* (Cmnd 6731, 1976).
[16]   Bell and McGillivray, *op. cit.*, p. 386.
[17]   See para. 4.2.3 above.
[18]   See the Pollution Prevention and Control (England and Wales) Regulations 2000, S.I. 2000 No 1973, reg. 8 and Sched. 1 (Part A and Part B Installations within each Section of the Sched). See also Wolf and Stanley, *op. cit.*, p. 291, for a further explanation.
[19]   See para. 4.2.1.3(1) above.

By s.6(1), Environmental Protection Act 1990, no person must carry on a prescribed process 'except under an authorisation granted by the enforcing authority and in accordance with the conditions to which it is subject'. Additionally, by Environmental Protection Act 1990, s.2(5), the Secretary of State for Environment, Food and Rural Affairs[20] may, by regulations, 'prescribe any description of substance as a substance the release of which into the environment is subject to control' under s.6 of the Act.

A prescribed process is a process[21] for the carrying on of which an authorisation is required by regulations.[22] The regulations in question, the Environmental Protection (Prescribed Processes and Substances) Regulations 1991,[23] contain two lists of processes, 'A' and 'B', in Sched. 1 thereto, the more seriously polluting processes falling within List A. Prescribed substances appear within Scheds 4–6, Environmental Protection (Prescribed Processes and Substances) Regulations 1991. Within Sched. 1, the prescribed processes are grouped within six chapters, according to sector:

1   fuel production processes, combustion processes and associated processes;[24]
2   metal production and processing;[25]
3   mineral industries;[26]
4   the chemical industry;[27]
5   waste disposal and recycling;[28] and
6   other industries.[29]

Schedule 4 contains prescribed substances released into the air; Sched. 5 contains prescribed substances released into water; and Sched. 6 contains prescribed substances released into land.

The exemptions appearing in the Environmental Protection (Prescribed Processes and Substances) Regulations 1991, reg. 4, apply to those processes in which the amount of a prescribed substance released is very small.

Authorisations are covered by Environmental Protection Act 1990, s.6 and Sched. 1. It is a criminal offence to carry on a prescribed process or to discharge a prescribed substance without prior authorisation.[30] This is punishable, on summary conviction, by a fine not exceeding £20,000 or by imprisonment for a term not exceeding three months (or both). On conviction on indictment, it is subject to an unlimited fine or to imprisonment for a term not exceeding two years or to both.[31]

---

20   See para. 4.2.1.2(1) above.
21   See Environmental Protection Act 1990, s.1(5).
22   *Ibid.*, s.2(1).
23   S.I. 1991 No. 472, as amended at least eight times.
24   Environmental Protection (Prescribed Processes and Substances) Regulations 1991, S.I. 1991 No. 472, Sched. 1, ch. 1.
25   *Ibid.*, Sched. 1, ch. 2.
26   *Ibid.*, Sched. 1, ch. 3.
27   *Ibid.*, Sched. 1, ch. 4.
28   *Ibid.*, ch. 5.
29   *Ibid.*, Sched. 1, ch. 6.
30   Environmental Protection Act 1990, s.23(1)(a).
31   *Ibid.*

The enforcing authority, as mentioned above, is either the Environment Agency[32] or the local authority.[33] List A processes are designated for central control, whilst List B processes are designated for local control.[34]

Section 7, Environmental Protection Act 1990, deals with the imposing of conditions on the grant of authorisations. By s.6(3), Environmental Protection Act 1990, on an application for an authorisation, the enforcing authority shall:

> ... either grant the authorisation subject to the conditions required or authorised to be imposed by section 7 ... or refuse the application.

In s.7 are stated the considerations that the enforcing authority may take into account in granting an authorisation and the objectives for which the conditions may be imposed. Among these objectives is that of ensuring that, in carrying on a prescribed process, the best available techniques not entailing excessive cost ('BATNEEC') will be used for preventing or reducing the release of prescribed substances.[35]

The enforcing authority has wide powers, including:

1   varying and revoking authorisations;[36]
2   serving enforcement notices[37] and prohibition notices;[38] and
3   power to take reasonable steps to remedy the harm.[39]

6.2.3   Integrated Pollution Prevention and Control

Section 2, Pollution Prevention and Control Act 1999, empowers, *inter alia*, the Secretary of State for Environment, Food and Rural Affairs, to make regulations for particular purposes, including:

1   establishing standards, objectives or requirements in relation to emissions;[40]
2   authorising the making of plans for the setting of overall limits, the allocation of quotas or the progressive improvement of standards or objectives;[41] and
3   authorising the making of schemes for the trading or other transfer of quotas so allocated.[42]

Most importantly, Sched. 1, Pt 1, para. 4, of the 1999 Act empowers the Secretary of State to prohibit persons from operating any installations or plant of any specified

---

32   See para. 4.2.1.3 above.
33   See para. 4.2.3 above.
34   Environmental Protection Act 1990, s.5.
35   *Ibid.*, s.7(2).
36   *Ibid.*, ss.10–12.
37   *Ibid.*, s.13.
38   *Ibid.*, s.14.
39   *Ibid.*, s.27.
40   Pollution Prevention and Control Act 1999, Sched. 1, Pt 1, para. 1(1).
41   *Ibid.*, Sched. 1, Pt 1, para. 1(2).
42   *Ibid.*, Sched. 1, Pt 1, para. 1(3).

description or otherwise carrying on any activities of any specified description, except under a permit in force under regulations and in accordance with any conditions to which the permit is subject.[43]

The Pollution Prevention and Control Act 1999 also empowers the Secretary of State for Environment, Food and Rural Affairs to make schemes for fees and charges in respect of applications for the grant, variation, subsistence, transfer or surrender of permits or testing substances and regulations may require charges to be so framed as to cover expenditure.[44] Pollution Prevention and Control (England and Wales) Regulations 2000[45] represent the exercise of these powers in England and Wales.

The basic provision of Pollution Prevention and Control (England and Wales) Regulations 2000 is that 'no person shall operate an installation or mobile plant except under, and to the extent authorised by, a permit granted by the regulator'.[46] An installation is a stationary technical unit where prescribed activities are carried out and any other location on the same site where any other directly associated activities are carried out.[47] Mobile plant is plant designed to move or to be moved, whether on roads or otherwise, and which is likewise used to carry out prescribed activities.[48]

The descriptions of these activities are immensely detailed, but cover:

1   combustion activities, gasification, liquefaction and refining activities, in the energy industries;[49]
2   the production and processing of metals and the surface treating of metals and plastic materials, in the production and processing of metals;[50]
3   the production of cement and lime, activities involving asbestos, manufacturing glass and glass fibre, the production of other mineral fibres, other mineral activities and ceramic production, in the mineral industries;[51]
4   in the chemical industry, producing both organic and inorganic chemicals, chemical fertilisers, plant health products and biocides, pharmaceutical products and explosives, manufacturing activities involving carbon disulphide or ammonia and the storage of chemicals in bulk;[52]
5   the disposal of waste by landfill, as well as the disposal of waste other than by incineration or landfill, the recovery of waste and the production of fuel from waste, in the waste management industry;[53] and

---

[43]   Pollution Prevention and Control Act 1999, Sched. 1, Pt 1, para. 1(4).
[44]   *Ibid.*, Sched. 1, Pt 1, para. 24.
[45]   S.I. 2000 No. 1973.
[46]   Pollution Prevention and Control (England and Wales) Regulations 2000, S.I. 2000 No. 1973, reg. 9(1).
[47]   *Ibid.*, reg. 2(1).
[48]   *Ibid.*, reg. 2(1).
[49]   *Ibid.*, Sched. 1, Pt 1, Ch. 1, Section 1.1.
[50]   *Ibid.*, Sched. 1, Pt 1, Ch. 1, Sections 2.1–2.3.
[51]   *Ibid.*, Sched. 1, Pt 1, Ch. 1, Sections 3.1–3.6.
[52]   *Ibid.*, Sched. 1, Pt 1, Ch. 1, Sections 4.1–4.8.
[53]   Pollution Prevention and Control (England and Wales) Regulations 2000, S.I. 2000 No. 1973, Sched. 1, Pt 1, Ch. 1, Sections 5.1–5.5.

6   other activities, that is, paper, pulp and board manufacturing activities; carbon
     activities; tar and bitumen activities; coating activities, printing and textile
     treatments; the manufacture of dyestuffs, printing ink and coating materials;
     timber activities; activities involving rubber; the treatment of animal and
     vegetable matter and food industries; and intensive farming.[54]

When determining the conditions to be attached to a permit, the regulator is obliged
to take account of certain general principles.[55] Installations and plant should be
operated in such a way that:

   (a) all the appropriate preventative measures are taken against pollution, in particular
        through application of the best available techniques [BAT]; and
   (b) no significant pollution is caused.[56]

Regulation 12, Pollution Prevention and Control (England and Wales) Regulations
2000, specifies the conditions that must be included in a permit, and the concept of
best available techniques ('BAT') is elaborated in reg. 3. There, 'best' is defined as
meaning the most effective techniques in achieving a high general level of protection
of the environment as a whole; 'available' means those techniques that have been
developed 'on a scale that allows implementation in the relevant industrial sector
under economically and technically viable conditions'; and 'techniques' includes
'both the technology used and the way in which the installation is designed, built,
maintained, operated and decommissioned'. Regulation 4 contains particular
provision for the case where the regulator has to decide whether or not a person
is a 'fit and proper person to carry out a specified waste management activity'.[57]
A person will not be fit and proper to carry out those activities if, for example, he
has not made adequate financial provision to discharge his obligations under the
permit.[58]

The IPPC regulator[59] has a large number of enforcement powers, including the
power to serve enforcement notices;[60] the power to serve a revocation notice,
revoking all or part of the activities authorised by the permit;[61] and the power to
serve a suspension notice if it considers that the operation of an installation involves
an imminent risk of serious injury.[62] There is also a range of criminal offences

---

[54]   *Ibid.*, Sched. 1, Pt 1, Ch. 1, Sections 6.1–6.9.
[55]   *Ibid.*, reg. 11(1).
[56]   *Ibid.*, reg. 11(2). There are additional principles in the case of mobile plant (see reg.
       11(3)).
[57]   See point 5 above.
[58]   See Pollution Prevention and Control (England and Wales) Regulations 2000, S.I. 2000
       No. 1973, reg. 4(3)(c).
[59]   Who may be either the Environment Agency (see para. 4.2.1.3 above) or the relevant local
       authority (see para. 4.2.3 above).
[60]   Pollution Prevention and Control (England and Wales) Regulations 2000, S.I. 2000 No.
       1973, reg. 24.
[61]   *Ibid*, reg. 21.
[62]   *Ibid.*, reg. 25.

involving permits,[63] the maximum penalty on summary conviction being a fine of up to £20,000 and/or imprisonment for a period of up to six months, with the maximum penalty on conviction on indictment being an unlimited fine and/or imprisonment for a period up to five years.[64]

6.2.4   *Environmental Impact Assessment and Strategic Environmental Assessment*

It is important at this stage to note a second overarching set of provisions which are designed to ensure that, when a public authority is considering plans for public and private projects, it takes account of their environmental impacts.[65] These provisions originate from the Environmental Impact Assessment Directive ('the EIA Directive'), which is examined in some detail in a later chapter of the study,[66] the legislative basis of the provisions being s.2(2) of the European Communities Act 1972. In England and Wales, the EIA Directive has been implemented by the Town and Country Planning (Environmental Impact Assessment) (England and Wales) Regulations 1999,[67] together with a number of other sets of regulations which cover projects falling outside the scope of the law on town and country planning.[68]

The requirement for an EIA in relation to specific projects has been considered in a number of important cases in this jurisdiction, including *R v. North Yorkshire County Council*, ex parte *Brown*[69] and *Berkeley v. Secretary of State for the Environment, Transport and the Regions (No. 1)*.[70] In the former case, the House of Lords rejected an argument by the County Council that there had been no necessity for an EIA in a case where it had simply imposed conditions on a planning permission of indefinite duration originally granted in 1947 for the working of Wensley Quarry, at Preston-under-Scar, in the Yorkshire Dales. The Council had imposed the conditions under s.22 of the Planning and Compensation Act 1991 and, in support of its argument, it had contended that s.22 involved merely the detailed regulation of activities for which the principal consent had already been given in 1947. Clearly unimpressed with the argument, Lord Hoffmann said:

> The procedure created by the Act of 1991 was not merely a detailed regulation of a project in respect of which the substantial environmental issues had already been considered. The purpose of the procedure was to give the mineral planning authority [that is, the Council] a

---

63   *Ibid.*, reg. 32.

64   *Ibid.*, regs 32(2) and 32(3).

65   See Stephen Tromans and Karl Fuller, *Environmental Impact Assessment: Law and Practice* (London: LexisNexis Butterworths, 2003).

66   See para. 12.2.4 below.

67   S.I. 1999, S.I. No. 293.

68   For example, the Environmental Assessment (Afforestation) Regulations 1988, S.I. 1988 No. 1207, and the Land Drainage Improvement Works (Assessment of Environmental Effects) Regulations 1988, S.I. 1988 No. 1217.

69   [2000] 1 AC 397.

70   [2001] 2 AC 603.

power to assess the likely environmental effects of old mining permissions which had been granted without, to modern ways of thinking, any serious consideration of the environment at all.[71]

Similarly, in the *Berkeley* case, the House of Lords dismissed the argument put on behalf of the Secretary of State that, where substantial compliance with the EIA Directive's requirements could be demonstrated by reference to public documents, the fact that no EIA had actually taken place did not invalidate a planning permission. In dismissing this argument also, Lord Hoffmann said:

> My Lords, I do not accept that this paper chase[72] can be treated as the equivalent of an environmental statement[73] ... The point about the environmental statement contemplated by the Directive is that it constitutes a single and accessible compilation, produced by the applicant at the very start of the application process, of the relevant environmental information and the summary in non-technical language.[74]

Although both the *Brown* and the *Berkeley* cases involved the interpretation of the predecessor regulations dating from 1988, they each illustrate the strictness with which the courts approach the question of whether there has been compliance with the requirements of the Directive.[75]

Against the background of a transposition of the EIA Directive, which has hardly been free from difficulty, the ODPM[76] consulted in spring 2004[77] on the implementation of the Strategic Environmental Assessment ('SEA') Directive.[78] This was transposed into UK law, on time, in July 2004, via four separate sets of regulations for each of the UK's four countries.[79] Attention has recently been focused on SEAs since they are needed for the development of arrays of coastal wave power devices for electricity generation. The lack of an SEA for the southwest coast of England has been a matter of considerable concern to developers (and others) looking to wave power as a form of renewables generation.[80]

---

71   [2000] 1 AC 397, 405.
72   That is, a close reading of the documentation actually submitted in the case.
73   That is, as required by Council Directive 85/337/EEC, (1985) OJ L175 40, Arts 5, 6(2) and Annex IV. See para. 12.2.4 below.
74   [2001] 2 AC 603, 617.
75   But see the argument on the failure to carry out an EIA in *R (on the application of The Mayor, Citizens of Westminster and others) v. The Mayor of London*, [2002] EWHC 2440. See para. 18.2 below.
76   See para. 4.2.1.2(1) above.
77   See Office of the Deputy Prime Minister, *Consultation Document on Implementation of SEA Directive (2001/42/EC)*, March 2, 2004 (available from www.odpm.gov.uk).
78   See para. 12.2.4 below.
79   See, for example, for England, Environmental Assessment of Plans and Programmes Regulations 2004, S.I. 2004 No. 1633. See also para. 1.3 above.
80   See 346 ENDS Report (2003).

6.3   Waste management regulation

6.3.1   The international and European Union background

United Kingdom waste management regulation operates within the broader context of public international law and Community law on the regulation of waste. That broader context is considered in Chapters 8 and 12 below.

6.3.2   Command and control

The regulation of waste management has, until very recently, been dominated by command and control regulation.[81] The key set of provisions appears in Part II of the Environmental Protection Act 1990, but account must also be taken of the relationship between Part II, Environmental Protection Act 1990, and both IPC (in Part I of the same statute) and IPPC (in the Pollution Prevention and Control Act 1999).[82]

The Environmental Protection Act 1990, ss.44A and 44B, place an obligation on the Secretary of State for Environment, Food and Rural Affairs to prepare a national waste strategy for England and Wales. This is currently contained in what was then the DETR's[83] *Waste Strategy 2000 for England and Wales*, of May 2000.[84] Three principal targets of the strategy are: to reduce, by 2005, the amount of industrial and commercial waste sent to landfill to 85 per cent of 1998 levels; to recover value from 45 per cent of municipal waste by 2010 and to recycle or compost at least 30 per cent of household waste by that date; and to recover value from two-thirds of municipal waste by 2015, with at least 33 per cent of household waste being recycled or composted by the same date.

Following concern that the targets in *Waste Strategy 2000* would not be met, the Strategy was reviewed by the Cabinet Office Strategy Unit[85] in November 2002 in its now-famous *Waste Not, Want Not* document.[86] The Strategy Unit criticised *Waste Strategy 2000* for giving too little attention to minimising waste (the *Strategy* not containing any waste minimisation targets)[87] and for failing to create '... the economic and regulatory framework and enough associated policy tools to deliver tangible improvements in waste minimisation, re-use and recycling'.[88] *Waste Not, Want Not* made over 30 recommendations for taking forward and monitoring by Defra.[89] The recommendations included raising landfill tax to £35 per tonne for

---

[81]   See Duncan Laurence, *Waste Regulation Law* (London: Butterworths, 1999).
[82]   See para. 6.2 above.
[83]   See para. 4.2.1.2(1) above.
[84]   Department of Environment, Transport and the Regions, *Waste Strategy 2000 for England and Wales* (Cm 4693, 2000).
[85]   See para. 4.2.1.2(1) above.
[86]   Cabinet Office Strategy Unit, *Waste Not, Want Not: A Strategy for Tackling the Waste Problem in England* (London: Cabinet Office Strategy Unit, 2002).
[87]   *Ibid.*, p. 30.
[88]   Cabinet Office Strategy Unit, *op. cit.*, p. 30.
[89]   *Ibid.*, pp. 116–39.

active waste[90] and keeping under review the case for an incineration tax.[91] They were subsequently implemented in the government's Waste Implementation Programme ('WIP'), which began in June 2003.[92]

6.3.2.1   Environmental Protection Act 1990, Part II

There are three main elements in the Environmental Protection Act 1990:

1   a waste management licensing system;
2   a statutory duty of care in relation to how waste is handled; and
3   a reorganisation of the functions of the regulatory authorities between Waste Regulation Authorities ('WRAs'), Waste Disposal Authorities ('WDAs') and Waste Collection Authorities ('WCAs').

The WRAs were formerly the County Councils[93] but, as a result of amendments made to the Environmental Protection Act 1990 by the Environment Act 1995, the WRA is now the Environment Agency.[94] Generally speaking, the WDA is the County Council is a non-metropolitan area and the District Council in a metropolitan area.[95] The WDA has a duty to arrange for the disposal of controlled waste (that is, household, industrial and commercial waste)[96] collected by the WCA in the area in question.[97] The WCA has a duty to arrange for the collection of household waste and of commercial waste when so requested.[98] Although, in certain circumstances, a WCA may charge for the collection of household waste, the collection of such waste is generally free of charge. The same is not true of industrial or commercial waste, the collection and disposal of which is subject to the payment of some reasonable amount.[99]

Crucial to the whole system, of course, is the definition of 'waste' itself. Although it is possible to restate this definition in brief compass, the definition in practice gives rise to a range of potentially extremely complex issues. 'Waste' is defined in Environmental Protection Act 1990, s.75(2), as:

---

90   *Ibid.*, p. 124.
91   *Ibid.*, p. 127.
92   See Department for Environment, Food and Rural Affairs, *WIP: One Year On*, June 2004, available from www.defra.gov.uk.
93   See para. 4.2.3 above.
94   See para. 4.2.1.3 above.
95   See para. 4.2.3 above. See also Wolf and Stanley, *op. cit.*, p. 203. The authors would like to acknowledge a particular debt to this work in this para. 6.3.
96   See Environmental Protection Act 1990, s.75(4), and the Controlled Waste Regulations 1992, S.I. 1992 No. 588 (see *Thanet District Council v. Kent County Council*, [1993] Env LR 391). 'Household waste' is defined in the Environmental Protection Act 1990, s.75(5); 'industrial waste' is defined in the Environmental Protection Act 1990, s.75(6); and 'commercial waste' is defined in the Environmental Protection Act 1990, s.75(7).
97   *Ibid.*, s.51.
98   *Ibid.*, s.45.
99   *Ibid.*, s.45(4).

... any substance or object in the categories set out in Schedule 2B to this Act which the holder discards or intends or is required to discard; and for the purposes of this definition – 'holder' means the producer of the waste or the person who is in possession of it; and 'producer' means any person whose activities produce waste or any person who carries out pre-processing, mixing or other operations resulting in a change in the nature or composition of this waste.[100]

Environmental Protection Act 1990, Sched. 2B, provides a non-exhaustive list of items that count as waste (for example, off-specification products, residues from industrial processes and pollution abatement processes, adulterated materials, etc.) when they have been discarded.[101]

Thus, for the purposes of Environmental Protection Act 1990, s.75, in determining whether an item is 'waste', it is crucial to decide whether it has been 'discarded'.[102] What constitutes discarding has been a matter of considerable controversy in the context of landfill tax,[103] since that tax is charged on disposals of material as waste.[104] In the context of Environmental Protection Act 1990, the ECJ has held that 'discard' has a special meaning that includes both the consignment of waste for disposal and the consignment of waste to a recovery operation.[105] This definition was followed by Carnwath, J. in *Mayer Parry Recycling Ltd v. Environment Agency (No. 1)*,[106] where it was held that a company, part of whose business involved receiving scrap metal and dealing with it so that steel manufacturers could use it for making other items, was managing 'waste'.[107]

Besides the concept of 'controlled waste', there is also that of 'special waste'. Special waste is controlled waste in respect of which the Secretary of State for Environment, Food and Rural Affairs has made an order under the Environmental Protection Act 1990, s.62.[108] It includes waste that is especially '... dangerous or

---

[100]   This incorporates the definition in the Waste Framework Directive, Council Directive 75/442/EEC, (1975) OJ L194 39, Art. 1(a). Waste falling within this definition is therefore often known as 'Directive Waste'.

[101]   See para. 12.2.5.2 below.

[102]   See, for example, J. Cheyne and M. Purdue, 'Fitting definition to purpose: the search for a satisfactory definition of waste', [1995] 7 JEL 149. See also the articles cited at para. 12.2.5.2n below.

[103]   See para. 15.2 below.

[104]   See para. 1.4.2.1(1) above.

[105]   See *Criminal proceedings against Euro Tombesi and others*, Joined Cases C–304/94, C–330/94, C–342/94 and C–224/95, [1997] ECR I–3561; *Inter-Environnement Wallonie ASBL v. Region Wallonne*, C–129/96, [1998] 1 CMLR 1057; and *ARCO and EPON*, Joined Cases C–418–419/97, [2002] QB 646. See further para. 12.2.5.2 below.

[106]   [1999] Env LR 489. See also *Parkwood Landfill Ltd v. C & E Commrs*, [2002] STC 1536.

[107]   With the result that it therefore required the appropriate licences under the regulations described in (a) below. In *Mayer Parry Recycling Ltd v. Environment Agency (No. 1)*, Carnwath, J. held that the material continued to be waste until the completion of the recovery process (as to the position when that process has been completed, see *Castle Cement v. Environment Agency*, [2001] EWHC Admin 224 (Stanley Burnton, J.); [2001] Env LR 46).

[108]   See Environmental Protection Act 1990, s.75(9).

difficult to treat, keep or dispose of' (for example, because it is explosive, flammable or carcinogenic, etc.).[109]

The exclusions from the definition of commercial waste in the Environmental Protection Act 1990, s.75, include waste from any mine or quarry.[110]

(1) Waste management licensing system

Environmental Protection Act 1990, ss.35–44, contain a framework for the management of waste. The detail is contained in the Waste Management Licensing Regulations 1994.[111] Both sets of provisions are intended to comply with the Waste Framework Directive ('the WFD').[112]

Details of the basic provisions of the 1990 Act are as follows:[113]

a. Section 35(1) gives the following definition of a waste management licence:

> ... a licence granted by a waste regulation authority[114] authorising the treatment, keeping or disposal of any specified description of controlled waste in or on specified land or the treatment or disposal of any specified description of controlled waste by means of specified mobile plant.

> Subsections (3) and (4) of s.33, Environmental Protection Act 1990, give the Secretary of State for Environment, Food and Rural Affairs the power to exempt certain activities from the licensing system. The activities thus exempted include those covered by other statutory controls (for example, IPC)[115] as well as those falling within an exhaustive list,[116] which includes the temporary storage of waste, on the site where it is produced, pending its collection.[117]

b. The licence is granted to the person who is in occupation of the land (that is, to 'the site operator') or, in the case of a mobile waste treatment plant, the person who operates the plant.[118]

c. The Environment Agency's circumscribed discretion in relation to the grant of licences is contained in Environmental Protection Act 1990, s.36. The Agency must refuse to issue a licence if planning permission is required in relation to the land and no such planning permission is in force.[119] Otherwise, the Agency

---

109 See Special Waste Regulations 1996, S.I. 1996 No. 972.
110 See Environmental Protection Act 1990, s.75(7), and the Mines and Quarries (Tips) Act 1969.
111 S.I. 1994 No. 1056.
112 See para. 12.2.5.1(1) below.
113 It should be noted, of course, that the grant of a waste management licence does not obviate the need to comply with town and country planning law (see Hughes, *op. cit.*, pp. 423–6).
114 That is, the Environment Agency (see para. 4.3.2(1) above).
115 Waste Management Licensing Regulations 1994, S.I. 1994 No. 1056, reg. 16 (see below).
116 *Ibid.*, reg. 17 and Sched. 3.
117 *Ibid.*, Sched. 3, para. 41(1).
118 Environmental Protection Act 1990, s.35(2).
119 *Ibid.*, s.36(2).

must not reject the application if it is satisfied that the applicant is a fit and proper person, unless:

... it is satisfied that its rejection is necessary for the purpose of preventing–
(a)   pollution of the environment;
(b)   harm to human health; or
(c)   serious detriment to the amenities of the locality ...[120]

d.   The grant of a licence may be made subject to such terms and conditions relating to the activities that the licence authorises and to the precautions to be taken and works to be carried out.[121]
e.   Section 43, Environmental Protection Act 1990, provides for there to be a right of appeal from decisions of the Environment Agency to the Secretary of State.
f.   By Environmental Protection Act 1990, s.33(1), it is an offence to deposit, treat, keep or dispose of controlled waste except in accordance with a waste management licence.[122] Breach of the section makes the wrongdoer liable to imprisonment for up to six months or a fine not exceeding £20,000 or both (on summary conviction); or to imprisonment for up to two years and/or a fine (on conviction on indictment).[123]
g.   Fees are payable to the Agency for the grant, modification, transfer, etc., of the waste management licence.[124] This is a practical example of the polluter pays principle.[125]

(2) Statutory duty of care as to handling of waste
Section 34 of the Environmental Protection Act 1990, imposes a duty on various categories of persons (including importers, producers and carriers) to take all such measures applicable to them in that capacity as are reasonable in the circumstances:

a.   'to prevent contravention by any other person' of Environmental Protection Act 1990, s.33;[126]
b.   to 'prevent the escape of waste from his control or that of any other person'; and
c.   on the transfer of the waste, to secure that the transfer is to an authorised person or for authorised transport purposes only and to ensure that sufficient written description of the waste is provided to anyone to whom the waste is transferred.[127]

---

[120]   *Ibid.*, s.36(3).
[121]   *Ibid.*, s.35(3).
[122]   *Ibid.*, s.33(1).
[123]   *Ibid.*, s.33(8) and (9). The penalties are harsher in relation to special waste (see s.33(9) and para. 6.3.2.6 below).
[124]   See Environment Act 1995, s.41, and the Environmental Licences (Suspension and Revocation) Regulations 1996, S.I. 1996 No. 508.
[125]   See para. 6.2.3 above.
[126]   See para. 6.3.2.1(1)(a) above.
[127]   Environmental Protection Act 1990, s.34(1) (see also Environmental Protection (Duty of Care) Regulations S.I. 1991 No. 2839).

The only exception to the duty of care is for occupiers of domestic property with regard to household waste produced on their property.[128]

Breach of the duty of care in s.34 is a criminal offence. It is punishable, on summary conviction, by a fine not exceeding the statutory maximum and, on conviction on indictment, by a fine.[129]

6.3.2.2 Integrated Pollution Control

Regulation 16(1)(a) of the Waste Management Licensing Regulations 1994[130] provides that the recovery or disposal of waste under an IPC authorisation, where the activity is, or forms part of, a process designated for central control,[131] is exempt from the need for a waste management licence.

Moreover, no condition can be attached to an IPC authorisation that regulates the final disposal of directive waste in or on land.[132]

6.3.2.3 Integrated Pollution Prevention and Control

The IPPC applies to the disposal of waste by incineration,[133] by landfill[134] or by other specified means.[135] It also applies to the recovery of waste[136] and the production of fuel from waste.[137]

Prior to June 2002,[138] landfill sites were either subject to the Waste Management Licensing Regulations 1994[139] or to the IPPC Regime. If a landfill site fell outside the wording of the relevant definition in the Pollution Prevention and Control (England and Wales) Regulations 2000,[140] then it would fall within the licensing system of Part II of the Environmental Protection Act 1990.[141]

The IPPC regime has recently been extended to all landfill sites pursuant to the measure discussed in the following para.[142]

---

[128] Environmental Protection Act 1990, s.34(2).
[129] *Ibid.*, s.34(6).
[130] S.I. 1994 No. 1056. See para. 6.3.2.1(1)(a) above.
[131] That is, under Environmental Protection Act 1990, s.2(4).
[132] *Ibid.*, s.28(1).
[133] See the Pollution Prevention and Control (England and Wales) Regulations 2000, S.I. 2000 No. 1973, Sched. 1, Pt 1, Ch. 1, section 5.1.
[134] *Ibid.*, Sched. 1, Pt 1, Ch. 1, section 5.2.
[135] *Ibid.*, Sched. 1, Pt 1, Ch. 1, section 5.3.
[136] *Ibid.*, Sched. 1, Pt 1, Ch. 1, section 5.4.
[137] *Ibid.*, Sched. 1, Pt 1, Ch. 1, section 5.5.
[138] See the Landfill (England and Wales) Regulations 2002, S.I. 2002 No. 1559, reg. 1(2).
[139] See para. 6.3.2(1)(a) above.
[140] See S.I. 2000 No. 1973, Sched. 1, Pt 1, Ch. 1, section 5.2.
[141] See para. 6.3.2.1(1)(a) above.
[142] See the Landfill (England and Wales) Regulations 2002, S.I. 2002 No. 1559, reg. 6(1). See para. 6.3.2.4 below.

6.3.2.4   Implementation of the Landfill Directive

The Landfill Directive[143] aims for a 65 per cent reduction, on the basis of 1995 waste arisings, in the amount of methane producing, biodegradable household and municipal waste that is disposed of in landfill sites.[144] Although the means by which the Waste and Emissions Trading Act 2003 seeks to achieve these targets are largely in the form of economic instruments,[145] Part 1 of the 2003 Act places a duty on the Secretary of State for Environment, Food and Rural Affairs to set the maximum amount of biodegradable municipal waste to be sent to landfill from each of the UK's constituent countries.[146] The UK, as an EU Member State which sent more than 80 per cent of its collected municipal waste to landfills in 1995, qualifies for the four-year derogation from the 2016 target.[147]

The need to comply with the terms of the Landfill Directive[148] has also given rise to the creation of a special regulatory regime for landfill sites. The regime is contained in the Landfill (England and Wales) Regulations 2002.[149]

The 2002 regulations apply to 'landfills', the concept of a landfill being defined, subject to certain exclusions,[150] as 'a waste disposal site for the deposit of the waste onto or into land'.[151] Specifically excluded from the scope of the regulations, among other things, is 'the use of suitable inert waste for redevelopment, restoration and filling-in work or for construction purposes'.[152] The regulations operate by modifying the conditions under which an IPPC permit relating to the disposal of waste in a landfill (referred to as a 'landfill permit')[153] may be granted under the Pollution Prevention and Control (England and Wales) Regulations 2000.[154]

Besides specifying the conditions to be contained in a landfill permit,[155] procedures for accepting waste at landfills[156] and the inspection, control and monitoring of landfills by the Environment Agency,[157] the regulations contain four key sets of provisions:

---

143   Council Directive 99/31/EC, (1999) OJ L182 1.
144   See para. 12.2.5.1(2)(b) below.
145   See para. 6.3.3 below.
146   Waste and Emissions Trading Act 2003, s.1.
147   See para. 12.2.5.1(2)(b) below.
148   Council Directive 99/31/EC, (1999) OJ L182 1.
149   S.I. 2002, No. 1559.
150   *Ibid.*, regs 3(4) and 4.
151   *Ibid.*, reg. 3(2). See *Blackland Park Exploration Ltd v. Environment Agency*, [2003] EWCA Civ 1795, [2003] All ER (D) 249 (Dec). There, the CA (Simon Brown, Mummery and Scott Baker, L.JJ.) held that the disposal of hazardous liquid industrial wastes through a borehole into strata around 1,000 metres below sea level was a deposit 'into land' within the 2002 regulations and that the site was therefore a 'landfill'.
152   S.I. 2002, No. 1559, reg. 4(b).
153   *Ibid.*, reg. 6(2).
154   S.I. 2000, No. 1973. See para. 6.2.3 above.
155   S.I. 2002, No. 1559, reg. 8.
156   *Ibid.*, reg. 12.
157   *Ibid.*, regs 13 and 14.

1 when considering the grant of planning permission for a landfill, the relevant planning authority must take into consideration specific matters relating to its location which indicate that the landfill 'does not pose a serious environmental risk';[158]

2 before granting a landfill permit, the Environment Agency must classify the landfill as a landfill for hazardous waste,[159] a landfill for non-hazardous waste[160] or a landfill for inert waste[161] and must ensure that the classification is given in the landfill permit;[162]

3 an absolute ban on the acceptance at landfills of certain types of waste, including any waste in liquid form (including waste waters but not including sludge);[163] infectious hospital and clinical wastes;[164] whole tyres (from July 2003);[165] and shredded tyres (from July 2006);[166]

4 a requirement for the landfill site operator to ensure that the charges it makes cover the costs of setting up and operating the landfill, of his financial provision in relation to his obligations under IPPC and the estimated costs of the eventual closure and after-care of the site.[167]

The provision summarised at 2 above thus bans co-disposal of hazardous and non-hazardous waste. Furthermore, in relation to the classifications within the provisions at 2 above, the site operator[168] must generally[169] ensure that the landfilled waste has been subject to prior volume-reducing or safety treatment.[170] He must also ensure that he accepts for landfilling only waste which will not have specified adverse effects to the environment or human health.[171] Finally, if the landfill permit relates to non-hazardous waste, the operator must landfill only municipal waste, other non-hazardous waste and stable, non-reactive hazardous waste.[172]

Besides placing the target-setting duty on the Secretary of State, Part 1 of the Waste and Emissions Trading Act 2003 requires the appropriate authority for each country

---

158 *Ibid.*, reg. 5 and Sched. 2.
159 *Ibid.*, reg. 7(2). See 12.2.5.1(3) below.
160 *Ibid.*, reg. 7(3).
161 *Ibid.*, reg. 7(4).
162 *Ibid.*, reg. 7(1).
163 *Ibid.*, reg. 9(1)(a). In *Blackland Park Exploration Ltd v. Environment Agency*, [2003] EWCA Civ 1795, [2003] All ER (D) 249 (Dec), Scott Baker, L.J. commented that the appellant company was said to be the only company in the UK bringing liquid waste onto a site for disposal by means of an injection well (see judgment, paras 1 and 4).
164 See S.I. 2002, No. 1559, reg. 9(1)(c).
165 *Ibid.*, reg. 9(1)(e).
166 *Ibid.*, reg. 9(1)(f).
167 See S.I. 2002, No. 1559, reg. 11.
168 See the Pollution Prevention and Control (England and Wales) Regulations 2000, S.I. 2000 No. 1973, reg. 2 (see S.I. 2002, No. 1559, reg. 2).
169 But see S.I. 2002 No. 1559, reg. 10(1).
170 *Ibid.*
171 S.I. 2002, No. 1559, reg. 10(2) and Sched. 1, paras 1 and 2.
172 *Ibid.*, No. 1559, reg. 10(3).

of the UK[173] to develop a national strategy to reduce the amount of biodegradable municipal waste sent to landfills.[174] In addition, where a local authority is a two-tier authority,[175] the WDA and the WCAs within its area must usually[176] develop a joint strategy for the management of waste from households and similar waste.[177] Finally, although the 2003 Act empowers WDAs to direct WCAs to deliver their waste in a separated state,[178] it also requires the WDA to make payments to the WCA to cover the cost of doing this.[179]

6.3.2.5   Measures relating to specific types of waste

Certain categories of waste have their own disposal rules.[180] For example, special provisions apply to the disposal of waste oils;[181] to the disposal of polychlorinated biphenyls ('PCBs') and polychlorinated terphenyls ('PCTs');[182] to the disposal of waste from the titanium dioxide industry;[183] to the agricultural use of sewage sludge;[184] and to the recovery and controlled disposal of batteries and accumulators which are spent and contain quantities of mercury, cadmium or lead.[185]

6.3.2.6   Measures relating to hazardous ('special') waste

Section 62 of the Environmental Protection Act 1990 provides that, where the Secretary of State for Environment, Food and Rural Affairs considers that a particular kind of controlled waste is particularly dangerous or difficult to treat, keep or dispose of, so that special provision is needed in relation to it, then he must make regulations

---

173   See para. 4.2.2 above (that is, the Secretary of State in England, the Scottish Ministers in Scotland, the National Assembly for Wales and the Department of the Environment in Northern Ireland).
174   Waste and Emissions Trading Act 2003, ss.17–20.
175   See para. 4.2.3 above.
176   Waste and Emissions Trading Act 2003, s.33.
177   *Ibid.*, s.32.
178   Environmental Protection Act 1990, s.48(1A) (as inserted by the Waste and Emissions Trading Act 2003, s.31(2)).
179   *Ibid.*, s.31(4)).
180   The categories listed here all originate in Community law. See para. 12.2.5.1(2) below.
181   See the Waste Management Licensing Regulations 1994, S.I. 1994 No. 1056 and the Special Waste Regulations 1996, S.I. 1996 No. 972.
182   See the Environmental Protection Act 1990 (Extension of Section 140) Regulations 1999, S.I. 1999 No. 396 and the Environmental Protection (Disposal of Polychlorinated Biphenyls and Other Dangerous Substances) (England and Wales) Regulations, S.I. 2000 No. 1043 (amended by the Environmental Protection (Disposal of Polychlorinated Biphenyls and Other Dangerous Substances) (England and Wales) (Amendment) Regulations 2001 (S.I. 2001 No. 3359).
183   This is a prescribed process under the Environmental Protection Act 1990, Part 1 (see para. 6.2.2 above).
184   See the Sludge (Use in Agriculture) Regulations 1989 (as amended), S.I. 1989 No. 1263.
185   See the Batteries and Accumulators (Containing Dangerous Substances) Regulations 1994, S.I. 1994 No. 232.

for dealing with such 'special waste'.[186] The regulations made under the section are the Special Waste Regulations 1996,[187] as amended.

6.3.2.7 Measures relating to the importation and exportation of waste

Although the Waste Shipment Regulation[188] is directly applicable in the UK, it has been transposed into UK law by the Transfrontier Shipment of Waste Regulations 1994.[189]

6.3.3 Economic instruments

The main economic instruments in UK waste management regulation have for long been landfill tax, packaging waste recovery notes ('PRNs') and the waste recycling credits scheme. To these will shortly be added the LATS, a form of tradeable landfill allowance.[190]

6.3.3.1 Landfill tax

An outline of the tax has been given at para. 1.4.2.1(1) above.[191] It is discussed in detail in Chapter 15 below.

6.3.3.2 Packaging waste recovery notes ('PRNs')

Again, an outline of the role of PRNs has been given above at para. 1.4.2.1(2). There is a more detailed discussion in Chapter 19.

6.3.3.3 Waste Recycling Credits Scheme

A note on the WRCS has been given at para. 1.4.2.1(3) above. It is not considered further in this study.

6.3.3.4 Landfill Allowances Trading Scheme ('the LATS')

The Landfill Allowances Trading Scheme ('the LATS'), which is now due to begin in

---

186 See para. 6.3.2.1 above.
187 S.I. 1996 No. 972.
188 Council Regulation EEC/259/93, (1993) OJ L30 1. See para. 12.2.5.1(4) below.
189 S.I. 1994 No. 1137.
190 See paras 1.4.2.1(4) above and 20.7 and 21.3.2 below.
191 See D.N. Pocklington and R.E. Pocklington, 'The United Kingdom Landfill Tax – Externalities and External Influences', [1998] JPL 529–545; Patricia Park, 'An Evaluation of the Landfill Tax Two Years on', [2000] JPL 3–13; and Inger Brisson and Jane Powell, 'The UK Landfill Tax', in *Environmental Policy: Objectives, Instruments and Implementation*, ed. by Dieter Helm (Oxford: Oxford University Press, 2000), pp. 260–80.

April 2005 in England and Wales,[192] is discussed in para. 20.7 below. There is again an outline of the LATS in para. 1.4.2.1(4) above.

6.3.3.5   The Waste Performance Reward Grant

In January 2004, Defra issued a consultation paper on the introduction, in 2005/2006, of a Waste Management Performance Reward Grant ('WPRG'), which is intended to provide incentives for local authorities to improve their recycling levels.[193]

6.4   Control of air and atmospheric pollution

6.4.1   The international and European Union background

Issues of air and atmospheric pollution are nothing if not transboundary in their scope. The measures discussed in this part of the chapter should therefore be understood within the public international and Community contexts discussed in Chapters 8 and 12 below.

6.4.2   Environmental regulation

If the IPC and IPPC regimes are put to one side,[194] then, with two main exceptions,[195] and unlike with waste management, the instruments for carrying into effect the UK's climate change programme[196] are primarily economic ones. That programme is based on the UK's commitment under the Kyoto Protocol to reduce its 1990 levels of all greenhouse gases by 12.5 per cent by 2010.[197]

The two main exceptions to the predominance of economic instruments in the area are the command and control regimes for regulating emissions contained in the Clean Air Act 1993 and the mechanism for ensuring air quality contained in Part IV of the Environment Act 1995. The former empowers local authorities[198] to bring criminal proceedings in respect of a range of prohibited activities involving the emission of smoke, dust and grit.[199] The latter places a duty on the Secretary of

---

[192] But not in Wales and Scotland, where the scheme began on 1 October 2004 (see 358 ENDS Report (2004) 44–45).

[193] See Department for Environment, Food and Rural Affairs, *Consultation Paper on the Design of the Waste Performance Reward Grant*, 7 January 2004 (available from www.defra.gov.uk).

[194] See paras 6.2.2 and 6.2.3 above.

[195] Note should also be taken of the statutory nuisance provisions to be found in the Environmental Protection Act 1990, Part III (see Wolf and Stanley, *op. cit.*, Ch. 9).

[196] See Department of Environment, Transport and the Regions, *Climate Change: the UK Programme*, 2000 (Cm 4913, 2000).

[197] See Chapter 8 below.

[198] See para. 4.2.3 above.

[199] See, for example, ss.1 (prohibition of dark smoke from chimneys), 2 (prohibition of dark smoke from industrial premises), 14 (height of chimneys for furnaces) and 15 (applications

State for Environment, Food and Rural Affairs[200] to prepare, publish and keep under review a National Air Quality Strategy.[201]

Setting aside the measures to which reference has just been made, the government's climate change policy rests on the combination of four economic instruments:[202]

1 climate change levy;
2 the UK Emissions Trading Scheme ('the UK ETS');
3 the EU Emissions Trading Scheme ('the EU ETS'); and
4 the Renewables Obligation ('the RO').

Each of these have been outlined at the beginning of the study[203] and are discussed in detail in the rest of the book.[204] Both climate change levy and the EU ETS rely heavily on IPPC concepts.

6.4.3   Energy regulation

6.4.3.1   Electricity

(1) Overview
The electricity supply industry[205] in England and Wales[206] is regulated by the Electricity Act 1989, which has been substantially amended by the Utilities Act 2000.[207] Besides creating the new regulator, Ofgem,[208] Utilities Act 2000 made the distribution of electricity a separate licensable activity from transmission,[209] abolished the concept of public electricity suppliers,[210] reformulated the relevant duties of the Secretary of State for Trade and Industry[211] and introduced standard conditions for the grant of licences in the industry.

---

for approval of height of chimneys for furnaces). See the discussion of the Clean Air Act 1993 in Wolf and Stanley, *op. cit.*, pp. 335–42.

200 See para. 4.2.1.2(1) above.
201 See Department of Environment, Transport and the Regions, *Air Quality Strategy for England, Wales, Scotland and Northern Ireland*, 2000 (Cm 4548, 2000).
202 See Benjamin J. Richardson and Kiri L. Chanwai, 'Taxing and trading in corporate energy activities: pioneering UK reforms to address climate change', (2003) 14(1) ICCLR 18–27.
203 See para. 1.4.2.2 above.
204 See Chapter 14 below (climate change levy); Chapter 20 (the UK ETS); Chapter 28 (the EU ETS); and paras 6.4.3.1(2)(a) above and 21.5 below (the RO).
205 The use of this term is common but, properly speaking, supply is only one of four elements in the sector (see para. 2.4.1 above).
206 *Ibid.* See also Tudway, *op. cit.*, Ch. 25.
207 Unless otherwise indicated, all references to the Electricity Act 1989 are to the Act as amended by the 2000 Act. See Roggenkamp *et al.*, *op. cit.*, paras 13.177–13.240.
208 See para. 4.2.1.3 above.
209 Both terms are defined in Electricity Act 1989, s.4(4). See also para. 2.4.1 above.
210 Defined in Electricity Act 1989, s.6(9), as originally enacted, as '… any person who is authorised by a licence … to supply electricity … to premises in his authorised area'.
211 See para. 4.2.1.2(1) above.

The Electricity Act 1989 is divided into three parts: Part I creates, under the supervision of Ofgem, a regulatory structure for the electricity supply industry. By s.4(1), the generation, transmission, distribution or supply of electricity is unlawful unless the activity in question is authorised by an appropriate licence.[212] This general rule is, however, subject to the power given by s.5 to the Secretary of State for Trade and Industry[213] to create exemptions from the rule by secondary legislation. Part II of the Act deals with the industry's structure and the privatisation of the parts which make it up; and Part III provides for measures to ensure the security of electricity installations, for dealing with civil emergencies and for making government loans in relation to matters of public security (for example, nuclear reprocessing). In Part I, ss.32–3, as amended, contain measures relating to the promotion of renewable sources of energy.[214]

Further changes to the regulation of the electricity industry are authorised by the Energy Act 2004, passed on 22 July 2004.[215] These will be in three main areas: the creation of a new body to ensure the decommissioning and clean-up of public sector civil nuclear sites; the promotion of renewable energy sources; and the creation of a single wholesale electricity market for Great Britain, to be called 'BETTA', which, although it will replace NETA,[216] will be based upon the latter, which itself replaced the Electricity Pool[217] as recently as 2001.

The Electricity Act 1989, as amended by Utilities Act 2000, applies both to England and Wales and to Scotland.[218] In Northern Ireland, the relevant legislation is contained in the Electricity (Northern Ireland) Order 1992,[219] as amended by the Energy (Northern Ireland) Order 2003.[220] The New Electricity Trading Arrangements did not apply to Scotland or Northern Ireland. The introduction of BETTA under the Energy Act 2004 will, as mentioned above, extend to Scotland but not to Northern Ireland.

(2) Renewable energy sources

(a) Renewables Obligation ('RO')
Section 32, Electricity Act 1989, as amended by Utilities Act 2000, empowers the Secretary of State to impose by order the so-called 'Renewables Obligation' ('the RO') on designated electricity suppliers.[221] The RO is considered in more detail

---

212   The licences held by particular companies in the industry are usefully listed in Utility Week, *The Electricity Supply Handbook 2004*, 57th edn (Sutton: Reed Business Information, 2004).

213   See para. 4.2.1.2(1) above.

214   See para. 6.4.3.1(2) below.

215   There had been thought to be some danger of the Bill being lost (see Taylor, *Financial Times*, 1 April 2004, p. 4).

216   See para. 2.4.1 above.

217   As described by Tudway, *op. cit.*, '[t]he Pool mechanism was a contractual arrangement between parties to the Pooling and Settlement Agreement, which set wholesale buying and selling prices for electricity' (*ibid.*, para. 25–49).

218   Electricity Act 1989, s.113(3).

219   S.I. 1992 No. 231 (N.I. 1).

220   S.I. 2003 No. 419 (N.I. 6).

221   Electricity Act 1989, s.32(1).

later in the study[222] but, since the nature of the mechanisms introduced by it fall for consideration in the next chapter,[223] it will be useful to include here a brief discussion of the primary legislation which creates the legislative framework for the secondary legislation to be discussed later. By s.32(3), the RO is that:

> … the designated electricity supplier must, before a specified day (or before each of several specified days, or before a day specified in each year), produce to [the Gas and Electricity Markets Authority][224] evidence of a specified kind showing –
> (a) that it has supplied to customers in Great Britain during a specified period such amount of electricity generated by using renewable sources as is specified in relation to such a supplier; or
> (b) that another electricity supplier has done so (or that two or more others have done so); or
> (c) that, between them, they have done so.

For the purposes of s.32, renewable sources are defined as sources of energy other than nuclear fuel or fossil fuel but including waste of which 'not more than a specified proportion is waste which is, or is derived from, fossil fuel'.[225]

Section 32B(1) stipulates that an order under s.32 may provide for Ofgem[226] to issue from time to time a so-called 'Green Certificate'[227] to the operator of a generating station or to an electricity supplier. The production by the supplier of a Green Certificate (or 'ROC') to Ofgem is 'sufficient evidence' of the facts certified for the purposes of s.32(3) above.[228] A Green Certificate certifies two matters:

> (a) that the generating station or, in the case of a certificate issued to an electricity supplier, a generating station specified in the certificate, has generated from renewable sources the amount of electricity stated in the certificate; and
> (b) that it has been supplied to customers in Great Britain (or the part of Great Britain stated in the certificate).

In practice, it has been inferred from s.32(3)(b) above that, once issued, Green Certificates can be traded between suppliers, since that subsection seems to assume that a supplier can comply with the RO by producing Green Certificates originally issued to another supplier.[229]

---

222 See para. 21.5 below.
223 See para. 7.3.1 below.
224 See para. 4.2.1.3 above.
225 Electricity Act 1989, s.32(8).
226 The section actually refers to 'the Authority', that is, to the Gas and Electricity Markets Authority ('GEMA'), whose functions are exercised through Ofgem. See para. 4.2.1.3 above.
227 This is the term used in the heading to s.32B but they are also known as Renewables Obligation Certificates ('ROCs'). That section has been amended by Energy Act 2004, s.116.
228 Electricity Act 1989, s.32B(3).
229 See para. 21.5 below.

Section 32C states that an order under s.32 may provide that, instead of producing a Green Certificate/Green Certificates, an electricity supplier may meet its RO, whether in whole or in part, by making a payment to Ofgem.[230] This power is then backed up by provisions on fixing the amount of the payment,[231] as well as a rule on the application of the revenue generated under s.32C. The rule as to the application of the revenue thereby received is contained in s.32(3), which provides that Ofgem 'must pay the amounts received to electricity suppliers in accordance with a system of allocation specified in the order'. The order in due course made by the Secretary of State under Electricity Act, ss.32–2C, is the Renewables Obligation Order 2002 ('ROO 2002'),[232] which came into force on 1 April 2002.[233]

Section 32A(1)(b) states that an order under s.32 may make general provisions in relation to the RO imposed by the order and, in particular, may specify that only electricity generated using certain types of renewable source is to count towards discharging the RO. In ROO 2002, this leads to the creation of the concept of 'eligible renewable sources',[234] which is made to exclude most hydro generating stations commissioned before April 2002.[235] The reasons for this exclusion, as well as the other main provisions of ROO 2002 are discussed in detail in a subsequent chapter.[236]

The provisions just described apply in England and Wales. In Scotland, the powers conferred by Electricity Act 1989, ss.32–2C, are exercisable by the Scottish Ministers. The Renewables Obligation (Scotland) Order 2002[237] was made in exercise of these powers and came into force on 1 April 2002. In Northern Ireland, the Department of Enterprise, Trade and Investment was given power to impose an RO by the Energy (Northern Ireland) Order 2003.[238] Like the system for England and Wales, each of the Scotland and Northern Ireland systems envisage the trading of Green Certificates (in Scotland called 'SROCs' and in Northern Ireland called 'NIROCs').

The combination of the RO and Green Certificates together replace the previous combination of Non-Fossil Fuel Orders ('NFFOs') and fossil fuel levy ('FFL').[239] Since some further reference will be made to the nature and purpose both of NFFOs and of FFL,[240] it may be useful briefly to describe the earlier instruments here, notwithstanding their subsequent abolition.[241]

---

[230]　Electricity Act 1989, s.32C(1), replaced in part by Energy Act 2004, s.115(2).

[231]　*Ibid.*, s.32C(2), replaced in part by Energy Act 2004, s.115(3).

[232]　S.I. 2002, No. 914 (as amended by the Renewables Obligation (Amendment) Order 2004, S.I. 2004 No. 924).

[233]　S.I. 2002, No. 914, Art. 1(1).

[234]　*Ibid.*, Art. 8(1).

[235]　*Ibid.*, Art. 8(2).

[236]　See para. 21.5 below. See also para. 12.2.6.3(2) below.

[237]　S.S.I. 2002 No. 163.

[238]　S.I. 2003 No. 419 (N.I. 6).

[239]　See Utilities Act 2000, s.66, and Sched. 8.

[240]　See paras 21.4.4n and 21.6.1 below.

[241]　The writers are indebted for this explanation to the valuable contemporary annotations to the Electricity Act 1989 by Patrick McAuslan and John F. McEldowney in [1989] 2 CLSA, pp. 29–52–29–54.

Section 32, Electricity Act 1989, as originally enacted, empowered the Secretary of State to order public electricity suppliers[242] – that is, by the making of a NFFO – to ensure that specified proportions of their electricity supplies were acquired from non-fossil fuel generators. Such generators could include, under the original provisions, not only water, wind and solar generators,[243] but also nuclear ones.[244] Between 1990 and 1998, five NFFOs were made,[245] each one being followed by a tendering process for electricity generation contracts in each category specified in the relevant NFFO.[246] Only the earliest of the NFFOs (the so-called 'NFFO 1')[247] included nuclear generators within the categories of non-fossil fuel generators. The shift is perhaps explained by a greater emphasis originally being placed upon diversity (that is, security) of supply rather than upon environmental concerns as such.[248]

Fossil fuel levy was imposed under Electricity Act 1989 to subsidise the additional costs on suppliers which NFFOs created, since it was foreseen that non-fossil fuel-generated electricity would tend to be more expensive than electricity generated at fossil-fuel-powered stations. The now-repealed s.33, Electricity Act 1989, accordingly empowered the Secretary of State, by regulations, to impose FFL both on suppliers who were subject to an NFFO and also on other persons who were licensed to supply electricity within the authorised areas[249] of such suppliers.[250] Although FFL, which was payable each month to the Director General of Electricity Supply,[251] was charged on suppliers, the levy was actually borne by electricity consumers, since it

---

[242] A term which is now superseded (see para. 6.4.3.1(1)n above).

[243] See Electricity Act 1989, s.32(9) (as originally enacted).

[244] This may be gathered from the wording of Electricity Act 1989, ss.32(8) and 32(9) (as originally enacted).

[245] That is, Electricity (Non-Fossil Fuel Sources) (England and Wales) Order 1990, S.I. 1990 No. 263 (as amended by S.I. 1990 No. 494) and Electricity (Non-Fossil Fuel Sources) (England and Wales) (No. 2) Order 1990, S.I. 1990 No. 1859 (together referred to as 'NFFO 1'); Electricity (Non-Fossil Fuel Sources) (England and Wales) Order 1991, S.I. 1991 No. 2490 ('NFFO 2'); Electricity (Non-Fossil Fuel Sources) (England and Wales) Order 1994, S.I. 1994 No. 3259 ('NFFO 3'); Electricity (Non-Fossil Fuel Sources) (England and Wales) Order 1997, S.I. 1997 No. 248 ('NFFO 4'); and Electricity (Non-Fossil Fuel Sources) (England and Wales) Order 1998, S.I. 1998 No. 2353 ('NFFO 5'). Contracts under NFFOs 1 and 2 have now terminated; those under NFFOs 3–5 will continue, however, the last contract being due to terminate in 2018. Contracts under NFFOs 3–5 are in fact replacement contracts, the originals having terminated when NETA commenced (see Tudway, *op. cit.*, para. 25–54).

[246] See, for example, 284 ENDS Report (1998) for the schemes which were awarded contracts under NFFO 5.

[247] That is, Electricity (Non-Fossil Fuel Sources) (England and Wales) Order 1990, S.I. 1990 No. 263 (as amended) and Electricity (Non-Fossil Fuel Sources) (England and Wales) (No. 2) Order 1990, S.I. 1990 No. 1859.

[248] See McAuslan and McEldowney, *op. cit.*, pp. 29–50–29–51 ('General Note').

[249] A superseded concept. See Electricity Act 1989, s.6(9) (as originally enacted).

[250] See Electricity Act 1989, s.33(1)(a) (as originally enacted). See also the Fossil Fuel Levy Act 1998.

[251] An office which is now abolished (see Utilities Act 2000, s.1(3), and para. 4.2.1.3 above).

was incorporated in their electricity bills. After deduction of administrative expenses, the revenue from the levy was distributed each month to public electricity suppliers in accordance with the regulations made under s.33, Electricity Act 1989.[252] Fossil fuel levy therefore operated as a mechanism for supporting non-fossil fuel-generated electricity.[253]

In parallel with the arrangements for England and Wales, three NFFOs were made for Scotland,[254] prior to their replacement, with appropriate savings, by ROO 2002. Confusingly, the three orders were together known as the 'Scottish Renewables Obligation' (or 'SRO').[255] As in the UK, FFL had been introduced to compensate the two companies licensed to generate, transmit, distribute and supply electricity in the post-privatisation structure.[256] Section 66, Utilities Act 2000, which had abolished FFL in England and Wales, did the same for Scotland.[257] In Northern Ireland, prior to March 1, 2004,[258] Northern Ireland Electricity plc was obliged to secure quantities of generation capacity from renewable energy sources under Art. 35 of the Electricity (Northern Ireland) Order 1992.[259] The concomitant Northern Ireland FFL[260] was abolished by Art. 58 of the Energy (Northern Ireland) Order 2003.

(b) Renewable energy guarantees of origin ('REGOs')
The Electricity (Guarantees of Origin of Electricity Produced from Renewable Energy Sources) Regulations 2003 are designed to transpose into the law of England and Wales and of Scotland[261] the provisions of Art. 5 of the Renewables Directive.[262] Article 5 imposes an obligation on Member States to ensure that the origin of electricity produced from renewable energy sources can be guaranteed as such by creating systems for granting guarantees of origin.

---

[252]   See the Fossil Fuel Levy Regulations 1990, S.I. 1990 No. 266, reg. 28.
[253]   Note that Utilities Act 2000, s.67, enables provision to be made for FFL to be continued in relation to the outstanding contracts entered into under NFFOs (see the Fossil Fuel Levy (Amendment) Regulations 2001, S.I. 2001 No. 1200 and other post-2000 S.I.s dealing with FFL).
[254]   That is, the Electricity (Non-Fossil Fuel Sources) (Scotland) Order 1994, S.I. 1994 No. 3275 (S. 190); Electricity (Non-Fossil Fuel Sources) (Scotland) Order 1997, S.I. 1997 No. 799 (S. 76); and Electricity (Non-Fossil Fuel Sources) (Scotland) Order 1999, S.I. 1999 No. 439 (S. 24).
[255]   See Tudway, *op. cit.*, para. 25–08.
[256]   That is, Scottish Power plc and Scottish and Southern Electricity plc.
[257]   Utilities Act 2000 applying, as it does, to England and Wales and Scotland but not to Northern Ireland.
[258]   See the Energy (2003 Order) (Commencement No. 2) Order (Northern Ireland) 2004, S.R. 2004 No. 71 (C. 1), Art. 2 and Sched.
[259]   S.I. 1992 No. 231 (N.I. 1). The 2003 Electricity Order does, however, have saving powers for existing contracts entered into under the pre-existing powers.
[260]   Which had been introduced by Art. 36, S.I. 1992 No. 231.
[261]   See S.I. 2003 No. 2562, reg. 1(2). The equivalent provision for Northern Ireland is the Electricity (Guarantees of Origin of Electricity Produced from Renewable Energy Sources) Regulations (Northern Ireland) 2003, S.R. 2003 No. 470.
[262]   That is, European Parliament and Council Directive 2001/77/EC (see para. 12.2.6.3 below).

The key provision of the Regulations is to be found in reg. 3, which provides for guarantees of origin, 'certifying that the electricity in respect of which the certificate is issued was electricity produced from renewable energy sources',[263] to be issued by Ofgem.[264] Each renewable energy guarantee of origin ('REGO') is issued in respect of one kilowatt hour of electricity produced from renewable energy sources.[265] The general rule, contained in reg. 4(1), is that the only person who may request Ofgem to issue a REGO is the producer of the electricity.[266] However, where electricity has been purchased under an arrangement originally made under an NFFO,[267] the only person who may request the issue of a REGO is generally the relevant NFFO purchaser.[268] Moreover, in order to issue a REGO under reg. 4, Ofgem must be satisfied that the electricity covered by the request was produced from renewable resources.[269] Renewable Energy Guarantees of Origin may be transferred to persons other than the original maker of the request by amendment to an electronic register of REGOs to be maintained by Ofgem.[270] In circumstances such that access to plants is denied to, or information is withheld from, the Authority, then it may refuse to issue REGOs[271] and it may also revoke them in certain circumstances.[272] Finally, reg. 9 contains provisions relating to the recognition and non-recognition of REGOs issued in Northern Ireland and other EU Member States. This reflects the stated aim of the Renewables Directive as being to promote an increase in the contribution of renewable energy sources to the production of electricity in the internal market.[273]

No amendment is made by the Regulations either to ROO 2002[274] or to the Climate Change Levy (General) Regulations 2001.[275] It would therefore appear that REGOs neither provide a means of complying with the RO[276] nor do they provide entitlement to exemption from climate change levy.[277]

---

[263] S.I. 2003 No. 2562, reg. 2(1).

[264] See para. 4.2.1.3 above.

[265] S.I. 2003 No. 2562, reg. 6(3). Renewable energy sources are defined as '... renewable non-fossil energy sources, that is, wind, solar, geothermal, wave, tidal, hydropower, biomass, landfill gas, sewage treatment plant gas and biogases' (see S.I. 2003 No. 2562, reg. 2(1)).

[266] S.I. 2003 No. 2562, reg. 4(1)(b).

[267] See para. 6.4.3.1(2)(a) above.

[268] S.I. 2003 No. 2562, reg. 4(1)(a). In Scotland, the reference is to the relevant SRO purchaser (see para. 6.4.3.1(2)(a) above).

[269] See S.I. 2003 No 2562, reg. 6(1).

[270] *Ibid.*, reg. 7 and Sched. 2.

[271] *Ibid.*, reg. 5.

[272] *Ibid.*, reg. 8.

[273] See para. 12.2.6.3(2) below.

[274] See para. 6.4.3.1(2)(a) above.

[275] S.I. 2001 No. 838. See paras 14.1–14.3 below.

[276] See para. 6.4.3.1(2)(a) above.

[277] See para. 14.4 below.

(3) Other aspects of electricity regulation

In addition to the regulatory framework discussed above, the UK's electricity supply industry is subject to at least four other regulatory regimes: competition law, planning law, IPC[278] and IPPC.[279]

Competition in the electricity supply industry is regulated both by general UK competition law and by the specific licensing provisions applicable to the electricity supply industry discussed above.[280] The general provisions are exercisable in relation to the industry concurrently, but subject to conditions,[281] by the Director General of Fair Trading and by Ofgem.[282] These general provisions include the power under the Fair Trading Act 1973, Part III, to take proceedings in relation to courses of conduct detrimental to consumers;[283] the power to refer monopoly situations to the Competition Commission;[284] and the power to take action in relation to courses of conduct restricting, distorting or preventing competition.[285]

So far as planning law is concerned,[286] Electricity Act 1989, ss.36 and 37 impose special conditions on the construction of certain power stations and overhead lines.[287] Section 36 currently provides that a generating station with a capacity greater than 50 MW must not be constructed, extended or operated, except in accordance with a consent granted by the Secretary of State for Trade and Industry.[288] Section 37 likewise provides that, subject to certain *de minimis* exceptions, the consent of the Secretary of State is likewise required for an overhead line to be installed or kept installed. Although the power under s.36 has been used to impose the moratorium on the construction of gas-fired power stations,[289] both s.36 and s.37 are coming under increasing scrutiny as presenting obstacles to the development of renewables generators, specifically wind farms. Under Electricity Act 1989, Sched. 8, notice of applications for consent under ss.36 and 37 must be served on the local planning authority. If the planning authority objects, then a public inquiry

---

278 See para. 6.2.2 above.
279 See para. 6.2.3 above.
280 See para. 6.4.3.1(1) above.
281 See Electricity Act 1989, ss.43(4)–43(6A).
282 *Ibid.*, s.43.
283 *Ibid.*, s.43(1).
284 *Ibid.*, s.43(2).
285 *Ibid.*, s.43(3).
286 These issues are well discussed in John Grady, 'Climate Change and Great Britain's Electricity Generation Policy', [2003] IELTR 105–15, p. 114. See also Jonathan Evans, 'The Implications of the United Kingdom's Long-Term Energy Policy for its Renewables Market', [2003] IELTR 233–8. It should be noted also that, on 13 May 2004, the Planning and Compulsory Purchase Act 2004, with the avowed purpose of 'speeding up the planning system', received the Royal Assent, although it is to be brought into effect in ongoing stages.
287 Other provisions relate to the fuelling of power stations, compulsory acquisition of land and street works (see the outline in Tudway, *op. cit.*, paras 25–75–25–77). There are also some relatively little-noted provisions in Electricity Act 1989, s.38 and Sched. 9 on the preservation of amenity and fisheries.
288 General planning law applies to generating stations with a capacity of less than 50 MW.
289 See Grady, *op. cit.*, p. 114. See also para. 21.4.3 below.

must be held.[290] In relation to objections from persons other than local planning authorities, the Secretary of State may order the holding of a public inquiry if he thinks it appropriate to do so.[291] Objections have been raised by the Ministry of Defence, alleging the impact of wind turbines on radar signals and dangers from the turbines to low-flying aircraft,[292] and by the Royal Society for the Protection of Birds ('the RSPB'),[293] given that the proposed locations of the wind farms are nearly all in areas of international importance for birds.[294] Wind farms also give rise to concerns over s.37, given that such installations are generally located in remote areas and that the cost of installing overhead wires is a fraction of that of laying wires underground.

More generally, the planning of overhead electricity cables as well as of nuclear power stations, as well as nuclear fuel installations, is subject to the EIA regime discussed above.[295] There have traditionally been few statutory provisions specifically relating to the decommissioning of nuclear power stations, the matter usually having been dealt with in site licences granted under the Nuclear Installations Act 1965. The provisions of Part 1 of the Energy Act 2004 creates a new statutory decommissioning framework. This is one of a number of reforms to the sector to which it is necessary to refer next.

Finally, it should, of course be noted that electricity generation is a combustion activity within the scope of the IPPC regime.[296] Likewise, combustion processes are caught by IPC.[297]

(4) Regulatory reform in the electricity sector
There are, at the time of writing, two sets of further reforms at various stages of development: those in the recently passed Energy Act 2004 and, insofar as they are not dealt with in that Act, certain other measures that are designed to transpose into UK law the provisions of the Electricity Acceleration Directive (the 'EAD').[298] It is

---

[290] See Electricity Act 1989, Sched. 8, para. 2(2).

[291] *Ibid.*, Sched. 8, para. 3(2).

[292] See, for example, Taylor, *Financial Times*, 1 March 2004, p. 4. But see the case for the Ministry of Defence presented in the letter from Williams, *Financial Times*, 5 March 2004, p. 18.

[293] See www.rspb.org.uk.

[294] See 350 ENDS Report (2004) 10–11. The ENDS correspondent reports that the RSPB objected to 27 wind farms between 1998 and 2003. It is also reported that the RSPB has confirmed its opposition to the 1,000MW London Array, proposed by Shell and Powergen, on the basis that the Thames Estuary 'supports large numbers of wintering red-throated divers' (*ibid.*).

[295] See para. 6.2.4 above.

[296] See Pollution Prevention and Control (England and Wales) Regulations 2000, S.I. 2000 No. 1973, Sched. 1, Section 1.1. See para. 6.2.3 above.

[297] See the Environmental Protection (Prescribed Processes and Substances) Regulations 1991, S.I. 1991 No. 472, Sched. 1, Ch. 1.

[298] European Parliament and Council Directive 2003/54/EC, (2003) OJ L176 37. The provisions of the EAD are discussed in detail in a subsequent chapter (see paras 12.2.6.3(1) and 12.2.6.3(2) below).

the DTI's intention to implement the latter by a mixture of statutory instrument under European Communities Act 1972, s.2(2) and other administrative action.[299]

The Energy Act 2004 affects the electricity sector in three main areas.[300] First, it provides for the establishment of the Nuclear Decommissioning Authority ('NDA'), for the creation of a new Civil Nuclear Police Authority and for the authorisation of government expenditure in relation to the restructuring of British Energy.[301] Next, the Act provides for a number of matters in relation to renewables generation. These provisions include measures relating to the development of offshore renewable energy sources in accordance with the rights conferred on the UK under the 1982 UN Convention on the Law of the Sea ('the UNCLOS')[302] as well as provisions for the mutual recognition of NIROCs.[303]

Finally, the Energy Act 2004 establishes the basis for the creation of a single wholesale electricity market for the whole of Great Britain (that is, England and Wales and Scotland but not Northern Ireland), to be called the British Electricity Trading and Transmission Arrangements ('BETTA'), based on NETA;[304] for a new licensing system applicable to electricity interconnectors, to bring them under the authority of Ofgem; and for the Secretary of State to have the power to appoint additional inspectors in relation to inquiries under Electricity Act 1989, ss.36 and 37.[305] The new licensing system reflects the fact that there has hitherto been no relevant legislation in Great Britain on the regulation and exemption of interconnectors.

The elements of the EAD which are not covered by the Energy Act 2004 provisions are, as mentioned above, to be implemented by a combination of statutory instrument and administrative action. They relate to fuel disclosure,[306] dispute settlement and reporting requirements.[307]

6.4.3.2 Gas

The counterpart of the Electricity Act 1989 in the context of the gas industry is the Gas Act 1986.[308] Like the 1989 Act, the 1986 Act has, however, been extensively

---

[299] For a detailed review of the changes, see Department of Trade and Industry, *Consultation: Implementation of EU Directive 2003/54 Concerning Common Rules for the Internal Market in Electricity*, February 2, 2004 (available from www.dti.gov.uk/energy).
[300] Within these broad heads are some interesting sub-areas, for example, a special insolvency regime for energy licensees in Chapter 3 of Part 3, following the coal power station insolvency debacle referred to in para. 21.4.2 below.
[301] See Energy Act 2004, Part 1. For the background, see paras 2.4.1 above and 21.4.3 below.
[302] See para. 8.3.2 below.
[303] Energy Act 2004, Part 2. For NIROCs, see para. 6.4.3.1(2)(a) above.
[304] The provisions relating to BETTA originally appeared, in January 2003, in a draft Electricity (Trading and Transmission) Bill.
[305] Energy Act 2004, Part 3.
[306] That is, a requirement for electricity suppliers to specify their fuel mix to customers (see para. 12.2.6.3(1) below).
[307] See Department of Trade and Industry, *op. cit.*, para. 1.8, where the changes are summarised.
[308] See Roggenkamp *et al.*, *op. cit.*, paras 13.114–13.166.

amended by subsequent legislation (in this case, by the Gas Act 1995, as well as by the Utilities Act 2000). However, the 1986 Act did not consolidate the whole of the relevant legislation and parts of the Gas Act 1965, as well as the Energy Act 1976 remain in force.[309]

The gas industry's regulator has, since the Utilities Act 2000, been Ofgem,[310] the 2000 Act having brought the two industries into line with each other in this respect. It should be noted, however, that, generally speaking, the liberalisation of the gas industry in Great Britain has preceded that of electricity supply. Although the 2000 Act substituted the new regulator and redefined the duties of the Secretary of State,[311] many of the most important liberalisation measures had already been taken in the Gas Act 1995.

The Gas Act 1986, as amended, is divided into three Parts, only the first of which is relevant in the present context. This Part contains elements of the regulatory structure for the industry, which is placed under Ofgem's supervision. By s.5, the transportation,[312] supply[313] or shipment[314] of gas is unlawful without the appropriate licence. However, the storage of gas is still regulated by the Gas Act 1965, whilst exploration for gas and gas production are subject to their own licensing regime.[315] The requirement for a licence for each of transportation, supply and shipment is, as for electricity, subject to the Secretary of State's power to grant exemptions under s.6A, Gas Act 1986. Schedule 2B, which is incorporated by s.8B, Gas Act 1986, contains the Gas Code, which relates to the rights and obligations of licence holders and consumers and related matters.

The DTI has identified a number of areas in which legislative action is required in order for UK law to comply with the provisions of the Gas Acceleration Directive (the 'GAD').[316] These include the need for a new licensing system for gas interconnectors, which appears in the Energy Act 2004, as well as a regulated TPA system and exemptions for LNG import terminals and storage facilities, to be effected by modifying the 1986 Act and the Petroleum Act 1998.

The legislation discussed above applies both to England and Wales and to Scotland.[317] The gas industry in Northern Ireland is regulated by the provisions of the Gas (Northern Ireland) Order 1996[318] and the Energy (Northern Ireland) Order 2003.[319]

---

309 Unlike gas levy, which was provided for by the Gas Levy Act 1981 and abolished by Finance Act 1998, s.153(2), for 1998–1999 onwards. The levy financed certain aspects of the denationalisation arrangements introduced by the Gas Act 1986 (see Tudway, *op. cit.*, para. 23–43).
310 Utilities Act 2000, s.1.
311 Gas Act 1986, s.4AA.
312 *Ibid.*, s.5(1)(a).
313 *Ibid.*, s.5(1)(b).
314 *Ibid.*, s.5(1)(c).
315 See, for example, Tudway, *op. cit.*, Ch. 23.
316 European Parliament and Council Directive 2003/55/EC, (2003) OJ L176 57. See paras 12.2.6.3(1) and 12.2.6.3(2) below.
317 Gas Act 1995, s.18(4).
318 S.I. 1996 No. 275 (N.I. 2).
319 S.I. 2003 No. 419 (N.I. 6).

Very similar points apply in relation to competition law issues as obtain in relation to electricity supply.[320] Thus, for instance, s.36A, Gas Act 1986 tracks very closely s.43 of the Electricity Act 1989 and, likewise, s.41E of the 1986 Act (which deals with references to the Competition Commission) tracks the wording of s.56C of the Electricity Act 1989.

The planning of installations, whether on- or offshore for the extraction of natural gas, of installations for its storage and of gas pipelines is subject to the EIA regime discussed above.[321] Gasification and associated processes are also, of course, subject to the IPC[322] and IPPC regimes.[323] Both off- and onshore oil and gas exploration and production are subject to the environmental controls outlined in the next para. below. The decommissioning of offshore oil and gas installations is covered by Part I of the Petroleum Act 1987.

6.4.3.3   Oil

Most of the regulatory framework for the oil industry is contained in the Petroleum Act 1998 and in delegated legislation made thereunder by the Secretary of State for Trade and Industry.[324]

However, clean air legislation confers powers on the Secretary of State for Transport[325] to regulate the composition and content of motor fuels[326] and health and safety at work legislation confers powers on the same Minister in relation to the transport and storage of petroleum.[327]

The five Parts of the Petroleum Act 1998 include provisions relating to the ownership of petroleum[328] reserves, the grant of licences to search, bore for and get petroleum,[329] the construction and use of submarine pipelines[330] and the abandonment of offshore installations.[331] Section 2 of the Act provides that:

---

[320]   See para. 6.4.3.2(3) above.
[321]   See para. 6.2.4 above.
[322]   See the Environmental Protection (Prescribed Processes and Substances) Regulations 1991, S.I. 1991 No. 472, Sched. 1, Ch. 1.
[323]   See the Pollution Prevention and Control (England and Wales) Regulations 2000, Sched. 1, Section 1.2. See para. 6.2.3 above.
[324]   See para. 4.2.1.2(1) above. See Roggenkamp *et al.*, *op. cit.*, paras 13.25–13.94. The discussion in the text is by reference to offshore (seaward) oil and gas only. For a useful discussion of the regulation of the onshore (landward) industry, see Hughes, *op. cit.*, pp. 402–4.
[325]   See para. 4.2.1.2(1) above.
[326]   See Clean Air Act 1993, ss.30(1), 30(3), 32(1) and 63(1).
[327]   See Health and Safety at Work Act 1974, ss.15, 43, 82 and Sched. 3.
[328]   The definition of petroleum in Petroleum Act 1998, s.1, emphasises the artificiality of dealing with it separately from natural gas (see para. 2.4.2 above). So defined, it 'includes any mineral oil or relative hydrocarbon and natural gas existing in its natural condition in strata; but does not include coal or bituminous shales or other stratified deposits from which oil can be extracted by destructive distillation'.
[329]   Petroleum Act 1998, Part I.
[330]   *Ibid.*, Part III.
[331]   *Ibid.*, Part IV.

[H]er Majesty has the exclusive right of searching and boring for and getting petroleum ... (including petroleum in Crown Land)[332] which for the time being exists in its natural condition in strata in Great Britain or beneath the territorial sea adjacent to the United Kingdom.

Subject to petroleum rights being vested in Her Majesty the Queen in this way, the Secretary of State for Trade and Industry may grant licences 'to such persons as he thinks fit' to 'search and bore for and get [the] petroleum' covered by the Act. Section 4 enjoins the Secretary of State to make regulations relating to applications for licences, including the model clauses to be incorporated in such licences.[333] The construction and use of submarine pipelines is similarly restricted, s.14 requiring that any company[334] wishing to construct or use a pipeline[335] in, under or over the UK's territorial sea or continental shelf[336] may do so only with the written authorisation of the Secretary of State.

Oil and gas exploration and production is subject to its own environmental control regime. Besides oil pollution offences such as that of discharging oil or an oily mixture into UK territorial waters (which can be committed wherever the pipeline in question is located),[337] environmental protection provisions are incorporated in the exploration and development licences granted under s.4.[338]

The planning of oil refineries, of installations for the extraction of petroleum and of oil pipelines is subject to the EIA regime discussed above.[339] Refining mineral oils and its associated activities are subject to the IPC[340] and IPPC regimes.[341]

The powers delegated as above to the Secretary of State for Transport have recently been exercised in order to ensure UK law's compliance with standards contained in Community law.[342] Thus the lead and sulphur content of fuels, as well as their transport and storage, is regulated by a number of statutory instruments,[343] each of which transpose Directives in these areas.

To these powers, Energy Act 2004, s.124 has added an enabling power for the Secretary of State to impose renewable transport fuel obligations on particular

---

332   Defined in Petroleum Act 1998, s.2(3).
333   Petroleum Act 1998, s.4(1)(e). See the Petroleum Licensing (Exploration and Production) (Seaward and Landward Areas) Regulations 2004, S.I. 2004 No. 352.
334   Authorisations may not be issued to individuals! (See Petroleum Act 1998, s.15(2).)
335   Defined in *ibid.*, s.26.
336   See Continental Shelf Act 1964, s.1(7).
337   See Prevention of Oil Pollution Act 1971, s.2.
338   See the Petroleum Licensing (Exploration and Production) (Seaward and Landward Areas) Regulations 2004, S.I. 2004 No. 352.
339   See para. 6.2.4 above.
340   See the Environmental Protection (Prescribed Processes and Substances) Regulations 1991, S.I. 1991 No. 472, Sched. 1, Ch. 1.
341   See Pollution Prevention and Control (England and Wales) Regulations 2000, S.I. 2000 No. 1973, Sched. 1, Section 1.2. See para. 6.2.3 above.
342   See para. 12.2.6.3(5) below.
343   See the Motor Fuel (Composition and Content) Regulations 1999, S.I. 1999 No. 3107; the Motor Fuel (Composition and Content) (Amendment) Regulations 2003, S.I. 2003 No. 3078; and the Carriage of Dangerous Goods by Road Regulations 1996, S.I. 1996 No. 2095.

types of transport fuel supplier. Unusually, such Orders have to be approved by an affirmative vote in the UK Parliament.

6.4.3.4   Coal

Most of the legislation regulating the coal industry dates from the denationalisation of the industry in the mid-1990s. The main statute is the Coal Industry Act 1994, which sets out a basic regulatory framework, although this should be read in the context of the general law of town and country planning.[344]

The 1994 Act establishes the Coal Authority as the industry's regulatory body,[345] provides for the licensing of coal-mining operations,[346] sets out the rights and obligations of coal mine operators[347] and contains various supplementary provisions relating to matters such as access to information.[348] The central idea in the 1994 Act is that, the interests of the British Coal Corporation[349] in unworked coal mines having vested in the Coal Authority,[350] the latter has the power to grant coal-mining licences to persons wishing to conduct mining operations on the land to which the application relates.[351]

In addition to the town and country planning law referred to above, the planning of both underground and open cast coal mines, of fossil fuel storage facilities and coke ovens are among the items subject to the EIA regime discussed above.[352] Finally, coal production falls within the scope both of the IPC/LAAPC[353] and of the IPPC regime.[354]

The abandonment of coal mines, with the potential for land contamination[355] that may result, is subject to the notification and other procedures in the Mines (Notice of Abandonment) Regulations 1998.[356]

---

[344]   Currently (December 2004) the subject of the massive reform referred to at para. 6.4.3.1(3) above. The impact of planning policy on coal mining is usefully discussed in Hughes, *op. cit.*, pp. 399–402. Town and country planning is outside the scope of the present volume.
[345]   Coal Industry Act 1994, Part I. See para. 4.2.1.3 above.
[346]   *Ibid.*, Part II.
[347]   *Ibid.*, Part III.
[348]   *Ibid.*, Part IV.
[349]   Dissolved on 27 March 2004 (see Coal Industry Act 1994, s.23(2) and the Coal Industry Act 1994 (Commencement No. 7) and Dissolution of the British Coal Corporation Order 2004, S.I. 2004 No. 144 (C.6)).
[350]   Coal Industry Act 1994, s.7(3). The vesting date was 31 October 1994, pursuant to Coal Industry Act 1994, s.7(1).
[351]   Coal is thus an exception to the general rule of English property law that 'land' includes all mines and minerals beneath it (see *Coke on Littleton*, para. 4a).
[352]   See para. 6.2.4 above.
[353]   See the Environmental Protection (Prescribed Processes and Substances) Regulations 1991, S.I. 1991 No. 472, Sched. 1, Ch. 3. For LAAPC, see para. 6.2.2 above.
[354]   See the Pollution Prevention and Control (England and Wales) Regulations 2000, S.I. 2000 No. 1973, Sched. 1, Section 3.5. See para. 6.2.3 above.
[355]   See para. 6.7 below.
[356]   S.I. 1998 No. 892.

6.5 Air passenger and road freight transport regulation

6.5.1 The international and European Union background

Since the environmental issues raised by air transport and by road freight transport are largely (though not exclusively)[357] questions of air and atmospheric pollution, they have a significant transboundary aspect. Again, the measures discussed in this part of the chapter should therefore be understood within the public international and Community contexts discussed in Chapters 8 and 12 below.

6.5.2 Environmental regulation

Given the origins of the environmental aspects of transport policy in policy on air and atmospheric pollution, the ministerial predilection for economic instruments (both existing and, as yet, only imaginary) in the transport arena should occasion little surprise. It is perhaps curious, however, that neither the air transport nor the road freight transport sector is currently subject to any form of environmental levy.[358] That said, there continues to be a seemingly endless discussion of the policy options, each of which are reviewed elsewhere in this study.[359]

For the present, such economic instruments as do exist are embodied in excise duty differentials,[360] income tax provisions on company cars and fuel benefits[361] and in the provisions of the Transport Act 2000. This last was the result of the government's July 1998 White Paper, *A New Deal for Transport: Better for Everyone*.[362] The Act ranges over at least four distinct areas, the one with the most obvious environmental significance appearing in Part III.[363] The provisions of Part III give to local authorities[364] powers to implement two important economic instruments: workplace parking levies and road user charging schemes. Each of these is discussed in detail in Chapters 17 and 18 below and there is also a brief introductory outline of both concepts in the opening chapter above.[365]

---

357 When, for example, loss of amenity through road traffic congestion is considered.
358 See para. 2.6 above.
359 See paras 2.6 above and 27.3 below (nationwide road-user charging scheme for heavy lorries); paras 2.6 above and 27.5 below (possibility of allowing airport charges to have an emissions-related element); and para. 2.6 above and 27.5 below (possibility of bringing the aviation industry within the scope of the EU ETS).
360 See para. 22.2.1 below.
361 See para. 23.2 below.
362 Cm 3950, 1998.
363 This is, of course, an artificial statement, since the Transport Act 2000's provisions on local transport (Part II) and railways (Part IV) also have profound environmental implications. Part IV establishes the Strategic Rail Authority, sets out its objectives and functions and generally 'makes provision for the better regulation of the railway industry'. Part II requires local transport authorities to prepare and publish local transport plans for 'safe, integrated, efficient and economic transport facilities in their areas' (*Explanatory Notes*, paras 10 and 14).
364 See para. 4.2.3 above.
365 See paras 1.4.2.4(1) and (2) above.

It should be recognised that, in addition to the economic instruments – actual or projected – just referred to, specific command and control measures deal with particular environmental issues raised by transport. For example, legislation imposes requirements on airport operators as to the noise and vibration caused by aircraft[366] as well as on airline operators, with regard to engine emissions and noise certificates.[367] Equally, in addition to generally-applicable measures relating to vehicle construction and use, such as the control of emissions[368] and noise,[369] there are specific provisions relating to environmental matters in the licensing of goods vehicle operators.[370] More generally, there is extensive legislation relating to the acquisition and use of land for the construction of highways[371] as well as legislation enabling a highways authority to make agreements with adjoining landowners to mitigate the injurious effects of highways by tree-planting, etc.[372]

6.5.3   Transport regulation

The main statutes regulating the civil aviation industry are the Civil Aviation Act 1982 and the Airports Act 1986, in each case together with the relevant secondary legislation. Transport Act 2000, Part I, establishes a public-private partnership for the provision of air traffic services.[373]

The road haulage industry is regulated both by the Road Traffic Acts and by a special goods vehicle licensing regime.[374] The basic provision of the latter is that, subject to certain exceptions, an operator's licence is required in order for any person to use a goods vehicle on a road, for the carriage of goods for hire or reward, or for or in connection with any trade or business carried on by him.[375]

The Civil Aviation Authority ('the CAA'),[376] which may charge for the statutory functions that it discharges in relation to civil aviation,[377] has, since 1971, been responsible for the operational, economic and technological regulation of the civil air transport industry.

The operation of airports is subject to the CAA's licensing system,[378] as is the carrying of passengers or cargo for remuneration.[379] An airport operator may only

---

[366]   See Civil Aviation Act 1982, s.77 and the Air Navigation Order 2000, S.I. 2000 No. 1562, Art. 108. These provisions mean that the scope for private individuals to bring legal action over aircraft noise is severely limited.
[367]   Air Navigation (Environmental Standards) Order 2002, S.I. 2002 No. 798, Art. 8.
[368]   See Road Vehicles (Construction and Use) Regulations 1986, S.I. 1986 No. 1078, reg. 3(2), Table.
[369]   *Ibid.*, reg. 97.
[370]   See para. 6.5.3 below. See Goods Vehicles (Licensing of Operators) Act 1995, s.34.
[371]   See, generally, Highways Act 1980.
[372]   See Highways Act 1980, s.253.
[373]   That is, National Air Traffic Services Ltd ('NATS'). See Transport Act 2000, Pt I.
[374]   That is, the Goods Vehicles (Licensing of Operators) Act 1995.
[375]   Goods Vehicles (Licensing of Operators) Act 1995, s.2(1).
[376]   See para. 4.2.1.3 above.
[377]   Civil Aviation Act 1982, s.11(1).
[378]   See, generally, the Air Navigation Order 2000, S.I. 2000 No. 1562.
[379]   See the Licensing of Air Carriers Regulations 1992, S.I. 1992 No. 2992, reg. 2.

make airport charges with the authorisation of the CAA[380] and then only when the airport has, having achieved a certain level of turnover, become subject to the economic regulation of the CAA.[381]

6.6   Regulation of mineral extraction

Like transport, issues arising from quarrying are hardly a discrete area of environmental regulation. However, there is a range of general measures relating to noise pollution[382] and, in enacting the aggregates levy provisions of Finance Act 2001, the government's professed aims were to introduce an economic instrument similar to landfill tax that would reflect the environmental costs of aggregates quarrying, in terms of noise, dust, traffic, visual impacts, blasting, etc.[383] Mineral development is subject to the general law on town and country planning,[384] responsibility therefore being with local authorities, and may require an EIA to be carried out.[385]

An outline of aggregates levy appears at para. 1.4.2.3 above and there is a detailed account in Chapter 13 below.

6.7   Regulation of contaminated land

The historical legacy of contaminated land is dealt with in Environmental Protection Act 1990, Pt II, and in the Contaminated Land (England) Regulations 2000.[386] A duty is imposed on local authorities[387] to inspect their areas to identify contaminated land,[388] details of which are entered on a public register and, although agreement as to the costs of clean up may be reached between those responsible, local authorities are empowered to serve remediation notices[389] on the appropriate person or persons,[390] the failure to comply with which without reasonable excuse is a criminal offence.[391]

---

[380]   See Airports Act 1986, s.38(1) and the Civil Aviation Authority (Economic Regulation of Airports) Regulations 1986, S.I. 1986 No. 1544.

[381]   See Airports Act 1986, s.37.

[382]   See, generally, Environmental Protection Act 1990, s 79(1)(g) and (ga); Noise Act 1996; Control of Pollution Act 1974, ss.60–67.

[383]   See paras 5.5 above and 11.3.2 below (see also para. 13.1 below).

[384]   See Mark Stallworthy, *Sustainability, Land Use and Environment* (London: Cavendish, 2002); Hughes *et al.*, *op. cit.*, Ch. 13.

[385]   See para. 6.2.4 above.

[386]   S.I. 2000 No. 227.

[387]   See para. 4.2.3 above.

[388]   See the definition in Environmental Protection Act 1990, s.78A(2) as 'any land that appears to the local authority to be in such a condition, by reason of substances in, on or under the land, that significant harm is being caused or there is a significant possibility of such harm being caused or pollution of controlled waters is being, or is likely to be, caused'.

[389]   Environmental Protection Act 1990, s.78E.

[390]   *Ibid.*, s.78F(2).

[391]   *Ibid.*, s.78M.

The command and control regulation control regulation in Environmental Protection Act 1990, Part IIA is backed-up by an economic instrument in the form of a generous but closely restricted tax deduction, outlined in para. 7.3.4.2 below and detailed in para. 24.5.1 below.

6.8   Concluding comments

Only the briefest of glances at the present chapter will show the reader how far the idea of environmental levies and their associated economic instruments has gained ground in the regulatory landscape of the UK within the last decade.

In seeking to place environmental taxation within the context of environmental regulation generally, this chapter has highlighted the regulatory nature of the taxes and other levies under consideration. Such an emphasis is both necessary and appropriate. An evaluation of the UK's environmental taxation provisions which was divorced from their regulatory context would be, at the very least, incomplete. It will be recalled from earlier in the study[392] that the government views environmental levies and other economic instruments very much in terms of their cumulative effect, that is, as part of a diet of measures together designed to deliver environmental improvements. Thus it is important to realise that, in some areas, their significance may be greater than in others. Hitherto, for example, economic instruments have played a relatively insignificant role in waste management regulation, where command and control has dominated. Equally, however, economic instruments dominate the regulation of air and atmospheric pollution. As mentioned at the beginning, by examining in this chapter the regulation of the industries whose products fall within the scope of the various environmental taxes and other economic instruments, we hope to begin the process of drawing attention to possible anomalies as between the respective purposes and effects of what we have referred to respectively, and pragmatically, as market regulation and environmental regulation.

What falls to us next is the task of examining the duality of environmental levies and other economic instruments. Though spoken of, quite correctly, as regulatory instruments, environmental levies and subsidies also fall to be considered in their taxation context. Accordingly, in Chapter 7, we shall begin an examination of the issues involved in describing and evaluating environmental levies and other economic instruments in taxation terms.

---

[392]   See Chapter 5 above.

Chapter 7

Taxation Context

7.1   Introduction

In previous chapters, discussion has been focused on the regulatory qualities of the UK's environmental taxes and other economic instruments. Attention is narrowed in the present chapter to an examination of the nature of these instruments in tax law. The division of the material is designed to highlight the complimentary but distinct nature of the two contexts.

The chapter is divided into two main parts. In the earlier of these, we explore the concepts of 'taxes' and 'tax subsidies'. In the course of the former discussion, we shall be building on the definition of a tax introduced in Chapter 1 and, by reference to case-law from different legal traditions, attempting to explain the relationship between taxes and other, related but distinct, concepts. Thus, we shall be considering the distinction between a tax and a charge, or fee, for government services, as well as the relationship between the tax concept and those of tolls, penalties and fines. The rise of environmental taxation has necessitated the revisiting of these basic distinctions, none of which have traditionally been seen, from the legal point of view, as important ones in the UK. Secondly, we introduce the concept, not of a tax, but of a 'tax expenditure', or 'tax subsidy', something not so far alluded to in the book. If the legal concept of a tax is crucial to understanding the significance of the various environmental levies, then the idea of a tax expenditure is no less important to understanding what is required, at least in part,[1] in order to 'green' a tax system. Such tax expenditures or tax subsidies, if designed to incentivise environmentally-friendly behaviour, can properly be characterised as economic instruments for environmental protection, since they represent a form of financial assistance from government.[2] We shall be particularising them in UK law in Part III, Section B below.

The second main part of the discussion in the present chapter is the complex task of analysing the legal nature of the various payments under consideration in the book. These comprise, not only the environmental levies and environmentally-inspired subsidies currently at work in the UK tax system, but also items such as payments to government agencies on the grant of licences. The discussion in the latter part of the chapter builds on the distinctions drawn in the earlier part, in attempting to confirm which of the current environmental levies can properly be described as taxes, and in describing the legal nature of 'green' subsidies within the non-environmental tax codes.[3]

---

[1]   Greening a tax system can, of course, involve bringing items within the tax base, just as it is taken to refer introducing environmentally-friendly exemptions and reliefs.

[2]   See para. 1.2.1.5 above.

[3]   See para. 1.4.3.1 above.

We turn first to the distinctions between each of charges – or fees – for government services, taxes, tolls, fines and penalties.

7.2   Taxes and tax subsidies

7.2.1   Taxes distinct from charges (or fees)

Tiley isolates the characteristics of a charge or fee, as distinct from a tax, as follows:[4]

1   with a charge, 'some service must be provided direct to the individual';[5]
2   a charge must be related to the service provided, not varied according 'to some other criterion ... such as the value of that person's property' or his ability to pay;[6] and
3   'it is no objection that a charge may result in a profit' to the levying authority, 'provided only that the profit is a reasonable one'.[7]

Re Eurig Estate,[8] it will be recalled, insists on there being three criteria for identifying a tax, as distinct from a charge. A levy is a tax if it is: compulsory,[9] levied by a public body and intended for a public purpose.[10]

   These points are made on the basis of Commonwealth authorities. They are much more precise than either the case-law of the ECJ discussed below on the meaning of the expression 'taxation' as it is used in Art. 90, European Treaty (ex 95),[11] or GATT 1994/WTO jurisprudence on the construction of the GATT 1994 equivalent of Art. 90, Art. III(2), GATT 1994.[12] However, the Commonwealth authorities are consistent with the proposition a levy will fall outside the scope of Art. 25, European Treaty (ex 12), which bans customs duties as between EU Member States, only if it

---

4   See John Tiley, Revenue Law, 4th edn (Oxford: Hart Publishing, 2000), p. 4.
5   Re Tax on Foreign Legations and High Commissioner's Residence, [1943] SCR 208 (Can) (Duff, C.J.).
6   Société Centrale d'Hypothesques v. Cité de Quebec, [1961] QLR 661. Thus, in Re Eurig Estate (1999) 165 DLR (4th) 1 (see para. 1.2.1.1 above), one of the reasons that the probate fee was held to be a tax was that it varied with the size of the estate, there being no link between the amount of the fee and the cost of the service for granting letters probate.
7   Minister of Justice for Dominion of Canada v. Levis City, [1919] AC 505.
8   (1999) 165 DLR (4th) 1.
9   And therefore enforceable by law (see (1999) 165 DLR (4th) 1, para. 17); see also, for example, Attorney-General of New South Wales v. Homebush Flour Mills Ltd, (1937) 56 CLR 390.
10   (1999) 165 DLR (4th) 1, para. 15. On the public purpose requirement, see Lower Mainland Dairy Products Sales Adjustment Committee v. Crystal Dairy Ltd, [1933] AC 168, 175–176 (Lord Thankerton); and Lawson v. Interior Tree, Fruit and Vegetable Committee of Direction, [1931] 2 DLR 193, 197–198 (Duff, J.).
11   See para. 7.2.2.3 below.
12   See para. 8.4.3 below.

can be shown that there was a genuine bargain between the trader and the state, that is, that the trader receives a specific, identifiable benefit in return for the sum paid, and that the sum paid is proportionate to the benefit received.[13]

7.2.2   Taxes defined

In order to further refine the definition of a tax on which the discussion has so far proceeded, four areas of law fall for consideration:[14]

1   UK case-law on the concept of a tax;
2   Commonwealth case-law on the concept of a tax;
3   Case law of the ECJ on the meaning of 'taxation' as it is used in the European Treaty; and
4   Jurisprudence of the WTO Dispute Settlement Body and the AB on the meaning of 'internal taxes or other internal charges' as the phrase is used in GATT 1994.

It will be apparent to the reader that these four areas of law have been chosen for their relevance to the UK, given its status as a member of each of the Commonwealth, of the EU and of the WTO.[15] It should, however, be noted that the scope of the concept of a tax has been given detailed consideration in many national jurisdictions, most notably, perhaps, those of the USA and Germany.[16] Of the areas listed above, 2 and 3 yield fairly detailed information on the characteristics of a tax, whilst 1 and 4 are, possibly for historical reasons, relatively uninformative.

7.2.2.1   UK case-law on the concept of a tax

In Chapter 1 above, we stated that the UK's constitutional arrangements have so far tended to discourage a narrow and prescriptive view of a tax as a legal concept.[17]

---

13   See para. 12.3.3.2 below. See *Commission v. Italy*, C–24/68, [1969] ECR 193; *W. Cadsky SpA v. Instituto Nazionale per il Commercio Esterio*, C–63/74, [1975] ECR 281; and *Ford España v. Spain*, C–170/88, [1989] ECR 2305.
14   That is, from the UK perspective.
15   See para. 8.4.1 below.
16   Not considered further in this volume, although see, for example, Paul Marchetti, 'Distinguishing Taxes from Charges in the Case of Privileges' (1980) 33 *NTJ* 233–6. It has recently been reported that the US Supreme Court is about to consider the nature of the tax concept in US law in a case involving the beef levy (see Waldmeir, *Financial Times*, 27 September 2004, p. 12). See also, in the context of the term 'taxes', as used in the OECD's model double taxation convention, Klaus Vogel, *Klaus Vogel on Double Taxation Conventions*, 3rd edn (The Hague: Kluwer Law, 1997), pp. 146–8 and 205.
17   See para. 1.2.1.1 above. The Bill of Rights 1689, Art. 4, simply states that: ' … levying Money for or to the Use of the Crown, by Pretence of Prerogative, without Grant of Parliament, for longer Time, or in other Manner than the same is or shall be granted, is illegal'. See also 8(2) *Hals*, para. 228, esp. n3 and the references cited therein; O. Hood Phillips and Paul Jackson, *Constitutional and Administrative Law*, ed. by Paul Jackson and Patricia Leopold, 8th edn (London: Sweet and Maxwell, 2001), paras 3–008–3010; and E.C.S. Wade and A.W. Bradley, *Constitutional and Administrative Law*, ed. by A.W. Bradley and K.D. Ewing, 11th edn (London: Longman, 1993), pp. 366–7.

Probably for this reason, such English law authorities as actually exist on the nature of a tax are unsatisfactory.[18]

Following devolution,[19] however, the exact legal meaning of the concept of a tax will need to be considered before much longer.[20] Three well-known cases illustrate the historically impressionistic approach of English courts to the concept.

1   In *Congreve v. Home Office*,[21] the CA declared unlawful and void the Home Secretary's threatened revocation of certain television licences. The CA clearly assumed that the television licence fee was a form of taxation, although the judges did not consider the issue in any detail.[22]

2   *Daymond v. South West Water Authority*,[23] concerned charges for household sewage and sewerage disposal services. The claimant, whose property was not connected to the main drainage system, and was situated 400 yards from the nearest sewer, received a demand for payment of charges for sewerage and sewage disposal services. He refused to pay the charges and sought a declaration that the demand was unlawful, since the water authority had no power under the Water Act 1973, s.30, to demand charges other than those for services performed, facilities provided or rights made available to him by the authority. The House of Lords[24] held that, where a provision stated only that a statutory body could demand, take and recover such charges for the services that it performed, then the body in question could charge only those who availed themselves of its services. It followed that the body was not empowered to charge the claimant for services of which he did not avail himself. Whilst emphasising that the conclusion of the majority was a matter of statutory construction, Viscount Dilhorne asked:

> Is it to be inferred that it was the intention of Parliament that [water authorities] ... should be at liberty to charge anyone they thought fit in Great Britain? That has only to be stated

---

18   See, for example, *Brewster v. Kidgill*, (1697) 88 ER 1239 (Holt, C.J.); *Baker v. Greenhill*, (1842) 114 ER 463, at 470 (Lord Denman, C.J.); *Attorney-General v. Wilts United Dairies Ltd*, (1921) 124 LT 319, 322–323 (Bailhache, J., overruled by CA and HL at [1922] WN 217, 218); *Government of India v. Taylor* [1955] AC 491; and *Aston Cantlow v. Wallbank*, [2001] 3 WLR 1323, para. 40 (Sir Andrew Morrit, V.C.), overruled by HL at [2003] 3 WLR 283 (esp. para. 133) (Lord Scott of Foscote).

19   See para. 1.3 above.

20   On the near-federal nature of the UK, see, for example, Chris Hilson, *Regulating Pollution – A UK and EC Perspective* (Oxford: Hart Publishing, 2000), p. 47.

21   [1976] QB 629.

22   The background to the case was interesting. The Home Secretary had announced, on 29 January 1975, that the colour television licence fee would be increased from £12 to £18 on April, 1 that year, and had made an order to that effect under the Wireless Telegraphy Act 1949, s.2(1). In order to avoid the increase in the fee, some 24,500 licence holders had applied for new £12 licences, before 1 April and before their existing licences had expired. The Home Secretary acted in purported exercise of a statutory discretion to revoke these overlapping licences.

23   [1976] AC 609.

24   Lords Wilberforce and Diplock dissenting.

to be rejected for it is, to my mind, inconceivable that Parliament should have intended to entrust such an extensive power of taxation to a non-elected body. Is it then to be inferred that it was intended to give them only power to charge those living in their area and those who came into it and made use of their services ... ? I think that such a limitation must be implied.[25]

Lord Kilbrandon took a similar approach:

For my part I do not consider ... [the relevant statutory wording] adequate, if any other meaning is open, to empower an ad hoc nonrepresentative body to impose what is in truth a tax, namely an impost under the head of charges for services ... upon persons who do not directly receive such advantages.[26]

3   *IRC v. Océ van der Grinten*[27] concerned withholding tax under the 1980 UK/ Netherlands Double Tax Treaty. The Inland Revenue had appealed against the Special Commissioner's[28] referral of a double taxation issue to the ECJ. The UK subsidiary of a Dutch parent company had declared dividends of £13 million in a particular year when the former UK imputation system (now abolished) was in force.[29] The Revenue argued that, pursuant to the double tax treaty, the parent was liable to pay tax at 5 per cent on the total of the dividends distributed, plus the tax credit received on those dividends. The parent argued that the 5 per cent was a withholding tax and, as such, was forbidden by the Parent-Subsidiary Directive.[30] In agreeing that the case should be referred to the ECJ,[31] Jacob, J. held that the concept of a withholding tax was a global question of Community law. His Lordship also commented reluctantly, at the specific request of counsel on both sides, on the issue of whether the 5 per cent was a 'tax' in UK law:

The [tax credit] stems from the distribution of profits by the subsidiary. It is the distribution of profits by way of dividend which causes the [tax credit] to be given to the parent. And the amount of the [tax credit] represents part of the profits distributed. So the abatement of the [tax credit] by the 5 per cent ... is in substance a reduction in what the parent gets compared with what it would get if there were no abatement. Putting it another way, as a result of the abatement the Crown ends up with more money and the taxpayer less. Most people would call the difference a 'tax' and I do too. Since the 'tax' stems from the distribution it seems equally right to call the deduction a 'tax on (or in respect of) the distribution'.[32]

---

25   [1976] AC 609, 640.
26   [1976] AC 609, 651.
27   [2000] STC 951.
28   See para. 4.2.1.5 above.
29   For the detail of the UK imputation system, inapplicable after 5 April 1999, see, for example, Chris Whitehouse, *Revenue Law – Principles and Practice*, 16th edn (London: Butterworths, 1998), paras 29.62–29.65.
30   Council Directive 90/435/EEC, (1990) OJ L225 6.
31   See *Océ van der Grinten v. Commissioners of Inland Revenue*, C–58/01. The ECJ does not, however, shed any light on the issue under consideration here.
32   [2000] STC 951, para. 14.

The reasons for Jacob, J.'s evident impatience at counsels' request was presumably that the concept of a tax had already been considered by the Special Commissioners[33] in some detail.

7.2.2.2   Commonwealth case-law on the concept of a tax

Commonwealth case-law, of course, is the source of the definition of a tax upon which reliance has so far been placed (that is, compulsory, levied by a public body and intended for a public purpose).[34] To these three may be added the fourth criterion, which determines the lawfulness or otherwise of the tax thus identified. This fourth criterion, it will be recalled, is that the tax must have been imposed under the authority of the legislature.[35]

In *Re Eurig Estate* itself, the province of Ontario had imposed an *ad valorem* probate fee, which varied according to the size of estates. The appellant was the executor of the estate of her late husband. She applied to the Ontario Court (General Division) for an order that she be issued letters probate without payment of the probate fee and for a declaration that the regulation which required that payment was unlawful. The basis of her argument was that the fee was in reality a tax and was invalid, either because it was an indirect tax (and thus outside the powers of the provincial legislature)[36] or because, as a direct tax, the constitutional requirements for its implementation had been violated. Originally unsuccessful before the Ontario Court, the appellant appealed to the Ontario Court of Appeal, where her appeal was dismissed. Eventually the case reached the Supreme Court of Canada.

Major, J., for the majority of the Supreme Court,[37] applied the three criteria listed above, in holding that the probate fee was a tax:

1   the fee was compulsory and therefore enforceable by law: although probate was not the foundation of the executor's title, but only 'the authentic evidence of it', that authentication was 'nonetheless a practical and legal necessity in most cases'.[38]
2   The fee was levied by a public body, since probate fees in Ontario were levied by the Ontario Court (General Division).[39]
3   The fees were intended for a public purpose, since the revenue obtained from probate fees was 'used for the public purpose of defraying the costs of court administration in general ... not simply to offset the costs of granting probate'.[40]

Having held that the probate fee was therefore clearly a tax, Major, J. held that, although it was a direct tax – and therefore within the competence of the provincial

---

33   [2000] STC(SC) 127.
34   (1999) 165 DLR (4th) 1. See para. 1.2.1.1 above.
35   (1999) 165 DLR (4th) 1, para. 16.
36   See para. 1.2.1.2 above.
37   Bastarache and Gonthier, JJ. dissented.
38   (1999) 165 DLR (4th) 1, para. 17.
39   *Ibid.*, para. 18.
40   *Ibid.*, para. 20.

legislature[41] – it had not been implemented in accordance with the constitutional requirements for a direct tax.[42]

To the three characteristics of a tax enumerated in *Re Eurig Estate* may be added the following subsidiary points:[43]

1   A levy is not prevented from being a tax by the fact that raising revenue is not the government's main reason for imposing it (see *Northern Suburbs Cemetery Reserve Trust v. Commonwealth*).[44]
2   For a levy to be capable of being a tax, it must be possible to identify 'the criteria by reference to which liability to pay the tax is imposed' and to demonstrate that the process of applying the criteria is not arbitrary or capricious (see *MacCormick v. Federal Commissioner of Taxation*).[45]
3   The three criteria for the existence of a tax should not be regarded as providing an exhaustive definition of a tax (see *Air Caledonie v. The Commonwealth*).[46] For example, in relation to the requirement that the levy must be made by a public body, it has been held that a levy on blank tapes, to be paid to a body 'set up by the music industry' to compensate performers, was nonetheless a tax (see *Australian Tape Manufacturers' Association v. The Commonwealth*).[47]

7.2.2.3   European Union case-law on the meaning of 'taxation'

The ECJ has had to consider the meaning of the expression 'taxation' as it occurs in Art. 90, European Treaty (ex 95). Article 90 provides:

> No Member State shall impose, directly or indirectly, on the products of other Member States any internal taxation of any kind in excess of that imposed directly or indirectly on similar domestic products.
>   Furthermore, no Member State shall impose on the products of other Member States any internal taxation of such a nature as to afford indirect protection to other products.

As interpreted by the ECJ, 'taxation' in this context is a broad term. For the purposes of Art. 90, taxation 'includes not only taxes levied by central government but any fiscal or parafiscal levy[48] imposed by or with the authority of any level of government, including levies charged by or for quasi-official bodies and allocated to a specific purpose'.[49]

---

41   *Ibid.*, para. 27. See para. 1.2.1.2 above.
42   *Ibid.*, paras 28–37.
43   See Tiley, *op. cit.*, p. 5, to which the summaries in these paragraphs are much indebted. The test was considered in *Airservices Australia v. Canadian Airlines International Ltd*, [1999] HCA 62 and in *Luton v. Lessels*, [2002] HCA 13.
44   (1993) 176 CLR 555 (High Court of Australia).
45   (1984) 158 CLR 622 (High Court of Australia).
46   (1988) 165 CLR 462, para. 6 (High Court of Australia).
47   (1993) 176 CLR 480 (High Court of Australia).
48   See para. 12.3.3.1 below.
49   See Paul Farmer and Richard Lyal, *EC Tax Law* (Oxford: Clarendon Press, 1994), p. 46.

The approach of the ECJ in relation to Art. 90 is illustrated by the following cases:

1. In *Deutschmann v. Germany*,[50] when concluding that the then Arts 12 (now 25) and 95, European Treaty, were mutually exclusive, the ECJ did not consider there to be a relevant distinction between taxes and charges for services in the context of Art. 90, European Treaty (ex 95). Advocate General Gand said:

    From the fact that Article 95 appears in a chapter entitled '*steuerliche Vorschriften*' ('Fiscal provisions') and from its closeness to Article 98, the Federal Republic draws the conclusion that it only applies to taxes in the strict sense ... Fiscal terminology, already uncertain in the national legal systems, is all the more so when it is transferred to the Common Market sphere; such an exegesis seems to us to be rather useless and of little help in solving the problem put to us [that is, as to whether Art. 95 of the European Treaty applied to fees charged on the grant of import licences].

2   In *Iannelli & Volpi v. Meroni*,[51] the ECJ had to consider two cases involving, inter alia, Art. 90, European Treaty (ex 95). One case involved subsidies of newsprint purchased by Italian newspaper publishers, the paper being produced in Italy with levies being paid by manufacturers and importers of paper and cardboard. The other was about charges levied in relation to the processing of citrus concentrates imported to Germany. One of the points that the court made about the scope of Art. 90, was:

    Since Article 95 of the Treaty refers to internal taxation of any kind the fact that a tax or levy is collected by a body governed by public law other than the State or is collected for its benefit and is a charge which is special or appropriated for a specific purpose cannot prevent its falling within the field of application of Article 95 of the Treaty.[52]

3   *Apple and Pear Development Council v. Lewis*[53] involved a research and development body established by statutory instrument, which body's activities were financed by charges on fruit growers. One issue was whether the charges were covered by Arts 28 and 29, European Treaty (ex Arts 30 and 34). The ECJ said:

    ... the charges, being measures of a fiscal nature or of equivalent effect, fall within the scope, not of those Articles, but of Articles 9 to 16 and 95 of the Treaty. Since the charge in question does not apply to imported produce and only affects produce intended for export in the same way as produce sold on the home market, it does not raise any problem in relation to the last-mentioned articles either.[54]

---

50   C–10/65, [1965] CMLR 259.
51   C–74/76, [1977] 2 CMLR 688. See also *AGF Belgium SA v. European Economic Community*, C–191/94, [1996] ECR I–1859; *Weyl Beef Products BV v. Commission*, Case T–197/97, [2001] 2 CMLR 22.
52   [1977] 2 CMLR 688, para. 19.
53   C–222/82, [1984] 3 CMLR 733.
54   [1984] 3 CMLR 733, para. 30.

Although unwilling to make fine distinctions between taxes and related concepts for the purpose of Art. 90, European Treaty, the ECJ has been obliged to distinguish, for the purposes of Art. 25, European Treaty (ex 12) between taxes and charges having an effect equivalent to customs duties. Customs duties[55] as well as charges having equivalent effect,[56] are of course prohibited as between Member States of the EU, by Art. 25, European Treaty (ex 12). A charge having an equivalent effect (a 'CEE') is defined in a key passage in the following terms:

> Any pecuniary charge, however small and whatever its designation and mode of application, which is imposed unilaterally on domestic or foreign goods by reason of the fact that they cross a frontier, and which is not a customs duty in the strict sense, constitutes a charge having equivalent effect ... even if it is not imposed for the benefit of the State, is not discriminatory or protective in effect and if the product on which the charge is imposed is not in competition with any domestic product ...
>
> It follows from Article [90] ... *et seq.* that the concept of a charge having equivalent effect does not include taxation which is imposed in the same way within a State on similar or comparable domestic products, or at least falls, in the absence of such products, within the framework of general internal taxation ...[57]

A charge may, however, escape Art. 25, if it can be shown that, rather than having an effect equivalent to a customs duty, the charge is a tax and therefore subject to the discipline of Art. 90. In *Commission v. France*,[58] a tax, in this context, was defined as a levy relating to:

> ... a general system of internal dues applied systematically to categories of products in accordance with objective criteria irrespective of the origin of the products ...[59]

in circumstances such that there is no identical or similar domestic product.

This is not the only way in which an item may escape Art. 25, however. A levy will not count as a CEE if either it counts as consideration for a service supplied by the importing State to the importer[60] or it was imposed pursuant to mandatory requirements of Community Law (that is, it was an administrative charge).[61]

Given the purpose for which the complementary Arts 90 and 25 fall to be interpreted, these distinctions are unsurprising. Overall, however, they are consistent with a number of the subsidiary points about taxes made in the Commonwealth cases, that is, that:

1   the fact that the levy is allocated to a specific, as opposed to a general, purpose does not prevent it from being a tax;

---

55   See, generally, Timothy Lyons, *EC Customs Law* (Oxford: Oxford University Press, 2001).

56   See *Commission v. Italy*, C–24/68, [1969] ECR 193 (the 'statistical levy case').

57   [1969] ECR 193, paras 9 and 11.

58   C–90/87, [1981] ECR 283.

59   [1981] ECR 283, para. 14.

60   *Bresciani v. Amministrazione Italiana delle Finanze*, C–87/75, [1976] ECR 129.

61   See *Commission v. Germany*, C–18/87, [1988] ECR 5427.

2 a levy may be a tax where it is imposed by, and paid to, a body set up under statutory powers, rather than directly to the government itself; and

3 it must not be arbitrary or capricious, that is, it must be possible to identify the criteria by reference to which liability to pay the tax is imposed.[62]

7.2.2.4   *GATT 1994/WTO jurisprudence on the meaning of 'internal taxes and other internal charges'*

The equivalent provision to Art. 90, European Treaty (ex 95) as regards the trade relations of EU Member States with third countries, is Art. III(2), GATT 1994.[63] Article III, which is considered in more detail later in the book,[64] reads (so far as relevant) as follows:

> 1. The contracting parties recognize that internal taxes and other internal charges ... affecting the internal sale, offering for sale, purchase, transportation, distribution or use of products, ... should not be applied to imported or domestic products so as to afford protection to domestic production.

> 2. The products of the territory of any contracting party imported into the territory of any other contracting party shall not be subject, directly or indirectly, to internal taxes or other internal charges of any kind in excess of those applied, directly or indirectly, to like domestic products. Moreover, no contracting party shall otherwise apply internal taxes or other internal charges to imported or domestic products in a manner contrary to the principles set forth in paragraph 1.

It will be apparent to the reader that, instead of the expression 'internal taxation of any kind', which is the expression used in the (later) European Treaty, Art. III(2), GATT 1994,[65] refers to 'internal taxes and other internal charges'. The use of the word 'charges' in Art. III(2), rather than the wider expression 'levies', emphasises the fact that the Art. was not drafted mindful of the classification of taxes and charges as two different types of 'levy'.[66] Given its purpose,[67] it is plain that one of the main purposes of Art. III(2), however, is to distinguish, not between taxes and other types of levy, but between internal taxes, on the one hand, and customs duties, on the other. In short, customs duties, which are subject to the discipline of GATT 1994, Art. II,[68] though clearly taxes,[69] are not internal taxes. What makes a

---

62   Arts 25 (ex 12) and 90 (ex 95), European Treaty, are discussed in detail in Stephen Weatherill and Paul Beaumont, *EU Law*, 3rd edn (London: Penguin, 1999), chs 13 and 14; and Farmer and Lyal, *op. cit.*, ch 3.

63   See para. 12.3.3.1(1) below.

64   See para. 8.4.3 below.

65   Originally, Art. III(2), GATT 1947.

66   See para. 1.2 above.

67   See para. 8.4.3 below.

68   *Ibid.*

69   See Kenneth W. Dam, *The GATT: Law and International Economic Organization* (London: University of Chicago Press, 1970), p. 115. As such, they are indirect taxes (see para. 1.2.1.2 above).

tax an internal tax, and therefore subject to GATT 1994, Art. III(2), rather than GATT 1994, Art. II, is revisited below.[70] The distinction is one on which GATT and WTO institutions seem to have been remarkably reluctant to rule definitively. What is at least clear is that: (1) the mere fact that it is collected or enforced 'at the time or point of importation' does not prevent the tax in question from being an internal tax;[71] and (2) that, although described as internal taxes, levies may nonetheless be in reality import (that is, customs) duties, if:

> (a) they are collected at the time of, and as a condition to, the entry of the goods into the importing country, and (b) they apply exclusively to imported products without being related in any way to similar charges collected internally on like domestic products.[72]

7.2.3   *Taxes distinct from tolls, penalties and fines*

Tolls[73] have existed for centuries, both at common law and under statute, in relation to roads and bridges, street trading, markets, fairs and shipping and navigation. So far as roads are concerned, the necessary powers for tolls to be charged, whether by a highway authority or a concessionaire, have for long been taken by legislation.[74] Historically, rights to take tolls or dues in relation to the use of a port or harbour have also been conferred by immemorial usage or grant as well as by statute, tolls in relation for example, to harbours being required to be reasonable in amount.[75] Where a toll road has been constructed pursuant to an order made under the New Roads and Street Works Act 1991, the Secretary of State is not required to impose restrictions on the levels of tolls charged.[76] Such distinctions between tolls and taxes as historically there might have been (that is, as to restrictions on amount and the need for statutory authority) are not, however, relevant to the projected lorry road user charge, which is straightforwardly characterised as a tax in the 2002 enabling legislation.[77]

---

[70]   See para. 8.4.3 below.

[71]   Note Ad Art. III, GATT 1994 (reproduced at World Trade Organization, *The Legal Texts: the Results of the Uruguay Round of Multilateral Trade Negotiations* (New York: Cambridge University Press, 1999), p. 479). See para. 8.4.3n below.

[72]   GATT 1947 Committee Report from the 1948 Havana Conference, quoted in John H. Jackson, *World Trade and the Law of GATT* (Indianapolis: Bobbs-Merrill, 1969), section 12.3 (esp. pp. 280–81), in turn referred to in the same author's *The World Trading System*, 2nd edn (Cambridge, MA: MIT Press, 1997), p. 397.

[73]   The word 'toll', or *tolnetum*, is defined to be a sum of money which is taken in respect of some benefit' (see Bramwell and Willes (counsel) in argument, in *Adey v. Trinity House*, (1852) 22 LJ QB 3). See also para. 12.2.6.4(2)n below.

[74]   For example, as in relation to the M6 toll road, north of Birmingham (see para. 27.3n below), constructed under the New Roads and Street Works Act 1991.

[75]   See *Lord Falmouth v. George* (1828) 130 ER 1071 (fishing boats paying a toll for use of a capstan at Sennen Cove in Cornwall).

[76]   See *Alliance Against the Birmingham Northern Relief Road v. Secretary of State for the Environment, Transport and the Regions and Midlands Expressway Ltd*, [1999] JPL 426. The Birmingham Northern Relief Road is the M6 toll road referred to above. The legal background to the construction of the road is examined in the judgment of Latham, J.

[77]   See Finance Act 2002, s.137(1).

Civil penalty provisions appear throughout the VAT code[78] and have been transposed into the codes for each of aggregates levy, climate change levy and landfill tax.[79] Environmental taxes are closer to civil penalties in terms of the function that they serve than are fiscal taxes, such as corporation tax and VAT.[80] This proposition seems to be particularly well illustrated by landfill tax. Even at the current main rate, the level of the tax considerably exceeds the externalities associated with landfill. At the ultimate rate of £35 per tonne, as currently projected under the landfill tax escalator,[81] the rate of the tax will clearly be a penal one. The significance of this point is that, unlike taxes, penalties are not deductible in calculating trading profits under Schedule D, Case I.[82] The same point applies to fines that are imposed by the courts for breaches of offences created by command and control environmental and health and safety legislation.[83]

7.2.4   Tax subsidies distinct from taxes

Having distinguished taxes from related but distinct concepts, it is necessary to describe the various legal concepts that are together categorised as tax subsidies or 'tax expenditures'.[84]

As indicated above, tax expenditures are not 'free-standing' concepts, such as taxes and charges, but are rather elements within the structure of taxes themselves. The concept of the tax expenditure was defined by Willis and Hardwick as:

> ... an exemption or relief which is not part of the essential structure of the tax in question but has been introduced into the tax code for some extraneous reason – eg in order to ease the burden for a particular class of taxpayers, or to provide an incentive to apply income in a particular way, or perhaps to simplify administration. The term is used to cover, not merely specific exemptions but also gaps in the charge as a result of which receipts or benefits which represent or are equivalent to income are not subjected to tax. The choice of the term 'tax expenditure' indicates that, because they are not inherent in the structure of the tax, these reliefs are equivalent in terms of revenue foregone to direct government expenditure and should in general be judged by the same criteria.[85]

Two of the different legal forms that tax expenditures may take (that is, exemptions and reliefs) have been mentioned already. Within the structure of the various UK taxes, however, there are a range of other forms that tax expenditures may take.

---

[78]   See, for example, ss.59–67, Value Added Tax Act 1994.
[79]   See para. 16.14 below.
[80]   See Kalle Määttä, *Environmental Taxes – From an Economic Idea to a Legal Institution* (Helsinki: Finnish Lawyers' Publishing, 1997), pp. 207–10.
[81]   See para. 15.2 below.
[82]   See para. 1.4.3.1 above.
[83]   See Susan Wolf and Neil Stanley, *Wolf and Stanley on Environmental Law*, 4th edn (London: Cavendish, 2003), pp. 60–64.
[84]   These terms are treated as being synonymous (see, for example, Määttä, *op. cit.*, p. 28).
[85]   J.R.M. Willis and P.J.W. Hardwick, *Tax Expenditures in the United Kingdom* (London: Heinemann, 1978), p. 1. The quotation is limited to income only because Willis and Hardwick's study was confined to income tax and CGT.

The distinctions between these various possibilities can be illustrated by reference to income tax.[86]

1 *Exemptions* remove certain items from the scope of the tax altogether (for example income from personal equity plans[87] or certain employee benefits in kind).[88]
2 *Permitted deductions* are deductible in calculating income from particular sources (for example trading expenses).[89] Within this category can be included the logically necessary concept of capital allowances. Within the UK system, capital expenditure is not deductible from income, allowance instead being made for certain categories of such expenditure through the elaborate system of capital allowances.[90]
3 *Charges on income* are deductible in calculating total income from all sources, though the range of payments capable of qualifying as such is now very narrow (certain interest payments by individuals are still of this kind).[91]
4 *Personal allowances* and reliefs can, in certain cases, be deducted from total income from all sources (for example the personal allowance).[92]
5 *Income tax reductions* give rise to credits against tax (for example the income tax reduction provided for by the enterprise investment scheme ('EIS')).[93] This last form of tax expenditure has become particularly popular with certain OECD governments.

It should be emphasised that these are the tax expenditures within the income tax code only; as mentioned above, the legal form that tax expenditures take depends on the tax code in question.[94]

---

[86] See para. 1.4.3.1 above. This part of the para. draws heavily on the useful list in *CCH Tax Handbook 2003–04* (Banbury: CCH, 2003), para. 1840.
[87] Although no new personal equity plans ('PEPs') can now be taken out, existing ones continue to benefit from provisions allowing the income therefrom to be exempt from income tax.
[88] See Income and Corporation Taxes Act 1988, ss.333 and 333A and regulations thereunder (although note that no new PEP subscriptions have been permitted since 5 April 1999, having been replaced by Individual Savings Accounts or 'ISAs').
[89] See Chapter 24 below.
[90] *Ibid.* Capital allowances are so complex that they are governed by their own statute, that is, the Capital Allowances Act 2001; in the late 1960s, they were, for a time, replaced by investment grants, a system which favoured loss-making businesses, such as the (then) nationalised industries. As discussed in Chapter 24 below, there is a tension in the capital allowances system between the provision of a highly stylised system of depreciation and their use as policy incentives.
[91] See, for example, Income and Corporation Taxes Act 1988, s.359 (interest on loan to buy machinery or plant).
[92] See Income and Corporation Taxes Act 1988, s.257(1): these have to be distinguished from the mass of credits and deductions which have been added to the PAYE system since 1997 and which do not relate to tax allowances at all, even though they may be income based, for example, working family tax credit and student loan repayments.
[93] See Income and Corporation Taxes Act 1988, ss.289–312.
[94] This legal form is in each case described below.

Tax expenditures in general have a double significance in the context of the present study. Within the environmental tax and charging codes,[95] they will tend to limit the effectiveness of the regulatory instrument, and we should therefore expect them to be difficult, although not impossible, to justify in terms of regulatory principle. Conversely, it might be expected that tax expenditures within the non-environmental tax codes[96] should relatively easily be justifiable in regulatory terms.

7.3 Status of the main levies and subsidies under consideration

7.3.1 Environmental levies

Within the scope of this book are at least ten extant economic instruments, or categories thereof, that are of an environmental nature and that involve the making of payments. Possible payments include the following:

1   payments of aggregates levy;
2   payments of climate change levy;
3   payments of landfill tax;
4   payments of workplace parking levies;
5   payments under road user charging schemes;
6   payments made under the UK ETS;
7   payments made under the RO;
8   payments made under the EU ETS;
9   payments made for licences and permits (for example IPPC permits) under command and control legislation; and
10 payments of airport landing charges.

There is no doubt that landfill tax, climate change levy and aggregates levy are taxes. It is unclear whether licensing schemes relating to workplace parking levies or to road user charging impose taxes or charges; it is thought that taxes are more likely to arise under the former than under the latter. The payments referred to below under the UK ETS and RO also seem closer to charges than to taxes. The importance of the distinction is not yet a constitutional one as it is in the context of the Commonwealth jurisdictions referred to above. However, it may have a distinction in assessing the acceptability of the levies in question.

Charges, it will be recalled, involve the provision of a service direct to the individual, a relationship between the service and the charge and no more than a reasonable profit accruing to the service provider. Payments of charges are requited, whereas payments of taxes are not.

Licensing schemes for workplace parking and road user charging schemes each involve the provision of a service direct to individuals. The service in the former case is the provision of licences to park a certain number of cars at, or in the vicinity of,

---

95   See Part III, Section A below.
96   *Ibid.*, Section B below.

particular premises, such licences being granted in favour of occupiers of premises. In the latter case, the service is the provision of licences for motor vehicles to enter a certain area at certain times, the registered keeper of the vehicle generally being the person in whose favour the licence is granted.

In the case of workplace parking levies, vehicles are classified as 'licensed units' and the amount payable is based on the charge per unit. Section 186, Transport Act 2000, allows for variations in the charges according to different days or times of day, different parts of the licensing area, different classes of motor vehicles or different numbers of licensed units. Although Transport Act 2000, s.171, provides for the basic elements that must be included in the order establishing the charging scheme, the local authority is free to determine the levels of charge. There would therefore appear to be greater scope for the level of the charge to be related to the service provided, as opposed to extraneous criteria, in the case of road user charging schemes.

The question of whether the profit accruing to the service provider is reasonable is, of course, a nebulous one. Suffice it to say that Transport Act 2000, Sched. 12, contains financial provisions common to both types of scheme. The net proceeds of both types of scheme are found by subtracting the expenses of establishing or operating the scheme from the gross proceeds;[97] net proceeds must be hypothecated (that is, earmarked)[98] such that they can only be spent in support of the authority's local transport plan for the first ten years of a scheme's life;[99] and, in accordance with the general rule that local authorities must spend net proceeds only on things that offer value for money, they are required to prepare a ten-year general plan for spending the proceeds.[100]

The UK ETS has already been outlined.[101] The relevant feature of the scheme for present purposes is that, where participants make more emissions than their targets for a particular year, it will be necessary for them to purchase extra allowances. Where there is a failure to do this then, as an agreement participant, it will be necessary to pay the full rate of the levy, while, as a direct participant, the incentive payment will be lost and a tighter target will be imposed for the following year. It is thought that, since the right to retain the specific benefit conferred – that is, either a tax reduction or an incentive payment – is purchased at a price determined by the market, such payments are not to be regarded as taxes but are more in the nature of fees or charges. However, any loss of the incentive payment seems properly characterised as a penalty, with the loss of the climate change levy reduction resulting in a payment of the levy (that is, in the payment of tax rather than a penalty).

With the RO, which has again been described above,[102] electricity suppliers are obliged to prove that they have supplied specified quantities of renewable source electricity to customers in Great Britain within a specified period.[103] In order to

---

[97] Transport Act 2000, Sched. 12, para. 2.
[98] See para. 11.2 below.
[99] Transport Act 2000, Sched. 12, para. 7.
[100] *Ibid.*, Sched. 12, paras 8–10.
[101] See para. 1.4.2.2(2) above.
[102] See para. 1.4.2.2(4) above.
[103] See Utilities Act 2000, s.62, and Renewables Obligation Order 2002, S.I. 2002 No. 914, Sched. 1.

comply with the RO, suppliers must produce ROCs or Green Certificates to Ofgem. Where a supplier cannot produce the requisite Green Certificates, it may either pay a penalty to Ofgem for each MWh it is short of its obligation or purchase ROCs from other suppliers willing to trade them. Ofgem pays the funds collected in penalties to suppliers in proportion to the amount of renewable energy they have sourced.[104] To the extent that the supplier has had to pay for Green Certificates, such payments seem best characterised as fees or charges, since the payment is made directly to avoid the penalty and its amount is determined by the market place. Payments to Ofgem are, of course, in the nature of penalties.

Whatever technical issues may be raised by the levies discussed above, it is clear that each of landfill tax, climate change levy and aggregates levy are taxes. They are payable to a public body, that is, HM Customs and Excise; they are compulsory, in the sense that satisfaction of the relevant conditions for liability and the absence of any exemption or relief gives rise to a liability to pay them, which liability is backed up by elaborate sanctions; and their proceeds are intended for a public purpose. This is so, even though part of the revenue raised by the three taxes is, by various means, hypothecated.[105] It will be recalled from the ECJ case-law that the fact that a levy is allocated to a specific, as opposed to a general, purpose does not prevent it from being a tax.[106] It is also unimportant to the status of these levies as taxes whether raising revenue is the government's primary purpose in imposing them.[107]

7.3.2   Non-environmental levies

There is no issue as to the status of the UK's seven main non-environmental taxes as such.[108]

7.3.3   Tax subsidies within environmental levies

The nature of the main exemptions within the environmental levies is summarised below. The exemptions in question are analysed in detail in the relevant paragraphs of the rest of the study.

1   *Landfill tax:*[109] The exemptions from the tax relate to: disposals of material removed from water; disposals of materials removed from contaminated land; the

---

[104]   Richardson and Chanwai draw attention to the fact that energy supply companies may pass on the costs of purchasing renewable energy to their customers and refer to the suggestion in recent retail energy prices that the RO has added 3 per cent to consumers' electricity bills (see Benjamin J. Richardson and Kiri L. Chanwai, 'Taxing and Trading in Corporate Energy Activities: Pioneering UK Reforms to Address Climate Change' (2003) 14(1) ICCLR 18–27).

[105]   See paras 11.2.2 and 21.3 below.

[106]   See para. 7.2.2.3 above.

[107]   See para. 7.2.2.2 above.

[108]   See para. 1.4.3.1 above.

[109]   See para. 1.4.2.1(1) above.

use of inert materials for site restoration; the disposal of inert waste at quarries; and the disposal of the remains of dead domestic pets.[110] The exemptions all operate by removing the operation in question from the scope of the concept of the taxable disposal.

Tax rates are differentiated according to whether the waste in question is general or inert waste.[111]

2   *Climate change levy:*[112] Excluded from the levy are direct supplies for domestic or non-business charitable use and any supply made before 1 April 2001.[113]

Exempted from the levy are: gas supplies for burning in Northern Ireland;[114] gas supplies for burning outside the UK; supplies for use in public transport, on the railways, or international shipping; supplies to producers of taxable commodities other than electricity; supplies other than for use as fuel, that is, the electrolytic processes, steam reformation, dual use functions and non-heating uses specified in regulations; supplies where the person supplied intends the commodity to be used as fuel in a recycling process; and supplies of renewable source and good quality combined heat and power ('CHP') electricity.[115]

Excluded supplies are outside the scope of climate change levy, whereas exempt supplies are ones that, if not exempted, would have been within the scope of the levy.

There are reduced rates of tax for horticultural producers installations within the IPPC regime that have entered into climate change agreements.[116]

The exemptions and reduced rate provisions were subsidies for which state aid clearance was required.[117]

3   *Aggregates levy:* The exemptions from aggregates levy operate either by exempting the aggregate itself or by exempting the spoil and other by-products of a process, that is, an exempt process.

Exempt aggregate includes aggregate consisting wholly or mainly of coal, lignite, slate or shale and the spoil or waste from an industrial combustion process or from the smelting or refining of metal.[118]

Exempt processes include the cutting of any rock to produce stone with one or more flat surfaces; any process by which certain substances (including anhydrite and ball clay) are extracted or separated from any aggregate; and any process for the production of lime or cement, from lime alone or lime and another substance.[119]

The levy is charged at a single rate.[120]

---

110   See para. 15.3 below.
111   See para. 15.1 below.
112   See para. 1.4.2.2(1) above.
113   See para. 14.3 below.
114   On a temporary basis only (see para. 14.2 below).
115   See para. 14.3 below.
116   See para. 14.1 below.
117   See para. 12.2.7 below.
118   See para. 13.4 below.
119   See para. 13.5 below.
120   See para. 13.2 below.

4   *Workplace parking levies:* Transport Act 2000, s.187, grants powers to
    set exemptions, reduced rates or limits on workplace parking charges by
    regulations.[121]
5   *Road user charging:* Transport Act 2000, s.172, provides the power for regulations
    to set exemptions from charges, reduced rates or limits on charges which will
    apply to all charging schemes. By s.172(2), any charging scheme will be able to
    set additional exemptions, reductions or limits as the authority wishes, subject to
    approval.

    Under the Central London scheme, a discount of 90 per cent of the standard
    £5 charge is available to residents living within the congestion charging zone,
    and there is also a range of exemptions for particular types of vehicle. Some of
    these are of a purely environmental nature (for example for electrically propelled
    vehicles) but they are not exclusively so.[122]
6   *The UK Emissions Trading Scheme:* Direct participants in the UK ETS are
    eligible to receive incentive payments, calculated in accordance with the rules
    of the Scheme. Agreement participants are entitled to a reduced rate of tax. Both
    mechanisms were in the nature of subsidies, for which state aid clearance was
    required.[123]

7.3.4   Environmental subsidies within non-environmental levies

Below are summarised the main environmental exemptions and reliefs within the
non-environmental tax codes. They are discussed in detail as indicated in the relevant
paras of the rest of the book.

7.3.4.1   Employee taxes

(1) Income tax
a. *Exemptions:* In general, the earnings of an office or employment, including the
    value of any benefits in kind (usually their 'cash equivalent'), are chargeable to
    income tax under the Income Tax (Earnings and Pensions) Act 2003.[124] However,
    a number of relevant items are outside the scope of, what was, until 5 April 2003,
    Schedule E, altogether. These include: the provision for employees of works
    bus services; financial or other support for bus services used by employees; the
    provision for employees' use of cycles or cyclists' safety equipment; payments
    made to employees for carrying other employees travelling on business; the
    provision of up to six free meals per annum for employees cycling to work; and
    the use of works minibuses for certain shopping trips from work.[125]

    Although not a case of exemption, the rules for bringing into charge the 'cash
    equivalent' of vehicles and fuel are not as favourable as once they were. The

---

[121]   There are as yet (December 2004) no such schemes but the first workplace parking levy
        scheme is likely to be introduced in Nottingham in April 2005 (see para. 17.1 below).
[122]   See para. 18.2 below.
[123]   See para. 12.2.7 below.
[124]   See para. 1.4.3.1 above.
[125]   See para. 23.3 below.

amount of the cash equivalent is now based on the car's carbon dioxide emissions, no allowance being made for high business mileage or the age of the car.[126]

b. *Permitted deductions:* These are very restricted and, significantly, do not include the expenses of ordinary commuting.[127]

(2) National insurance contributions

Since the tax base of National insurance contributions ('NICs') is the employed earner's earnings, albeit in separate legistaion from that applicable to income tax, the general rule is that a payment in kind, or by way of the provision of services, board and lodging or other facilities is to be disregarded in the calculation of earnings.[128]

On this basis, and without more, all of the items within the income tax exemptions that are not 'earnings' for NICs purposes would fall outside the scope of NICs. However, the position is not as simple as this, since certain items that do not qualify as 'earnings' under the general definition are nonetheless brought within the scope of NICs. These include non-cash vouchers, the range of exemptions for which is apt not to include, for example, vouchers provided to employees for use on works buses.[129]

7.3.4.2 Business taxes

1 *Income tax and corporation tax:* The scope of capital allowances in respect of expenditure on machinery and plant covers at least two areas of 'green' expenditure: expenditure on cars which are either electrically propelled or have low carbon dioxide emissions and on energy-saving plant and machinery of a type specified by the government.[130] One hundred per cent allowances are available in both cases. Additionally, 100 per cent allowances are available against rental income for the conversion of space above shops into small self-contained flats.[131]

Relief for acquisitions as trading stock, or as capital assets of a trade or a property business, for expenditure on remedial works on contaminated land, is also available. Tax relief is available on a notional 150 per cent of expenditure, which may explain why it is very closely circumscribed.[132]

2 *Value added tax:* A 5 per cent VAT rate is available for grant-aided installation in dwellings of energy saving materials and heating equipment.[133]

3 *Stamp duties:* There is an exemption from stamp duty land tax ('SDLT') for both conveyances and transfers and leases of land. This is subject to a value cap for residential properties but is complete for non-residential ones until 2006.[134]

---

126 See para. 23.2 below.
127 See para. 23.1 below.
128 See Social Security (Contributions) Regulations 2001, S.I. 2001 No. 1004, reg. 25 and Sched. 3, Part II, para. 1.
129 See para. 23.3 below.
130 For details, see paras 21.3.1, 24.3 and 24.4 below.
131 See para. 24.5.2 below.
132 See para. 24.5.1 below.
133 See para. 21.3.1 below.
134 See para. 24.5.3 below.

7.4 Concluding comments

Throughout the study thus far, the pattern of the overall discussion has been one of a gradual move from the general to the specific. The material in the present and previous chapter has been designed to shown how environmental taxes and other economic instruments can be analysed both in environmental and energy law terms and in tax terms. Environmental taxation law occupies a difficult terrain across what in the UK, at least, have traditionally been regarded as entirely distinct areas of law.

Chapter 8

International Aspects

8.1   Introduction

This chapter seeks to complete the conspectus of the contexts of environmental taxation law by locating the UK's environmental levies and subsidies within the framework both of international environmental law and the multilateral rules on the taxation of international trade.

The discussion unfolds in three main stages, followed by two brief paras dealing with subsidiary issues. Paragraph 8.2 provides a general introduction for the non-specialist to the nature and sources of public international law. Some idea of this background material is necessary for a full appreciation of the topics considered elsewhere in the chapter, since the principal instrument in both the international environmental and the international trading contexts is the treaty, or international agreement. A significant part is played also, however, at least in the environmental context, by more controversial concepts, such as customary international law and so-called 'soft law', including declarations and recommendations. The principal characteristics of these sources of law are also delineated in para. 8.2. This background material is of some significance, in terms of the study as a whole, since it helps to explain the extent and shape of the environmental and taxation rules applicable within the EU and its Member States. Such internal EU issues will form the subject matter of Chapter 12 below. The main theme of the present chapter, however, is the UK's environmental and taxation ties, via the law and institutions of the EU, with the world beyond the EU's borders.

International agreements have, of course, been reached on a number of the most serious environmental problems, most importantly air and atmospheric pollution. Chief among these, arguably, is the 1992 United Nations Framework Convention on Climate Change and its eponymous Kyoto Protocol of 1997. The details of the main agreements in this sector are explored in para. 8.3, where they are briefly related to the economic instruments justified in the UK by reference to those agreements.[1] Although there is some discussion of the main international agreements concluded in order to control pollution by waste, this is relatively brief, since these have been by no means as influential in the creation of the UK's waste tax as have been the initiatives of the European Commission. Discussion of these initiatives is accordingly postponed to the consideration of internal EU issues in Chapter 12 below.

The third main stage of the discussion, contained in para. 8.4, seeks to locate the UK, together with its existing green levies and subsidies, within the framework of the 1994 revision of the General Agreement on Tariffs and Trade ('GATT 1994'). In doing so, it seeks to elucidate the nature of the relationship between the international environmental agreements referred to above and the system of multilateral trade agreements based on GATT 1994. Since UK environmental regulation, as it relates to

---

[1]   See Chapter 6 above for details of the UK's transposition of its international and EU obligations into domestic law.

air and atmospheric pollution, has its origins in international environmental law then, to the extent that regulatory instruments comprise environmental levies and subsidies, those already in existence must be consistent with the relevant provisions of GATT 1994, while the discussion of those as yet only envisaged must take place with those provisions in mind. Besides explaining the place of the UK's existing environmental taxes and other economic instruments in the environmental and trade contexts, the present chapter therefore paves the way for the discussion of new directions in these areas in Part IV of the book.

Prior to summing up on Part II of the book, in Chapter 9, we briefly highlight two further relevant areas of public international law: those relating to energy and to air transport. The international aspects of each of these areas are not major themes in Parts III and IV but they do embody some key factors regarding the possibilities for future policy directions, especially, in the case of air transport, in relation to international tax or emissions trading schemes for airlines. The fact that air transport falls outside the provisions of the Kyoto Protocol is one of the Protocol's most striking features.

We begin this chapter, however, with a brief examination of the nature and sources of public international law, with special relevance to examples drawn from international environmental law and the law of international trade.

8.2   Public international law

8.2.1   Preliminary

Within the scope of public international law[2] are the intersecting areas of international environmental law and world trade law.[3] The expression 'international environmental law',[4] invariably used but not universally accepted,[5] is a convenient way of describing the whole body of public and private international law relating to environmental issues,[6] much of which originates in a series of UN-sponsored international agreements. Alongside the international agreements on the environment is the GATT 1994-based system of multilateral agreements on international trade that are administered by the World Trade Organization ('the WTO').[7] The extent, if any, to which the objectives of the two sets of treaties are reconcilable is one of the great questions of our time.[8] The crux of the legal argument is that, at present, international

---

2   See, generally, I. Brownlie, *Principles of Public International Law*, 5th edn (Oxford: Oxford University Press, 1999).

3   That is, insofar as the latter relates to the rules and institutions of the world trading system, as opposed to the rules and principles of the conflict of laws in relation, for example, to international contracts.

4   See Philippe Sands, *Principles of International Environmental Law*, 2nd edn (Cambridge: Cambridge University Press, 2003).

5   See, for example, Brownlie, *op. cit.*, ch. 12.

6   See Patricia Birnie and Alan Boyle, *International Law and the Environment*, 2nd edn (Oxford: Oxford University Press, 2002), pp. 1–2.

7   See para. 4.4.1 above.

8   The WTO's position is discussed in Michael Moore, *A World Without Walls – Freedom, Development, Free Trade and Global Governance* (Cambridge: Cambridge University

policy neither endorses any general environmental exception to the principle of free trade, nor does it seek to give free trade priority over environmental protection.[9]

Public international law is usually divided into 'hard law' and 'soft law' components. As set out in Art. 38 of the 1945 Statute of the International Court of Justice, the sources of international law, that is, the 'hard' law component, are:

> ... international conventions, whether general or particular, establishing rules expressly recognised by the contesting states;[10] (b) international custom, as evidence of a general practice accepted as law; (c) the general principles of law recognised by civilised nations; (d) subject to the provisions of Article 59,[11] judicial decisions and the teachings of the most highly qualified publicists of the various nations, as subsidiary means for the determination of rules of law.

None of these forms of hard law is necessarily such as to create rights upon which private persons can rely, nor can they invariably be used as the basis for legal action against the state or other public authority, in the way that European legislation may sometimes be used.[12] They do, however, create rules and principles that are enforceable between one state and another in respect of the relations between them. It is useful briefly to describe the forms of hard law in turn, although it should

---

Press, 2003). Moore, a former Prime Minister of New Zealand, was the Director-General of the WTO from 1999 to 2002. For a somewhat different view, see *The Case Against the Global Economy and for a Turn Toward the Local*, ed. by J. Mander and E. Goldsmith (San Francisco: Sierra Club Books, 1996). The problems of evaluating the arguments on each side and of the absence of data are summarised in, for example, Gerd Winter, 'The GATT and Environmental Protection: Problems of Construction' (2003) 15 JEL 113–40 (esp. pp. 113–15).

9   Birnie and Boyle, *op. cit.*, p. 698. The chapter of Birnie and Boyle entitled 'International Trade and Environmental Protection' was contributed by Thomas J. Schoenbaum. As Schoenbaum points out (Birnie and Boyle, *ibid.*), the preamble to the 1994 Marrakesh Agreement Establishing the World Trade Organization recognises that expanding the production of, and trade in, goods and services, must allow for: '... the optimal use of the world's resources in accordance with the objective of sustainable development, seeking both to protect and preserve the environment and to enhance the means for doing so in a manner consistent with their respective needs and concerns at different levels of economic development ...'. See also the Rio Declaration, Principle 12, referred to in para. 8.2.6 below. The bald statement of the legal issue in the text barely does justice to the range of economic, political and even moral issues surrounding it. For an insight into the strength of the various arguments, see the websites of the myriad charitable and political organisations engaged in the debate.

10 The concept of international conventions includes treaties, 'acts', 'agreements', 'covenants', 'pacts', 'protocols', etc.

11 It is by virtue of Art. 59 that the decisions of the ICJ have no binding force except as between the parties and in respect of the case under consideration.

12 Stuart Bell and Donald McGillivray, *Ball and Bell on Environmental Law: The Law and Policy relating to the Protection of the Environment*, 5th edn (London: Blackstone Press, 2000), p. 91. For the direct effect of EU law and actions for damages against Member States, see, for example, Stephen Weatherill and Paul Beaumont, *EU Law*, 3rd edn (London: Penguin, 1999), ch. 11.

perhaps be stressed that, among the five, treaties and custom are the main sources of international law in general.[13]

8.2.2   Treaties

Treaties are defined in the 1969 Vienna Convention on the Law of Treaties[14] as 'international agreements concluded between states in written form and governed by international law, whether embodied in a single instrument or in two or more related instruments and whatever their particular designation'.[15] The basic principle of the law of treaties is that a treaty cannot be applied until it has been ratified and come into force.[16] The process of ratifying, and hence, of the coming into force of a treaty, can – unless the two take place at the same time[17] – be a slow and painful one. As an extreme example, the 1997 Kyoto Protocol[18] to the 1992 United Nations Framework Convention on Climate Change[19] (always referred to as 'the Kyoto Protocol') provided that the Protocol would enter into force on the ninetieth day after the date on which not less than 55 parties to the 1992 Convention, including those that accounted in total for at least 55 per cent of the total 1990 carbon dioxide emissions of those parties, had deposited their instruments of ratification, acceptance or accession.[20] The law and procedure applicable to making, operating and terminating treaties[21] are contained in the 1969 Vienna Convention. The most recent multilateral environmental agreements ('MEAs') have been signed by the EU, as the relevant regional economic integration organisation.[22] This means that both the EU and its Member States may be parties to the MEA in question. In the case of the Kyoto Protocol, the burden of the EU's commitment to an 8 per cent reduction in greenhouse gas emissions by 2010 is thus shared between the EU Member States under a so-called 'burden-sharing agreement'.[23]

---

[13]   Anthony Aust, *Modern Treaty Law and Practice* (Cambridge: Cambridge University Press, 2000), p. 10.
[14]   The 1969 Vienna Convention was drafted by the International Law Commission, a body established in 1947 by the UN General Assembly, with the object of promoting the progressive development of international law and its codification (see Aust, *op. cit.*, pp. 6–7).
[15]   1969 Vienna Convention, Art. 2(1)(a).
[16]   *Ibid.*, Art. 26.
[17]   Aust, *op. cit.*, p. 75.
[18]   (1998) 37 ILM 22.
[19]   (1992) 31 ILM 851.
[20]   Kyoto Protocol, Art. 25(1).
[21]   Although the 1969 Vienna Convention does not apply to oral agreements, this does not affect the legal force of such agreements, 'or the application to them of any of the rules in the Convention to which they would be subject under international law independently of the Convention, such as customary international law' (see 1969 Vienna Convention, Art. 3 and Aust, *op. cit.*, pp. 7 and 16).
[22]   See Bell and McGillivray, *op. cit.*, p. 87.
[23]   See now Council Decision 02/358/EC, (2002) OJ L130 1. Article 4 of the Kyoto Protocol allowed parties thereto to fulfil their commitments thereunder jointly; EU Member States accordingly entered into the burden-sharing agreement of 16–17 June 1998 (Doc. 9702/98

Recent developments in the interpretation of GATT 1994 have indicated that GATT 1994 should be interpreted, not in accordance with GATT 1994 interpretation norms, but with the provisions of Arts 31–33 of the 1969 Vienna Convention.[24] The principles contained in these articles, headed 'Interpretation of Treaties', may briefly be summarised as follows:

1 Article 31(1) provides for the interpretation of treaties 'in good faith in accordance with the ordinary meaning to be given to the terms of the treaty in their context and in the light of its object and purpose'; Art. 31(2) then states what the context comprises; and Art. 31(3) specifies what, together with the context, must be taken into account, that is: 'any subsequent agreement between the parties regarding the interpretation of the treaty or the application of its provisions;[25] any subsequent practice in the application of the treaty which establishes the agreement of the parties regarding its interpretation;[26] and any relevant rules of international law applicable in the relations between the parties'.[27]

2 Article 32 allows recourse to supplementary means of interpretation (including *travaux préparatoires*) to confirm the meaning resulting from the application of Art. 31 or to determine the meaning where interpretation under Art. 31 leaves the meaning ambiguous or obscure or leads to a manifestly absurd or unreasonable result.

3 Generally speaking, by Art. 33, a treaty is equally authoritative in each language in which it has been authenticated.

8.2.3   Customary international law

Customary international law, the second source of hard law referred to in Art. 38 of the ICJ statute, consists of two elements: '(1) a general convergence in the practice of states from which one can extract a norm (standard of conduct); and (2) *opinio juris*, the belief by states that the norm is legally binding on them'.[28] Such a definition needs only to be stated for it to be evident that the ascertainment of the relevant custom in a particular case is potentially fraught with difficulty, involving, as it may, both research and the exercise of careful judgment.[29] As to (1), the practice of states is thought to cover any act or statements by a state from which views about

---

of 19 June 1998, Annex I). The burden-sharing agreement actually allows Greece, Spain, Iceland, Portugal and Sweden to increase their greenhouse gas emissions, these being offset by greater than expected reductions in other Member States, notably the UK! (see Maurice Sunkin, David Ong and Robert Wight, *Sourcebook on Environmental Law*, 2nd edn (London: Cavendish, 2002), pp. 99, 143 and 148). See further P.G.G. Davies, 'Global warming and the Kyoto Protocol' (1998) 47 ICLQ 446–61.

24 See Birnie and Boyle, *op. cit.*, p. 704, and paras 8.4.2(2) and 8.4.2(3) below.
25 1969 Vienna Convention, Art. 31(3)(a).
26 *Ibid.*, Art. 31(3)(b).
27 *Ibid.*, Art. 31(3)(c).
28 Aust, *op. cit.*, p. 10. See Malcolm Shaw, *International Law*, 5th edn (Cambridge: Cambridge University Press, 2003), pp. 65–92.
29 See Birnie and Boyle, *op. cit.*, p. 16.

customary law may be inferred[30] whilst, as to (2), the belief that the practice is obligatory by virtue of the existence of a rule of law requiring it, is a subjective requirement.[31] To customary international law may be owed, in the environmental context, the precautionary principle or 'precautionary approach', the polluter pays principle[32] and the preventive principle. Principle 15 of the 1992 Declaration of the United Nations Conference on Environment and Development, made at the UN Conference on Environment and Development held in Rio de Janeiro in 1992 ('the Rio Declaration')[33] states that:

> In order to protect the environment, the precautionary approach shall be widely applied by states according to their capabilities. Where there are threats of serious or irreversible damage, lack of full scientific certainty shall not be used as a reason for postponing cost-effective measures to prevent environmental degradation.

In his dissenting opinion in a famous ICJ decision,[34] one judge described the precautionary approach as one that was gaining increasing support as part of the international law of the environment.[35] As to the polluter pays principle, this is reflected in Principle 16 of the Rio Declaration, which runs as follows:

> National authorities should endeavour to promote the internalisation of environmental costs and the use of economic instruments, taking into account the approach that the polluter should, in principle, bear the cost of pollution, with due regard to the public interest and without distorting international trade and investment.

The 1990 International Convention on Oil Pollution Preparedness, Response and Co-operation, as well as the 1992 Convention on the Transboundary Effects of Industrial Accidents, referred to the polluter pays principle as '... a general principle of international environmental law'.[36] The principle has for long been promoted by the OECD[37] and is specifically referred to in Art. 174(2), European Treaty (ex 130r) as one of the bases

---

[30]   See M. Akehurst, 'Custom as a Source of International Law' (1974–75) 47 BYIL 1.
[31]   See *Nicaragua v. US*, [1986] ICJ Rep. 14, 108–109.
[32]   See Bell and McGillivray, *op. cit.*, p. 92.
[33]   UN Doc. A/CONF151/26/Rev. 1, reproduced in Sunkin, Ong and Wight, *op. cit.*, p. 69.
[34]   *Request for an Examination of the Situation in Accordance with Paragraph 63 of the Court's Judgment in the 1974 Nuclear Tests case*, [1995] ICJ Rep. 288.
[35]   [1995] ICJ Rep. 288, 342 (Judge Weeramantry).
[36]   In the US, this has been taken to surprising lengths, under the 'Carter Act' (Comprehensive Environmental Response, Compensation, and Liability Act 1980), for instance resulting in a UK investment trust, Fleming American Investment Trust plc, which had owned a factory nearly a century before, in a previous guise, having to contribute towards the clean-up costs.
[37]   See, for example, Organisation for Economic Co-operation and Development, *Environmentally-Related Taxes in OECD Countries: Issues and Strategies* (Paris: OECD, 2001), p. 16, which contains the OECD's definition of the polluter pays principle, as originally expressed by the organisation in 1972. The OECD there emphasises that, in the context of environmentally related taxation, the principle is a non-subsidisation principle, 'meaning simply that governments should not as a general rule give subsidies to their industries for pollution control' (*ibid.*).

of Community policy on the environment. Finally, the preventive principle, that is, the prohibition of any activity which actually causes, or will cause, environmental damage or pollution,[38] is reflected in Principle 2 of the Rio Declaration:

> States have, in accordance with the Charter of the United Nations and the principles of international law, the sovereign right to exploit their own resources pursuant to their own environmental and developmental policies, and the responsibility to ensure that activities within their jurisdiction or control do not cause damage to the environment of other states or of areas beyond the limits of national jurisdiction.[39]

8.2.4 General principles of law

General principles of law recognised by civilised nations are the third source of law referred to in Art. 38 of the Statute of the ICJ. It is unclear whether the reference here is to the principles of international law (for example, as to the freedom of the seas) or simply to the principles of domestic legal systems (for example, as to the admissibility of evidence in legal proceedings). It appears, however, that this source is most frequently used, when it is used at all, in order to reason from analogy in relation, for example, to rules of procedure, evidence and jurisdiction. 'General principles of law' thus have little specific application to environmental issues but are of general significance in international law.[40]

8.2.5 Judicial decisions and jurisprudential writing

Finally, Art. 38 of the ICJ statute refers to 'judicial decisions and the teachings of the most highly qualified publicists of the various nations, as subsidiary means for the determination of rules of law'. This is expressly subject to Art. 59 of the ICJ statute, which provides that the decisions of the ICJ have no binding force, 'except as between the parties and in respect of the case under consideration'. Subject to this qualification, the judicial decisions referred to here are those of the ICJ itself; of other international courts (including arbitral tribunals);[41] of the European Court of Human Rights;[42] of the International Tribunal for the Law of the Sea[43] and of the

---

38 See Sunkin, Ong and Wight, *op. cit.*, p. 49.
39 See also Principle 21 of the 1972 Stockholm Declaration of the United Nations Conference on the Human Environment, reproduced in Sunkin, Ong and Wight, *op. cit.*, p. 63.
40 See Birnie and Boyle, *op. cit.*, p. 20.
41 See Brownlie, *op. cit.*, pp. 19–24 and Robert Jennings, 'The Judiciary, International and National, and the Development of International Law' (1996) 45 ICLQ 1–12. But see the comment in *Bowett's Law of International Institutions*, ed. by Philippe Sands and Pierre Klein, 5th edn (London: Sweet and Maxwell, 2001), para. 13–042, where the learned editors note that: 'What the Court has not done is refer to judgments of other international courts, no doubt bearing in mind its position as the "principal judicial organ of the United Nations". Whether this approach is tenable over the long term, given the increased specialisation of various areas of international law, remains unclear'.
42 See para. 4.3.7 above.
43 Set up under the 1982 Law of the Sea Convention, Annex VI (not discussed elsewhere in this book).

national courts of particular states.[44] With regard to the writings of publicists, besides including the writings of eminent international lawyers, these may also include reports of international codification bodies, such as, for example, the International Law Commission.

8.2.6   Soft law

Besides the various sources of hard law, there is what is usually referred to as 'soft law'. The idea behind the terminology is to show that the relevant document is not law as such but that specific attention must be accorded to it because of the influence that it exerts on the international scene. Bell and McGillivray draw together a number of examples of soft law. It is characterised, they say, by the fact that '… it contains general norms rather than specific rules … [and] provides a guide as to how disputes might be resolved rather than hard-and-fast rules applying to specific situations'.[45] Their list of examples consists of the following:

1   declarations;
2   principles;
3   recommendations; and
4   standards.[46]

Two of these merit some comment in the present context. Besides reflecting the agreed aspirations of the international community, declarations also contribute to the creation of customary international law and the consolidation of existing customs. The Rio Declaration, setting out 27 Principles of international environmental law, which together constitute the main outlines of the concept of sustainable development,[47] is a key example of this form of soft law. Principles 2 (the preventive principle), 15 (the precautionary principle) and 16 (the polluter pays principle) have already been mentioned. Other Principles of some importance in the present context are Principles 12 and 17. Principle 12 reads as follows:

> States should co-operate to promote a supportive and open international economic system that would lead to economic growth and sustainable development in all countries, to better address the problems of environmental degradation. Trade policy measures for environmental purposes should not constitute a means of arbitrary or unjustifiable discrimination or a disguised restriction on international trade. Unilateral actions to deal with environmental challenges outside the jurisdiction of the importing country should be avoided. Environmental measures addressing transboundary or global environmental problems should, as far as possible, be based on an international consensus.

---

44   See para. 4.2.1.5 above.
45   Bell and McGillivray, *op. cit.*, p. 94.
46   That is, other than those that do have the force of law, for example, Community regulations that impose standards (see, for example, John F. McEldowney and Sharron McEldowney, *Environmental Law and Regulation* (London: Blackstone Press, 2001), esp. pp. 12–14).
47   See Birnie and Boyle, *op. cit.*, pp. 44–47.

Principle 17 provides for environmental impact assessments ('EIAs'):[48]

> Environmental impact assessment, as a national instrument, shall be undertaken for proposed activities that are likely to have a significant adverse impact on the environment and are subject to a decision of a competent national authority.

As to recommendations, also of some significance in the present study are OECD recommendations on the development of environmental policy. These include the 1991 OECD Council Recommendation on the use of economic instruments in environmental policy,[49] which sought to make a general case for the more consistent and extended use of economic instruments. Amongst its main exhortations were the following:

> that Member countries:
> i.   make a greater and more consistent use of economic instruments as a complement or a substitute to other policy instruments such as regulations, taking into account national socio-economic conditions;
> ii. work towards improving the allocation and efficient use of natural and environmental resources by means of economic instruments so as to better reflect the social cost of using these resources;
> iii. make effort to reach further agreement at international level on the use of environmental policy instruments with respect to solving regional or global environmental problems as well as ensuring sustainable development ...

The special importance of such a Council Recommendation is that it is a unanimous recommendation from OECD member governments to themselves.[50]

8.3   International environmental law

Having illustrated the main sources of public international law by reference to the environment and international trade, we now turn to look in somewhat more detail at MEAs in two sectors: air and atmospheric pollution and the disposal of waste. The discussion of each of these topics will then be related to the economic instruments which have been deployed in the UK to assist in the achievement of that country's obligations under international law.

---

48   See paras 6.2.4 above and 12.2.4 below.
49   *Recommendation of the Council on the Use of Economic Instruments in Environmental Policy*, 31 January 1991 – C(90)177/Final. See Organisation for Economic Co-operation and Development, *Environmental Policy: How to Apply Economic Instruments* (Paris: OECD, 1991). See para. 1.2.1.5(2) above.
50   See Bowett, *op. cit.*, para. 11–007; also, David Williams, *EC Tax Law* (Longman: London, 1998), p. 11.

8.3.1   Air and atmospheric pollution

8.3.1.1   Introduction

Reference has already been made to the 1992 United Nations Framework Convention on Climate Change, as well as to the 1997 Kyoto Protocol thereto.[51] These are considered in somewhat more detail in the present para., together with (briefly):

1   the 1979 Geneva Convention for the Control of Long-Range Transboundary Air Pollution ('the 1979 Geneva Convention')[52] and its related Protocols;
2   the 1985 Vienna Convention for the Protection of the Ozone Layer;[53] and
3   the 1987 Montreal Protocol on Substances that Deplete the Ozone Layer.[54]

Before describing the treaty sources of air and atmospheric pollution regulation, it is important to note the principle of customary international law applied in the famous *Trail Smelter* arbitration.[55]

8.3.1.2   The Trail Smelter *arbitration*

Still the only example of an international adjudication on transboundary air pollution, the *Trail Smelter* arbitration saw the application by the arbitral tribunal of the principle that:

> ... no state has the right to use or permit the use of its territory in such a manner as to cause injury by fumes in or to the territory of another or the properties or persons therein, when the case is of serious consequence and the injury is established by clear and convincing evidence.[56]

The arbitral tribunal had been established to determine whether damage had been caused in the state of Washington by smoke emissions from a smelter located in Canada, seven miles from the US border. If it had, then the tribunal had to decide the level of compensation that had to be paid and the measures that had to be taken in order to prevent further damage. Having determined the causation question on scientific evidence, and in reliance on the principle extracted above, the tribunal went on to lay down a regime for the operation of the smelter in the future, in reliance on a precursor to the preventive principle.[57] The nature of the *Trail Smelter* principle, applicable as it is only in claims between states, has meant that it has been of limited utility,[58] although it seems to be a good basis for a general principle that customary

---

51   See paras 8.1 and 8.2.2 above.
52   (1979) 18 ILM 1442.
53   (1987) 26 ILM 1529.
54   *Ibid.*, 1550. This is generally taken to be a particularly successful agreement.
55   (1939) 33 AJIL 182 and (1941) 35 AJIL 684.
56   (1941) 35 AJIL 684, 716.
57   See para. 8.2.3 above.
58   See Birnie and Boyle, *op. cit.*, p. 505.

international law forbids one state from significantly harming another's environment through transboundary pollution.

8.3.1.3   The 1992 United Nations Framework Convention on Climate Change

Together with the Convention on Biological Diversity,[59] the 1992 United Nations Framework Convention on Climate Change ('the Framework Convention')[60] was opened for signature at the United Nations' Conference on Environment and Development at Rio de Janeiro in 1992 (the conference known as 'the Earth Summit'). Having entered into force in March 1994, the Framework Convention has now been signed by 166 parties, 188 parties having ratified it, accepted it, approved it or acceded to it.[61] Article 2 of the Framework Convention states its objective as being the:

> ... stabilisation of greenhouse gas concentrations in the atmosphere at a level that would prevent dangerous anthropogenic interference with the climate system. Such a level should be achieved within a time frame sufficient to allow eco-systems to adapt naturally to climate change, to ensure that food production is not threatened and to enable economic development to proceed in a sustainable manner.

Among other things, by Art. 4 of the Framework Convention, the states parties undertake to:

1   develop, periodically update and publish national inventories of anthropogenic emissions by sources and removals by sinks of all greenhouse gases not controlled by the Montreal Protocol;[62]
2   'formulate, implement, publish and regularly update national and, where appropriate, regional programmes containing measures to mitigate climate change';[63]
3   'promote and cooperate in the development, application and diffusion ... of technologies, practices and processes that control, reduce or prevent anthropogenic emissions' as above in all relevant sectors, including the energy, transport, industry, agriculture, forestry and waste management sectors;[64]
4   'promote sustainable management and conservation of all greenhouse gases not controlled by the Montreal Protocol, including biomass, forests and oceans as well as other terrestrial, coastal and marine eco-systems';[65]

---

59   (1992) 31 ILM 818.
60   (1992) 31 ILM 851.
61   That is, as at 23 September 2004 (see www.unfccc.de/resource/convkp.html). Acceptance and approval has the same effect as ratification and accession has the same effect ratification but is not preceded by signature (see Aust, *op. cit.*, p. xxxiii).
62   Framework Convention, Art. 4(1)(a). For the Montreal Protocol, see para. 8.3.1.5 below.
63   Framework Convention, Art. 4(1)(b).
64   *Ibid.*, Art. 4(1)(c).
65   *Ibid.*, Art. 4(1)(d).

5   so far as feasible for the states parties so to do, to take climate change considerations
     into account in policy-making:

> ... and employ appropriate methods, for example impact assessments, formulated and
> determined nationally, with a view to minimising adverse effects on the economy, on
> public health and on the quality of the environment, of projects or measures undertaken by
> them to mitigate or adapt to climate change; ...[66]

6   promote and cooperate in scientific, technological, technical, socioeconomic and
     other research related to the climate system;[67] and
7   'promote and cooperate in education, training and public awareness related to
     climate change and encourage the widest participation in this process'.[68]

By Art. 4(2) of the Framework Convention, developed country parties, as well as
certain other parties,[69] commit themselves to taking the lead in modifying longer-term
trends in anthropogenic emissions consistent with the objective of the Convention
and specifically to 'adopt national policies and take corresponding measures on the
mitigation of climate change, by limiting ... anthropogenic emissions of greenhouse
gases and protecting and enhancing ... greenhouse gas sinks and reservoirs'.[70]

Under Art. 4(2)(b) of the Framework Convention, the same parties must submit
within six months of the Framework Convention coming into force, and periodically
thereafter, detailed information on the matters referred to above, 'with the aim of
returning individually or jointly to their 1990 levels these anthropogenic emissions
of carbon dioxide and other greenhouse gases not controlled by the Montreal
Protocol'.[71] This information, continues Art. 4(2)(b), will be periodically reviewed
by the Conference of the Parties.

Financial resources to enable the developing country parties to meet their obligations
and to assist them in coping with the effects of climate change are to be provided by
the developed country parties[72] under Arts 2(3) and 2(4). By Art. 4(8), the parties
agree to give full consideration to what actions are necessary under the Framework
Convention to assist developing country parties that may be, for example, small
island countries, countries with low-lying coastal areas, countries prone to natural
disasters, etc.

Article 7 establishes the Conference of the Parties as the supreme body of the
Framework Convention, with the function of reviewing its implementation, etc, and
(by Art. 8) a secretariat is to be established, together with a subsidiary body for
scientific and technological advice and a subsidiary body for implementation.[73]

---

[66]   *Ibid.*, Art. 4(1)(f).
[67]   *Ibid.*, Art. 4(1)(g).
[68]   *Ibid.*, Art. 4(1)(i).
[69]   *Ibid.*, Annex I.
[70]   *Ibid.*, Art. 2(a).
[71]   *Ibid.*, Art. 2(b).
[72]   *Ibid.*, Annex II.
[73]   *Ibid.*, Arts 9 and 10.

8.3.1.4   The Kyoto Protocol

The 1997 Kyoto Protocol[74] to the Framework Convention, although notoriously not yet in force,[75] sets out the detail of the legally-binding greenhouse gas ('GHG') emissions reduction targets. Article 3 provides, in part:

> 1. The Parties included in Annex I[76] shall, individually or jointly, ensure that their aggregate anthropogenic carbon dioxide equivalent emissions of the greenhouse gases listed in Annex A do not exceed their assigned amounts, calculated pursuant to their quantified emission limitation and reduction commitments inscribed in Annex B and in accordance with the provisions of this Article, with a view to reducing their overall emissions of such gases by at least 5 per cent below 1990 levels in the commitment period 2008 to 2012.
>
> 2. Each Party included in Annex I shall, by 2005, have made demonstrable progress in achieving its commitments under this Protocol …

The GHGs listed in Annex A are six in number:

1   carbon dioxide (CO_2);
2   methane (CH_4);
3   nitrous oxide (N_2O);
4   hydrofluorocarbons (HFCs);
5   perfluorocarbons (PFCs); and
6   sulphur hexafluoride (SF_6).

Annex B contains differentiated targets for the Annex I parties with, for example, the EU having a reduction target of 8 per cent, the US 7 per cent and Japan 6 per cent. In this connection, it should be noted that, under the burden-sharing agreement referred to above,[77] the EU's 8 per cent commitment is translated into a commitment on the part of the UK to reduce its 1990 levels of all greenhouse gases by 12.5 per cent by 2010.[78]

Besides detailing the targets, however, the Kyoto Protocol also includes provisions for the parties to use economic instruments to achieve their targets (the so-called 'Kyoto flexible mechanisms'), including the following:

1   emissions trading schemes;
2   joint implementation ('JI'); and
3   the Clean Development Mechanism ('the CDM').[79]

---

74   (1998) 37 ILM 22.
75   That is, as at December 2004. See para. 20.2 below.
76   That is, to the Framework Convention (see para. 8.3.1.3 above). See also para. 4.4.3n above (Annex I parties).
77   See para. 8.2.2n above.
78   See Chapter 12 below and para. 8.2.2n above.
79   See Scott Barrett, 'Political Economy of the Kyoto Protocol', in *Environmental Policy: Objectives, Instruments, and Implementation*, ed. by Dieter Helm (Oxford: Oxford University Press, 2000), pp. 111–41.

Emissions trading schemes are provided for by Art. 17, which states:

> The Conference of the Parties shall define the relevant principles, modalities, rules and guidelines, in particular for verification, reporting and accountability for emissions trading. The Parties included in Annex B may participate in emissions trading for the purposes of fulfilling their commitments under Article 3. Any such trading shall be supplemental to domestic actions for the purpose of meeting quantified emission limitation and reduction commitments under that Article.

Joint implementation, provided for by Art. 6, allows the parties listed in Annex I to the Framework Convention, for the purpose of meeting their commitments under Art. 3, to:

> ... transfer to, or acquire from, any other such Party emission reduction units ['ERUs'] resulting from projects aimed at reducing anthropogenic emissions by sources or enhancing anthropogenic removals by sinks of greenhouse gases in any sector of the economy.

The exercise of this power is subject to a number of provisos, including those that any such project must have the approval of the parties involved and that the acquisition of ERUs be supplemental to domestic actions for meeting Art. 3 commitments.[80] Theoretically at least, such JI projects are a cost-efficient way for global targets to be achieved since, at the margin, it is cheaper for some countries to abate their greenhouse gases compared to other countries.[81]

The CDM, which is defined in Art. 12, has the stated purpose of assisting non-Annex I parties to achieve sustainable development and to contribute to the Framework Convention's ultimate objective,[82] whilst also assisting Annex I parties to meet their commitments under Art. 3.[83] This is to be achieved through the method elaborated in Art. 12(3), by which non-Annex I parties benefit from project activities resulting in certified emission reductions, that is, 'credits', which credits may then be used by Annex I parties to contribute to compliance with part of their quantified emission limitation and reduction commitments under Art. 3.

A major source of GHG emissions, of course, is international civil aviation. Of the six GHGs listed in Annex A to the Kyoto Protocol, the most relevant to aviation is carbon dioxide. Article 2(2) of the Protocol gives the Annex I parties responsibility for limiting or reducing GHG emissions from aviation bunker fuels, international aviation emissions being excluded from the Protocol's emissions reduction targets. This is to be achieved by working through the ICAO,[84] the relevant UN specialised agency.

---

[80]   Kyoto Protocol, Arts 6(1)(a) and 6(1)(d).
[81]   See A.D. Ellerman, H.D. Jacoby and A. Decaux, *The Effects on Developing Countries of the Kyoto Protocol and Carbon Dioxide Emissions Trading* (Washington DC: World Bank Policy Research Paper 2019, 2000). See also Zhong Xiang Zhang, 'Greenhouse Gas Emissions Trading and the World Trading System' (1998) 32 JWT 219–39.
[82]   See para. 8.3.1.3 above.
[83]   Kyoto Protocol, Art. 12(2).
[84]   See para. 4.4.3 above.

8.3.1.5   Other relevant agreements

It remains briefly to consider the other treaties referred to above, that is, the 1979 Geneva Convention and its related Protocols; the 1985 Vienna Convention for the Protection of the Ozone Layer;[85] and the 1987 Montreal Protocol on Substances that Deplete the Ozone Layer.[86]

The 1979 Geneva Convention for the Control of Long-Range Transboundary Air Pollution ('the 1979 Geneva Convention')[87] came into force in 1983. It is 'the only major regional multilateral agreement devoted to the regulation and control of transboundary air pollution',[88] and was negotiated through the United Nations Economic Commission for Europe ('UNECE').[89] It applies to pollution having 'adverse effects in the area under the jurisdiction of another state at such a distance that it is not generally possible to distinguish the contribution of individual emission sources or groups of sources'.[90] The contracting parties promise only to 'endeavour to limit and, as far as possible, gradually reduce and prevent air pollution including long-range transboundary air pollution'.[91] However, each contracting party also '... undertakes to develop the best policies and strategies including air quality management systems, and, as part of them, control measures compatible with balanced development, in particular by using the best available technology which is economically feasible and low and non-waste technology'.[92] The institutions of the 1979 Geneva Convention comprise an executive body, made up of environmental advisers to UNECE governments[93] and a secretariat provided by the UNECE.[94] So far, five protocols related to the 1979 Geneva Convention have entered into force,[95] with three others having been adopted.[96]

---

[85]   (1987) 26 ILM 1529.
[86]   *Ibid.*, 1550.
[87]   (1979) 18 ILM 1442.
[88]   Birnie and Boyle, *op. cit.*, p. 508.
[89]   See www.unece.org.
[90]   1979 Geneva Convention, Art. 1(b).
[91]   *Ibid.*, Art. 2.
[92]   *Ibid.*, Art. 6.
[93]   *Ibid.*, Art. 10.
[94]   *Ibid.*, Art. 11.
[95]   That is, the 1984 Geneva Protocol on Long-term Financing of the Co-operative Programme for Monitoring and Evaluation of the Long-Range Transmission of Air Pollutants in Europe, (1985) 24 ILM 484; the 1985 Helsinki Protocol on the Reduction of Sulphur Emissions or Their Transboundary Fluxes, (1988) 27 ILM 707; the 1988 Sofia Protocol Concerning the Control of Emissions of Nitrogen Oxides or Their Transboundary Fluxes, (1989) 28 ILM 212; the 1991 Geneva Protocol Concerning the Control of Emissions of Volatile Organic Compounds or Their Transboundary Fluxes, (1992) 31 ILM 568; and the 1994 Oslo Second Protocol on the Further Reduction of Sulphur Emissions, (1994) 33 ILM 1540.
[96]   The Protocol on Heavy Metals and the Protocol on Persistent Organic Pollutants (see www.unece.org/env).

The 1985 Vienna Convention for the Protection of the Ozone Layer[97] and its 1987 Montreal Protocol[98] were negotiated under the auspices of the United Nations Environment Programme ('UNEP').[99] In advance of firm scientific proof as to harm to the ozone layer caused by chlorofluorocarbons ('CFCs'), the Convention seeks to address increasing concerns over the possibility. Parties are to take 'appropriate measures', including the adoption of legislation and administrative controls, to protect human health and the environment '... against adverse effects resulting or likely to result from human activities which modify or are likely to modify the ozone layer'.[100] The 1987 Montreal Protocol to the Convention sets clear targets for reducing and ultimately eliminating consumption of ozone damaging chemicals.

8.3.1.6   UK domestic law

In order to meet its commitments under the Kyoto Protocol, as translated into the European context by the EU's burden-sharing agreement,[101] the UK has introduced three economic instruments: climate change levy, the UK ETS and the Renewables Obligation ('RO'), a structure for trading in renewable energy supply obligations. Climate change levy has been outlined in para. 1.4.2.2(1) above and is discussed in detail in Chapter 14 below. The UK ETS was introduced in para. 1.4.2.2(2) above and is discussed in detail in Chapter 20 below. The RO was outlined in paras 1.4.2.2(4) and 6.4.3.1(2)(a) above and is discussed in para. 21.5 below. The EU ETS, also having Kyoto-based objectives, has been outlined at para. 1.4.2.2(3) above and is discussed in detail in Chapter 28 below.

8.3.2   International transportation and dumping of waste

The regulation of waste by international agreement has concentrated on two principal aspects:

1   the transport of, and trade in, waste between countries; and
2   the disposal of waste outside national jurisdictions, that is, on the high seas.[102]

In addition, whilst the definition of the concept of waste in EU law is a relatively wide one, public international law on waste concentrates on hazardous waste. The reasons for these emphases are not hard to discern: sensitivities concerning national sovereignty and the problems of opportunities for arbitrage. As to the former, any treaty designed to regulate the transport and disposal of waste within the jurisdictions of states would constrain unacceptably the freedom of action of national governments.[103] As to arbitrage, costs of waste disposal rose with the tightening of

---

[97]   (1987) 26 ILM 1529.
[98]   (1987) 26 ILM 1550.
[99]   See www.unep.org.
[100]   1985 Vienna Convention, Art. 2 (see para. 12.2.6.2 below).
[101]   See para. 8.3.1.4 above.
[102]   See Sunkin, Ong and Wight, *op. cit.*, p. 343.
[103]   *Ibid.*, p. 344.

disposal regulation in the US and Europe. The result of this was that hazardous wastes were historically either dumped at sea or exported to less developed countries where, because of lower standards, disposal costs were less.[104] International measures taken to address these problems lie outside the scope of the present study. Since, however, they would be a relevant element in assessing the feasibility of some form of global economic instrument, a brief indication of their nature and scope is justifiable here.

There are four key measures:[105]

1   the 1972 Convention on the Prevention of Marine Pollution by Dumping of Wastes and Other Matter ('the London Dumping Convention');[106]
2   the 1989 Basel Convention on the Control of Trans-boundary Movements of Hazardous Wastes and their Disposal ('the 1989 Basel Convention');[107]
3   the 1999 Basel Protocol on Liability and Compensation for Damage Resulting from Trans-boundary Movements of Hazardous Wastes and their Disposal ('the Basel Protocol');[108] and
4   the 1991 Bamako Convention.[109]

Revised in 1993, the London Dumping Convention ('the LDC') will eventually be replaced by a 1996 Protocol, although the latter is not yet in force.[110] The LDC operates within the general principles laid down in the 1982 UN Convention on the Law of the Sea ('the UNCLOS'). Under the LDC, the states parties are obliged to take all practicable steps to prevent the pollution of the sea by the dumping of waste and other hazardous matter, such dumping being prohibited or regulated according to which of three lists a particular substance belongs. Besides the UNCLOS and the LDC, there are a number of agreements relating to specific regions, that is, to the North Sea, the Baltic, the North East Atlantic, the Mediterranean and the South Pacific.

The 1989 Basel Convention, although not accepted by African states, attempts to establish a global regime for the control of international trade in hazardous and other wastes.[111] It confirms the sovereign right to ban imports, whether on a unilateral, bilateral or regional basis, but the exercise of this right must be notified to the other parties through the Convention's Secretariat;[112] it promotes disposal at source and

---

104   See Bell and McGillivray, *op. cit.*, p. 467.
105   But note also (not discussed in the text) the 1989 Lomé IV Convention made between the EU and a group of African, Caribbean and Pacific countries (prohibits export of hazardous wastes from the EU to these countries) and the 1991 and 1994 Decisions by OECD Member States prohibiting trade in hazardous wastes, whether for disposal or recycling and/or recovery, between themselves and non-OECD countries (OECD Decisions I/22 (1992) and II/12 (1994)).
106   (1972) 11 ILM 1294.
107   (1989) 28 ILM 657.
108   Reproduced in Sunkin, Ong and Wight, *op. cit.*, pp. 369–80. Not yet in force. Described in Birnie and Boyle, *op. cit.*, pp. 435–6.
109   (1991) 30 ILM 775.
110   See Birnie and Boyle, *op. cit.*, p. 420.
111   1989 Basel Convention, Art. 1.
112   *Ibid.*, Preamble and Art. 4(1)(a).

embodies the principle of minimising the generation of hazardous waste;[113] and, most significantly, in the conduct of international trade in hazardous wastes, the 1989 Basel Convention requires the prior, informed and written consent of both transit and import states.[114] The Basel Protocol, adopted in 1999 but not yet in force,[115] provides for liability (strict and otherwise) for environmental damage arising from international trade in hazardous wastes.

As mentioned above, African states did not accept the 1989 Basel Convention. Rather than being parties to a compromise between regarding trade in waste as an emerging market opportunity or a growing environmental threat, they decided, in the Bamako Convention, to ban imports of hazardous wastes into Africa from non-parties altogether,[116] and to regulate trade in waste among African states themselves.

Whilst none of the above is of direct relevance to the existing pattern of economic instruments in UK waste regulation,[117] it may become relevant if moves for global regulation of waste disposal using economic instruments[118] gather momentum. Noting the problems surrounding international waste regulation mentioned above, however, it is not surprising that 'soft' forms of public international law have had a part to play in the international regulation of waste disposal. In this connection, the influence of the OECD has been decisive. Reference has already been made to the 1991 OECD Council Recommendation on the use of economic instruments in environmental policy.[119] Specifically in relation to waste management, it made five recommendations, covering financing (user) charges, emission charges, product charges and deposit-refund systems.[120]

8.4   International trade law

The legal implications of the intersection of the various types of agreement discussed above, of their consequent implementation and reflection in national policies and their relationship with the regulation of world trade, are extremely complex.[121] This part

---

[113]   1989 Basel Convention, Art. 4.
[114]   *Ibid.*, Arts 4 and 6.
[115]   See S.D. Murphy, 'Prospective Liability Regimes for the Transboundary Movement of Hazardous Wastes' (1994) 88 AJIL 24–75.
[116]   Bamako Convention, Art. 4(1). The prohibition is confined to non-parties to the Bamako Convention and membership of the Convention is restricted to Member States of the Organisation of African Unity ('OAU'), since March 2001 the African Union (not discussed elsewhere in this book).
[117]   See para. 6.3 above.
[118]   Such as the Austrian waste tax.
[119]   See para. 8.2.6 above.
[120]   OECD Council Recommendation C(90)177/Final, January 31, 1991, paras 41–45 (see para. 1.2.1.5 above).
[121]   See Ole Kristian Fauchald, *Environmental Taxes and Trade Discrimination* (London: Kluwer Law International, 1998). For a brief, although extremely useful account, see Michael J. Trebilcock and Robert Howse, *The Regulation of International Trade*, 2nd edn (London: Routledge, 1999), ch. 15. The classic account of GATT 1994/WTO in English is John H. Jackson's *The World Trading System*, 2nd edn (Cambridge, MA: MIT Press,

of the chapter cannot, of course, attempt an exhaustive analysis of this conjunction. The aim of this next main part of the chapter is a much more modest one: it is simply to show why and how, as an EU Member State, the UK's environmental taxes and other economic instruments, introduced pursuant to the international agreements discussed above, are subject to the discipline of GATT 1994[122] and its associated Uruguay Round agreements.[123]

As a preliminary to that discussion, a brief overview of the most relevant GATT 1994 articles may be useful. Besides reminding the reader of the outlines of world trade law, these should also be useful in drawing attention to the close correspondence between the GATT discipline and the rules for intra-EU trade contained in the European Treaty.[124]

a. GATT 1994, Art. I, provides for general most-favoured nation treatment;
b. GATT 1994, Art. II, provides for schedules of concessions (or 'bindings') with regard to customs duty rates and the freezing of rates of other duties and charges. Article II has some significance in the context of environmental levies and is considered in greater detail below;
c. GATT 1994, Art. III, bans protective and discriminatory internal tax and protective quantitive regulations. Article III is deeply significant in the context of environmental taxes and is subjected to detailed analysis below. Similar prohibitions, as Lyons points out,[125] are to be found in Arts 90 and 28, European Treaty (ex 95 and 30);[126]
d. GATT 1994, Art. VI 'permits the imposition of anti-dumping and countervailing duties', a possibility with which the basic EC measures are consistent. Again, this is of potential significance in relation to economic instruments and is considered below;
e. GATT 1994, Art. XVI, contains provisions limiting state subsidies, which are again mirrored in the European Treaty (see Arts 87 and 88 (ex 92 and 93));[127] and
f. GATT 1994, Art. XX, contains general exceptions to GATT, including those which are sometimes described as the 'environmental exceptions', although it

---

1997), although, possibly since it predates many of the developments discussed in para. 8.4.2 below, it somewhat understates environmental issues in the context of the system it describes.

[122] See Kirsten Borgsmidt, 'Ecotaxes in the Framework of Community Law' [1999] EELR 270–81.

[123] GATT 1994 and its associated Uruguay Round agreements, as well as decisions of the Appellate Body from 1996 onwards, are available from the WTO website, that is, www. wto.org. For a hard copy version of the agreements, see World Trade Organization, *The Legal Texts: the Results of the Uruguay Round of Multilateral Trade Negotiations* (New York: Cambridge University Press, 1999).

[124] Grateful acknowledgement is made to a similar survey in Timothy Lyons's *EC Customs Law* (Oxford: Oxford University Press, 2001), p. 14, to which the present paragraph is indebted.

[125] See Lyons, *op. cit.*, p. 14.

[126] See paras 12.3.3.1 below and 12.4 below.

[127] See para. 12.2.7 below.

should be noted that the term 'environment' does not appear. The relevance of Art. XX to the present area is both obvious and intricate and is revisited in detail below.

The discussion in para. 8.4 begins by examining how the GATT 1994 rules constrain the freedom of action of member countries, including Member States of the EU, with regard to fiscal policy (see para. 8.4.1 below). Paragraph 8.4.2 then considers in detail the so-called 'environmental exceptions' to GATT 1994 that are contained in Art. XX. Article XX is the focus of a developing jurisprudence, the significance of which to the international development of environmental taxes is widely-agreed to be absolutely crucial.[128] The analysis of the case law on Art. XX is then followed by a detailed examination (in para. 8.4.3 below) of Arts II and III, GATT 1994, and the possibility of the incorporation in the design of environmental taxes the concept of the border tax adjustment. Together with para. 8.4.4, on GATT 1994 anti-subsidy rules, para. 8.4.3 provides a context for the examination of the shape of the UK's environmental levies and subsidies in para. 8.4.5.

8.4.1   Restrictions on fiscal policy under GATT 1994

The necessity for such of the UK's tax law rules as touch upon international trade to comply with the relevant rules of GATT 1994 derives both from the UK's membership of the EU and from the UK's and the EU's membership of the WTO.

When the European Economic Community Treaty (the Treaty of Rome) was concluded in 1957, the EEC's founding members were already bound by GATT 1947.[129] What is now Art. 131, European Treaty,[130] espoused objectives corresponding closely to those of GATT 1947, and Art. 307, European Treaty (ex 234) provided that rights and obligations arising from agreements concluded before January 1958 between one or more Member States, on the one hand, and one or more non-Member States (that is, 'third countries'), on the other, were not to be affected by the provisions of the EEC Treaty. Even though the Community was not a GATT 1947 contracting party, the ECJ subsequently held that the Community was bound by GATT 1947 by a process of substitution for the Member States.[131] In *International*

---

[128]   See, for example, Geert van Calster, 'Topsy-Turvy: the European Court of Justice and Border (Energy) Tax Adjustments – Should The World Trade Organization follow suit?', in *Critical Issues in Environmental Taxation*, ed. by Janet Milne *et al.* (Richmond: Richmond Law and Tax, c.2003), pp. 311–41 esp. pp. 332–5.

[129]   See the Preamble to GATT 1947. In 1957, there were six Member States of the EEC: France, the Federal Republic of Germany, Italy, the Netherlands, Belgium and Luxembourg (see, for example, Weatherill and Beaumont, *op. cit.*, pp. 1–5 for the early history of the EEC). Four of the six had been original signatories to GATT 1947, whilst Italy had signed GATT 1947 in 1950 and West Germany in 1951 (see www.wto.org/english/thewto_e/gattmem_e.htm).

[130]   Articles 131–135, European Treaty (ex 110–15), set out the EU's (incomplete) common commercial policy ('the CCP').

[131]   See Dominic McGoldrick, *International Relations Law of the European Union* (London: Longman, 1997), pp. 194–5.

Fruit Company NV v. Produktschap voor Groenten en Fruit,[132] the ECJ reached the conclusion that '... in so far as, under the EEC Treaty, the Community has assumed competences previously exercised by the member-States in the sphere of application of [GATT 1947] ..., the provisions of [GATT 1947] ... have the effect of binding the Community'.[133]

The ECJ began by noting that, '... at the time of concluding the Treaty instituting the European Economic Community, the member-States were bound by the undertakings of [GATT 1947]'.[134] It then inferred from the EEC Treaty, and especially from Arts 131 (ex 110) and 307 (ex 234), European Treaty, a desire on the part of the Member States to abide by the terms of GATT 1947.[135] Referring to the Community's assumption of the functions inherent in its tariff and trade policy, the Court said that this marked the Member States' readiness to bind the Community by the obligations which they had contracted under GATT 1947.[136] Finally, and especially since the establishment of the common customs tariff ('the CCT'),[137] '... the Community, acting through its institutions, ... [had] appeared as a participant in the tariff negotiations and as a party to the agreements ... concluded within the framework of [GATT 1947] ...'.[138] The *International Fruit* case illustrates, therefore, that, although Member States remained parties to GATT 1947, the EU took on the principal role in conducting the relationship between the Member States and the other contracting parties.

The position just summarised was subsequently developed following the EU's becoming an original member of the WTO in 1995.[139] In *Re the Uruguay Round Treaties*,[140] the ECJ explained in detail the competences of each of the EU and its Member States in relation to the WTO agreement, which had been signed in Marrakesh in April 1994.[141] The Court interpreted Art. 133, European Treaty (ex 113) as according competence to the EU to conclude multilateral agreements on trade in goods, to the exclusion of its Member States, the EU and its Member States having joint competence, subject to certain exceptions,[142] in relation to trade in services (under the General Agreement on Trade in Services ('GATS')) as well as trade-related aspects of intellectual property rights (under the TRIPS agreement).[143]

---

[132] C–21–24/72, [1975] 2 CMLR 1 (decision, December 1972).

[133] [1975] 2 CMLR 1, para. 18.

[134] *Ibid.*, para. 10.

[135] *Ibid.*, paras 12 and 13.

[136] *Ibid.*, paras 14 and 15.

[137] That is, on 1 July 1968. For a valuable discussion of the CCT, see Timothy Lyons, *op. cit.*, esp. (in the present context) pp. 58–60.

[138] [1975] 2 CMLR 1, paras 16 and 17.

[139] See Asif Qureshi, *The World Trade Organization* (Manchester: Manchester University Press, 1996), pp. 164–91. For further details, see the WTO website: www.wto.org.

[140] Opinion 1/94, [1995] 1 CMLR 205.

[141] The Uruguay Round of GATT 1947 was concluded in December 1993, the WTO Agreement being signed in April 1994 in Marrakesh.

[142] That is, those provisions of GATS and TRIPS that relate to cross-frontier supplies of services ('GATS') and the means of enforcement of intellectual property rights ('TRIPS'), each of which fell within Art. 133, European Treaty (ex 113) and therefore the exclusive competence of the EU.

[143] That is, Trade-Related Aspects of Intellectual Property Rights.

It followed that when, following the conclusion of the Uruguay Round, not only the EU but also its Member States,[144] had signed the WTO agreement, incorporating GATT 1994, GATS and TRIPS, the latter should not (in the Court's opinion) have done so. Indeed, the fact that Arts 131–135, European Treaty (ex 110–15), relating to the common commercial policy ('the CCP'), deny individual Member States any freedom of action in relation to GATT 1947/GATT 1994 has been reaffirmed in a succession of ECJ decisions.[145]

On its accession to the EU in 1973, the UK's trade policy was thus already subject to the rules of GATT 1947. Following the signature of the WTO agreement in 1994, by the EU and its Member States, that policy is now constrained by their membership of the WTO.[146] The restraints placed on the EU and its Member States by GATT 1994, Williams tells us, are fundamental, being '... a form of basic law to that constitution'.[147] As a member of the WTO, the EU has a duty to ensure that the UK, as an EU Member State, complies with the requirements of GATT 1994. As will be seen in Chapter 12, the rules of the European Treaty mirror the rules of GATT in many respects.[148] Nonetheless, the obligation of the UK as an EU Member State to comply with the EU's international obligations is independent of, and separate from, the duties imposed on it as a result of the creation of the single market.[149]

Thus, in the design and implementation of the UK's environmental taxes, it is necessary for policy makers to keep in mind the potential legal implications of the UK's and the EU's membership of the WTO and the subjection of its tax law to the discipline of GATT 1994.

---

[144] Of which there were at that time only 12.

[145] See, for example, *Re OECD Local Costs Standard*, Opinion 1/75, [1975] ECR 1355; *International Fruit Company NV v. Produktschap voor Groenten en Fruit*, C–21–24/72, discussed in the text above; *Diamantarbeiders v. Indiamex*, C–37–38/73, [1973] ECR 1600, *Donckerwolcke*, C–41/76, [1976] ECR 1921. See the discussion in Stefano Inama and Edwin Vermulst, *Customs and Trade Laws of the European Community* (The Hague: Kluwer Law International, 1999), para. 1.1. For a qualification to the point made in the text, see Inama and Vermulst, *op. cit.*, para. 1.3.5.

[146] See David Williams, *EC Tax Law* (Longman: London, 1998), p. 9.

[147] Williams, *op. cit.*, p. 10.

[148] This similarity was an important part of the ECJ's reasoning in *International Fruit Company NV v. Produktschap voor Groenten en Fruit*, C–21–24/72, discussed in the text above. In referring to the Members States' desire to abide by the terms of GATT 1947, the Court said that the Members States' readiness '... to respect the undertakings of the General Agreement results as much from the provisions of the EEC Treaty itself as from the declarations made by the member-States when they presented the Treaty to the Contracting Parties of the General Agreement in accordance with Article XXIV of the latter' ([1975] 2 CMLR 1, para. 12). See, further, Lyons, *op. cit.*, pp. 11–20.

[149] Williams, *op. cit.*, p. 71. Thus, Art. 90, European Treaty (ex 95), which applies to intra-Community trade, is both similar to, and crucially different from, Art. III, GATT 1994 (see below in this chapter and para. 12.3 below).

8.4.2 Environmental exceptions to GATT 1994 restrictions

Before considering in detail the WTO provisions of most relevance to environmental taxes and other economic instruments (that is, those on customs duties, internal taxes and subsidies), it is useful to refer to the general exceptions to the GATT discipline, contained in GATT 1994, Art. XX ('General Exceptions'). Interestingly, the terms 'environment' and 'environmental' do not appear; instead, the relevant wording runs as follows:

> Subject to the requirement that such measures are not applied in a manner which would constitute a means of arbitrary or unjustifiable discrimination between countries where the same conditions prevail, or a disguised restriction on international trade, nothing in this Agreement shall be construed to prevent the adoption or enforcement by any contracting party of measures:[150]
>
> ...
>
> (b) necessary to protect human, animal or plant life or health;
>
> ...
>
> (g) relating to the conservation of exhaustible natural resources if such measures are made effective in conjunction with restrictions on domestic production or consumption; ...

Four general points might be made about this wording:[151]

1   clearly, it does not suggest any straightforward means of reconciling the tensions between free trade and environmental protection referred to above;
2   the burden of demonstrating that one of the exceptions in Art. XX is applicable in a particular case falls upon the party seeking to use it as a defence;[152] mainly because Art. XX is construed strictly, the burden has not often been discharged;
3   Article XX, GATT 1994, is the subject of a developing jurisprudence, which may see some resolution of the international environmental law/world trade law tension referred to in para. 8.2.1 above; and
4   If Art. XX, GATT 1994 applies, then it disapplies the relevant GATT 1994 rule, for example, Art. III or Art. VI.

Three cases help to particularise the point made in 2 above.[153] They represent a developing jurisprudence, so 1 below must be read subject to 2 and 3.

1   In *Restrictions on Imports of Tuna*[154] ('the *Tuna-Dolphin I* case'), the US had imposed restrictions on imports of yellowfin tuna because of concerns that they

---

[150]   This introductory wording is referred to as 'the chapeau' of the Article.
[151]   For a fuller treatment, see, for example, Birnie and Boyle, *op. cit.*, pp. 701–2.
[152]   See *Canada – Administration of the Foreign Investment Review Act*, GATT, BISD 30 Supp 140 (1984), para. 5.20.
[153]   Each decision is notoriously long, if not complex. Great assistance has been derived in the following summaries from the summaries in Trebilcock and Howse, *op. cit.*, ch. 15; Birnie and Boyle, *op. cit.*, pp. 701–14; and Bell and McGillivray, *op. cit.*, pp. 105–8.
[154]   (1991) 30 ILM 1598.

were caught using methods dangerous to dolphins, a protected species under the US Marine Mammal Protection Act. A GATT dispute settlement panel upheld Mexico's complaint that this violated GATT 1947, Art. XI(1) and rejected a justification based on Art. III, GATT 1947.[155] The panel also held that the exemptions in GATT 1947, Arts XX(b) and XX(g) did not apply. In relation to Art. XX(b), the panel held that 'necessary' did not simply mean 'needed' but that no other reasonable alternative existed,[156] whilst, under Art. XX(g), it held that 'relating to' and 'in conjunction with' meant 'primarily aimed at'.[157] Since the restrictions were imposed to force other countries to change their environmental policies, they were not 'necessary', within Art. XX(b), nor were they 'primarily aimed at' conserving exhaustible natural resources. The decision in the *Tuna-Dolphin I* case was not, however, adopted by the GATT Council.[158]

Note that the dispute settlement panel accepted that dolphins were an 'exhaustible natural resource' within Art. XX(g), GATT 1947.

2   *Standards for Reformulated and Conventional Gasoline*[159] ('the *US Gasoline Standards* decision') concerned the reformulated and conventional gasoline programmes created under the US Clean Air Act 1990. Under both programmes, changes were required in the composition of gasoline sold to consumers, with 1990 being used as the baseline year. The US Environmental Protection Agency ('EPA') distinguished, in their baseline establishment rules, between foreign and domestic producers and refiners. Domestic refiners were allowed to establish individual 1990 baselines, whilst foreign refiners had to use instead the EPA's statutory baselines. The WTO Appellate Body ('the AB') held that the scheme was caught by the *chapeau* of Art. XX as 'unjustifiable discrimination' and a 'disguised restriction on international trade'.[160] In so doing, the AB applied a two-stage test:

In order that the justifying protection of Article XX may be extended to it, the measure at issue must not only come under one or another of the particular exceptions – paragraphs (a) to (j) – listed under Article XX; it must also satisfy the requirements imposed by the opening clauses of Article XX. The analysis is, in other words, two-tiered: first, provisional justification by reason of characterization of the measure under Article XX(g); second, further appraisal of the same measure under the introductory clauses [that is, the *chapeau*] of Article XX.[161]

The AB held that the second clause of Art. XX(g) appeared to '... refer to governmental measures like the baseline establishment rules being promulgated

---

155   See para. 8.4.3 below for a full treatment of Art. III, GATT 1994.
156   (1991) 30 ILM 1598, para. 5.28.
157   *Ibid.*, para. 5.31.
158   The reasoning in *Tuna-Dolphin I* was also followed by the panel in *Restrictions on Imports of Tuna*, (1994) 33 ILM 839 (the *Tuna-Dolphin II* case), which was again not adopted by the GATT Council.
159   (1996) 35 ILM 603 (see Birnie and Boyle, *op. cit.*, p. 701).
160   (1996) 35 ILM 603, 633.
161   *Ibid.*, 626.

or brought into effect together with restrictions on domestic production or consumption of natural resources ... The clause is a requirement of *even-handedness* in the imposition of restrictions, in the name of conservation, upon the production or consumption of exhaustible natural resources'.[162] Furthermore, when considered in the light of the introductory clauses of Art. XX, it was clear that the baseline establishment rules involved arbitrary or unjustified discrimination. The US had failed adequately to explore means of mitigating the administrative problems that it had relied on in imposing the statutory baselines on foreign refiners; it had also failed to count the costs for foreign refiners of denying them individual baselines.[163]

Note that the wide view of 'exhaustible natural resources' taken by the dispute settlement panel in the *Tuna-Dolphin I* case was accepted by the AB, the latter agreeing that 'clean air' was an exhaustible natural resource within Art. XX(g).[164] Furthermore, the AB found that the baseline establishment rules related to the conservation of natural resources,[165] given their primary aim, and having regard to their purpose and effect.[166]

3 Finally, in *Import Prohibition of Certain Shrimp and Shrimp Products*,[167] ('the *Shrimp-Turtle* case'), a national US measure required countries exporting shrimp to the US to show either that their fishing environments did not pose a threat of the incidental taking of sea turtles[168] in the course of shrimp harvesting, or that their fishing industry was regulated to standards comparable to those in force in the US. Countries exporting to the US to which either possibility applied were so certified; those to which neither possibility applied were banned from exporting shrimp to the US. Applying the two-tiered test in the *US Gasoline Standards* decision, the AB held that the measure was not justified under Art. XX, GATT 1994. Although the measure qualified for provisional justification under Art. XX(g), it failed to meet the requirements of the *chapeau* thereto.[169]

a. In finding that the measure was provisionally justified under Art. XX(g), the AB took a similarly wide view of exhaustible natural resources as had been

---

162 *Ibid.*, 624–5.
163 *Ibid.*, 632. The panel observed that: 'There was more than one alternative course of action available to the United States in promulgating regulations implementing the ... [Clean Air Act]. These included the imposition of statutory baselines without differentiation as between domestic and imported gasoline. This approach, if properly implemented, could have avoided any discrimination at all. Among the other options open to the United States was to make available individual baselines to foreign refiners as well as domestic refiners' (see (1996) 35 ILM 603, 629).
164 See (1996) 35 ILM 603, 613–14, the Panel already having accepted this point also.
165 See the opening words of Art. XX(g).
166 (1996) 35 ILM 603, 623.
167 (1999) 38 ILM 118.
168 Five species of sea turtles fell within the regulations: loggerhead (*Caretta caretta*), Kemp's ridley (*Lepidochelys kempi*), green (*Chelonia mydas*), leatherback (*Dermochelys coriacea*) and hawksbill (*Eretmochelys imbricata*).
169 See para. 8.4.2 above.

taken in the two previous decisions. It rejected an argument put forward
by India, Pakistan and Thailand as joint appellees that 'exhaustible natural
resources' referred to finite resources such as minerals, rather than biological
or renewable resources, the AB noting that living species were capable of
depletion, exhaustion and extinction.[170] Referring to the UNCLOS,[171] to the
Convention on Biological Diversity,[172] to the Rio Declaration[173] and to other
international environmental agreements,[174] it found that the relevant species
of sea turtle were 'exhaustible' because they were recognised as endangered
species.[175] In referring to these instruments, the AB was following the
general rule in the 1969 Vienna Convention, Art. 31(3), allowing account
to be taken of relevant rules of international law applicable in the relations
between the parties.[176]

As regards the remaining requirements for the measure to fall within Art.
XX(g), the AB found that the measure related to[177] the conservation of
exhaustible natural resources:

Focusing on the design of the measure here at stake, it appears to us that ... [it] is not
disproportionately wide in its scope and reach in relation to the policy objective of
protection and conservation of sea turtle species. The means are, in principle, reasonably
related to the ends.[178]

Finally, the AB found that the measure was an 'even-handed' one, the
measure being made effective in conjunction with restrictions on domestic
production or consumption.[179]

b.   Although the AB was satisfied as to the provisional justification for the
measure under Art. XX(g), it then went on to find that the measure failed
to meet the requirements of the *chapeau* to Art. XX. The *chapeau*, it will
be recalled, makes the application of the General Exceptions in Art. XX
subject to the requirement that the measure in question must not be applied
'... in a manner which would constitute a means of arbitrary or unjustifiable
discrimination between countries where the same conditions prevail'.

---

170   (1999) 38 ILM 118, para. 128.
171   See para. 8.3.2 above.
172   See para. 8.3.1.3 above.
173   See para. 8.2.3 above.
174   Including, not mentioned elsewhere in this chapter, the 1973 Washington Convention
      on International Trade in Endangered Species of Wild Fauna and Flora, (1973) 12 ILM
      1085 and the 1979 Bonn Convention on the Conservation of Migratory Species of Wild
      Animals, (1980) 19 ILM 15.
175   (1999) 38 ILM 118, para. 132. Bell and McGillivray, *op. cit.*, pp. 107–8, regard it as
      significant that the sea turtles' exhaustibility depended on the fact that they were already
      recognised as endangered, not because action was required to prevent endangering them.
176   See para. 8.2.2 (point 1) above and Birnie and Boyle, *op. cit.*, p. 704.
177   See the opening words of Art. XX(g), GATT 1994.
178   (1999) 38 ILM 118, para. 141.
179   *Ibid.*, paras 143–4: see the second part of Art. XX(g), GATT 1994.

The AB concluded that the measure in question was rigid and inflexible;[180] that the certification processes followed by the US were 'singularly informal and casual', to the point that they could result in the negation of rights of WTO members;[181] and that the measure lacked transparency and fairness.[182] For all three reasons, the measure amounted to *arbitrary* discrimination between countries where the same conditions prevailed.[183]

Besides being arbitrary, the measure constituted a means of *unjustifiable* discrimination:

i.   The measure lacked flexibility, since the US inquired only into whether exporting country used TEDs (that is, 'turtle excluder devices') not whether they authorised comparable methods;[184]
ii. the measure also had the effect of banning imports of shrimp to the US which had been caught using TEDs, where the shrimp had originated in the waters of countries not certified under the measure;[185]
iii. thirdly, the US had failed to engage the appellees, as well as other countries exporting shrimp to the US, '... in serious, across-the-board negotiations with the objective of concluding bilateral or multilateral agreements for the protection and conservation of sea turtles, before enforcing the import prohibition against the shrimp exports of those other [WTO] Members'.[186] Among other things, the US had failed to take account of Principle 12 of the Rio Declaration.[187] Finally,
iv. the US had made differing levels of effort in transferring the TED technology to other countries. Far greater efforts had been made to transfer the technology to 14 wider Caribbean/western Atlantic countries than to other exporting countries, including the appellees.[188]

In the light of the above decisions, it may well be the case that Art. XX, GATT 1994 will come to play a greater role in rendering certain types of environmental tax lawful which would otherwise fall foul of its discipline. The most obvious candidate would be a carbon tax, which, as a direct tax, is not, on present learning, capable of incorporating a BTA,[189] even on environmental grounds.

8.4.3   GATT 1994 provisions on customs duties and internal taxes

Articles II and III, GATT 1994, distinguish between customs duties and import charges, on the one hand, and internal taxes and charges, on the other. The distinction

---

180   *Ibid.*, para. 177.
181   *Ibid.*, para. 181.
182   *Ibid.*, para. 183.
183   *Ibid.*, para. 184.
184   *Ibid.*, para. 164.
185   *Ibid.*, para. 165.
186   (1999) 38 ILM 118, para. 166.
187   See para. 8.2.6 above.
188   (1999) 38 ILM 118, para. 175.
189   See para. 1.2.1.2 above.

forms the basis of two different sets of rules. Article III is of the greater relevance in the present context, since it allows the imposition of levies that are indirect – as opposed to direct[190] – on domestic and imported products alike. Such blanket imposition is necessary in the case of environmental levies, as with other levies, to guarantee both the competitiveness of domestic products as well as the 'tax base' of the environmental levy in question.

In relation to customs duties and import charges, Art. II(1)(a), GATT 1994, requires WTO members to accord to the commerce of other members a treatment no less favourable than that provided for in the agreed schedules of concessions annexed to the WTO agreement. Thus, Art. II(1)(b), GATT 1994, reads:

> The products described in ... the Schedule relating to any contracting party, which are the products of territories of other contracting parties, shall, on their importation into the territory to which the Schedule relates, and subject to the terms, conditions or qualifications set forth in that Schedule, be exempt from ordinary customs duties in excess of those set forth and provided therein.[191]

The obligation contained in Art. II(1) is generally referred to as the 'tariff-concession obligation', the tariff commitments of each country or regional trading organisation in the schedules to the WTO agreement being referred to as 'bindings' or 'concessions'. These schedules range from the voluminous, in the cases of the EU and the US, to the relatively brief, in the case of less developed countries.[192] In accordance with the CCP provisions of the European Treaty discussed above, Member States of the EU are represented at the WTO by the European Commission,[193] which has the sole right to speak for its Member States at virtually all WTO meetings, including tariff negotiations.[194] Furthermore, just as the European Treaty bans Member States from unilaterally imposing customs duties on goods from third countries,[195] so also does it outlaw charges having equivalent effect to customs duties on imports from non-member countries.[196]

So far as internal taxes and charges are concerned,[197] Art. III, GATT 1994 subjects them to the so-called 'national treatment obligation'. Article III(1) articulates the general policy goal that internal taxes and charges '... should not be applied to imported or domestic products so as to afford protection to domestic production'.[198] Referring to the goal mentioned in Art. III(1), Art. III(2) then goes on to require that:

---

[190] *Ibid.*
[191] GATT 1994, Art. II(1)(b).
[192] The maximum tariffs are contained in Part I of each country's or customs territory's four-part goods schedule (see www.wto.org).
[193] See para. 4.3.2 above.
[194] For further details, see Williams, *op. cit.*, pp. 62–4.
[195] *Sociaal Fonds voor de Diamantarbeiders v. SA Ch Brachfeld & Sons and Chougal Diamond Co*, C–2/69 and 3/69, [1969] ECR 211.
[196] *Aprile Srl*, in *Liquidation v. Amministrazione Delle Finanze dello Stato*, C–125/94, [1995] ECR I–2919, para. 34.
[197] See para. 7.2.2.4 above.
[198] Besides internal taxes and charges, GATT 1994, Art. III(1) refers to 'laws, regulations and requirements'.

The products of the territory of any contracting party imported into the territory of any other contracting party shall not be subject, directly or indirectly, to internal taxes or other internal charges of any kind in excess of those applied, directly or indirectly, to like domestic products. Moreover, no contracting party shall otherwise apply internal taxes or other internal charges to imported or domestic products in a manner contrary to the principles set forth in [Art. III(1)].

Exceptions to the national treatment obligation are limited, and are set out in Art. III(8). They consist of an exception for government purchases,[199] as well as one for the payment of subsidies exclusively to domestic producers.[200] These are in addition to the general exceptions in GATT 1994, Art. XX referred to above. It is plain from Art. III, GATT 1994, that the design of new internal taxes and charges in WTO member countries must comply with the national treatment obligation.

Turning back to Art. II(2)(a), GATT 1994, we find it stated that the provisions of Art. II (which covers customs duties and import charges) do not prevent WTO members from imposing at any time on the importation of any product 'a charge equivalent to an internal tax, in respect of the like domestic product, or in respect of an article from which the imported product has been manufactured or produced, whether in whole or in part'.[201] However, Art. II(2)(a) also says that any such charge must be imposed consistently with the provisions of Art. III(2).[202] When taken together, it is apparent from the wording of Arts II and III that the distinction between a customs duty and an internal tax or other charge does not depend on when or where the levy in question is imposed.[203] Instead, it depends on whether the levy on imported products is also borne by like domestic products. If it is, then the levy falls within Art. III; if it is not, then the levy falls within Art. II.[204] In other words, an internal tax or charge, provided it complied with Art. III(2), could be applied at the border with the third country. Subject to the application of the border tax adjustment ('BTA') rules, Art. III(2) is not prima facie infringed, therefore, by an internal tax or other charge, imposed by an EU/WTO member, which is designed simply to ensure parity of tax treatment between domestic products and third country products and which is imposed at the border.

Whilst the BTA concept is a simple one to articulate, its application to environmental taxes involves a number of difficult questions.[205] Briefly, a BTA is designed to put

---

[199] GATT 1994, Art. III(8)(a).
[200] *Ibid.*, Art. III(8)(b).
[201] This division is then underlined by a note to GATT 1994, Art. III in Annex I, which states that '... [a]ny internal tax or other internal charge ... which applies to an imported product and to the like domestic product and is collected or enforced in the case of the imported product at the time or point of importation, is nevertheless to be regarded as an internal tax or other internal charge ... and is accordingly subject to the provisions of Article III'.
[202] See above in this para.
[203] Birnie and Boyle, *op. cit.*, p. 729.
[204] See WTO Secretariat, *Taxes and Charges for Environmental Purposes – Border Tax Adjustment* (WT/CTE/W/47), para. 55.
[205] See 1.2.1.2 above. See also Paul Demaret and Raoul Stewardson, 'Border Tax Adjustments under GATT and European Law and General Implications for Environmental Taxes' (1998) 28 JWT 5–65.

into effect the general principle of international indirect taxation[206] that goods should be taxed where they are used or consumed.[207] The definition of a BTA used by the WTO was originally used by the OECD and defines a BTA as:

> ... any fiscal [measure] which put[s] into effect, in whole or in part, the destination principle (that is, which enable[s] exported products to be relieved of some or all of the tax charged in the exporting country in respect of similar domestic products sold to consumers on the home market and which enable imported products sold to consumers to be charged with some or all of the tax charged in the importing country in respect of similar domestic products).[208]

The phenomenon of BTAs is assumed throughout GATT 1994, not only in Arts II and III, but also, as will be seen, in Art. XVI, GATT 1994.[209] It rests in turn on a second assumption that is made in GATT 1994, that is, that of the distinction between direct and indirect taxes.[210] As explained at the beginning of the book, only indirect taxes are eligible for BTA.[211] Thus, only internal indirect taxes imposed by an EU/WTO member state, which are designed to ensure parity of tax treatment between domestic products and third country products (whether or not they are imposed at the border), are capable of satisfying GATT 1994 requirements.

The availability of BTA for environmental taxes, under GATT/WTO rules, is a major technical consideration in their design. The idea of a carbon tax, as mentioned above, has famously been controversial in this context. How the UK's environmental levies have sought to address the technical challenge involved is considered below.

8.4.4   GATT 1994 anti-subsidy rules

The WTO's anti-subsidy rules are to be found in GATT 1994, Arts VI (anti-dumping and countervailing duties) and XVI (subsidies), as well as in the 1994 WTO Agreement on Subsidies and Countervailing Measures ('the Subsidies Agreement').[212] Here, no less than in the Art. III context, the distinction between direct and indirect taxes,[213] and the consequent availability of BTA, are crucial technical questions.

Article VI, GATT 1994, permits WTO members to deal with the problem of dumping, that is, the introduction of the products of one country into the commerce

---

206   See para. 1.2.1.2 above.
207   *Ibid.*
208   WT/CTE/W/47, para. 28.
209   The 1970 WTO Working Party on Border Tax Adjustments agreed that the main provisions of GATT relating to BTA codified practices that existed in commercial treaties when it was drafted (see WTO Secretariat, *Taxes and Charges for Environmental Purposes – Border Tax Adjustment* (WT/CTE/W/47), para. 29).
210   See para. 1.2.1.2 above.
211   WT/CTE/W/47, para. 33–35. See para. 1.2.1.2 above.
212   See, generally, Konstantinos Adamantopoulos and Marìa J. Pereyra-Friedrichsen, *EU Anti-Subsidy Law and Practice* (Bembridge: Palladian, 2001), and A. Leigh Hancher, Tom Ottervanger and Piet Jan Slot, *EC State Aids*, 2nd edn (London: Sweet and Maxwell, 1999), ch. 5.
213   See para. 1.2.1.2 above.

of another country at less than the normal value of the products, by the imposition of countervailing duties. Article XVI, GATT 1994, deals with the related, but distinct, phenomenon of subsidisation by Governments of exporting countries; it is divided into a Section A, entitled 'Subsidies in General', and a Section B, headed 'Additional Provisions on Export Subsidies'. Article XVI(A)(1) obliges WTO members to notify to the Ministerial Conference the granting or maintaining of '... any subsidy, including any form of income or price support, which operates directly or indirectly to increase exports of any product from, or to reduce imports of any product into, its territory'. Article XVI(B)(4) bans the direct or indirect grant of '... any form of subsidy on the export of any product other than a primary product which subsidy results in the sale of such product for export at a price lower than the comparable price charged for the like product to buyers in the domestic market'.

The Subsidies Agreement, which dates from the Uruguay Round, deals with two major areas: the regulation of subsidies that impact on international trade and advice to WTO members on the best ways of protecting their domestic industries from those subsidies.[214] The Subsidies Agreement is incorporated in Community law via the 1997 Council Regulation on Protection against Subsidized Imports from Countries not Members of the European Community (usually referred to as 'the Basic Regulation' or 'the Countervailing Duty Regulation').[215] The Basic Regulation is intended to provide greater transparency and effectiveness in the application by the European Community of the rules laid down in the Subsidies Agreement, as regards subsidised imports into the EU. It does not affect the virtually identical obligations of the EU and its Member States under the Subsidies Agreement, which continues to govern claims by third countries in respect of governmental subsidies in the EU as a whole or any of its Member States.[216]

In practice, these provisions have presented far fewer problems for environmental taxes than have those of Art. III, GATT 1994. The Interpretative Note Ad Article XVI, GATT 1994 makes it plain that:

> The exemption of an exported product from duties or taxes borne by the like product when destined for domestic consumption, or the remission of such duties or taxes in amounts not in excess of those which have accrued, shall not be deemed to be a subsidy.

Moreover, Art. VI(4), GATT 1994, states that the exemption of exported products from taxes borne by like domestic products, as well as the refund of such taxes, cannot be subject to anti-dumping or countervailing duties. The principles of Arts VI and XVI are then underlined in the footnote to Art. 1.1 of the Subsidies Agreement, which reads:

> In accordance with the provisions of Article XVI of GATT 1994 (Note to Article XVI) and the provisions of Annexes I through III of this Agreement, the exemption of an

---

[214] See Raymond Luja, 'WTO Agreements versus the EC Fiscal Aid Regime: Impact on Direct Taxation' (1999) 27 *Intertax* 207–25, esp. pp. 207–11.

[215] Council Regulation EC/2026/97, (1997) OJ L288 1.

[216] The granting of subsidies by EU Member States to their national industries is dealt with by the state aid rules (see Arts 87–89, European Treaty (ex 92–94)), insofar as such subsidies affect trade between Member States (see para. 12.2.7 below).

exported product from duties or taxes borne by the like product when destined for domestic consumption, or the remission of such duties or taxes in amounts not in excess of those which have accrued, shall not be deemed to be a subsidy.

Thus, when designing environmental taxes or, indeed, other economic instruments, it is necessary for the UK to take into account the anti-subsidy rules.

8.4.5   UK environmental levies and subsidies in the GATT 1994 context

8.4.5.1   GATT 1994 aspects of the UK's environmental taxes

The design of each of aggregates levy and of climate change levy is conceptually somewhat similar. One area in which this conceptual similarity is apparent is the attempted assimilation of the international dimension of each tax to the discipline both of GATT 1994 and of the European Treaty.[217]

Aggregates levy and climate change levy are unusual among 'the existing UK environmental levies, as having a cross-border dimension. Although conceived within the GATT 1994/European Treaty discipline, landfill tax is designed in such a way as to obviate the need for dealing specifically with cross-border issues. A substantially similar point might obviously be made about the concepts of workplace parking levies and road user charging schemes. Landfill tax, it will be recalled, is charged on a disposal of material as waste by way of landfill at a landfill site.[218] By Finance Act 1996, s.66, a site is a landfill site at any given time if one of five alternative types of licence is in force in relation to the land, which authorises disposals on the land.[219] Since a landfill site, as so defined, must necessarily be within the UK,[220] there is no need within the landfill tax code for any special provisions relating to the taxation of imports of waste from, and exports of waste to, other Member States of the EU or to third countries.[221] Moreover, the structure of rates and exemptions does not differentiate between waste generated within the UK or outside it. Given these design features of the tax, there would therefore seem to be no issue in relation to any of the arts of GATT 1994 discussed above.

However, by contrast with the landfill tax code, each of the aggregates levy and climate change levy codes contain provisions relating to imports and exports of the products in question. As such, certain aspects of each may be difficult to justify under the GATT 1994 discipline, although the true position is unclear. On one view, the structure of the taxes means that, in the absence of special factors, no GATT 1994 issue is likely to arise, irrespective of Art. XX. This view is based on a close reading of the respective tax codes in the light of the GATT 1994 provisions. It can be summarised as follows.

---

[217]   See Chapter 12 below.
[218]   See paras 1.4.2.1(1) above and 15.2 below.
[219]   See para. 15.2 below.
[220]   Generally, this is a landfill permit (see para. 6.3.2.4 above).
[221]   Such international trade in waste is in any event restricted by the requirements of the 1989 Basel Convention, the European Waste Shipment Regulation and other international agreements (see paras 8.3.2 above and 12.2.5.1(4) below).

The aggregates levy provisions relating to the importation of aggregates appear in Finance Act 2001, ss.16(2) and 19(1). Together, these two subsections impose a charge to aggregates levy whenever taxable aggregate is subjected to commercial exploitation in the UK, such exploitation being made to include the two situations envisaged in s.19(1), that is, when:

...

(b) ...[taxable aggregate] becomes subject to an agreement to supply it to any person; [or]

(c) it is used for construction purposes ...

Whatever the other ambiguities of the wording of s.19 may be,[222] a number of points seem clear, no less from the section itself, as from the structure of the tax as a whole. First of all, it is apparent that, when applied to imports of aggregates, the levy is an internal tax within Art. III, GATT 1994, rather than a customs duty or charge with equivalent effect under GATT 1994, Art. II. This is because the concept of subjecting taxable aggregate to commercial exploitation in the UK makes no distinction between aggregate originating within the UK or that originating outside it. The tax is charged on imported and domestic products alike. Secondly, it seems unlikely that aggregate imported into the UK from a third country would be subject, even indirectly, to a charge to the levy in excess of the amount applied to domestically-produced aggregate. Thirdly, aggregates levy is clearly an indirect tax,[223] in respect of which the BTA built into its structure is designed to ensure parity of treatment between domestic aggregate and third country aggregate.[224] As to the second of these factors, unlikely as it may be that there is any indirect discrimination, it is not inconceivable that, whilst the tax appears on its face to be non-discriminatory, various circumstances in the market place or elsewhere might have the effect in a particular case of 'tilting the scales against the imported product'.[225]

A similar conclusion, as regards the treatment of imports of taxable commodities from third countries, seems possible in the case of climate change levy. With climate change levy, all supplies of taxable commodities which are not excluded or exempt from the levy count as taxable supplies,[226] irrespective of whether the supplier is resident in the UK.[227] In the absence of special factors, climate change levy is thus again a tax borne by domestic and imported like products alike, and lawful under GATT 1994 as an internal indirect tax for ensuring parity of tax treatment between domestic and third country products.[228]

---

[222]   See paras 13.2 and 13.3 below.

[223]   See para. 1.4.2.3 above.

[224]   See para. 12.4 below.

[225]   See Jackson, *op. cit.*, p. 216.

[226]   Finance Act 2001, Sched. 6, para. 2(2).

[227]   For the definition of 'resident in the UK' for these purposes, see Finance Act 2000, Sched. 6, para. 156, discussed at para. 16.7n below.

[228]   Under Finance Act 2000, Sched. 6, para. 40(2), the person liable to account for the levy charged on the supply is the person to whom the supply is made. Where either of these requirements is missing, then Customs have wide powers to appoint a resident tax representative (see para. 16.7 below). For the formalities associated with imports of

As regards the treatment of exports within the aggregates levy code, Finance Act 2001, s.30(1)(a), provides for the making of relevant regulations for the purpose of conferring entitlement to tax credits on the exportation of aggregate 'in the form of aggregate'. The relevant regulations accordingly appear in the Aggregates Levy (General) Regulations 2002,[229] reg. 13:

> (1) This regulation applies to a person who has commercially exploited taxable aggregate and who has accounted for the aggregates levy chargeable on that commercial exploitation.
> (2) Such a person is entitled to a tax credit in respect of any aggregates levy accounted for in respect of that commercial exploitation where the taxable aggregate in question –
>   (a) is exported or removed from the United Kingdom without further processing;
> ...[230]

There is judicial authority for the proposition that this provision is cast in terms of a tax credit, rather than an exemption, to reflect the fact that it cannot be ascertained whether aggregate otherwise taxable is in fact exempt until it is known what has happened to the aggregate in question.[231] In any event, it would seem plain from the provisions just discussed that there is no unlawful subsidy under GATT 1994, since the effect of the tax credit is merely to remit taxes in amounts not exceeding those that have accrued.[232] Moreover, the remission on exportation of duties and taxes borne by a like product when destined for domestic consumption does not entitle a third country to impose anti-dumping or countervailing duties.[233]

Within the climate change levy code, the exportation of commodities is covered by the wording of Finance Act 2000, Sched. 6, para. 11. This provides that, where a taxable commodity[234] is caused to be exported from the UK, then the supply is exempt from the levy, provided that the recipient has previously notified the supplier that he intends to cause the relevant commodity to be exported from the UK and has

---

     taxable commodities, see HM Customs and Excise Notice CCL 1, *Climate Change Levy* (March 2002), pp. 13–14 (available from www.hmce.gov.uk).

[229] S.I. 2002 No. 671. See para. 16.9 below.

[230] See also HM Customs and Excise Notice AGL 1, *Aggregates Levy* (March 2003), pp. 21–2 for the formalities (available from www.hmce.gov.uk).

[231] See *R (on the application of British Aggregates Associates and others) v. C & E Commrs*, [2002] EWHC 926 (Admin), [2002] 2 CMLR 51, para. 29 (Moses, J.). See paras 12.1, 12.3.3.1 below.

[232] See para. 8.4.4 above.

[233] *Ibid.*

[234] Finance Act 2000, Sched. 6, para. 3 (see para. 14.1 below). It should be noted that, although a distinction between goods and services is not part of the structure of climate change levy, the commodities which fall within its scope are also subject to customs duties as goods under the CCT (for the categories into which they fall, see the 'TARIC' web-pages on the EU website, www.europa.eu.int). This is consistent with the classification of the supply of any form of power, heat, refrigeration or ventilation as a supply of goods in the VAT legislation (see Value Added Tax Act 1994, Sched. 4, para. 3). TARIC, it should be noted, is the single register representing the collected tariffs on goods imported into, and exported from, the EU (see Williams, *op. cit.*, p. 62n).

no intention of causing it to be brought back into the UK thereafter.[235] Again, so far as third countries are concerned, there would seem to be no unlawful subsidy, since the exemption is expressly permitted by GATT 1994 provisions on subsidies and countervailing duties.[236]

On the basis of the foregoing, and in the absence of special factors, there would appear to be little possibility for conflict between the GATT 1994 provisions and those of the domestic legislation. However, it is instructive to reflect on certain other features of the two taxes, insofar as they relate to imports, in the light of possible bases for green taxes generally.[237] As we noted much earlier in this study,[238] the OECD distinguishes between three possible forms of environmental tax or charge: those based on the emission of pollutants, including (in the case of air) noise, into various media ('emission taxes'); those based on the cost of collective treatment of effluent or waste ('user taxes'); and those based on products (including raw materials, intermediate or final products) that are harmful to the environment when used in production processes ('product taxes').[239] There is clearly an environmental link between the first and third of these and, indeed, the OECD acknowledges that 'product ... taxes can act as a substitute for emission ... taxes when charging directly for emissions is not feasible'.[240] Since the scope of landfill tax is geographically limited, the fact that it appears to be a hybrid of an emissions tax and a user tax[241] does not seem to raise any issue. However, the other two green taxes discussed above, not being geographically limited, appear to give cause for some concern, if only at the margins. This is because both climate change levy and aggregates levy may be regarded as a hybrid of an emissions tax and a product tax. The reason why this is significant is that, whilst BTA is lawful for taxes on products under GATT 1994, Art. III(2), it is not lawful to apply it to taxes on emissions, since such taxes are regarded within GATT 1994 as a tax on the producer.[242] Both points are well made by a close reading of Arts II(2)(a) and III, GATT 1994. Article II(2)(a), the outline of which has already been mentioned, provides:

> Nothing in this Article shall prevent any contracting party from imposing at any time *on the importation of any product* ... [authors' emphases] a charge equivalent to an internal tax imposed consistently with the provisions of paragraph 2 of Article III in respect of the like domestic product or in respect of an article from which the imported product has been manufactured or produced in whole or in part.

---

235 See HM Customs and Excise Notice CCL 1, *Climate Change Levy*, above, p. 10, and HM Customs and Excise Notice CCL1/3, *Reliefs and Special Treatments for Taxable Supplies* (available from www.hmce.gov.uk).
236 See para. 8.4.4 above.
237 See Chapter 5 above.
238 See para. 1.2.1.5(2) above.
239 See OECD Council Recommendation C(90)177/Final, paras 2–4. See also para. 1.2.1.5(2) above.
240 OECD Council Recommendation C(90)177/Final, para. 4.
241 See para. 1.2.1.5(2) above.
242 See Birnie and Boyle, *op. cit.*, p. 730.

If each of the taxes is properly seen as a product tax, then the two articles are clearly not infringed. However, if they are properly seen as taxes on emissions, that is, on resource use, then they fall foul of Art. II(2)(a).[243] The position is, in the view of the present writers, unclear.[244]

8.4.5.2   GATT 1994 aspects of UK tax subsidies

In Chapter 7, when examining the taxation, as opposed to the regulatory, context of the UK's green levies and subsidies, we divided the discussion of subsidies into tax subsidies within the environmental levies and environmental subsidies within the non-environmental taxes.[245]

Tax subsidies within the green levies were characterised as having been introduced into the structure of the levy in question for reasons, not of regulatory efficiency, but of sectoral competitiveness. By contrast, green subsidies within the non-environmental tax codes were characterised as having been introduced into the structure of the tax for reasons of regulatory efficiency.

For present purposes, it is proposed to treat the two types of subsidy together, whilst noting their difference of function. The question here is whether, in either case, the subsidies in question are such as to infringe the relevant provisions of the Subsidies Agreement.[246] In common with the rest of the discussion in the present chapter, the analysis is intended to be read not simply in connection with existing subsidies, upon which it does not purport to offer a conclusive view, but also with those that have already been, or may yet be, proposed.[247]

The Subsidies Agreement begins by defining the concept of a 'subsidy',[248] with 'specificity'[249] as the fundamental condition of actionability. It then divides subsidies into three categories, each of which is subject to different rules:

1   prohibited subsidies;[250]
2   actionable subsidies;[251] and
3   non-actionable subsidies.[252]

---

[243]   A related, but different, problem does not however seem to arise with climate change levy, although it might have done with a more traditionally-conceived carbon tax. If a tax were to be imposed on energy consumed in the production process of a product, then it would be extremely doubtful whether the tax would be lawful under art II(2)(a), GATT 1994. Article II(2)(a) permits the imposition of a tax under Art. III, GATT 1994, only on an article 'from which', that is, not 'with the help of which', the 'imported and the like domestic product were produced' (see Birnie and Boyle, *op. cit.*, p. 731). See para. 1.2.1.4n above.

[244]   See, generally, WT/CTE/W/47, above.

[245]   See paras 8.3.3 and 8.3.4 above.

[246]   See para. 8.4.4 above.

[247]   See Chapter 27 below.

[248]   Subsidies Agreement, Art. 1.

[249]   *Ibid.*, Art. 2.

[250]   *Ibid.*, Art. 3.

[251]   *Ibid.*, Art. 5.

[252]   *Ibid.*, Art. 8.

For the purposes of the Subsidies Agreement, a subsidy includes a financial contribution by a government or public body within the territory of a WTO member where, among other things, 'government revenue that is otherwise due is foregone or not collected (for example, fiscal incentives such as tax credits)'.[253] However, a benefit must thereby be conferred.[254]

Article 2 of the Subsidies Agreement distinguishes specific subsidies from those which are non-specific. Non-specific subsidies are ones which are generally available to all enterprises or industries in the WTO member country. Specific subsidies are those access to which is, formally or in fact, confined to certain specific enterprises, industries, groups of enterprises or industries, or to enterprises in a specific geographical region.[255] Only specific subsidies are actionable under the Subsidies Agreement.[256]

Where a subsidy is prohibited, the complaining WTO member may seek the removal of the subsidy through the WTO dispute settlement mechanism, a procedure which may result in its being authorised to take appropriate, proportionate countermeasures.[257] There is no need for the complaining member to demonstrate adverse effects. Prohibited subsidies include the following:

(e) The full or partial exemption, remission or deferral specifically related to exports, of direct taxes or social welfare charges paid or payable by industrial or commercial enterprises.

(f) The allowance of special deductions directly related to exports or export performance, over and above those granted in respect to production for domestic consumption, in the calculation of the base on which direct taxes are charged.

(g) The exemption or remission, in respect of the production and distribution of exported products, of indirect taxes in excess of those levied in respect of the production and distribution of like products when sold for domestic consumption.[258]

If a subsidy, rather than being prohibited, is one which is actionable, the WTO complaining member must follow a very similar procedure to the one just described.[259] However, the complaining member must obviously demonstrate that the conditions for actionability are made out. These are that the subsidy in question has adverse effects consisting of injury to the complaining member's domestic industry, the nullification or impairment of benefits under GATT 1994, or serious prejudice to its interests.[260] It follows from this that, even if a subsidy is actionable, it will not have a relevant adverse effect on international trade if its economic effects are confined

---

253 *Ibid.*, Art. 1, esp. Art. 1.1(a)(1).
254 *Ibid.*, Art. 14, which excludes certain items from the scope of a 'benefit'.
255 This neat summary of the intricacies of Art. 2, Subsidies Agreement, is adopted from Hencher et al., *op. cit.*, para. 5–012.
256 Subsidies Agreement, Art. 8.1(a).
257 *Ibid.*, Art. 4, esp. Art. 4.8–4.12.
258 *Ibid.*, Annex I, paras (e)–(g).
259 *Ibid.*, Art. 7.
260 *Ibid.*, Art. 5.

within national borders.[261] The procedure in relation to an actionable subsidy may again result in the complaining WTO member being authorised to take proportionate countermeasures.[262]

Finally, if a subsidy is non-actionable, then it cannot be challenged under the dispute settlement procedure, provided however that it has been duly notified in advance to the WTO's Committee on Subsidies and Countervailing Measures.[263] Non-actionable subsidies are ones which are either non-specific or specific but meet the conditions of Art. 8 of the Subsidies Agreement. Article 8 comprises three categories:

1   assistance for certain research activities, subject to stringent conditions;
2   assistance to disadvantaged regions within the territory of a Member given pursuant to a general framework of regional development and non-specific (within the meaning of Article 2) within eligible regions ... [subject to the satisfaction of detailed criteria]; ...[264]
3   assistance to promote adaptation of existing facilities to new environmental requirements imposed by law and/or regulations which result in greater constraints and financial burdens on firms ... [subject to the satisfaction of detailed criteria].[265]

Under Arts 10–23 of the Subsidies Agreement, read in conjunction with GATT 1994, Art. VI, WTO members may impose countervailing duties in conjunction with invoking the dispute settlement mechanism in cases of prohibited and actionable subsidies, although not subsidies that are not actionable.[266]

From the foregoing, it is apparent that, as regards all of the tax subsidies identified in Chapter 7 above:

1   all are subsidies within Art. 2 of the Subsidies Agreement;
2   if access to any of them is, formally or in fact, confined to certain specific enterprises, industries, groups of enterprises or industries, or enterprises in a specific geographic region, then they are specific subsidies;
3   one of them appear to be prohibited subsidies;
4   any specific subsidies that have a relevant adverse effect beyond the borders of the UK will entitle a complaining WTO member (for example, the EU, via the European Commission) to take steps under the dispute settlement mechanism; and
5   any that are specific but non-actionable will be incapable of challenge by the EU or any other WTO member.

---

261   See Hancher *et al.*, *op. cit.*, para. 5–013.
262   Subsidies Agreement, Art. 7.
263   *Ibid.*, Art. 25.
264   *Ibid.*, Art. 8.2(b).
265   *Ibid.*, Art. 8.2(c).
266   *Ibid.*, note 35 (Art. 10).

8.5   International air transport law

We referred at the beginning of the chapter to the need to allude, if only briefly, to international agreements governing air transport. This is in order to provide a context for the discussion in Chapter 27 below[267] of proposals to introduce new economic instruments in relation to the noise and carbon emissions externalities caused by air transport. The key multilateral treaty, as mentioned in Chapter 4 above, is the 1944 Chicago Convention on International Civil Aviation,[268] as amended and clarified by subsequent policy guidance issued by the International Civil Aviation Organization ('the ICAO').[269] The UK, in common with most other developed countries, including the EU Member States,[270] is a party to the Chicago Convention.

The Chicago Convention is significant in the context of the present book for two reasons. One of these, of course, is the exclusion already mentioned of international aviation emissions from the scope of the 1997 Kyoto Protocol targets.[271] The other is that Art. 24 (a) of the Chicago Convention provides that:

> Fuel, lubricating oils, spare parts, regular equipment and aircraft stores on board an aircraft of a contracting State, on arrival in the territory of another contracting State and retained on board on leaving the territory of that state shall be exempt from customs duty, inspection fees or similar national or local duties and charges.

The subsequent ICAO policy guidance referred to above strongly recommends that 'any environmental levies on air transport which States may introduce should be in the form of charges rather than taxes[272] and that the funds collected should be applied in the first instance to mitigating the environmental impact of aircraft engine emissions'; that there should be no fiscal purpose to the charges; that such charges should be related to costs; and that they should not discriminate against air transport as compared with other transportation modes.[273]

Although the Chicago Convention is the fundamental multilateral agreement on civil aviation, it should be noted that there is also a network of bilateral agreements, made between pairs of states and based on the UK/US 'Bermuda II' agreements, and having the chief purpose of shielding national airlines from competition.[274] Such bilateral agreements reflect the ICAO's traditional policy guidance[275] of

---

[267] See para. 29.8 below.
[268] International Civil Aviation Organization Doc. 7300/8, *Convention on International Civil Aviation*, 8th edn (Montreal: ICAO, 2000), available from www.icao.int. See also para. 4.4.3 above.
[269] See, especially, International Civil Organization Doc. 8632, *ICAO's Policies on Taxation in the Field of International Air Transport*, 3rd edn (Montreal: ICAO, 2000) and ICAO Council Resolution on Environmental Charges and Taxes, 9 December 1996 (available from www.icao.int).
[270] Before enlargement in 2004.
[271] See para. 8.3.1.4 above.
[272] See para. 7.2.1 above.
[273] See 1996 Council Resolution, paras 4 and 5.
[274] See Rosa Greaves, *EC Transport Law* (Harlow: Pearson Education, 2000), pp. 65–7.
[275] See International Civil Organization Doc. 8632, above.

recommending 'the reciprocal exemption from all taxes levied on fuel taken on board by aircraft in connection with international air services, ..., and also ... [the reduction or elimination of] taxes related to the sale or use of international air transport'.[276]

8.6 International energy law

The Energy Charter Treaty ('the ECT')[277] is, in origin, the most recent of the multilateral agreements referred to in this chapter, having been opened for signature at Lisbon in December 1994.[278] Its purpose is to 'establish a legal framework in order to promote long-term co-operation in the energy field, based on complementarities and mutual benefit ...'.[279] Although the scope of the ECT is limited to one sector, it creates a range of legal obligations and rights within the energy sector relating to investment and trade, while also creating a number of rights and obligations which relate to the environment. The ECT is an extremely innovative document, making explicit reference to the philosophy of economic liberalism, while also holding out the possibility of at least in part having direct effect in signatory countries.[280]

The basic tenor of the ECT investment provisions, which appear in Part III thereof, is to ensure that investors receive a basic minimum standard of fair treatment from the contracting parties. These include a commitment to accord to the investments of investors 'fair and equitable treatment'.[281] By Art. 13, a contracting party may not nationalise or expropriate the investment of another contracting party, except subject to certain conditions, one of which is the payment of 'prompt, adequate and effective compensation'. In the present context, Art. 13, which deals with expropriation, has a twofold importance. First, it certainly covers windfall and other confiscatory taxation, such the windfall tax imposed on the UK's privatised utilities in 1997.[282] Secondly, it might, according to Waelde, cover expropriation by exorbitant environmental

---

[276]   See 1996 Council Resolution, recital d.

[277]   See para. 4.4.4 above. The text of the Energy Charter Treaty is available from the website referred to in that para.

[278]   This paragraph is heavily indebted to *Energy Law in Europe: National, EU and International Law and Institutions*, ed. by M. Roggenkamp *et al.* (Oxford: Oxford University Press, 2001), ch. 4 (authored by Craig Bamberger, Jan Linehan and Thomas Waelde).

[279]   Energy Charter Treaty, Art. 2.

[280]   See the painstaking examination of the Energy Charter Treaty's provisions in Thomas Waelde, 'International Investment under the 1994 Energy Charter Treaty' (1995) 29 *JWT* 5–72.

[281]   Energy Charter Treaty, Art. 10.

[282]   See para. 2.4 above. The windfall tax was created by Finance (No.2) Act 1997, ss.1–5 and Scheds 1–2. The problem was that most of the original allottees had sold out and so the 'clawback' was against investors who had not received the alleged 'benefit'. See Dieter Helm, *Energy, the State and the Market: British Energy Policy since 1979*, revised edn (Oxford: Oxford University Press, 2004), pp. 288–90, and Thomas Waelde, 'Renegotiating Previous Governments' Privatisation Deals: The 1997 UK Windfall Tax on Utilities and International Law' (1997) 2 *Journal of the Centre for Energy, Petroleum and Mineral Law and Policy* (available from www.dundee.ac.u/cepmlp).

regulation.[283] However, Art. 13 must be read subject to Art. 21 on taxation, which provides that nothing in the ECT creates rights or imposes obligations with regard to the domestic tax laws of the contracting parties and, if there is any inconsistency between Arts 13 and 21, Art. 21 is to prevail.

The ECT's trade provisions, which are set out in Art. 29 of the ECT, are designed, in essence, to bring the trade of contracting parties who are not parties to GATT 1994 into line with the provisions of GATT 1947.[284] As regards environmental matters, Art. 19, ECT, contains merely hortatory commitments of good environmental practice.

It is a pity that it is not possible here, for reasons of space, to go into greater detail on the ECT; it has been described as 'arguably the most innovative of the modern international economic treaties'.[285] The ECT:

> ... breaks away from the pattern of multilateral trade agreements by making Governments directly accountable to aggrieved investors before non-national tribunals for important duties specified in the [ECT]. It also pushes the concept of state responsibility further than in traditional international law by formulating a concept of State responsibility for regulating private enterprises.[286]

8.7 Concluding remarks

This chapter has covered two main sets of multilateral agreements in detail: the MEAs, mainly UN-sponsored, which cover matters of international environmental law, and the GATT 1994-based system of multilateral trade agreements administered by the World Trade Organization. In addition, we have briefly considered the interaction of each of these two main sets of agreements with the sectorally-specific multilateral agreements on international aviation law (most importantly the 1944 Chicago Convention on International Civil Aviation) and international energy law (as embodied in the 1994 Energy Charter Treaty).

The heart of the discussion in the present chapter has been the problem of reconciling the objectives of international trade law with those of international environmental law. In the EU as a whole, as we shall see,[287] meeting the GHG emissions reduction targets in the Kyoto Protocol has meant, *inter alia*, the introduction of the EU Emissions Trading Scheme ('the EU ETS'), which specifically envisages the linking of the EU ETS with the Kyoto flexible mechanisms of JI and the CDM from 2008. The EU ETS is more easily assimilated to GATT 1994 than would be an EU-wide carbon/energy tax, given the problems that the latter presents for the GATT 1994 concept of the BTA.[288]

---

[283] See Thomas Waelde, 'Sustainable Development and the 1994 Energy Charter Treaty: Between Pseudo-Action and the Management of Environmental Investment Risk', in *International Economic Law with a Human Face*, ed. by Freidl Weiss, Erik Denters and Paul de Waart (The Hague: Kluwer Law International, 1998), pp. 223–70.

[284] See Roggenkamp *et al.*, *op. cit.*, pp. 188–9.

[285] *Ibid.*, p. 208.

[286] *Ibid.* See Energy Charter Treaty, Arts 7(6) and 22.

[287] See Chapter 28 below.

[288] See para. 1.2.1.2 above.

The UK, for its part, has decided to meet its own emissions reduction targets through three economic instruments, that is, the RO, the UK ETS and climate change levy. The last of these avoids the problems associated with the BTA by being structured as a product, rather than an emissions, tax.

Setting aside the three instruments just referred to, the two other main environmental taxes have also clearly been designed with the GATT 1994 discipline in mind. Aggregates levy, although obviously not inspired by Kyoto, sacrifices adherence to the possibility, under GATT 1994, of rebating exports. Landfill tax, whose scope does not extend to imports and exports of waste, neatly sidesteps technical considerations under GATT 1994 altogether.

The key question for the present is whether it would be possible to design a future environmental tax which was lawful, not because it satisfied general GATT 1994 norms, but because it was held to be exonerated from the general GATT 1994 discipline by Art. XX.

One of the most problematic aspects of Kyoto is the exclusion from its scope of aircraft emissions. The ECT may act as a 'brake' on increases in rates of energy taxes

Chapter 9

Conclusions on Part II

We began this part by alluding to the intricate web of departmental involvement in the design and implementation of the UK's environmental taxes and other economic instruments. In fact, as has also been explained, the range of relevant institutional actors at the central government level is even wider than this would suggest, since it includes advisory bodies, such as the Advisory Committee on Business and the Environment ('ACBE') and non-departmental public bodies and agencies, such as Ofgem and the Environment Agency. In summary, the involvement of the various departments, bodies and agencies of central government in the instruments under discussion in the book is as follows.

The biggest of the environmental taxes, climate change levy, has been, subject to one particular aspect,[1] designed, implemented and administered by HM Customs and Excise. Customs is responsible for registering holders of electricity supply licences, that is, the privatised utilities and, in the case of a non-resident supplier, the consumer, as the persons liable to account for the tax. The Department is also responsible for making the credits and repayments referred to in Chapter 16 below.[2] Furthermore, Customs is empowered to enter and search premises for, as well as to copy and remove, documents, as well as to take samples. Customs' wide-ranging powers in relation to the levy also include those of charging interest on overdue tax, imposing civil penalties for incorrect returns and taking criminal proceedings in cases of fraudulent evasion of the tax. Their enforcement powers include arrest, distress and diligence and the provision of security. The existence of each of these powers, which are discussed in detail in Chapter 16 below, underline the oft-made point that economic instruments are not necessarily any less exhaustive of administrative time and effort than are the more traditional command and control ones. They also underscore the importance to economic instruments of 'backstop' penalty regimes. The mechanism by which Customs is held to account is a combination of Parliamentary scrutiny (via the various Select Committees)[3] and scrutiny by the courts (via judicial review).[4] Although, as will been seen in Chapter 21 below, there have been a range of Select Committee reports relating, among other things, to the operation of climate change levy, there has, to the authors' knowledge, been not a single judicial review case involving Customs' administration of the levy.

It was mentioned in Part I of the book that a key characteristic of climate change levy is its operation in combination with three other economic instruments, that is, the UK Emissions Trading Scheme ('the UK ETS'), the EU Emissions Trading Scheme ('the EU ETS') and the Renewables Obligation ('the RO'). Unlike the levy, however,

---

[1]   See below in this chapter.
[2]   See para. 16.9 below.
[3]   See para. 4.2.1.1(3) above.
[4]   See para. 4.2.1.5 above.

various aspects of these instruments are managed by various different departments, public bodies and agencies. Furthermore, the exception to Customs' control of climate change levy administration referred to above consists of the responsibility of Defra for the administration of the system of climate change agreements.[5] Of the three other instruments, Defra is responsible for administering the UK ETS and for the production of the UK's national allocation plan ('NAP') under the EU ETS. Interestingly, however, the regulator of the EU ETS, in relation to installations located in England and Wales, will be, not Defra, but instead the Environment Agency.[6] Finally, Ofgem is responsible for monitoring the operation of the RO and for issuing the Green Certificates which are used as a way of demonstrating compliance with it. Each of the Environment Agency and Ofgem are accountable, in the sense of having to give an account of their activities in an annual report.[7] In addition, they are subject, of course, to Select Committee scrutiny and to judicial review. As to the latter, it is perhaps rather surprising, in view of the astonishing complexity of the climate change agreement system, that there have been no reported judicial review proceedings relating to it. This is especially so when it is considered that, as was pointed out in the Standing Committee debates on Finance Bill 2000, the effect of the climate change agreement system was to give the Secretary of State a discretion to fix the rate of the levy as between different businesses.[8] What is truly surprising, given the enormous complexity both of the levy and its interaction with its associated economic instruments, is that is the least-litigated of the UK's environmental taxes. There have, in fact, to the authors' knowledge, been no reported cases involving the levy.

By contrast with the intricate regime surrounding climate change levy, aggregates levy is, has already been mentioned, a self-standing economic instrument. The sole regulatory authority for the tax is therefore HM Customs and Excise. The peculiarities of the tax mean that Customs has some unusually wide powers, however, in relation to registration for the tax. These include a power to determine the boundaries of a site, which is a point discussed in Chapter 16 below.[9] This is in addition to the provision for joint and several liability to the tax from among a class.[10] As with climate change levy, Customs is also tasked with dealing with tax credits and repayments of tax.[11] Customs is also empowered, as for climate change levy, to enter and search premises for documents and samples. Again, Customs' powers in relation to aggregates levy include those of charging interest on overdue tax, imposing civil penalties for incorrect returns and taking criminal proceedings where appropriate. Their enforcement powers for the levy again include arrest, distress and diligence and the provision of security. In a contrast with climate change levy, however, Customs did fall prey to an – albeit unsuccessful – judicial review action in relation

---

5   See paras 1.4.2.2(1) and 4.2.1.2(1) above.
6   See para. 28.4 below.
7   See para. 21.5 below.
8   HC, Standing Committee H, 8th Sitting, 18 May 2000, c.2.30 pm.
9   See para. 16.2 below.
10 *Ibid.*
11 See para. 16.9 below.

to aggregates levy.[12] There has also been litigation on Customs' exercise of their power to exclude registrables from registration.[13]

Like aggregates levy, landfill tax has, until very recently, been a self-standing environmental tax. However, as discussed below, the need to comply with the Landfill Directive has shifted the policy emphasis away from the internalisation of externalities, to the need to meet the targets laid down in the Directive.[14] With this shift has come an attempt to coordinate the tax with other economic instruments, in the form of packaging waste recovery notes ('PRNs') and the Landfill Allowances Trading Scheme ('the LATS'). Thus, whilst Customs remains in charge of landfill tax, the responsibility for the design and implementation of economic instruments in waste management as a whole lies with the Environment Agency, as regards PRNs, and, as regards the LATS, with a combination of Defra and (so far as England at least is concerned) local authorities, via their WDAs.[15] The same comments as to the accountability and responsibilities of Customs apply in relation to landfill tax as in relation to the other taxes referred to above. It is striking that, despite its relative longevity, none of the litigation on landfill tax[16] has involved judicial review. All the reported cases have been tax disputes between Customs, on the one hand, and the taxpayer on the other. This may be because the legislation creating the tax – that is, Finance Act 1996 – contains fewer provisions conferring a discretion on the Commissioners.

We have already commented on the UK's quasi-federal structure. Its significance in relation to the material discussed above in this Part II is that of fixing the appropriate level of fiscal intervention. There was some discussion of this point in Chapter 5 above, in the context of the technical justifications for the instruments under discussion. Indeed, on reflection, it is rather strange that, having referred to the fact some environmental problems (for example climate change), are best dealt with at an international level, the 2002 Treasury paper does not take account of the unilateral action typified by each of the UK ETS and climate change levy.[17] In seeking to address possibly the largest environmental issue of all, that is, the emissions responsible for global warming, on a unilateral basis, each of these two instruments may be said, in Steven Sorrell's words, to have turned 'an early start into a false one'.[18] At the same time, it must be stressed that the problem is the political one of acceptance, easier to achieve unilaterally, perhaps, than on an EU basis, a point that is vividly illustrated by the EU carbon/energy tax proposal[19] and which may yet have consequences for

---

12 That is, *R (on the application of British Aggregates Association and Others) v. C & E Commrs*, [2002] EWHC 926 (Admin), [2002] 2 CMLR 51 (see para. 11.3.2 below).
13 See para. 13.2 below.
14 See para. 6.3.2.4 above.
15 See para. 1.4.2.1(4) above.
16 See para. 15.3 below.
17 See para. 5.5 above.
18 See Steve Sorrell, 'Turning and Early Start into a False Start: Implications of the EU Emissions Trading Directive for the UK Climate Change Levy and Climate Change Agreements', in Organisation for Economic Co-operation and Development, *Greenhouse Gas Emissions Trading and Project-based Mechanisms* (Paris: OECD, 2004), pp. 129–51.
19 See para. 28.1n below.

the EU ETS. However, just as some environmental problems are best dealt with at the international level, so also are others best dealt with locally or regionally. The current inability of the devolved administrations in the UK to levy environmental taxes comparable to those levied, for example, in Spain, will no doubt for this reason come under greater scrutiny in the future. Equally, the same issue may arise in relation to local authorities, although their general failure to exercise the environmental taxation powers contained in the Transport Act 2000 (that is, in relation to the imposition of workplace parking levies and congestion charging)[20] does not suggest that, at least in the absence of a more general overhaul of local government finance, this is presently a particularly fruitful line of inquiry.

If the level of governance at which environmental levies are sought to be imposed is one factor affecting their political acceptability, so also is their status as taxes. In Chapters 6 and 7 above, for the purposes of explaining the relationship between environmental taxes and other types of environmental regulation, we looked in turn at the regulatory context and taxation context of environmental levies. It goes without saying that this separation is in a way artificial. However, it will also have been clear that, even where regulation is command and control based, in fixing standards, the range of possible payments, for example, for licences, is still considerable and encompasses payments which might fairly be regarded as taxes. That is why Chapter 7 has taken some time to separate out taxes, properly so called, from concepts which may at first appear similar but which are crucially different.

The question of whether a particular payment under a regulatory regime is a tax has a double significance for the rest of the book. The first is that, as a tax, a payment is susceptible to fairly well-established criteria for separating out taxes which are well designed from those which are not. Of course, some allowance is necessary in this context for the nature of the taxes under discussion as environmental, rather than as fiscal, taxes. These are matters to which we shall allude in Chapter 29. However, evaluation is not the only significance of designating particular payments as taxes and thereby segregating them from concepts which may at first appear similar. There is also an analytical and legal reason. This is that the nature of a payment as a tax and, in the cases mentioned at the beginning of the book, its allocation to one or other of the categories of 'direct' or 'indirect' taxes,[21] has a universal significance in the law of constitutions and in the law of treaties. We have already hinted at the potential significance of a precise definition of a tax in the devolution of powers to the regions of the UK. However, it also has a significance for the issues discussed in Chapter 8; not, perhaps, for the use of the term in GATT 1994 but certainly for the pricing of air transport and the ban in the 1944 Chicago Convention on taxes on aircraft fuel.[22] We shall return to this issue in Chapter 27.

---

[20]   See para. 1.4.2.4 above.
[21]   See para. 1.2.1.2 above.
[22]   See paras 8.5 above and 12.2.6.4(3) below.

PART III
PRACTICE

Chapter 10

Introduction to Part III

In Part II, we examined the institutional, theoretical and regulatory contexts in which the UK has evolved its policies on environmental taxes and their associated economic instruments. In this Part, we examine the practical operation of each instrument, both in terms of an analysis of the legal shape that they have ultimately taken and in terms of the findings of Parliamentary Select Committees on the way in which they have thus far operated.

The discussion in Part III is split into two Sections, the earlier (A) dealing with the UK's environmental taxes and their associated economic instruments, the latter (B) covering the 'greening' of the UK tax system. Such a division of the material is sanctioned, for example, by the approach of the OECD in successive reports.[1]

Within Section A of Part III, we have made a fivefold division of the material. Division 1 of Section A consists of two chapters, the earlier (Chapter 11) dealing with the process by which each of the post-1997 environmental taxes came to be designed and implemented, as well as the process in which the pre-1997 tax, landfill tax, has been monitored. We have not included an account of the design and implementation of the UK ETS in Chapter 11, thinking it more natural for that history to be traced in the Chapter which deals with the UK ETS.[2] Likewise, the design and implementation of the EU ETS is analysed as part of the general discussion of the EU ETS in Chapter 28 below.[3] Chapter 12, the later part of Division 1, is devoted to analysis of those areas of Community law which have had the most immediate impact on the design and implementation of the UK's environmental taxes and their associated economic instruments.

Section A, Division 2, is devoted to the detailed analysis of the tax codes relating to each of aggregates levy, climate change levy and landfill tax. These three taxes have broadly similar administrative structures, so it was decided to deal with these together, in Chapter 16.

The third Division of Part III, Section A, is devoted to the two local levies introduced in earlier chapters, that is, workplace parking levies and road user charging schemes.

The environmental tax or charge, especially the Pigouvian tax,[4] is the archetype of the economic instrument. However, as has been often stated, each of landfill tax and climate change levy operate as one of a combination of economic instruments. These other economic instruments are the subject matter of Division 4 of Part III. Thus, packaging waste recovery notes ('PRNs') are discussed in Chapter 19, while

---

[1]   See, for example, Organisation for Economic Co-operation and Development, *Environmental Taxes and Green Tax Reform* (Paris: OECD, 1997).

[2]   See para. 20.2 below.

[3]   See para. 28.2 below.

[4]   See para. 5.4 above.

the Landfill Allowances Trading Scheme ('the LATS') is discussed in para. 20.7. Each of these operate in conjunction with landfill tax. The UK Emissions Trading Scheme ('the UK ETS'), which is discussed in Chapter 20, operates in conjunction with climate change levy and with the Renewables Obligation ('the RO'), the detail of which has already been considered in Chapter 6 above.[5] As from January 2005, these three instruments will operate alongside the EU ETS, the detailed discussion of which is allocated to a separate chapter in Part IV.[6]

Whilst much has been written about the theoretical possibilities presented by environmental taxes and other economic instruments, data on how they have operated in practice, especially in relation to their broader regulatory context, is altogether much scarcer. In the UK, a valuable source of information are the reports of Parliamentary Select Committees, and these are analysed, so far as they relate to the instruments under discussion, in Chapter 21. Although Select Committee evaluation of the various instruments has tended to concentrate on their efficiency and effectiveness in relation to their avowed environmental goals, this is not the only aspect on which it is possible to comment in the light of experience. We return to some of the possible criticisms in Chapter 29 below. One of the key justifications for environmental taxes, however, has been their potential for delivering the so-called 'employment double dividend'. How this has translated into reality in the UK over the last seven years or so is also discussed in Chapter 21.

Section B of Part III, by far the shorter of the two Sections, is devoted to non-environmental levies and to the measures taken so far by the UK government, both to remove hidden subsidies within those levies which tend to encourage environmentally-unfriendly behaviour, and to introduce tax incentives for more environmentally-friendly behaviour. We have made a three-way partition of the material, the second and third of which, employee taxation (Chapter 23) and business taxation (Chapter 24) are essentially self-explanatory. Excise duties, relating, as they do, to motor vehicles, are not susceptible to quite so straightforward a thematic allocation and have therefore been considered separately, in Chapter 22. The elements of these three chapters are predictable enough and, whilst they are mainly concerned with the statutory removal of subsidies and the provision of incentives, they are not entirely confined to this territory. An important issue also covered in Chapters 23 and 24, one to which brief reference was made in Chapter 1 above, is that of the interaction of environmental levies with non-environmental ones.

---

5   See para. 6.4.3.1(2) above.
6   See Chapter 28 below.

PART III, SECTION A
ENVIRONMENTAL LEVIES AND
OTHER ECONOMIC INSTRUMENTS

Division 1
General

Chapter 11

Design and Implementation

11.1   Introduction

The key HM Treasury policy document, from 2002, on the use of economic instruments in environmental regulation,[1] has already been discussed in some detail in a previous chapter.[2] One topic that has not yet been discussed in detail, however, is the outline of the process of policy development which is presented in the 2002 paper. Chapter 7 of the 2002 paper is devoted to providing illustrations of the process by which policy has been developed in relation to environmental taxes and other economic instruments. The sole purpose of the present chapter is to examine the discussion in the 2002 paper by reference to the stages by which the UK's environmental taxes have passed into law. Its concern, in other words, is to demonstrate how the economic theories discussed in an earlier chapter of the book[3] have actually been translated into practice.

The discussion in the present chapter is divided into three main parts. Paragraph 11.2 offers an overview of the design and implementation process deployed when new regulation, in whatever form, is proposed. This is based on the concept of the 'regulatory impact assessment' (the 'RIA'). Regulatory impact assessment, which is not to be confused with environmental impact assessment ('EIA'),[4] is the process by which a government department compares the benefits of regulatory proposals with their costs, on the basis that, if the latter are thought to be excessive in relation to the former, the proposals in question should be abandoned.[5] Subsequently, in para. 11.3, we trace the policy development process in relation to the design and implementation of the two post-1997 environmental taxes, that is, climate change levy and aggregates levy.[6]

The story of landfill tax is rather different from the other two taxes, given that, when the present government came to power in 1997, the tax was already in existence.[7] What has instead happened with landfill tax, is that an ongoing process of review and legislative amendment has taken place that has seen the original policy objectives of the tax transformed from the relatively modest one, of internalising the externalities caused by the landfilling of waste, to its enlistment in the much more ambitious

---

[1]   HM Treasury, *Tax and the Environment: Using Economic Instruments* (London: HMSO, 2002). This is referred to below as 'the 2002 Treasury paper' or, simply, 'the paper'.

[2]   See Chapter 5 above.

[3]   *Ibid.*

[4]   See para. 6.2.4 above and para. 12.2.4 below.

[5]   See Chris Hilson, *Regulating Pollution – A UK and EC Perspective* (Oxford: Hart Publishing, 2000), p. 49.

[6]   This is derived in part from Jeremy de Souza and John Snape, 'Environmental Tax Proposals: Analysis and Evaluation' (2000) ELR 74–101.

[7]   See para. 15.1 below.

project of meeting the targets imposed by the Landfill Directive.[8] This process of reorienting the objectives of the tax is traced in paragraph 11.4 below.

We restrict the coverage in this chapter to the development of the three environmental taxes referred to above. The background to the design and implementation of the seven other main economic instruments is, however, discussed in various places elsewhere in the book.[9]

11.2   Regulatory impact assessment

11.2.1   Overview

The explanation of the development process in the 2002 paper is cast in terms of a general description of the stages of the RIA, with outline accounts of how that process was in fact followed through in relation to the design and implementation of each of aggregates levy and climate change levy.[10] These outline accounts are elaborated in the discussion below.

Regulatory impact assessment replaced the previous Conservative Government's two separate 'compliance cost assessment'[11] and 'regulatory appraisal' procedures.[12] Although a unified process, the inspiration behind RIA is essentially the same as those earlier procedures, however,[13] since it is 'an assessment of the impact of policy options in terms of the costs, benefits and risks of a proposal'.[14] In accordance, therefore, with RIA procedures in the field, the stages involved in the development and implementation of environmental taxes and other economic instruments are described in the 2002 paper as follows:[15]

---

[8]   See para. 6.3.2.4 above and para. 12.2.5.1(2)(b) below.

[9]   See paras 19.1 below (background to PRNs); 20.7 below (background to the LATS); 20.2 below (background to the UK ETS); 28.2 (background to the EU ETS); 21.5 (background to the RO); 17.1 (workplace parking levies); and 18.1 (road user charging schemes).

[10]   The 2002 paper also contains a summary of the pesticides tax consultation, which is discussed in para. 21.9.5 below, and which has not as yet (December 2004) resulted in a pesticides tax being imposed.

[11]   See Department of Trade and Industry, *Checking the Cost of Regulation: A Guide to Compliance Cost Assessment* (London: HMSO, 1996).

[12]   See Cabinet Office Deregulation Unit, *Regulation in the Balance: A Guide to Regulatory Appraisal Incorporating Risk Assessment* (London: Cabinet Office, 1996).

[13]   See Cabinet Office, *Better Policy Making: A Guide to Regulatory Impact Assessment*, 2003.

[14]   See *Better Policy Making: A Guide to Regulatory Impact Assessment*, para. 1.1. In August 1998, the Prime Minister, the Rt Hon. Tony Blair, MP, 'announced that no policy proposal which has an impact on business, charities or voluntary bodies, should be considered by Ministers without an RIA being carried out' (*ibid.*, para. 1.2). See para. 4.2.1.2(1) above.

[15]   In the chapter below, press releases and other documents issued during the development and implementation process have been referenced where possible to *Simon's Weekly Tax Intelligence* ('SWTI'), as a ready source of reference. Reports of Parliamentary debates in *Hansard* are available on the web from 1989 onwards, at www.parliament.uk.

1   the establishment of the long-term goal or environmental objective for the economic instrument in question;
2   the institution of a long consultation period;
3   an announcement by an 'early signal' that the government is minded to intervene, or further intervene, in the market;
4   the active collection of evidence;
5   an announcement, again by an 'early signal', of the government's choice of economic instrument;
6   the consideration of how best to use the revenue raised;
7   a willingness to consider voluntary alternatives;[16]
8   the easing of the adjustment to the new instrument by introducing other fiscal measures which facilitate investment in new technology;
9   the provision of compensation and relief for the groups most seriously affected by the instrument;
10 a commitment to ongoing monitoring and evaluation of the instrument;
11 a commitment to future policy flexibility; and
12 a commitment, if possible, to working internationally.[17]

It should be emphasised that the 12 stages listed above can, and indeed do, run in parallel: stage 3 above, for instance, leaves scope for additional discussion with interested parties[18] on the possibilities for voluntary arrangements as an alternative to a tax (see stage 7 above).[19]

11.2.2   Hypothecation and recycling of revenues

Stage 6 above had been a matter of considerable debate prior to the introduction in 2001 and 2002, respectively, of climate change levy and aggregates levy, part of the revenue from each of which has been 'earmarked' or 'hypothecated' for particular purposes since the introduction of each respective tax. This was because HM Treasury, in common with finance ministries generally in developed countries, had traditionally resisted the notion of the earmarking of taxes.[20] Hypothecated taxes were very much the exception in the UK's taxation landscape.[21] Even the LTCS (the tax rebate which had operated since 1996 in conjunction with landfill tax)[22] represented what O'Riordan characterised as a clever ruse 'to accommodate the "sustainability reality" of quasi-hypothecation to the "other world" reality of Treasury-speak that hypothecation has not taken place'.[23]

---

16 See 2002 paper, ch. 6.
17 *Ibid.*, Table 7.1 (first column), pp. 42–3.
18 The government's term is 'stakeholders' (see the Foreword to the 2002 paper by The Rt Hon. Gordon Brown, MP, Chancellor of the Exchequer). See para. 4.2.1.2(2) above.
19 See 2002 paper, para. 7.5.
20 See Martin Daunton, *Just Taxes: The Politics of Taxation in Britain, 1914–1979* (Cambridge: Cambridge University Press, 2002), pp. 6 and 35.
21 Notoriously so, after the case of the road fund tax of the 1920s (see para. 21.2 below).
22 See para. 21.3.2 below.
23 See Timothy O'Riordan, 'Editorial Introduction to the Hypothecation Debate', in *Ecotaxation*, ed. by Timothy O'Riordan (London: Earthscan, 1997), pp. 37–44, p. 40.

11.3   Design and implementation of the two post-1997 taxes

11.3.1   Climate change levy

The 2002 Treasury paper[24] traces the proposals for climate change levy back to the March 1998 report of the Advisory Committee on Business and the Environment ('ACBE').[25] This decided in favour of a policy framework over the long term within which, without harming competitiveness, businesses in the UK could produce carbon savings. However, as Helm points out,[26] the idea of a carbon tax – as opposed to an energy tax – was much older, and economic instruments had been advocated in a Government White Paper as long ago as 1990.[27]

Subsequently, in his March 1998 Budget speech, the Chancellor of the Exchequer, the Rt Hon. Gordon Brown MP, announced that Sir Colin Marshall, then the chairman of British Airways and President of the CBI,[28] would head 'a Government review into economic instruments to improve the industrial and commercial use of energy'.[29] If, as has subsequently been claimed, this announcement was the prelude to a merely 'paper' exercise, then the manner in which it was expressed – in terms of energy consumption, rather than carbon emissions – may well suggest a foregone conclusion.[30] Nonetheless, Marshall forthwith invited all interested parties to submit their views, before 1 August 1998, on how industry and commerce could best contribute to tackling climate change. He also met a wide range of what the government refers to as 'stakeholder groups' to discuss the issues involved.[31] Finally, with the support of a task force of senior civil servants from HM Treasury, the former DETR,[32] from the DTI[33] and from Customs,[34] he published a 64-page report in November of the

---

Hypothecation can take a number of subtly different forms (*ibid.*, pp. 45–51, referring to *Charging for Government: User Charges and Earmarked Taxes in Principle and Practice*, ed. by R.E. Wagner (London and New York: Routledge, 1991) and M. Wilkinson, 'Paying for Public Spending: Is There a Role for Earmarked Taxes?' (1994) 15 FS 119–35.

[24]   See 2002 paper, Table 7.1.

[25]   Advisory Committee on Business and the Environment, *Climate Change: A Strategic Issue for Business*, March 1998 (London: Department of Transport, Environment and the Regions, 1998). See para. 4.2.1.4 above.

[26]   See Dieter Helm, *Energy, the State, and the Market: British Energy Policy since 1979*, revised edition (Oxford: Oxford University Press, 2004), p. 354.

[27]   See Department of the Environment, *This Common Inheritance: Britain's Environmental Strategy*, 1990 (Cmnd 1200, 1990), pp. 271–8.

[28]   Sir Colin had become Lord Marshall of Knightsbridge by the time of the publication of his report in November 1998. As Helm points out, 'Marshall had not been previously known for his environmental expertise, but he did have wide industrial experience, and therefore could be regarded as better able to 'sell' an unpopular tax to industry' (see Helm, *op. cit.*, p. 354).

[29]   *Hansard* HC, 17 March 1998, cols 1108–9.

[30]   See Helm, *op. cit.*, pp. 354–5.

[31]   See 2002 paper, para. 7.26.

[32]   See para. 4.2.1.2(1) above.

[33]   *Ibid.*

[34]   See para. 4.2.1.2(2) above.

same year.[35] The evidence collected in this period of just over four months, and set forth in the Marshall Report, seems to represent the extent of the consultation and active evidence collection involved in the production of the Marshall Report. A total of around 140 individuals, companies and other organisations had responded to the original request for views.[36]

The Marshall Report concluded that 'the leading option for a tax would appear to be a downstream tax on the final use of energy by industrial and commercial consumers', rather than an upstream carbon tax on energy suppliers, with the rates of tax reflecting 'at least in broad terms' the carbon content of different fuels.[37] However, Marshall cautioned that the design of any tax should ensure that combined heat and power ('CHP') was not disadvantaged and that the tax should aim, wherever possible, 'to increase incentives for the take-up of renewable sources of energy'.[38] The report also recommended the full 'recycling' of tax revenues to business, possibly through carbon trust-type schemes, and that any measures taken should be 'subject to detailed consultation about their design'.[39]

When the Chancellor made his Budget speech in March 1999, he confirmed that the government had decided to implement Marshall's recommendations from April 2001.[40] The Chancellor also announced that the levy would be brought in, after further consultation with the industry, on a revenue-neutral basis, with 'no overall increase in the burden of taxation on business'.[41] To this end, the introduction of the levy would be accompanied by a reduction in the main rate of employers' NICs[42] from 12.2 per cent to 11.7 per cent.[43]

The Chancellor's announcement was followed by a further period of consultation with various interested parties, during which the Rt Hon. Patricia Hewitt, at that time the Financial Secretary to the Treasury,[44] was asked to make a statement in the House of Commons,[45] on 24 June 1999, on the climate change levy generally. The Minister described in broad terms the purpose of the levy, briefly explained why the government expected the levy to involve no increase in the overall burden of tax on business, and recognised a need for the government to give special consideration to the position of energy-intensive industries. The purpose of the levy was to encourage energy efficiency in business, which in turn would enable the government to meet its target for reducing greenhouse gas emissions laid down by the Kyoto Protocol.[46] There would be no increase in the overall burden of tax on business, since the revenue

---

[35] Lord Marshall, *Economic Instruments and the Business Use of Energy* (London: HM Treasury, 1998).
[36] *Ibid.*, Table A.3, p. 34.
[37] *Ibid.*, p. 21.
[38] *Ibid.*, p. 3.
[39] *Ibid.*, p. 2.
[40] See 2002 paper, para. 7.27.
[41] See *Hansard* HC, 9 March 1999, col. 181.
[42] See para. 1.4.3.1 above.
[43] See *Hansard* HC, 9 March 1999, col. 181.
[44] See para. 4.2.1.2(2) above.
[45] See para. 4.2.1.1(1) above.
[46] As modified by the EU burden-sharing agreement (see para. 8.3.1.4 above).

raised by the levy would fund a 0.5 per cent cut in the main rate of employer's NICs. The position of energy-intensive industries would be reflected in lower rates of levy for those energy intensive sectors which agreed targets with the government for improving the efficiency of their energy consumption.[47]

Mrs Hewitt had revealed the fact that negotiations between the government and representatives of energy-intensive industry sectors were already ongoing when she made her Parliamentary statement on 24 June 1999. On 27 July 1999, a matter of some four weeks later, the Treasury announced that three Ministers, including Mrs Hewitt, had met representatives of these sectors.[48] Mrs Hewitt stressed that it was the Government's wish to co-operate, not only with industry, but with other interested parties, in order to make sure that the "environmental effectiveness" of the projected climate change levy was maximised, while attempting to ensure that the levy did not compromise the competitiveness of industry. In the same announcement, Mrs Hewitt indicated that the exemption for coal and gas used in chemical reactions would be extended to electricity, following representations on the point.[49]

In his Pre-Budget Statement of 9 November 1999,[50] the Chancellor of the Exchequer announced further structural amendments to climate change levy. These amendments had two main strands, one of which was the introduction of climate change agreements[51] and the measures needed to implement them.[52] The other strand was described as a recognition of the need to increase the environmental effectiveness of the levy. This was to be achieved in two ways: by the exemption from the levy of electricity generated from renewable sources of energy and good quality CHP stations;[53] and by the creation of a further bolster to the levy in the form of the introduction, from the tax year 2001/2002,[54] of enhanced income tax and corporation tax relief for expenditure on energy-saving investments.[55] The 2002 Treasury paper referred to above sees the exemptions for renewable source and CHP electricity[56] as evidence of the government's willingness to compensate sectors that would otherwise have been particularly badly affected by the introduction of the

---

[47]   See *House of Commons Written Answers*, vol. 110, cols 461, 462, 24 June 1999. On the same occasion, Mrs Hewitt gave a somewhat elliptical explanation for the imposition of the levy on CHPs, as originally envisaged (see para. 14.4 below).

[48]   HM Treasury Press Release, 27 July 1999 (reproduced in [1999] SWTI 1348). The Rt Hon. Michael Meacher, MP, the Environment Minister, and Mr John Battle, MP, the Trade and Industry Minister, were the other ministers involved.

[49]   [1999] SWTI 1348, 1349.

[50]   See *Hansard* HC, 9 March 1999, col. 889.

[51]   See para. 1.4.2.2(1) above and para. 14.6 below.

[52]   See HM Treasury Press Release, 9 November 1999 (reproduced in [1999] SWTI 1818).

[53]   As envisaged at the time, 'renewable energy sources' would included solar power, wind, wave and tide and hydroelectricity; solid renewables, such as wood, straw and waste; and gaseous renewables such as landfill and sewage gas (see HM Customs and Excise, *Budget 99: A Climate Change Levy – A Consultation Document* (HM Customs and Excise: London, 1999), p. 4).

[54]   That is, from April 2001.

[55]   See HM Treasury Press Release, 9 November 1999 (reproduced in [1999] SWTI 1816).

[56]   The exemptions were enacted as Finance Act 2000, Sched. 6, paras 15 and 19 (see paras 14.4 and 14.5 below).

levy and underscores the point by stating that the levy was designed to be revenue-neutral and that it was not designed to apply to the domestic sector because of issues of fuel poverty.[57] Likewise, the same document sees the enhanced income tax and corporation tax relief, in the form of generous capital allowances, as evidence of the government's commitment to support investment in new technology in order to ease adjustment to the levy. These measures were eventually enacted, to coincide with the introduction of the levy, in the Finance Act 2001.[58]

Published with the 1999 Pre-Budget Report was a detailed description of the design of climate change levy.[59] Useful though this description was, practitioners had to wait until 26 November 1999 for Customs to publish drafts of the clauses creating the levy, designed to be included in Finance Bill 2000.[60] On 11 November 1999, as the draft clauses were being awaited, Mr Stephen Timms MP, the new Financial Secretary to the Treasury, had emphasised that any expectations that particular sectors or firms had, for being exempted from the levy, would be disappointed.[61] Just before Christmas 1999, it was confirmed that the Secretary of State and representatives of energy-intensive sectors had reached climate change agreements.[62]

Climate change levy passed into law in the Finance Act 2000, although it was expressly provided that any supply made before 1 April 2001 was excluded from its scope.[63] Thereafter, although the rates of the levy have been reviewed annually as part of the Budget process, they have so far remained unchanged.[64] The impact of the levy is being monitored by Customs, with details of the environmental outcomes being published, not only in pre-Budget report documentation but that relating to the Budgets themselves.[65] The government's commitment to future policy flexibility, with the climate change levy is reflected, says the 2002 Treasury paper, in two developments that have occurred since its introduction in April 2001: the introduction of the general exemption for electricity produced in good quality CHP stations[66] or from coal mine methane[67] and from certain secondary recycling processes;[68] and the introduction of the UK ETS in April 2002,[69] with its link to the levy via climate change agreements.[70]

---

[57] See 2002 paper, Table 7.1.
[58] See Capital Allowances Act 2001, ss.45A–45C, as inserted by Finance Act 2001, s.65, and Sched. 17, para. 2.
[59] Reproduced in [1999] SWTI 1794–1798.
[60] See *House of Commons Written Answers*, vol. 339, col. 255, 26 November 1999.
[61] *Ibid.*, vol. 337, col. 768, 11 November 1999.
[62] See Brown, *Financial Times*, 22 December 1999, p. 3.
[63] Finance Act 2000, Sched. 6, para. 10.
[64] See 2002 paper, Table 7.1. The levy rates as finally enacted are given in paras 14.1 and 14.2 below.
[65] 2002 paper, Table 7.1.
[66] Finance Act 2002, s.123(1), inserting new para. 20A in Sched. 6, Finance Act 2000.
[67] *Ibid.*, s.126, inserting new paragraph 19(4A) in Sched. 6, Finance Act 2000.
[68] Business Brief 18/02, *Climate change levy – new exemption for certain recycling processes*, 8 July 2002.
[69] See para. 1.4.2.2(2) above.
[70] See 2002 paper, Table 7.1. See para. 14.6 below.

11.3.2   Aggregates levy

Whereas the goal of climate change levy has an international significance, that of aggregates levy is much more specific, possibly even parochial. Also, whereas climate change levy is one of three economic instruments designed to tackle climate change, aggregates levy is the key regulatory instrument in relation to mineral extraction.[71]

With some linguistic imprecision, the 2002 Treasury paper describes the goal of aggregates levy as being 'to tackle the environmental costs of aggregate extraction including noise, dust, visual intrusion and biodiversity loss'.[72] At the same time, the government has stressed that it is essential that there continues to be an adequate supply of aggregates.[73]

The government approached the issue of whether or not to create, as it was originally called, an 'aggregates tax', with considerable dedication, if not zeal.[74] In his very first Budget speech, barely eight weeks after the current government was first returned in 1997, the Chancellor announced that he was considering the imposition of a tax on aggregates extraction, together with one on water pollution:[75]

> The extraction of aggregates – including stone, sand and gravel – involves significant environmental costs and damage to the landscape, which may go beyond that recognised in the scope and level of the landfill tax. Too little is also being done to discourage water pollution. The environmental case for charges on polluters needs to be examined carefully. After a period of consultation, I will return with any proposals in those two areas in my next Budget.[76]

This was the early signal that intervention in the market for primary aggregates was likely, although it is not identified as such in the 2002 paper.[77] The DETR accordingly commissioned a research project from London Economics[78] which, beginning in September 1997, undertook an investigation of the environmental costs and benefits of the supply of aggregates.[79]

The London Economics report was published by the DETR in April 1998.[80] It indicated, as many had expected that it would, that aggregates extraction had considerable environmental costs which were not already subject to regulation.[81] It also identified a number of areas in which further work would be necessary, since it had only been possible to take a small sample of sites and the work, which involved

---

71   The other being town and country planning regulation.
72   See para. 5.3 above, quoting 2002 paper, Table 7.1, p. 42.
73   See, for example, HM Treasury, *Pre-Budget Report 1998*, 3 November 1998, para. 5.60.
74   See para. 13.1 below.
75   See para. 21.9 below.
76   The aggregates tax proposals obviously progressed; the water tax proposals did not (see para. 21.9 below).
77   See 2002 paper, Table 7.1, p. 42.
78   A private sector specialist economics consultancy (see www.londecon.co.uk).
79   See Phase Two, para. 1.1.
80   That is, *The Environmental Costs and Benefits of the Supply of Aggregates*, above.
81   That is, mainly the town and country planning legislation (see para. 6.6 above).

a fairly novel methodology, had been carried out in a relatively short time.[82] The DETR therefore asked David Pearce[83] and Susana Mourato, of University College, London, to review the work (which became 'Phase One' of the research) and to identify possible improvements to it. Although they did suggest a number of ways of improving the methodology, in June 1998 Pearce and Mourato concluded that, given Phase One's objectives, the methodology was appropriate.[84] The government then commissioned further research from London Economics, the nature and scope of which has been discussed in an earlier chapter.[85] The results of this research, referred to above as 'Phase Two', were accordingly published, again by the DETR, on 9 March 1999.[86] The work undertaken consisted of a combination of:

1　an assessment of local impacts through a survey of just under 10,000 people within five miles of 'representative' sites, interpreted as involving an average environmental cost of 70p per tonne; and
2　a separate national survey of public attitudes to measure the cost of quarrying in areas designated for special protection from development, for example, National Parks and Areas of Outstanding Natural Beauty, interpreted as involving external costs of £6 per tonne.

The DETR concluded that 'the costs of quarrying Britain's construction aggregates – in terms of local and national environment – are significant', putting the annual cost at £250m.

This considerable amount of work was designed to establish the long-term goal of a prospective tax and, as mentioned in the 2002 paper, was part of the process of evidence collection.

Meanwhile, in June 1998, Customs had issued a consultation paper on a potential aggregates tax, proposing a conceptually similar tax to landfill tax, which would be charged on a volume basis and applied at the point of first sale, use or transfer of the material away from the site of extraction.[87] This was the beginning of a consultation period that was eventually to last for the best part of two years, no doubt justifying the 2002 paper in its claim that there was an extensive consultation with the industry from 1998 onwards.[88] Five months later, the Quarry Products Association ('the QPA')[89] offered an alternative to a tax in the form of a set of voluntary measures. The

---

[82]　That is, something over seven months.
[83]　According to Helm, *op. cit.*, p. 346n, David Pearce had been a significant influence on *This Common Inheritance* (see para. 11.3.1n above) and especially its advocacy of economic instruments.
[84]　See *Environmental Costs and Benefits of the Supply of Aggregates: A Review of the London Economics Report* (see para. 5.5n above).
[85]　*Ibid.*
[86]　The DETR said that this research was 'one factor' to be taken into account by the Chancellor of the Exchequer.
[87]　See HM Customs and Excise, *Consultation on a Proposed Aggregates Tax*, 15 June 1998.
[88]　See 2002 paper, Table 7.1, p. 42.
[89]　See para. 2.5 above.

Chancellor referred both to the offer of voluntary measures and to the consultation process in his Pre-Budget Report of 3 November 1998, reiterating that, in the light of the responses to the consultation and of the proposals made by the aggregates industry, the government was still considering the imposition of a tax.[90] At this stage, however, the Pre-Budget Report simply confirmed that Ministers had offered to consider the industry's proposals for some alternative to the imposition of a tax. Whatever these proposals were, they would, however, 'have to amount to a deliverable package of measures which would permanently secure equivalent or greater benefits than a tax'.[91] Nonetheless, the Pre-Budget Report referred with approval to the fact that the industry had latterly taken certain steps to curtail its activities in the National Parks, and indicated that there might be scope to build on that positive approach in dealing with the remaining environmental impacts of the industry's activities.[92]

This somewhat conciliatory approach was still in evidence on Budget Day 1999[93] when, referring to the publication of the Phase Two report,[94] the Economic Secretary to the Treasury,[95] said:

> ... [B]efore coming to a final decision on whether to proceed with a tax, the Government would first like to pursue the possibility of an enhanced voluntary package of environmental improvements with the industry. Should the industry not be able to commit to an acceptably improved offer, or fail to deliver an agreed package of voluntary measures, the Government would [*sic*] introduce a tax.[96]

The Economic Secretary accordingly announced that Ministers would be meeting representatives of the quarrying industry to begin these negotiations. In the atmosphere of all this co-operation, it was therefore somewhat surprising when, in April 1999, Customs issued a summary of the responses to its June 1998 consultation, together with a set of draft legislative provisions.[97] Undeterred, the QPA submitted a revised package of voluntary measures in July 1999, but these again proved to be unacceptable to the government. In the Pre-Budget Report of November 1999, the Chancellor looked for more:

> The Quarry Products Association submitted a revised package of voluntary environmental measures to the Government in July. The Government welcomed this package, which shows some improvement on their original package, notably in the areas of aggregates transport and air quality. In particular there is an extra £20 million a year from seven major companies to promote and develop recycling. But this continues to fall short of what is necessary to match the overall environmental and economic effects of a tax on primary

---

90   See *Pre-Budget Report 1998*, para. 1.33.
91   *Ibid.*, para. 5.63.
92   *Ibid.*
93   That is, 9 March 1999.
94   See above in this para.
95   See para. 4.2.1.2(2) above.
96   See HM Treasury News Release HMT 8, *Reducing the Environmental Impact of Quarrying*, 9 March 1999.
97   Customs and Excise News Release, 30 April 1999, reproduced in [1999] SWTI 859, 881.

aggregates. So the Government is minded to introduce a tax in the next Budget unless the industry can further improve on this package.[98]

For one reason or another, the QPA failed subsequently to improve on their July 1999 package. The relevant part of the 2000 Budget Report reads as follows:

> Since the [1999] Pre-Budget Report, there have been further discussions about the content of the industry's voluntary package. But the industry has made delivery of the voluntary package conditional on undertakings from the Government on procurement policy which were unacceptable. The Government has therefore decided to introduce an aggregates levy[99] which will come into effect from April 2002.[100]

The 2002 paper is adamant that the consultation process that had ended so unsuccessfully for the industry is evidence of the government's willingness to consider a voluntary approach to the problem of environmental damage caused by aggregates extraction.[101] The postponement of the commencement of the levy for another two years from the date on which its introduction was announced constituted, according to the 2002 paper, an early signal of the choice of economic instrument, comparable to the way in which climate change levy had been announced in the 1999 Budget, for implementation in 2001.[102]

The outline of the tax offered in the 2000 Budget Report shows how much importance the government attached to the ideas of recycling revenue,[103] of making a commitment to supporting new technology to ease adjustment and of compensating the groups who would be hardest hit by the tax:[104]

> To further the Government's aim of shifting the burden of taxation from 'goods' to 'bads' the revenues from the levy will be fully recycled to the business community through a 0.1 percentage point reduction in employers' NICs and a new Sustainability Fund. The Government will be consulting shortly on how this fund can best be used to deliver local environmental improvements.[105]

Accordingly, in June 2000, Customs issued draft clauses for comment, and, in August 2000, the government consulted on the best uses for the sustainability fund which it was proposing to sent up with the benefit of receipts from the tax. In the Pre-Budget

---

[98] See *Stability and Steady Growth For Britain*, 1999 (Cm 4479, 1999), para. 6.95. Despite the ominous tone of the Pre-Budget Report, the quarrying industry remained confident that an aggregates tax would *not* be introduced, after all (see, for example, Daniel Lyons and Richard Mackender, 'The Environment and the Pre-Budget Report', TJ, 29 November 1999, pp. 11–12).
[99] Note the change of name (see para. 13.1 below).
[100] HM Treasury, *Budget Report 2000*, 21 March 2000, para. 6.90.
[101] See 2002 paper, Table 7.1, p. 42.
[102] *Ibid.*, Table 7.1, p. 42.
[103] A point which is reflected in the relevant statutory provisions also (see Finance Act 2001, s.44).
[104] See 2002 paper, Table 7.1, p. 42.
[105] HM Treasury, *Budget Report 2000*, 21 March 2000, para. 6.94.

Report of November 2000, the Chancellor announced that the government would allocate £35 million to the aggregates levy sustainability fund ('ALSF'), and that it would be introduced in April 2002, along with the levy itself.

In July 2000, the QPA, having failed to stop the levy, had submitted a proposal for reductions in tax rates if the industry reached defined environmental standards. In the 2001 Budget, the Chancellor announced that he was interested in principle in the idea of a differentiated rate of tax,[106] a point which might be taken to substantiate the claim made in the 2002 paper that rates of tax would be reviewed annually, as part of the Budget process.

The levy duly began operation on 1 April 2002, pursuant to the Treasury power contained in Finance Act 2001[107] to provide for its commencement date by statutory instrument.[108] An application by the British Aggregates Association[109] for judicial review of the legislation, based, among other things, on alleged breaches of Community law and of the European Convention on Human Rights, was dismissed by Moses, J. on 19 April 2002.[110]

11.4   Review of landfill tax

When the present government came to power in May 1997, landfill tax had already been in operation for seven months.[111] The goals of the tax, as originally articulated, were:

- [t]o ensure that landfill waste disposal is properly priced, which will promote greater efficiency in the waste-management market and in the economy as a whole; and
- To apply the 'polluter-pays' principle and promote a more sustainable approach to waste management in which we produce less waste, and reuse or recover value from more waste.[112]

As reformulated in the 2002 paper, the tax's goals are 'to internalise the environmental costs of landfill', exemplified by methane emissions, nuisance and groundwater

---

106 HM Treasury, *Budget 2001: Investing for the Long Term – Building Opportunity and Prosperity for All*, 7 March 2001 (see [2001] SWTI 475, 479).

107 See Finance Act 2001, s.16(6).

108 Finance Act 2001, Section 16, (Appointed Day) Order 2002, S.I. 2002 No. 809.

109 See para. 2.5 above.

110 See *R (on the application of British Aggregates Association and Others) v. C & E Commrs*, [2002] EWHC 926 (Admin), [2002] 2 CMLR 51. See paras 12.1, 12.3.3.1 and 12.3.5 below.

111 See Jane Powell and Amelia Craighill, 'The UK Landfill Tax', in O'Riordan, *op. cit.*, pp. 304–18; Bob Davies and Michael Doble, 'The Development and Implementation of a Landfill Tax in the UK', in Organisation for Economic Co-operation and Development, *Addressing the Economics of Waste* (Paris: OECD, 2004), pp. 63–80; and Inger Brisson, 'The UK Landfill Tax', in *Environmental Policy: Objectives, Instruments, and Implementation*, ed. by Dieter Helm (Oxford: Oxford University Press, 2000), pp. 260–80.

112 See HM Customs and Excise, *'Landfill Tax'*, *A Consultation Paper*, 1995, quoted in Powell and Craighill, *op. cit.*, p. 307.

pollution; to give better price signals for alternatives to landfill; and to assist meeting waste targets in the most efficient way.[113] The last of these marks a significant shift in policy, since, in order to 'steer' behaviour, the level of the tax will be vastly in excess of the externalities which it was originally introduced to address.[114] The tax was originally, and continues to be, chargeable on taxable disposals of material as waste, by way of landfill, at landfill sites.[115]

The 2002 Treasury paper places some emphasis on the fact that, despite[116] landfill tax's origins as a creation of the previous administration, the government continues to review, gather evidence and consult[117] on the operation of the tax.[118] The paper also stresses that, not only is revenue from the tax recycled, via the landfill tax credit scheme (that is, the LTCS),[119] but that the rates of the tax are reviewed annually as part of the Budget process and, specifically, as part of the Prime Minister's Strategy Unit[120] report on waste policy.[121] In relation to the former, it draws attention to the consultation launched in April 2002 on changes to the LTCS, especially on its funding priorities and on the way in which it is itself funded.[122] The consultation resulted in the changes to the LTCS made in April 2003, although these were of course subsequent to the publication of the Treasury paper itself.[123] The Chancellor's announcement in the 1999 Budget Report of the escalator in the rates of tax with effect from April 2000[124] is taken in the Treasury paper as an example of an early signal of the choice of an economic instrument, in this case to deal with the obligations imposed by the Landfill Directive.[125] A similar point might also be made of the second stage of the escalator, that is, the announcement in the 2002 Pre-Budget Report of the increase ('revenue neutral to business as a whole') in the rate of landfill tax to £35, by instalments of at least £3 per annum, starting in 2005.[126] Davies and Doble have the following comment:

> From a starting point of seeking to internalise externalities and incentivize sustainable waste management, policy considerations have changed the focus ... [the increases in the rate of

---

113 See para. 5.3 above.
114 See Davies and Doble, *op. cit.*, p. 78.
115 See para. 1.4.2.1(1) above and para. 15.2 below.
116 Or perhaps 'because of'.
117 The first consultation on the tax instituted by the government began as early as January 1998 (see 2002 paper, Table 7.1, p. 42).
118 See 2002 paper, Table 7.1, p. 42.
119 See para. 4.2.1.2(2)n above.
120 See para. 4.2.1.2(1) above.
121 See 2002 paper, Table 7.1, p. 43.
122 See HM Treasury News Release 35/02, *Consultation on Landfill Tax Credit Scheme Published*, 17 April 2002.
123 See the Landfill Tax (Amendment) Regulations 2003, S.I. 2003 No. 605. See also para. 21.3.2 below.
124 The 2002 paper misattributes this to the 1999 Pre-Budget Report (see Table 7.1, p. 42), whereas it was actually announced eight months earlier in the 1999 Budget itself (see *Pre-Budget Report 1999*, para. 6.86). See also para. 15.1 below.
125 See para. 6.3.2.4 above and para. 12.2.5.1(2)(b) below.
126 See *Pre-Budget Report 2002*, para. 7.51 (see [2002] SWTI 1562, 1577).

tax] ... have been driven by an acceptance that landfill tax must be increased to achieve behavioural change, through closing the cost gap on methods of diversion from landfill and ultimately to contribute to the incentive to achieve diversion to meet EU Landfill Directive targets on municipal waste.[127]

It is difficult to avoid the conclusion that landfill tax rates are such as to suggest that the tax is, or will be, more in the nature of an environmental penalty than an environmental tax. Such is the extent of the policy change that has resulted from the ongoing review referred to above.

11.5   Concluding comments

It will be appreciated from the discussion in this chapter that, whilst certainly taxes as traditionally understood,[128] the UK's three main environmental taxes have features which set them apart from most other taxes.

The key unusual feature which is common to all of them is their potential for steering behaviour[129] but there is also the hypothecation or, in the case of landfill tax, quasi-hypothecation, of the revenue raised by them.

What is striking about the application of the RIA approach to designing and implementing aggregates levy and climate change levy is that has introduced an element of bargaining into the decision as to whether to introduce new environmental taxes. The relative bargaining strengths of the governmental and industrial actors may, however, mean that taxes are more likely to be imposed on some sectors of industry than on others. The quarrymen's trade association was, after all, conspicuously less successful in negotiating the imposition of aggregates levy than were the energy-intensive sectors of industry in negotiating the introduction of climate change levy.

Finally, the soon-to-be penal levels of landfill tax raise the question of whether the tax stands to be transformed from a tax to a penalty. It is certainly no longer a Pigouvian tax,[130] properly so-called. This may conceivably be important, since taxes are usually deductible in the calculation of profits for the purposes of income tax and corporation tax, whereas fines are not.[131] In any event, there has certainly been a considerable shift in the policy which landfill tax is being used to advance since its introduction in 1996.

---

[127]   Davies and Doble, *op. cit.*, p. 77.
[128]   See Chapter 7 above.
[129]   Which, since they are consciously designed so to do, means that they do not, apparently, infringe the neutrality principle (see John Tiley, *Revenue Law*, 4th edn (Oxford: Hart Publishing, 2000), p. 11. See para. 1.2.1.5(2) above.
[130]   See para. 5.4n above.
[131]   See *McKnight v. Sheppard*, [1999] STC 669.

Chapter 12

Community Law Aspects

12.1   Introduction

This chapter seeks to examine how the design of the UK's environmental levies and subsidies is shaped by its status as a Member State of the EU. In practice, this may be the single most significant constraint on the design and implementation of the UK's environmental taxes and other economic instruments.

Throughout the study, and whatever the particular context in which we have sought to locate the UK's environmental taxes and other economic instruments, the rules and institutions of Community law have never been far away. In Chapter 4, we indicated how the ECJ has a role to play in determining the compatibility of UK tax law with the European Treaty.[1] In Chapter 4 also, as part of our analysis of the institutional framework of the UK's environmental taxes and other economic instruments, we described those Community institutions with some responsibility or other for environmental law and policy and for tax law and policy.[2] The UK's obligations as an EU Member State underlay the technical justifications for the green levies and subsidies discussed in Chapter 5. The overview of UK environmental and market regulation offered in Chapter 6 stressed the Community origins of much of the material, whilst reserving discussion of its detail until now.[3] In Chapter 7, we reviewed the insights provided by the jurisprudence of the ECJ into the concept of a tax, as distinct from related concepts.[4] Finally, in Chapter 8, we considered the impact of international environmental law and international trade law on the environmental taxation law of the UK as an EU Member State. The arrangement of the material in the present chapter reflects the principal concerns of these earlier chapters.

The current chapter is divided into two main parts. Of these, the more extensive is paragraph 12.2, which begins by examining those aspects of Community environmental law that have shaped the domestic environmental regulation discussed in Chapter 6 above. These are the Integrated Pollution Prevention and Control ('IPPC') regime; the Community's Environmental Action Programmes; and Community procedures for evaluating the likely environmental impacts of certain construction projects. The first of these provides a convenient basis for much UK domestic legislation (including climate change levy), while the second, in setting out the main environmental priorities of the Community, supplies the technical justification for both Community and domestic legislation in specific areas.[5] Subsequent parts of 12.2 are devoted, respectively, to Community waste management regulation and to

---

1   See para. 4.3.7 above.
2   See para. 4.3 above.
3   See Chapter 6 above, generally, esp. paras 6.2.1, 6.3.1 and 6.4.1.
4   See para. 7.2.2.3 above.
5   See paras 12.2.1–12.2.3 below.

Community regulation of air and atmospheric pollution.[6] An important element of the former is the definition of 'waste', as used in the Waste Framework Directive ('the WFD').[7] That definition had formerly been seen as being entirely separate from the one applicable for landfill tax; a discussion of the WFD definition has, however, been included, following a recent decision of the Court of Appeal which seems to recognise the possibility of a link between the definition of 'waste' for landfill tax purposes and the objectives of the WFD.[8] The discussion of air and atmospheric pollution control includes an overview both of Community energy law and of Community transport law.[9] Each of these overviews is necessary, not only for the environmental aspects of energy and transport law, but also for the implications of the structures thereby created for both the design of UK environmental taxes and environmental instruments and the prospects for their ultimate success or failure.[10] Since both environmental and energy policy usually involve the outlay of state resources, paragraph 12.2 ends with a brief discussion of the significance of state aids, both in combating air and atmospheric pollution and in the energy and transport sectors more generally.[11]

The latter part of the present chapter[12] mainly concerns the impact of Community tax law rules on the development of the UK's environmental taxes and economic instruments. The discussion unfolds, of course, against the background of the fiscal veto[13] and the essentially 'national' character of the instruments covered in the present study. After an overview of the relevant issues, the discussion examines the division of taxing powers between the Community and the Member States;[14] the ban on fiscal barriers to trade, including the prohibition on discriminatory internal taxation[15] and the prohibition on customs duties and, more importantly here, charges having equivalent effect.[16] Woven into the discussion in paragraph 12.3 are discussions of the cases in which various national attempts at creating green levies have received judicial attention. Most often, the relevant tribunal has been the ECJ but, as has already been mentioned,[17] the validity of the UK's aggregates levy in terms of Community law has been tested and upheld before a UK national court[18] and the relevant aspects of this decision are incorporated into the discussion as appropriate. Paragraph 12.3 ends

---

[6]   See P.G.G. Davies, *European Union Environmental Law* (Aldershot: Ashgate, 2004), chs 7 and 8.
[7]   Council Directive 75/442/EEC, (1975) OJ L194 39.
[8]   See *Parkwood Landfill Ltd v. Customs and Excise Commissioners*, [2002] STC 1536. See also paras. 6.3.2.1 above and 15.3 below.
[9]   See paras. 12.2.6.3 and 12.2.6.4 below. This mirrors the discussion of UK energy and transport law and policy in paras. 6.4 and 6.5 above.
[10] See Chapter 21 below.
[11] See para. 12.2.7 below.
[12] See para. 12.3 below.
[13] See para. 4.3.1 (point 1) above.
[14] See para. 12.3.2 below.
[15] See para. 12.3.3.1 below.
[16] See para. 12.3.3.2 below.
[17] See para. 8.4.5.1 above.
[18] In *R (on the application of British Aggregates Associates and Others) v. C & E Commrs*, [2002] EWHC 926 (Admin), [2002] 2 CMLR 51. See para. 12.3.3.1 below.

with a discussion of recent measures to harmonise excise duties on fuel products[19] and a review of the application of the law on state aids, as introduced earlier in the chapter, to the specific context of environmental levies and subsidies.[20]

The final part of the chapter briefly discusses a hypothetical issue for environmental levies, in the context of the Community rules on free movement of goods in Arts 28–31, European Treaty (ex 30, 34, 36 and 37).[21] Partly in view of the relative familiarity of those rules, partly also because of the hypothetical nature of the issue there discussed and partly because of the overall length of the chapter, the explanation of the relevant Treaty Arts and case law in that part has been kept to the bare minimum necessary to elucidate the issue concerned.

Evidently, the inclusion of such a wide range of material has not made for a concise chapter. The writers are firmly of the view, however, that, it is not possible to assess the true significance of the levies and subsidies that form the subject matter of the present study without an appreciation of their European regulatory and taxation context.

12.2   Regulatory aspects

12.2.1   General

Community environmental law constitutes a regional regime of international environmental law which, since 1 May 2004, has applied directly to 25 European countries.[22] By Art. 2 of the European Treaty, the Community is tasked, among other things, with promoting '… a harmonious, balanced and sustainable development of economic activities … [as well as] … a high level of protection and improvement of the quality of the environment', Art. 3(1)(l) stipulating that the activities of the Community are to include 'a policy in the sphere of the environment'.[23]

The listing of a high level of protection and improvement of the quality of the environment as an independent goal in Art. 2, rather than as an incidental requirement of economic growth, was an achievement of the Treaty of Amsterdam which, although it was signed in 1997, only came into effect on 1 May 1999.[24] Provisions on environmental protection had first appeared in the European Treaty as a result of the amendments made to it by the 1986 Single European Act ('the SEA 1986') but this had not deterred the Community from adopting environmental legislation under the Treaty powers conferred on the Council for the harmonisation of laws affecting the establishment or functioning of the common market,[25] and under the powers, also

---

19   See para. 12.3.4 below.
20   See para. 12.3.5 below.
21   See para. 12.4 below.
22   See Philippe Sands, *Principles of International Environmental Law*, 2nd edn (Cambridge: Cambridge University Press, 2003), p. 733; Joanne Scott, *EC Environmental Law* (London: Longman, 1998); and *European Environmental Law*, ed. by Ludwig Kramer (Aldershot: Ashgate/Dartmouth, 2003).
23   The expression 'the environment' is not, however, defined. See para. 1.2.1.5(1) above.
24   See Paul Craig and Gráinne de Búrca, *EU Law: Text, Cases and Materials*, 3rd edn (Oxford: Oxford University Press, 2003), pp. 29–42, esp. pp. 30 and 32.
25   That is, Art. 94, European Treaty (ex 100).

conferred by the Treaty on the Council, for adopting measures necessary for attaining one of the Treaty objectives but for which the Treaty had not provided the necessary powers.[26] Amsterdam also saw the inclusion of a new article, now Art. 6, European Treaty (ex 3c), of a requirement that '[e]nvironmental protection requirements ... be integrated into the definition and implementation of ... Community policies and activities ...', in particular with a view to promoting sustainable development'. Thus, as we shall see, Community policy on both energy and transport has a strongly 'environmental' dimension.

In its current form, Art. 174(1), European Treaty (ex 130r) states that the Community policy on the environment must contribute to the pursuit of four objectives:

- preserving, protecting and improving the quality of the environment;
- protecting human health;
- prudent and rational utilisation of natural resources; [and]
- promoting measures at international level to deal with regional or worldwide environmental problems.[27]

Article 174(2) elaborates these four objectives by stating that Community policy on the environment is to '... aim at a high level of protection taking into account the diversity of situations in the various regions of the Community'. Accordingly, continues Art. 174(2), Community environmental policy is to be based on the precautionary principle, the preventive principle, the proximity principle and the polluter pays principle.[28] The application of the precautionary principle to Community policy on the environment was yet another achievement of the Amsterdam Treaty,[29] although the inclusion of the other three principles dated back to the SEA 1986.[30] Article 174(3) provides that Community action on the environment must take account of available scientific and technical data, environmental conditions in the various regions of the Community, the potential benefits and costs of action or lack of action and the economic and social development of the Community as a whole and the balanced development of its regions.

By Art. 175, European Treaty (ex 130s), decisions on any action to be taken by the Council to achieve the four objectives in Art. 174(1) are usually to be taken by qualified majority voting.[31] One of the exceptions to this, of course, is where the decision relates to a provision primarily of a fiscal nature, in which case the decision must be unanimous.[32] Art. 176, European Treaty (ex 130t) specifically allows Member States to maintain or introduce more stringent protective measures, provided they are compatible with the Treaty and notified to the Commission.

---

[26]   That is, *ibid.*, Art. 308 (ex 235).
[27]   See the commentary on the four objectives in Maurice Sunkin, David Ong and Robert Wight, *Sourcebook on Environmental Law*, 2nd edn (London: Cavendish, 2002), pp. 18–27.
[28]   See para. 8.2.6 above. The proximity principle is also known as 'the rectification-at-source principle'.
[29]   See Sands, *op. cit.*, p. 271.
[30]   *Ibid.*, p.743.
[31]   *Ibid.*, Art. 175(1).
[32]   *Ibid.*, Art. 175(2). See para. 4.3.1 (point 1) above.

Unlike the common commercial policy,[33] the common agricultural and fisheries policy[34] and the common transport policy,[35] environmental policy as provided for by Arts 174–176 is not the exclusive policy of the Community. Thus, the Community and the Member States enjoy a shared competence in relation to environmental regulation. However, it should be noted that, under Art. 95, European Treaty (ex 100a), the Council may, by qualified majority,[36] adopt measures for the approximation of provisions relating to the establishment and functioning of the internal market. Such proposals may, of course, concern environmental protection.[37]

12.2.2 *Integrated Pollution Prevention and Control ('IPPC')*

The Integrated Pollution Prevention and Control Directive ('the IPPC Directive')[38] was adopted in 1996, on the basis of what is now Art. 175, European Treaty (ex 130s). Member States were given a period of three years in which to implement its provisions, its transcription into the law of England and Wales being effected by the Pollution Prevention and Control Act 1999 and the Pollution Prevention and Control (England and Wales) Regulations 2000.[39] Both of these measures have been discussed in detail in Chapter 6 and our concern here is only with the IPPC Directive itself,[40] although it is perhaps worth remembering the point made there that, *inter alia*, the IPPC regime has been taken to form the regulatory basis both of climate change levy and the EU ETS.

Article 1 of the IPPC Directive states the Directive's purpose as being to reduce emissions from industrial activities into the air, water and land, including waste, in order to achieve a high level of protection of the environment taken as a whole. By Art. 3 of the IPPC Directive, Member States are required to take the necessary measures to make sure that the competent authorities ensure that installations[41] are operated in such a way that:

1   all appropriate preventive measures against pollution are taken, especially through the application of best available techniques ('BATs');
2   no significant pollution is caused;
3   waste production is avoided or, where waste is produced, it is recovered or disposed of in such a way as to avoid or reduce environmental impact where recovery is not possible;

---

33   *Ibid.*, Art. 3(1)(b).
34   *Ibid.*, Art. 3(1)(e).
35   *Ibid.*, Art. 3(1)(f). See para. 12.2.6.4 below.
36   That is, under Art. 251, European Treaty (ex 189b).
37   Art. 95(3), European Treaty (ex 100a).
38   Council Directive 96/61/EC, (1996) OJ L257 26.
39   S.I. 2000 No. 1973.
40   See para. 6.2.3 above.
41   Defined as 'a stationary technical unit where one or more activities [covered by the IPPC Directive] ... are carried out, and any other directly associated activities which have a technical connection with the activities carried out on that site and which could have an effect on emissions and pollution' (see the IPPC Directive, Art. 2(3)).

4   energy is used efficiently;
5   necessary measures are taken to avoid accidents and limit their consequences; and
6   on definitive cessation of activities, necessary measures are taken to avoid any pollution risk and to return the site of operation to a satisfactory state.

The installations that fall within the IPPC regime are listed in Annex I to the Directive. They include installations in the energy industries,[42] industries involved in the production and processing of metals, the mineral industry,[43] the chemical industry and the waste management industry.[44] The purpose of the Directive is to be achieved by the competent authorities in Member States issuing permits to the operators of installations carrying out the targeted activities.[45] Permits must contain conditions guaranteeing that the installation complies with the requirements of the Directive and must otherwise be refused.[46]

The BAT concept is specifically defined in Art. 2(11) of the IPPC Directive. 'Best' means 'most effective in achieving a high general level of protection of the environment as a whole'; 'available' refers to techniques:

... developed on a scale which allows implementation in the relevant industrial sector, under economically and technically viable conditions, taking into consideration the costs and advantages, whether or not the techniques are used or produced inside the Member State in question, as long as they are reasonably accessible to the operator ... [47]

Finally, 'techniques' includes 'both the technology used and the way in which the installation is designed, built, maintained, operated and decommissioned'.[48]

As originally conceived by the EC Commission, the IPPC Directive had been intended to ensure 'integrated permitting' for industrial processes but, following objections from certain Member States, the emphasis shifted to the preventative nature of the control mechanism and away from integrating the permitting system.[49] It is to be amended following the passage of the EU Emissions Trading Directive ('the EU ETS Directive').[50]

---

[42]   That is, certain combustion installations; mineral oil and gas refineries; coke ovens; and coal gasification and liquefaction plants.
[43]   That is, installations for producing cement clinker.
[44]   That is, installations for the disposal or recovery of hazardous waste; for the incineration of municipal waste; and for the disposal of non-hazardous waste.
[45]   See Arts 4 and 5, IPPC Directive.
[46]   See Art. 8, IPPC Directive.
[47]   *Ibid.*, Art. 2(11).
[48]   *Ibid.*
[49]   See Stuart Bell and Donald McGillivray, *Environmental Law*, 5th edn (London: Blackstone, 2000), p. 380.
[50]   See Sands, *op. cit.*, p. 754n and Chapter 28 below.

12.2.3   *Environmental Action Programmes*

A phenomenon of no less importance to the EU aspects of environmental regulation than the primary and secondary European legislation discussed in this chapter are the six EU Environmental Action Programmes launched to date.[51] Proposed by the Commission[52] and approved by the Council,[53] they chart the development of Community policy on the environment over a 30-year period since 1973. The policy initiatives contained in the Programmes, together with the tools for their realisation (such as the increased use of economic instruments) have not only shaped Community environmental policy but have proved influential on the wider international stage.[54]

The current Sixth Environmental Action Programme, with its central themes of sustainable development and the integration of environmental policies, had a long gestation period, finally being adopted by the European Parliament[55] and Council in July 2002.[56] The Programme pays special attention to four priority areas for action:

1   tackling climate change by stabilising 'the atmospheric concentrations of greenhouse gases at a level that will not cause unnatural variations of the earth's climate'. The key priority for the sixth Programme is thus the ratification and implementation of the Kyoto Protocol to reduce greenhouse gas emissions by 8 per cent from 1990 levels by 2008–2012;[57]
2   protecting nature and biodiversity, with a view to protecting and restoring the functioning of natural systems and halting the loss of biodiversity by 2010;
3   achieving a quality of environment where the levels of man-made contaminants do not give rise to significant impacts on or risks to human health; and
4   ensuring sustainable use of natural resources and management of wastes by ensuring that 'the consumption of renewable and non-renewable resources does not exceed the carrying capacity of the environment'.

The Programme considers the EU's 8 per cent commitment to the reduction of greenhouse gases as being a first step to a 70 per cent cut (see 1 above). Objective 3 includes a commitment to ensuring by 2020 that chemicals are produced only in ways that do not have a significant negative impact on health and the environment. The objective of ensuring sustainable use of natural resources, etc. (see 4 above) includes the objective of producing 22 per cent of electricity from renewable sources by 2010.

---

[51]   The first Programme was for the period 1973–1976; the second was for the period 1977–1981; the third was for the period 1982–1986; the fourth was for the period 1987–1993; and the fifth was for the period 1993–1997, subsequently extended (see Sunkin, Ong and Wight, *op. cit.*, p. 28). Details of the Sixth Programme are given in the text.

[52]   See para. 4.3.2 above.

[53]   See para. 4.3.1 above.

[54]   See Sands, *op. cit.*, pp. 753–4.

[55]   See para. 4.3.3 above.

[56]   Decision 1600/02/EC, (2002) OJ L242 1.

[57]   See paras 8.2.2n and 8.3.1.4 above.

The Programme also identifies a number of international priorities (for example, achieving mutual supportiveness between trade and environmental needs, including the sustainability impact assessment of multilateral trade agreements)[58] and the implementation of Community environmental law and policy in the ten new Member States.

12.2.4   Environmental Impact Assessment and Strategic Environmental Assessment

Areas in which EU law and policies continue to prove influential on the wider international stage are those of Environmental Impact Assessment ('EIA') and Strategic Environmental Assessment ('SEA').[59] These concern the necessity of evaluating the likely environmental impacts of projects for the construction of, for example, thermal and nuclear power stations, waste disposal installations and, crucially, projects for the construction of renewables generators.[60] For the purposes of the present study, it is only necessary to sketch the broad outlines of each one.

The Directive on the Assessment of the Effects of Certain Public and Private Projects on the Environment ('the EIA Directive')[61] 'was [says Sands] the first international instrument to provide details on the nature and scope of environmental assessment, its use, and participation rights in the process'.[62] The Directive on the Assessment of Certain Plans and Programmes on the Environment ('the SEA Directive')[63] is, also, again according to Sands, 'the first international instrument to impose binding obligations, requiring member states to ensure that "an environmental assessment is carried out of certain plans and programmes which are likely to have significant effects on the environment"'.[64] The EIA Directive was to be transcribed into the laws of Member States by 3 July 1988[65] and the SEA Directive was so transcribed into UK by the deadline of 21 July 2004.[66]

Under the EIA Directive, where public and private projects are likely to have significant environmental effects, an Environmental Impact Assessment ('EIA') is required,[67] as part of the process of which the developer must supply specified information, consult with the relevant authorities, provide information to, and

---

[58]   The environmental aspects of world trade regulation are, of course, among the most pressing of contemporary issues (see para. 8.2.1 above).

[59]   See Sands, *op. cit.*, pp. 807–813 and Sunkin, Ong and Wight, *op. cit.*, pp. 777–82. Dealt with in the UK context in para. 6.2.4 above.

[60]   See para. 21.5 below.

[61]   Council Directive 85/337/EEC, (1985) OJ L175 40, subsequently heavily amended.

[62]   See Sands, *op. cit.*, p. 807. Sands also says, however, that EIAs emerged internationally after the 1972 Stockholm Declaration of the United Nations Conference on the Human Environment (see para. 8.2.3n above) and that they were first established in the domestic law of the US under the National Environmental Protection Act of 1972 (see Sands, *op. cit.*, p. 800).

[63]   Council Directive 01/42/EC, (2001) OJ L197 30.

[64]   See Sands, *op. cit.*, p. 812.

[65]   Council Directive 85/337/EEC, Art. 13.

[66]   Council Directive 01/42/EC, Art. 1.

[67]   Council Directive 85/337/EEC, Arts 1(1) and 2(1).

consult with, the public, and provide information to other Member States likely to be affected.[68] The projects covered are divided into 12 categories, which include ones in the extractive industry, the energy industry and the mineral industry.[69] There are certain exceptions for, for example, projects serving national defence purposes.[70]

The SEA Directive is aimed at the plans and programmes producing the projects covered by the EIA Directive. It applies to plans and programmes prepared for 10 specific sectors, including energy, transport, waste management and town and country planning or land use.[71] In essence, Member States must assess any plans or programmes[72] that contain a framework for future development consents and which are likely to have significant environmental effects.[73] The idea of the SEA Directive is that, as required by Art. 6, European Treaty (ex 3c)[74] and, as envisaged by the Sixth Environmental Action Programme,[75] environmental considerations will be integrated ever more firmly into the policy-making of the Member States.

The EIA Directive was implemented in England and Wales by the Town and Country Planning (Environmental Impact Assessment) (England and Wales) Regulations 1999,[76] as well as by the other measures already discussed in Chapter 6 above.[77] The implementation of the SEA Directive was effected by secondary legislation on which the ODPM[78] consulted in spring 2004.[79] As discussed in Chapter 6 above, the absence of an SEA has already been controversial in relation to the deployment of arrays of wave power devices off the coast of South West England.[80]

12.2.5   Waste management

12.2.5.1   General

Community law and policy on waste management is contained in a 1990 Community Strategy for Waste Management[81] and in legislation of four kinds: (1) a framework

---

68 *Ibid.*, Arts 5–10.
69 Council Directive 85/337/EEC, Annex II.
70 *Ibid.*, Art. 1(4).
71 Council Directive 01/42/EC, Art. 3.
72 Defined to mean all plans and programmes subject to preparation and/or adoption by an authority at national, regional or local level or which are prepared by an authority for adoption, through a legislative procedure by Parliament or government and which are required by legislative, regulatory or administrative provisions (see Art. 2(a), Directive 01/42/EC).
73 See Sunkin, Ong and Wight, *op. cit.*, p. 781.
74 See para. 12.2.1 above.
75 See para. 12.2.3 above.
76 S.I. 1999 No. 293.
77 See para. 6.2.4 above.
78 See para. 4.2.1.2(1) above.
79 See the Environmental Assessment of Plans and Programmes Regulations 2004, S.I. 2004 No. 1633 (see 'Strategic assessment guidance to planners at odds with EC views', 346 ENDS Report (2003)).
80 See para. 6.2.4 above.
81 Council Resolution of 7 May 1990 on Waste Policy, (1990) OJ C122 2.

directive on waste management regulation within the Community; (2) various directives on the disposal of specific wastes; (3) a specific directive on the management of hazardous waste; and (4) a regulation on the shipment of waste.[82] There are also Community measures prohibiting the disposal of certain wastes into the marine environment[83] and measures limiting the emission into the atmosphere of particular waste gases.[84]

The Community Strategy for Waste Management was reviewed by the Commission in 1996, who, in a Communication of July that year,[85] formulated a hierarchy of preferred ways of dealing with waste, first preference being given to waste prevention, increased recovery being next favoured and safe disposal being the least-favoured option. In addition to promoting reuse and recycling of waste, the Communication promotes the increased use of economic instruments, along with rules for restricting the use of dangerous substances in products, lifecycle analyses and eco-audits and means of increasing consumer awareness, such as eco-labelling. Although the Council greeted the Communication favourably, the Parliament expressed its dissatisfaction with the progress of Community waste policy and called (unsuccessfully) on the then Environment Commissioner to produce a new Action Programme on waste management.[86]

Before discussing the four kinds of legislation referred to above, it is necessary to refer briefly to the fact that waste management is specifically within the scope of the IPPC Directive, already discussed.[87] Without prejudice to specific provisions of other directives on waste management,[88] the installations in respect of which Member States are enjoined to introduce an IPPC permit system include the waste management installations listed in para. 5 of Annex 1 to the IPPC Directive. These are installations for the disposal or recovery of hazardous waste[89] with a capacity exceeding 10 tonnes per day;[90] installations for the incineration of municipal waste, provided their capacity exceeds three tonnes per hour;[91] installations for the disposal of non-hazardous waste, with a capacity exceeding 50 tonnes per day;[92] and landfills receiving more than 10 tonnes per day or with a total capacity exceeding 25,000

---

[82]   See Sands, *op. cit.*, pp. 786–92, from which the fourfold classification is taken; see also Sunkin, Ong and Wight, *op. cit.*, pp. 380–418.

[83]   As to which the reader is referred, for example, to Sands, *op. cit.*, pp. 768–83.

[84]   See para. 12.2.6 below.

[85]   COM (1996) 399 final.

[86]   See the account in Sunkin, Ong and Wight, *op. cit.*, p. 389.

[87]   See para. 12.2.2 above.

[88]   That is, Council Directive 75/442/EEC (the Waste Framework Directive), Art. 11, and Council Directive 91/689/EEC (the Directive on Hazardous Waste), Art. 3. See below in this para. for further details of these Directives, including OJ references thereto.

[89]   As defined, not only in Council Directive 75/442/EEC, but also in Council Directive 91/689/EEC (see above) and in Council Directive 75/439/EEC, (1975) OJ L194 23.

[90]   See Council Directive 96/61/EC, (1996) OJ L257 26, Annex I, para. 5.1.

[91]   *Ibid.*, Annex I, para. 5.2.

[92]   That is, as defined in Council Directive 75/442/EEC (that is, the Waste Framework Directive). See Council Directive 96/61/EC, (1996) OJ L257 26, Annex I, para. 5.3.

tonnes, excluding landfills of inert waste.[93] The transcription of these requirements into UK law has already been discussed in Chapter 6 above.[94]

The Directive has been transcribed into UK law by Part I of the Waste and Emissions Trading Act 2003 and by the Landfill (England and Wales) Regulations 2002, each of which have been discussed in Chapter 6 above.[95]

(1) Framework of waste management regulation

The Waste Framework Directive ('the WFD')[96] directs Member States to take appropriate measures to encourage 'the prevention or reduction of waste production and its harmfulness';[97] to encourage 'the recovery of waste by recycling, re-use, reclamation, etc.;[98] and to encourage the use of waste as energy.[99] The definition of the concept of waste which is used in the WFD, that is, that of specified substances that the owner discards, or intends to discard or is required to discard, is potentially of some importance in the context of the present study and is considered further below.[100]

The objectives of preventing and reducing waste production and its harmfulness are to be attained by three means in particular, namely: the development of clean technologies, which are 'more sparing in their use of natural resources' than older ones; the development of products whose manufacture, use and disposal are such as to minimise waste; and 'the development of appropriate techniques for the final disposal of dangerous substances contained in waste'.[101]

Article 4 of the WFD enjoins Member States to take measures necessary to ensure that waste is disposed of or recovered without endangering human health and without harming the environment and to prohibit the abandonment, dumping or uncontrolled disposal of waste. Subsequently, in Art. 5(1), they are required to take appropriate measures (taking account of BATNEEC) to create 'an integrated and adequate network of disposal installations', to enable the Community to become self-sufficient in waste disposal. Article 5(2) provides that such waste disposal networks must enable waste to be disposed of in one of the nearest appropriate installations and by means of the most appropriate methods, so as to ensure 'a high level of protection for the environment and public health'.[102] By Art. 7, the competent authorities in the Member States are required to draw up waste management plans and are permitted

---

93   See Council Directive 96/61/EC, (1996) OJ L257 26, Annex I, para. 5.4.
94   See paras 6.2.3 and 6.3.2.3 above.
95   See para. 6.3.2(4) above. See also para. 20.7 below (the LATS).
96   Council Directive 75/442/EEC, (1975) OJ L194 39. See Sands, *op. cit.*, pp. 787–9; also Sunkin, Ong and Wight, *op. cit.*, pp. 383–4.
97   Council Directive 75/442/EEC, Art. 3(1)(a).
98   *Ibid.*, Art. 3(1)(b).
99   *Ibid.*
100   See para. 12.2.5.2 below.
101   Council Directive 75/442/EEC, Art. 3(1)(a).
102   But see *Chemische Afvalstoffen Dusseldorp BV and Others v. Minister van Volkshuisvesting, Ruimtelijke Ordening en Milieubeheer*, C–203/96, [1998] 3 CMLR 873, 912 (para 30), where the ECJ held that the principles of self sufficiency and proximity did not apply to waste for *recovery*.

to take such measures as are necessary to prevent the movement of waste except in accordance with those plans. The current national waste strategy for England and Wales appears in *Waste Strategy 2000: England and Wales*.[103]

The WFD also contains provisions requiring undertakings carrying out disposal or recovery operations[104] to obtain a permit from the competent authority;[105] requiring collectors, transporters and those arranging for the disposal of waste to be registered;[106] and providing that the polluter pays principle[107] means that the holder of the waste (that is, its producer or the person in possession of it)[108] must bear the cost of disposing of it.[109]

(2) Disposal of specific wastes
As mentioned above, special disposal rules apply to specific categories of waste.[110] Thus, legislation has been adopted relating, for example, to the disposal of waste oils;[111] to the disposal of polychlorinated biphenyls ('PCBs') and polychlorinated terphenyls ('PCTs');[112] to waste from the titanium dioxide industry;[113] to sewage sludge;[114] and to the recovery and controlled disposal of spent batteries and accumulators, which contain quantities of mercury, cadmium or lead.[115] The UK measures implementing the directives on these matters have already been referred to in Chapter 6 above.[116] Two directives in specific areas are of particular importance in the context of the present study, however, that is, the Packaging and Packaging Waste Directive;[117] and the Landfill Directive.[118]

(a) Packaging and packaging waste
The Packaging and Packaging Waste Directive, which was adopted in response to the ECJ's decision in the *Danish Bottles* case,[119] has two main objectives: (a) to provide

---

103 See para. 6.3.2 above.
104 See para. 12.2.5.2 below, the disposal and recovery operations in question being specified in Annexes IIA and IIB to the Directive.
105 Council Directive 75/442/EEC, Arts 9 and 10.
106 *Ibid.*, Art. 12.
107 See para. 8.2.3 above.
108 Council Directive 75/442/EEC, Art. 1(c).
109 *Ibid.*, Art. 15.
110 See Sands, *op. cit.*, pp. 791–2; also Sunkin, Ong and Wight, pp. 384–8.
111 Directive on the Disposal of Waste Oils, Council Directive 75/439/EEC, (1975) OJ L194 23.
112 Directive on the Disposal of Polychlorinated Biphenyls and Polychlorinated Terphenyls, Council Directive 96/59/EC, (1996) OJ L243 31.
113 Directive on Waste from the Titanium Dioxide Industry, Council Directive 92/112/EEC, (1992) OJ L409 11.
114 Directive on Sewage Sludge, Council Directive 86/278/EEC, (1986) OJ L181 6.
115 Directive on Batteries and Accumulators, Council Directive 91/157/EEC, (1991) OJ L78 38.
116 See paras 6.3.2.5–6.3.2.6 above.
117 European Parliament and Council Directive 94/62/EC, (1994) OJ L365 10.
118 Council Directive 99/31/EC (1999) OJ L182 1.
119 *Commission v. Denmark*, C–302/86, [1989] 1 CMLR 619 (see para. 12.4 below).

a high level of environmental protection by harmonising national measures on the management of packaging and packaging waste; and (b) to ensure the functioning of the internal market by avoiding obstacles to trade.[120] Since June 1996, Member States have been obliged to take measures to ensure that systems are set up to provide for:

a. the return and/or collection of used packaging and/or packaging waste from the consumer, other final user, or from the waste stream in order to channel it to the most appropriate waste management alternatives;
b. the reuse or recovery including re-cycling of the packaging and/or packaging waste collected ...[121]

In addition, there are obligations on Member States to take preventative measures,[122] including the imposition of a requirement that packaging may be marketed only if it complies with requirements set out in the Directive,[123] as well as set specific recovery and recycling targets which must be reached within time periods laid down in the Directive.[124] By Art. 4(1) of the Directive, preventative measures may include the creation of national programmes to prevent the formation of packaging waste.[125]

The Directive covers primary, secondary and tertiary packaging, applying (as it does) to packaging usually acquired by the purchaser (primary); packaging removed by the retailer near the point of sale (secondary); and packaging designed to enable transportation and handling (tertiary).[126]

The Directive was transcribed into UK law by the Producer Responsibility Obligations (Packaging Waste) Regulations 1997,[127] introduced in Chapter 6 above[128] and examined in detail in Chapter 19 below.

(b) Waste sent to landfill
Article 1 of the Landfill Directive states that its aim is to provide for, 'by way of stringent operational and technical requirements', the prevention or, so far as possible, for the reduction of the negative environmental effects of the 'landfilling of waste, during the whole lifecycle of the landfill'. A landfill is defined in Art. 2(g) as a 'waste disposal site for the deposit of waste onto land or underground' but the definition excludes certain unloading and storage facilities. Landfill sites already in operation at the time of transposition of the Directive must comply with its terms by 2009.[129]

---

120 European Parliament and Council Directive 94/62/EC, Art. 1(1). For the implementation of the Directive in UK law, see Chapter 19 below, *passim*.
121 European Parliament and Council Directive 94/62/EC, Art. 7(1)(a).
122 *Ibid.*, Art. 4.
123 *Ibid.*, Art. 9(1).
124 *Ibid.*, Art. 5.
125 In the UK, the programme appears in *Waste Strategy 2000: England and Wales* (see para. 6.3.2 above).
126 European Parliament and Council Directive 94/62/EC, Art. 3(1).
127 S.I. 1997 No. 648.
128 See para. 6.3.3.2 above.
129 See Council Directive 1999/31/EC (1999) OJ L182 1, Art. 14, from which this date is derived.

Member States must apply the Directive to landfill as defined in Art. 2(g).[130] Article 4 requires each landfill to be classified as to whether it is for hazardous waste, for non-hazardous waste or for inert waste. By Art. 5(1), Member States must set up national strategies for the reduction of biodegradable waste going to landfills. Article 5(2) fixes the targets, based on 1995 arisings, as: a reduction of 25 per cent by 2006; of 50 per cent by 2009; and of 65 per cent by 2016.

Biodegradable waste is defined in Art. 2(m) as 'any waste that is capable of undergoing anaerobic or aerobic decomposition, such as food and garden waste, and paper and paperboard'. Such national strategies should include recycling, composting, biogas production and materials or energy recovery[131] and, by Arts 7–9, competent authorities are to issue permits for operating landfill sites, the conditions of issue of which are to ensure compliance with the Directive. The costs involved in setting up, operating, closing and caring for landfill sites thereafter must, as far as possible, be covered by the price charged by the site operator for the disposal of waste in the site[132] and Member States must not only control and monitor site operations[133] but also provide for site closure and after-care procedures.[134] As to the last of these, by Art. 13(c), 'after a landfill has been definitely closed, the operator shall be responsible for its maintenance, monitoring and control in the after-care phase for as long as may be required by the competent authority, taking into account the time during which the landfill could present hazards'.

The Directive has been transcribed into UK law by Part I of the Waste and Emissions Trading Act 2003, which has been discussed in Chapter 6 above.[135]

(3) Disposal of hazardous waste
The Directive on Hazardous Waste,[136] which repealed Council Directive 78/319/ EEC on toxic and dangerous wastes,[137] applies to the wastes featuring on a list to be drawn up by the Commission and which is subject to periodic review as well as to any other waste considered by a Member State to possess certain properties.[138]

The definition of hazardous waste is thus a complex one and recognises that some waste only becomes hazardous in certain circumstances.[139] The objective of the Directive is 'to approximate the laws of the Member States on the controlled management of hazardous waste'[140] but it does not apply to domestic waste.[141] The competent authorities in each of the Member States are enjoined to draw up, 'either

---

[130] Council Directive 1999/31/EC.
[131] *Ibid.*, Art. 5(1).
[132] *Ibid.*, Art. 10.
[133] *Ibid.*, Art. 12.
[134] *Ibid.*, Art. 13.
[135] See para. 6.3.2 above and para. 20.7 below (the LATS).
[136] Council Directive 91/689/EEC, (1991) OJ L377 20.
[137] (1978) OJ L84 43. Certain of the provisions of the 1978 directive are incorporated into the 1991 Directive by reference (see Sands, *op. cit.*, pp. 789–91).
[138] As to which, see Council Directive 91/689/EEC, Annex III.
[139] See Sunkin, Ong and Wight, p. 385.
[140] Council Directive 91/689/EEC, Art. 1(1).
[141] *Ibid.*, Art. 1(5).

separately or in the framework of their general waste management plans, plans for the management of hazardous waste and shall make these plans public ...'.[142]

By incorporating references to Council Directive 78/319/EEC, Art. 4 of the Directive[143] requires those who produce and transport hazardous waste to maintain detailed records, to be retained for at least three years, as to the nature of the waste. Furthermore, under Art. 2, Member States are required to take 'the necessary measures to require that on every site where tipping (discharge) of hazardous waste takes place the waste is recorded and identified';[144] to take the necessary measures to ensure that, except in prescribed circumstances,[145] different categories of hazardous waste are not mixed with each other nor with non-hazardous waste;[146] and Member States must supply to the Commission details of establishments and undertakings carrying out the disposal or recovery of hazardous waste on behalf of third parties.[147] By Art. 5, Member States are required to take 'the necessary measures to ensure that, in the course of collection, transport and temporary storage, waste is properly packaged and labelled in accordance with the international and Community standards in force'.

The implementation of the Directive into UK law has been effected by the Special Waste Regulations 1996,[148] discussed in Chapter 6 above.[149]

(4) Shipment of waste

Council Regulation on the Shipment of Waste ('the Waste Shipment Regulation')[150] covers shipments of waste between Member States, shipments of waste within Member States and waste exports to, and waste imports from, third countries.[151]

Generally speaking, waste may be shipped between Member States for disposal or recovery, subject to prior notification, authorisation and possible conditions, although the rules and procedures differ according to whether disposal or recovery is involved.[152]

So far as shipments within Member States are concerned, Member States are simply required to establish an appropriate system for the supervision and control of waste shipments which takes account of the need for coherence with the Community system.[153]

---

[142] *Ibid.*, Art. 6(1). See *Commission v. United Kingdom*, C–35/00, [2002] ECR I–953, and *Commission v. Italy*, C–466/99, [2002] ECR I–851. The UK's plan is contained in *Waste Strategy 2000: England and Wales* (see para. 6.3.2 above).

[143] That is, Council Directive 91/689/EEC.

[144] *Ibid.*, Art. 2(1).

[145] *Ibid.*, Art. 2(3).

[146] *Ibid.*, Art. 2(2).

[147] *Ibid.*, Art. 8(3).

[148] S.I. 1996 No. 972 (as amended).

[149] See para. 6.3.2(6) above.

[150] Council Regulation 259/93/EEC, (1993) OJ L30 1.

[151] See Sands, *op. cit.*, pp. 699–703 and Sunkin, Ong and Wight, *op. cit.*, pp. 386–7. See also the Transfrontier Shipment of Waste Regulations 1994, S.I. 1994 No. 1137. As to the legal effects of Regulations, see Craig and de Búrca, *op. cit.*, pp. 189–93.

[152] Council Regulation EEC/259/93, Arts 3–8.

[153] *Ibid.*, Art. 13.

With regard to exports to third countries,[154] exports to ACP countries are simply banned.[155] As regards non-ACP third countries, exports of waste for disposal are prohibited, except where the country of destination is an EFTA country which is also party to the 1989 Basel Convention.[156] In this case, although both the EFTA country and the Member State of origin has the option of imposing a ban,[157] exports are generally permitted, subject to notification and authorisation.[158] As to exports for recovery, these are prohibited, except where the country of destination falls into one of three categories: (a) it is a country to which the OECD Council Decision on the control of the transfrontier movements of wastes for recovery operations applies;[159] (b) it is a party to the 1989 Basel Convention or an agreement thereunder; or (c) it is another country, which is a party to a compatible pre-existing bilateral agreement with the Member State in question.[160]

Finally, waste imports from third countries are again divided between those for disposal and those for recovery. Imports of waste for disposal are banned except when they are: (a) from EFTA countries which are also parties to the 1989 Basel Convention (b) from other countries which are parties to the Convention; or (c) from countries with which certain bilateral agreements have been concluded between the EC/its Member States.[161] In each of these three situations, there is a system of notification and authorisation.[162] As regards imports of waste for recovery, these are again prohibited, except where they are from countries to which the OECD Council Decision applies[163] or from other countries which fall into one or other of two other categories: (a) countries which are also parties to the 1989 Basel Convention or agreements thereunder;[164] or (b) countries with which certain bilateral agreements have been concluded between the EU/its Member States.[165] Somewhat differing control procedures are applicable to each of these non-prohibited categories.[166]

12.2.5.2   The concept of waste

Characteristic of Community law on waste are the problems that have arisen from the way in which waste is defined in the WFD. For obvious reasons, these problems,

---

154   See para. 8.3.2 above.
155   Council Regulation EEC/259/93, Art. 18(1). But note that a Member State may return to an ACP country processed waste which that country has chosen to have processed in the EC (see Council Regulation EEC/259/93, Art. 18(2)).
156   Council Regulation EEC/259/93, Art. 14(1).
157   *Ibid.*, Art. 14(2).
158   *Ibid.*, Art. 15.
159   See para. 8.3.2 above.
160   See (as to all three possibilities) Council Regulation EEC/259/93, Art. 16.
161   *Ibid.*, Art. 19(1).
162   *Ibid.*, Art. 20.
163   *Ibid.*, Art. 21(1)(a).
164   Council Regulation EEC/259/93, Art. 21(1)(b).
165   *Ibid.*
166   *Ibid.*, Art. 22.

which have generated a considerable academic literature,[167] also underlie the very similar questions arising both from the transcription of the WFD into domestic law, as discussed in Chapter 6 above,[168] and from the interpretative problems arising in relation to landfill tax, especially as to the situations in which material is disposed of as waste, to be discussed in Chapter 15 below.[169] Having been slow to accept it, the Court of Appeal has now acknowledged the existence of a possible link between the objectives of the WFD and the definition of waste for landfill tax purposes.[170] The Directives referred to above adopt the same definition of waste as that used in the WFD itself.[171]

The definition of 'waste' in Art. 1 of the WFD consists of two elements.[172] The first is 'any substance or object ... which the holder discards or intends or is required to discard'.[173] The second is that the substance or object in question must fall into the categories contained in Annex I to the WFD.[174] Annex I lists 16 categories of substances and objects (including residues,[175] off-specification products,[176] date-expired products,[177] unusable parts,[178] adulterated materials,[179] etc.) and concludes, rather unhelpfully, with a sweeping-up clause which refers to '[a]ny materials, substances or products which are not contained in the [preceding] ... categories'.[180] The definition in Art. 1 is then supplemented by a provision requiring the Commission to draw up a list, to be reviewed periodically, of wastes belonging to the categories listed in Annex I to the WFD.[181] The list is known as the European Waste Catalogue ('the EWC')[182] and, whilst it is useful for fleshing-out the definition, the ECJ has stressed in *Criminal proceedings against Euro Tombesi and Others*,[183] that 'the fact that a substance is mentioned in ... [the EWC] does not mean that it is waste in all circumstances. An entry is only relevant when the definition of waste has been satisfied'.[184]

---

167 See, for example, Stephen Tromans, 'EC Waste Law – A Complete Mess?' (2001) 13 JEL 133–56 and Ilona Cheyne, 'The Definition of Waste in EC law' (2002) 14 JEL 61–73. The latter contains a survey of the relevant literature at n2 thereto.

168 See para. 6.3.2.1 above.

169 See para. 15.2 below.

170 See David Pocklington, 'Industry Soundings', (2003) 15(3) ELM 207–9, 208. See also para. 6.3.2(1) above.

171 See, for example, Council Directive 91/689/EEC, Art. 1(3) (Directive on Hazardous Waste); Council Regulation EEC/259/93, Art. 2(a) (the Waste Shipment Regulation); and Council Directive 99/31/EC, Art. 2(a) (the Landfill Directive).

172 See the analysis by Cheyne, *op. cit.*, p. 64.

173 See Council Directive 75/442/EEC, Art. 1(a).

174 *Ibid.*, Art. 1(a).

175 *Ibid.*, Annex I, paras Q1, Q8, Q9, Q10 and Q11.

176 *Ibid.*, Annex I, para. Q2.

177 *Ibid.*, Annex I, para. Q3.

178 *Ibid.*, Annex I, para. Q6.

179 *Ibid.*, Annex I, para. Q12.

180 See Council Directive 75/442/EEC, Annex I, para. Q16.

181 *Ibid.*, Art. 1(a).

182 See Commission Decision 00/532/EC, (2000) OJ L226 3.

183 See Joined Cases C–304/94, C–330/94, C–342/94 and C–224/95, [1997] ECR I–3561.

184 [1997] ECR I–3561, I–3589.

What is not clear from the WFD is whether the disposal operations listed in Annex IIA thereto or the 'operations which may lead to recovery' listed in Annex IIB to the WFD are also situations in which substances and objects are 'discarded'. The disposal operations listed in Annex IIA include tipping above or under ground (for example, landfill),[185] release of solid waste into a water body other than the sea[186] and incineration, whether on land or at sea.[187] The operations which may lead to recovery listed in Annex IIB include solvent reclamation or regeneration,[188] recycling or reclamation,[189] oil re-refining or other re-uses of oil[190] and use principally as a fuel or other means to generate energy.[191] Cheyne argues that the essential characteristic of all of the 16 categories in Annex I to the WFD is that the objects and substances within them are not identified according to the amount of environmental damage that they might inflict, but with regard to the likelihood of their holders wishing to get rid of them.[192] The common implication in all of the 16 categories in Annex I is the idea of discarding of the substance or object in question, which makes it all the more surprising that there is no specific definition of the word 'discard', even though it is expressly used in Art. 1(a) of the WFD.

The ECJ has now considered these questions in a series of important decisions.[193]

1   In the co-joined cases *Criminal proceedings against Euro Tombesi and Others*,[194] the ECJ held that the concept of waste 'is not to be understood as excluding substances and objects that are capable of economic reutilisation, even if the relevant materials may be the subject of a transaction or quoted on public or private commercial lists'.[195]

The joined cases involved six defendants who had been charged with various offences under Italian law, concerning the unauthorised transportation of waste scrap metal, the unauthorised discharge of marble rubble, the unauthorised burning of toxic waste, etc. In each case, the national courts essentially sought to ascertain, under Art. 234, European Treaty (ex 177), whether the concept of waste referred to in the WFD, and thus the Waste Shipment Regulation[196] and

---

185   See Council Directive 75/442/EEC, Annex IIA, para. D1.
186   *Ibid.*, Annex IIA, para. D6.
187   *Ibid.*, Annex IIA, paras D10 and D11.
188   *Ibid.*, Annex IIB, para. R1.
189   *Ibid.*, Annex IIB, paras R2–R4.
190   *Ibid.*, Annex IIB, para. R8.
191   *Ibid.*, Annex IIB, para. R9.
192   See Cheyne, *op. cit.*, p. 64.
193   Other important cases, not discussed below, are as follows: *Criminal proceedings against Vessoso and Zanetti*, C–206–207/88, [1990] ECR I–1461; *Zanetti and Others*, C–359/88, [1990] ECR I–1509; *Commission v. Germany*, C–422/92, [1995] ECR I–1097; *Lirussi and Bizzaro*, C–175,177/98, [2001] ECR I–6881. See also the UK cases discussed at para. 6.3.2.1 above.
194   Joined Cases C–304/94, C–330/94, C–342/94 and C–224/95, [1997] ECR I–3561. Judgment was actually given in the cases on 25 June 1997.
195   See Sands, *op. cit.*, p. 788n; [1997] ECR I–3561, 3602 (paras 54 and 55).
196   That is, Council Regulation 259/93/EEC. See para. 12.2.5.1(4) above.

the Directive on Hazardous Waste,[197] must be taken to exclude substances or objects capable of economic re-use.[198] In answering the question in the negative, the ECJ further particularised matters as follows:

In particular, a deactivation process intended merely to render waste harmless, landfill tipping in hollows or embankments and waste incineration constitute disposal or recovery operations falling within the scope of the ... Community rules. The fact that a substance is classified as a re-usable residue without its characteristics or purpose being defined is irrelevant in that regard. The same applies to the grinding of a waste substance.[199]

2   In *Inter-Environnement Wallonie ASBL v. Region Wallonne*,[200] the ECJ underlined the approach that it had taken in *Euro Tombesi*.[201]

The ECJ held that the scope of the term 'waste' turned on the meaning of the term 'discard';[202] that 'discard' included both disposal and recovery of a substance or object;[203] that the concept of waste did not in principle exclude 'any kind of residue, industrial by-product or other substance' arising from a production process;[204] that, besides applying to the disposal and recovery of waste by specialist undertakings, the WFD also applied 'to disposal and recovery of waste by the undertaking which produced them, at the place of production';[205] and that, even though substances directly or indirectly forming part of an industrial process might constitute waste, there was nonetheless a distinction between waste recovery and the normal industrial treatment of non-waste products, however difficult to draw that distinction might be.[206]

The foregoing issues had arisen in the context of a reference to the ECJ by the Belgian Conseil d'Etat under Art. 234, European Treaty (ex 177), in which Inter-Environnement Wallonie had asked the Belgian court to annul a decree of the Walloon Regional Council purporting to exempt from a permit system for waste installations those installations which formed an integral part of an industrial production process, on the basis that the decree conflicted with the Community law on waste.

3   In the joined cases *ARCO* and *EPON*,[207] the ECJ confirmed that the definition of waste turned on the meaning of 'discard',[208] a term which was not to be

---

197   That is, Council Directive 91/689/EEC. See para. 12.2.5.1(3) above.
198   See [1997] ECR I–3561, 3599 (para 41).
199   See [1997] ECR I–3561, 3602–3603 (paras 54 and 55).
200   Case C–129/96, [1998] 1 CMLR 1057.
201   See para. 12.2.5.2 (point 1) above.
202   [1998] 1 CMLR 1057, 1082 (para 26).
203   *Ibid.* (para 27).
204   *Ibid.* (para 28).
205   *Ibid.* (para 29).
206   [1998] 1 CMLR 1057, 1083 (paras 32–34).
207   *ARCO Chemie Nederland Ltd and Others v. Minister van Volkshuisvesting, Ruimtelijke Ordening en Milieubeheer, Vereniging Dorpsbelang Hees and Others v. Directeur van de Dienst Milieu en Water Van de Provincie Gelderland*, C–418–419/97, [2002] QB 646. Judgment was actually given on 15 June 2000 (see [2002] QB 646, 671).
208   [2002] QB 646, 677 (paras 36 and 46).

interpreted restrictively;[209] that 'discard' included in particular the disposal and recovery of a substance or object;[210] and that the question of whether a particular substance or object was waste had to be decided in the light of all the circumstances, regard being had to the aim of the WFD and the need not to undermine its effectiveness.[211]

In *ARCO*, a Dutch company planned to ship 'LUWA-bottoms', a by-product of one of its manufacturing processes, to Belgium, for use in cement manufacture. The relevant authority had granted permission on the basis that the substance was 'waste' within the terms of the Waste Shipment Regulation.[212] In *EPON*, the competent Dutch authority had granted an application by EPON, a Dutch electricity generating company, for authorisation to use pulverised wood chips from the construction industry as fuel in one of its power stations. The national court doubted whether each of the LUWA-bottoms and the wood chips were raw materials or waste and referred two questions for a preliminary ruling to the ECJ. In reaching the decision summarised above, the ECJ set out the circumstances that have to be considered in deciding whether an object or substance which undergoes the operations set out in Annexes IIA and IIB is waste:

a. it may, but does not necessarily, follow from the fact that disposal or recovery methods are described in Annexes IIA and IIB that any substance treated by one of those methods is to be regarded as waste;[213] and

b. other factors that may constitute evidence that a substance has been discarded but which are not conclusive on the point are the facts that:
    i.   the substance in question is 'commonly regarded as waste;'[214]
    ii. use as fuel is 'a common method of recovering waste';[215]
    iii. the substance is a production residue, that is, 'a product not in itself sought for use as fuel';[216]
    iv. the substance is a residue 'for which no use other than disposal can be envisaged';[217] and
    v.   the substance is a residue 'whose composition is not suitable for the use made of it or where special precautions must be taken when it is used owing to the environmentally hazardous nature of its composition'.[218]

c. However, the fact that a substance is the result of a complete recovery operation for the purpose of Annex IIB to the WFD does not necessarily exclude that substance from classification as waste.[219]

---

209 *Ibid.*, 676 (para 40).
210 *Ibid.*, 677 (para 47).
211 *Ibid.*, 680 (para 73).
212 That is, Council Regulation 259/93/EEC. See para. 12.2.5.1(4) above.
213 That is, because it has been discarded. See [2002] QB 646, 677 (para 49).
214 [2002] QB 646, 680 (para 73).
215 *Ibid.*
216 *Ibid.*, 681 (para 84).
217 *Ibid.* (para 86).
218 *Ibid.* (para 87).
219 *Ibid.*, 682 (paras 94–95).

d. Finally, the concept of waste 'is not to be understood as excluding substances and objects which are capable of being recovered as fuel in an environmentally responsible manner and without substantial treatment'.[220]

4 The recent decision of the ECJ in *Criminal proceedings against Paul van de Walle and others*[221] has again reiterated the point that the term 'discard' should not be interpreted restrictively. In emphasising the point, the Court drew attention to the fact that category Q4 in Annex I to the WFD refers to 'materials spilled, lost or having undergone other mishap, including any materials, equipment, etc., contaminated as result of the mishap'.[222]

What had happened was that, because of defects in a petrol station's storage tanks, hydrocarbon oil had leaked into the ground. Despite the accidental nature of the spillage, the ECJ held that the defendant had 'discarded' the oil, since, although the leakage was involuntary:

It is clear that accidentally spilled hydrocarbons which cause soil and groundwater contamination are not a product which can be re-used without processing. Their marketing is very uncertain and, even if it were possible, implies preliminary operations would be uneconomical for their holder. Those hydrocarbons are therefore substances which the holder did not intend to produce and which he discards, albeit involuntarily, at the time of the production or distribution operations which relate to them.[223]

On the basis that the oil had been discarded, it was next a question of whether the contaminated earth was 'waste' within the provisions of the WFD. Disregarding distinctions based, for example, on whether the earth had been excavated, the ECJ held that, since the oil could not be separated from the soil, the soil itself had become waste:

That is the only interpretation which ensures compliance with the aims of protecting the natural environment and prohibiting the abandonment of waste pursued by the Directive. It is fully in accord with the aim of the Directive and heading Q4 of Annex I thereto, which, as pointed out, mentions any materials, equipment, etc., contaminated as a result of [materials spilled, lost or having undergone other mishap] among the substances or objects which may be regarded as waste.[224]

Under the WFD, and in the circumstances of the case, the independent manager of the station was the 'holder' of the waste, since he had stocked the oil when it had become waste and had therefore to be treated as having produced it.[225] However, if the leakage been attributable to some contractual breach by Texaco, the lessee of the station, then Texaco might instead be regarded as the holder.[226]

---

220 *Ibid.*, 679 (para 65).
221 Case C–1/03 (7 September 2004).
222 *Ibid.* (para 43).
223 *Ibid.* (para 47).
224 *Ibid.* (para 52).
225 *Ibid.* (para 59).
226 *Ibid.* (para 60).

It has been pointed out by Macrory that a consequence of the decision in *Van de Walle* might be to render the UK's contaminated land rules otiose.[227]

12.2.6 Control of air and atmospheric pollution

12.2.6.1 General

The discussion in paragraph 12.2.5 completes our survey of Community waste management regulation. Despite the intractable problems surrounding the definition of 'waste', a feature of Community law and policy on waste management is that it forms an identifiable body of law, gathered mainly around the WFD. With Community law and policies relating to air and atmospheric pollution, the position is somewhat different. Easy appreciation of the corpus of Community regulation relating to air and atmospheric pollution is hampered partly by the piecemeal nature of regulatory development in the area and partly by its relationship with Community regulation of the energy and transport sectors. The following discussion begins with an overview of Community regulation on air and atmospheric pollution.[228] Thereafter, it focuses on those aspects of Community energy law and policy that have implications for the control of atmospheric pollution.

12.2.6.2 Pollution of air and atmosphere

Introduced in 1996, the Air Quality Framework Directive ('the AQFD')[229] lays down the basis for common objectives on ambient air quality in order to prevent harmful effects on the environment and on human health. In due course, the AQFD will be supplemented by a dozen 'daughter' directives, at least three of which have already been adopted.[230] On 8 March 1999, the Commission published its European Climate Change Programme ('the ECCP'),[231] which was intended to initiate a new community approach to the implementation of the Kyoto Protocol. In addition, Directive 2001/81/EC, on National Emissions Ceilings for Certain Atmospheric Pollutants, which aims to lay down a strategy for combating acidification, eutrophication and photochemical air pollutants, has been adopted.[232]

---

227 See 356 ENDS Report (2004) 44 (see also para. 6.7 above).

228 See Sands, *op. cit.*, pp. 336–9, 755–8; Sunkin, Ong and Wight, *op. cit.*, pp. 142–71; and Susan Wolf and Neil Stanley, *Wolf and Stanley on Environmental Law*, 4th edn (London: Cavendish, 2003), pp. 344–6.

229 Council Directive 92/62/EC, (1996) OJ L296 55. See para. 21.5 below. This is singled out by HM Treasury in *Tax and the Environment: Using Economic Instruments* (London: HMSO, 2002), p. 33, as an example of good regulation.

230 That is, Directive 1999/30/EC, (1999) OJ L163 41 (limitation of values for sulphur dioxide, nitrogen dioxide, oxides of nitrogen and particulates and lead in the ambient air) (see para. 22.4 below); Directive 00/69/EC, (2000) OJ L313 12 (limitation of values for benzene and carbon monoxide in the ambient air); and Directive 02/3/EC (2002) OJ L67 14 (limitation of values for ozone in the ambient air).

231 COM(00) 88 final, available from www.europa.eu.int. See also 322 ENDS Report (2001).

232 Council Directive 92/62/EC, (2001) OJ L309 22. See para. 22.4 below.

Aside from attempts to create a unified approach to controlling air and atmospheric pollution, there a number of other initiatives, in origin somewhat older, which are directed either at controlling emissions from specific pollution sources or setting air quality standards generally (that is, irrespective of the pollution source). Examples of the former are the 2001 Large Combustion Plant Directive,[233] which replaces the Large Combustion Plant Directive of 1988,[234] and the vehicle emissions standards discussed below;[235] an example of the latter is the current Regulation banning the sale and use of most ozone depleting substances, including CFCs and HCFCs.[236] The latter is designed to enable the Community to meet its obligations under the 1985 Vienna Convention for the Protection of the Ozone Layer and its 1987 Montreal Protocol.[237]

The Large Combustion Plant Directive should now, of course, be read in conjunction with the IPPC Directive.[238] The measures implementing the provisions just discussed in UK law are discussed in Chapter 6 above.[239]

12.2.6.3 *Energy law and policy*

(1) Introduction
There is only brief mention of energy in the European Treaty itself.[240] However, two directives, each adopted under the co-decision procedure,[241] contain common rules for the electricity and gas industries:[242]

a. European Parliament and Council Directive 2003/54/EC Concerning Common Rules for the Internal Market in Electricity ('the Electricity Acceleration Directive' or 'EAD');[243] and
b. European Parliament and Council Directive 2003/55/EC Concerning Common Rules for the Internal Market in Natural Gas ('the Gas Acceleration Directive' or 'GAD').[244]

In addition, a new regulation, again adopted under the co-decision procedure, is concerned with increasing cross-border trade in electricity. This is Regulation EC/1228/2003 on Conditions for Access to the Network for Cross-Border Exchanges

---

[233] Directive 01/80/EC, (2001) OJ L309 1. See Sands, *op. cit.*, pp. 336–9.
[234] Council Directive 88/609/EEC, (1988) OJ L336 1. See para. 22.4 below.
[235] See para. 12.2.6.4(2) below.
[236] Council Regulation 2037/2000/EC, (2000) OJ L244 1.
[237] See para. 8.3.1.5 above.
[238] See para. 12.2.2 above.
[239] See para. 6.4 above.
[240] That is, Arts 3(1)(u) and 154 (ex 129b), European Treaty.
[241] See para. 4.3.3 above.
[242] See Peter Cameron, *Competition in Energy Markets* (Oxford: Oxford University Press, 2002); Carlos Ocana *et al.*, *Competition in Energy Markets* (OECD/IEA, 2001); and *Completing the Internal Energy Market*, COM (01) 125 final.
[243] 03/54/EC, (2003) OJ L176 37.
[244] (2003) OJ L176 57.

in Electricity.[245] Finally, the production of electricity from renewable energy sources is promoted by European Parliament and Council Directive 2001/77/EC ('the Renewables Directive'),[246] with electricity production from combined heat and power being covered by European Parliament and Council Directive 2004/8/EC ('the Cogeneration Directive').[247]

The EAD and GAD have comparable market-opening objectives. The overall objective of the EAD is to provide for common rules for the generation, transmission and distribution of electricity, whilst that of the GAD is to provide for common rules for the transmission, distribution, supply and storage of natural gas. The Directives repeal[248] the 1996 Electricity Directive[249] and the 1998 Gas Directive[250] respectively, although they retain the most important features of their predecessors. The background to the policy development process culminating in the adoption of the Acceleration Directives is the opening up of energy markets to competition, a process which has already been discussed above.[251] Specifically, the Acceleration Directives were designed to improve on the provisions of the 1996 and 1998 Directives with regard to the pace and level of market-opening and guarantees as to fair and non-discriminatory network access. Although most Member States had in fact opened their markets further than required by the 1996 and 1998 Directives, market distortions remained because of those that had not done so; moreover, there continue to be wide variations between Member States as to standards for third party access ('TPA').[252] Whilst market opening is one of the main objectives of each of the EAD and the GAD, both Directives also imposes public service obligations ('PSOs') and measures for consumer protection.[253] These include the obligation to impose adequate safeguards to protect vulnerable customers;[254] the obligation to ensure that electricity supplies disclose details of energy sources in bills and promotional materials;[255] and the implementation of appropriate measures, including economic incentives, to achieve environmental protection.[256]

Prior to considering the four instruments in more detail, it is useful to refer briefly to Community law and policy on the most controversial of means of electricity-generation: nuclear power.[257] Whilst nuclear generation is not in itself a central

---

[245]   (2003) OJ L176 1.
[246]   (2001) OJ L283 33.
[247]   (2004) OJ L52 50.
[248]   In relation to the EAD, with effect from 1 July 2004 (see Directive 03/54/EC, Art. 29). In relation to the GAD, also with effect from 1 July 2004 (see Directive 03/55/EC, Art. 32(2)).
[249]   European Parliament and Council Directive 96/92/EC, (1996) OJ L27 20.
[250]   European Parliament and Council Directive 98/30/EC, (1998) OJ L204 2.
[251]   See para. 2.4 above.
[252]   See Cameron, *op. cit.*, para. 8.11.
[253]   See Directive 03/54/EC, Art. 3 and Directive 03/55/EC, Art. 3.
[254]   See Directive 03/54/EC, Art. 3(5) and Directive 03/55/EC, Art. 3(3).
[255]   *Ibid.*, Art. 3(6). This provides for cross-border transfers at 'cost'. See also www. electricitylabels.com.
[256]   *Ibid.*, Art. 3(7) and 03/55/EC, Art. 3(4).
[257]   See Cameron, *op. cit.*, paras 2.13–2.16.

concern of the present study, and although it does not count as a renewable energy source,[258] factors such as the operating costs of nuclear reactors, as well as the costs of their decommissioning, explain many of the policy choices which governments have made in relation to nuclear power and the environment.[259] The creation of a common market in nuclear ores and fuels was a prime aim of the 1957 Euratom Treaty[260] and, to this end, Art. 2 thereof committed the Community, among other things, to the promotion of research and the dissemination of technical information; to the establishment of uniform safety standards; to the facilitation of investment; to ensuring a regular and equitable supply of ores and nuclear fuels; and, internationally, to fostering progress in the peaceful uses of nuclear energy.[261] Additionally, by Art. 30, Euratom Treaty, basic standards were to be laid down for the protection of workers' health and that of the general public against dangers arising from ionising radiations. With nuclear power accounting for about a third of the EU's electricity-generating capacity, there is currently much concern about ensuring security of supply and the EC Commission has recently launched a package of measures on the protection of the environment from radioactive waste.[262]

Apart from electricity and gas, some consideration is given in what follows to coal and hydrocarbons and hydrocarbon-based fuels. The emphasis throughout is on explaining market structures. Except as regards state aids, which receive relatively detailed discussion,[263] Community competition law is not separately discussed. The reader should note, however, that, generally speaking,[264] Community energy markets are subject to general Community competition law rules.[265] In relation to both gas and electricity, the thrust of the Commission's competition policy has been 'to prevent private arrangements or practices that restrict the emergence of competition or that foreclose national markets against new entrants'.[266]

[258] See Directive 01/77/EC, Art. 2(a).
[259] See paras 21.4.4, 21.5 and 21.6 below.
[260] See para. 4.3 above.
[261] The Euratom Treaty was designed to reduce the dependence of European Countries on energy imports from the regions affected by the 1956 Suez crisis, at the same time countering the dominance at that time of the USSR and the US in nuclear power (see Cameron, *op. cit.*, para. 2.14).
[262] See Christiane Trüe, 'Legislative Competences of Euratom and the European Community in the Energy Sector: the Nuclear Package of the Commission', (2003) 28 EL Rev 664–85.
[263] See para. 12.2.7 below.
[264] There is, however, Council Directive 90/377/EEC, (1990) OJ L185 16 concerning a community procedure to improve the transparency of gas and electricity prices charged to industrial end-users. This has survived the EAD and GAD (unlike the Transit Directive (Council Directive 90/547/EEC, (1990) OJ L313 30), which has been repealed with effect from 1 July 2004 (see Directive 03/54/EC, Art. 29)).
[265] As to which, see, for example, Stephen Weatherill and Paul Beaumont, *EU Law*, 3rd edn (London: Penguin), chs 22–25; and Craig and de Búrca, *op. cit.*, chs 21–25.
[266] Cameron, *op. cit.*, para. 7.67.

(2) Electricity
The EAD envisages a fully open internal electricity market, with customers free to choose their suppliers and all suppliers free to deliver to their customers.[267] It lays down common rules in three areas of the electricity industry: generation, transmission and distribution.

The generation, or production,[268] of electricity is covered by Art. 6, EAD, which provides that an authorisation, as opposed to a tendering, procedure is to be the norm for the construction of new generating capacity. The authorisation procedure, which must be 'conducted in accordance with objective, transparent and non discriminatory criteria', may, among other things, lay down criteria relating to the protection of the environment.[269]

Transmission, which is defined as 'the transport of electricity on the extra high-voltage and high-voltage interconnected system with a view to its delivery to final customers or to distributors, but not including supply',[270] is dealt with in Arts 8–12, EAD. Article 8 provides for the designation and supervision of Transmission System Operators ('TSOs'), who are responsible for operating, ensuring the maintenance of and, if necessary, for developing a Member State's transmission system and its interconnections with other systems. By Art. 11(3), EAD, a Member State may require the TSO, '... when dispatching generating installations, to give priority to generating installations using renewable energy sources[271] or waste or producing combined heat and power'. Dispatching is not relevant to the UK, as has been discussed in Chapters 2 and 6[272] above.

The operation of the distribution system is dealt with in Arts 13–17, EAD. 'Distribution' is defined as '... the transport of electricity on high-voltage, medium voltage and low voltage distribution systems with a view to its delivery to customers, but not including supply'.[273] Member States generally have a single TSO and several distribution system operators ('DSOs').[274] The designation and supervision of DSOs is required by Art. 13.

Each of the activities referred to above is regulated in the UK by the Electricity Act 1989, as amended.[275] Although further liberalisation measures are currently proposed,[276] the UK electricity market was already the most liberalised in Europe, even before the EAD's adoption.[277]

---

[267] Directive 03/54/EC, recital 4.
[268] *Ibid.*, Art. 2(1).
[269] *Ibid.*, Art. 6(2)(c).
[270] *Ibid.*, Art. 2(3).
[271] Defined as 'renewable non-fossil energy sources (wind, solar, geothermal, wave, tidal, hydropower, biomass, landfill gas, sewage treatment plant gas and biogases) ...' (see Directive 03/54/EC, Art. 2(30)).
[272] See para. 6.4 above.
[273] See Directive 03/54/EC, Art. 2(5).
[274] See Cameron, *op. cit.*, para. 4.17.
[275] See para. 6.4 above.
[276] That is, through the introduction of BETTA in Part 3 of the Energy Act 2004 (see para. 6.4 above).
[277] See, for example, Cameron, *op. cit.*, paras 1.19, 4.116 and 5.86; IEA, *op. cit.*, pp. 37–43.

The crucial central provision of the EAD is the removal of two of the three types of TPA to the transmission and distribution networks as contained in the now repealed 1996 Directive. Access to the transmission and distribution networks is now to be based simply on published tariffs, applicable to all eligible customers objectively and without discrimination between users of the systems.[278] This is regulated TPA,[279] which allows producers, on the one hand, and eligible customers, on the other, to contract with each other direct for electricity supply on the basis of the published tariffs. This should encourage access to the market by new entrants on the basis that non-discriminatory access is possible.[280] Eligible customers are:[281]

a. before 1 July 2004, customers falling within specified market-opening percentages and permitted by Member States as being eligible to participate in the opening of the market;[282]
b. from 1 July 2004, at the latest, all non-household customers;[283] and
c. from 1 July 2007, all customers.[284]

Customers are defined to mean both wholesale and final customers of electricity,[285] the former referring to those who purchase electricity for resale,[286] the latter referring to those who purchase it for their own use.[287] The basis on which the UK has interpreted the concept of eligible customers has already been discussed in an earlier chapter.[288] Regulated TPA is already the basis for network access in the UK.

The expressed purpose of the Renewables Directive[289] is the promotion of an increase in the contribution of renewable energy sources to the production of electricity in the internal market.[290] By Art. 3 of the Renewables Directive, each Member State is to set national indicative targets for consumption of electricity produced from renewable energy sources, taking into account the reference values set out in the Directive and ensuring compatibility with obligations under the Kyoto Protocol.[291]

'Renewable energy sources' are defined as 'renewable non-fossil energy sources (wind, solar, geothermal, wave, tidal, hydropower, biomass, landfill gas, sewage treatment plant gas and biogases)'.[292] The definition thereby excludes nuclear

---

278 Directive 03/54/EC, Art. 20(1).
279 See Cameron, *op. cit.*, paras 4.24 and 8.12.
280 *Ibid.*, para. 8.11.
281 See Directive 03/54/EC, Arts 2(12) and 21(1).
282 *Ibid.*, Art. 21(1)(a).
283 *Ibid.*, Art. 21(1)(b), that is, purchasers of electricity other than for their own household use, including producers and wholesale customers (see Directive 03/55/EC, Art. 2(11)).
284 Directive 03/54/EC, Art. 21(1)(c).
285 *Ibid.*, Art. 2(7).
286 *Ibid.*, Art. 2(8).
287 *Ibid.*, Art. 2(9).
288 See para. 12.2.6.3(1) above.
289 *Ibid.*
290 Directive 01/77/EC, Art. 1.
291 Directive 01/77/EC, Art. 3(2).
292 *Ibid.*, Art. 2(a).

power from the definition of renewables. Since Member States are obliged only to take 'appropriate steps' to meet the indicative targets, failure to meet those targets is not by itself a breach of the terms of the Directive. However, Art. 3(4) of the Directive reserves to the Commission the right to set mandatory targets. Article 5 of the Directive requires Member States to ensure that the origin of electricity produced from renewable energy sources can be guaranteed as such by creating systems for granting guarantees of origin. Article 6(1) of the Directive then requires Member States to review their existing rules on the construction and operation of renewable power plants with a view to reducing the regulatory and non-regulatory barriers to renewables production.

The remaining provisions of the Renewables Directive concern transmission and distribution issues (that is, 'grid' issues). For instance, Art. 7(1) of the Directive requires Member States to take the necessary measures to ensure that TSOs and DSOs in their territory 'guarantee' the transmission and distribution of electricity produced from renewable energy sources. This same provision also allows Member States to afford renewables generators priority access to the transmission and distribution systems.

In the UK, the government has decided that 10 per cent of electricity supplies should come from renewable sources by 2010.[293] However, there has been until recently no system for guaranteeing the 'renewable' origin of electricity supplies and the UK transmission and distribution system is not structured in such a way as to enable renewables generators to be accorded priority.[294]

If the purpose of the Renewables Directive is the promotion of renewable source electricity in the internal market, the professed objective of the Cogeneration Directive is 'to increase energy efficiency and improve security of supply by creating a framework for promotion and development of high efficiency cogeneration of heat and power based on useful heat demand and primary energy savings in the internal energy market'.[295] 'Cogeneration' is defined in Art. 3(a) of the Cogeneration Directive as 'the simultaneous generation in one process of thermal energy and electrical and/or mechanical energy', a definition which covers the following cogeneration technologies:

    (a) combined cycle gas turbine with heat recovery;
    (b) steam backpressure turbine;
    (c) steam condensing extraction turbine;
    (d) gas turbine with heat recovery;
    (e) internal combustion engine;
    (f) microturbines;
    (g) stirling engines;
    (h) fuel cells;
    (i) steam engines;
    (j) organic Rankine cycles;

---

[293]   See para. 6.4 above.
[294]   See paras 2.4 and 6.4.3.1(2)(b) above and paras 21.5.4–21.5.6 below.
[295]   Directive 04/8/EC, Art. 1.

(k) any other type of technology or combination thereof falling under the definition laid down in Article 3(a).[296]

Article 6 of the Cogeneration Directive requires Member States both to analyse national potential for high-efficiency cogeneration and to monitor periodically their progress towards increasing the proportion of their electricity produced from this source,[297] while Art. 5 requires Member States to institute, as for renewable source electricity, a system of guarantees of origin, for the purpose of ensuring that electricity generated from high-efficiency cogeneration can be guaranteed as such.[298] Such a system of guarantees must be instituted within six months following the Commission's establishment of 'harmonised efficiency reference values for separate production of electricity and heat'.[299] Additionally, obligations on both the Commission and on Member States to report on the progress of the matters covered by the Directive are imposed by Arts 10 and 11, while Art. 8 requires Member States both to ensure that TSOs and DSOs in their territory 'guarantee' the transmission and distribution of electricity produced from high-efficiency cogeneration,[300] and, until the cogeneration producer becomes an eligible customer under Art. 21(1) of the EAD,[301] to ensure the due publication of tariffs for the purchase of electricity 'to back-up or top-up electricity generation'.[302]

Regulation EC/1228/2003 on Conditions for Access to the Network for Cross-Border Exchanges in Electricity is directed towards the further development of cross-border trade in electricity, whose level is currently rather modest. To this end, the Regulation establishes a compensation mechanism for cross-border flows of electricity; for the setting of harmonised principles on cross-border transmission charges; and for the allocation of available capacities of interconnections between national transmission systems.[303] Article 7 of the Regulation permits new electricity interconnectors to be exempt from TPA and from the regulatory control of tariffs.

In general terms, the UK presently complies with the three measures discussed above. Such legislative changes as are necessary, in relation to a regulatory and exemption regime for new interconnectors, have been made through the Energy Act 2004.[304]

(3) Gas

The Gas Acceleration Directive is rather similar to the EAD. Like the EAD, the GAD looks forward to a fully open internal market.[305] Reflecting the differences

---

[296] *Ibid.*, Annex I.

[297] *Ibid.*, Art. 6(1).

[298] *Ibid.*, Art. 5(1).

[299] *Ibid.*, Art. 4(1).

[300] *Ibid.*, Art. 8(1).

[301] See above.

[302] *Ibid.*, Art. 8(2).

[303] Regulation EC/1228/2003, Art. 1. See Cameron, *op. cit.*, paras 8.17–8.21; IEA, *op. cit.*, p. 43.

[304] See para. 6.4.3.1(4) above.

[305] Directive 03/55/EC, recital 4.

between the structures of the gas and electricity industries,[306] however, it establishes common rules in four areas: transmission, distribution, supply and storage.[307] Within the scope of natural gas, for the purposes of the GAD, is liquefied natural gas ('LNG'), biogas[308] and gas from biomass,[309] plus any other type of gas that can technically and safely be injected into, and transported through, the natural gas system.[310]

Gas transmission, storage and LNG is dealt with in Arts 7–10, GAD. Transmission is itself defined as 'the transport of natural gas through a high pressure pipeline network ... with a view to its delivery to customers, but not including supply'.[311] As with electricity, there is provision for the designation of TSOs,[312] who are tasked with various duties including the operation, maintenance and development, under economic conditions, of secure, reliable and efficient transmission, storage and/or LNG facilities,[313] due regard being paid to the environment.[314]

The distribution and supply of gas forms the subject matter of Arts 11–15, GAD. 'Distribution' is itself defined as 'the transport of natural gas through local or regional pipeline networks with a view to its delivery to customers, but not including supply'.[315] The provisions on distribution and supply are in substance very similar to those covering transmission, storage and LNG. As with the EAD, the designation and supervision of DSOs is required.[316]

As in the case of the EAD, the central provisions of the GAD are those that relate to market access. Access to transmission, distribution and LNG facilities is again to be based simply on published tariffs, applicable to all eligible customers objectively and without discrimination between users of the systems.[317] Again, the model is

---

306 See para. 2.4 above.
307 Directive 03/55/EC, Art. 1(1). Unlike the EAD, the GAD does not provide for common rules for operations relating to gas production although, like the EAD, it does contain rules for the granting of licences for the construction of natural gas facilities, including facilities for production, transmission, distribution and storage (see Directive 03/55/EC, Art. 4). Production itself is dealt with separately in the Hydrocarbons Licensing Directive (see para. 12.2.6.3(5) below).
308 That is, 'gas formed by anaerobic digestion of organic materials, for example, whey or sewage sludge' (Porteous).
309 That is, 'the mass of living organisms forming a prescribed population in a given area of the earth's surface' (Porteous).
310 Directive 03/55/EC, Art. 1(2).
311 *Ibid.*, Art. 2(3). 'Supply' is defined as the sale, including resale, of natural gas, including LNG, to customers (see Directive 03/55/EC, Art. 2(7)).
312 Directive 03/55/EC, Art. 7.
313 LNG facilities are defined in Directive 03/55/EC, Art. 2(11), as terminals used for the liquefaction of natural gas, or the importation, offloading and re-gasification of LNG, including ancillary services and temporary storage necessary for the re-gasification process and subsequent delivery to the transmission system but not including those parts of LNG terminals used for storage.
314 Directive 03/55/EC, Art. 8(1).
315 *Ibid.*, Art. 2(5).
316 That is, by Directive 03/55/EC, Art. 11.
317 *Ibid.*, Art. 18(1).

regulated TPA,[318] which allows producers, on the one hand, and eligible customers, on the other, to contract with each other direct for electricity supply on the basis of the published tariffs. Eligible customers are defined in exactly the same way as for purposes of the EAD, that is:[319]

a. before 1 July 2004, customers falling within specified market-opening percentages and permitted by Member States as being eligible to participate in the opening of the market;[320]

b. from 1 July 2004, at the latest, all non-household customers;[321] and

c. from 1 July 2007, all customers.[322]

Parallel with the EAD, in the GAD, customers are defined to mean both wholesale and final customers of natural gas and natural gas undertakings that purchase natural gas,[323] the former referring to those who purchase natural gas for resale (other than TSOs and DSOs),[324] final customers being those who purchase it for their own use.[325] The basis on which the UK has interpreted the concept of eligible customers for the purpose of the gas markets has already been discussed in an earlier chapter.[326] Regulated TPA is already the basis for network access in the UK and the DTI considers that, whilst the UK is in broad compliance with the provisions of the GAD, certain further measures needed to be taken.[327]

(4) Coal
The European coal industry was formerly subject to the system of regulated competition provided for by the 1951 European Coal and Steel Community Treaty ('the ECSC Treaty').[328] Since July 2002, however, the industry has been brought within the provisions of the European Treaty, although it has the benefit of a special set of rules on state aid.[329] At the time of the expiration of the ECSC Treaty, on 23 July 2002, the Commission reported that, within the countries which formed the then 15 Member States, coal output was down to 83 million tons, as against 485 million tons in 1953.[330] However, as the Commission also reported, the ten accession countries also have big, and not yet fully restructured, coal industries.

---

318 See Cameron, *op. cit.*, paras 4.88 and 8.13.
319 See Directive 03/55/EC, Arts 2(28) and 23(1).
320 *Ibid.*, Art. 23(1)(a).
321 *Ibid.*, Art. 23(1)(b), that is, purchasers of natural gas other than for their own household use (see Directive 03/55/EC, Art. 2(26)).
322 *Ibid.*, Art. 23(1)(c).
323 *Ibid.*, Art. 2(24).
324 *Ibid.*, Art. 2(29).
325 *Ibid.*, Art. 2(27).
326 See para. 6.4 above.
327 See para. 6.4.3.2 above.
328 See Cameron, *op. cit.*, paras 2.11–2.12.
329 See para. 12.2.7.3 below.
330 Press Release IP/02/898, *Fifty Years at the Service of Peace and Prosperity: The European Coal and Steel Community (ECSC) Treaty Expires*, Brussels, 19 June 2002.

(5) Oil
The sourcing, composition, storage and distribution of both hydrocarbon oils and
hydrocarbon-based motor fuels is highly regulated within the Community.[331]
European Parliament and Council Directive 94/22/EC ('the Hydrocarbons Licensing
Directive')[332] contains the rules relating to authorisations for exploring for and
extracting both oil and natural gas.[333] The three main objectives of the Hydrocarbons
Licensing Directive are to ensure that all entities possessing the necessary capabilities
can gain access to authorisations; to ensure that 'authorisations are granted according
to objective published criteria'; and to ensure that all entities taking part in the
authorisation procedure know in advance the conditions under which authorisations
are to be granted.[334]

Standards for the composition, storage and distribution of motor fuel are contained
in various directives. Both the lead content and the sulphur content of fuels is
regulated. Lead content is subject to the Fuel Quality Directive[335] and sulphur
content to the Sulphur Content Directive.[336] Subject to derogations, the Fuel Quality
Directive banned leaded petrol from the market from 1 January 2000[337] and made
provision for progressive improvements in the environmental quality of unleaded
petrol and diesel fuel. The Sulphur Content Directive, which amended Council
Directive 93/12/EEC on the Sulphur Content of Certain Liquid Fuels,[338] provides
for a gradual reduction in the sulphur content of liquid fuels 'to reduce the harmful
effects of [sulphur dioxide] ... emissions on man and the environment'.[339]

Directive 94/63/EC[340] contains standards for the storage of petrol and its distribution
from terminals to service stations. Such standards are concerned with controlling
volatile organic compound emissions. The transposition of the Directives referred to
above into UK law has been referred to in Chapter 6 above.[341]

12.2.6.4 Transport law and policy

(1) Introduction
The legislative basis of the Community's common transport policy ('the CTP') is
contained in Arts 70–80, European Treaty (ex 74–84).[342] Article 70 states only that

---

[331] Only the briefest indication of the scope of this regulation is given here. See, further,
Patricia D. Park, *Energy Law and the Environment* (London: Taylor and Francis, 2002),
ch. 5.
[332] (1994) OJ L164 3.
[333] See para. 12.2.6.3(3)n above.
[334] See Cameron, *op. cit.*, para. 3.38.
[335] European Parliament and Council Directive 98/70/EC, as amended, (1998) OJ L350 58.
[336] Council Directive 1999/32/EC, as amended, (1999) OJ L121 13.
[337] Directive 98/70/EC, Art. 3(1).
[338] (1993) OJ L74 81.
[339] Council Directive 1999/32/EC, Art. 1(1).
[340] (1994) OJ L365 24.
[341] See para. 6.4 above.
[342] Although 'a common policy in the sphere of transport' is referred to in Art. 3(1)(f),
European Treaty.

the objectives of the Treaty are to be pursued by the Member States 'within the framework of a common transport policy', however.[343]

Without stating the goals of the CTP, Art. 71 nonetheless provides that measures taken for the purpose of implementing Art. 70 are to be enacted under the co-decision procedure.[344] Article 72 then prevents the introduction of, or any increase in, discriminatory measures without the unanimous approval of the Council. Article 73, to be discussed below,[345] deals with state aids in relation to transport, while Art. 74, which relates to the method and extent to which Member States may intervene in the commercial activities of carriers, requires account to be taken of carriers' economic circumstances where measures are taken in relation to transport rates and conditions. Article 78 specifically allows state aid in the transport sphere in respect of parts of Germany to reflect the economic disadvantages caused by the division of Germany after the Second World War.

Article 75, which reflects the principles of Arts 23–31, European Treaty (ex 9, 10, 12, 28, 29, 30, 34, 36 and 37),[346] provides for the abolition of discrimination in rates and conditions for the transportation of goods. Article 76 continues this theme by reflecting the anti-discrimination provision of Art. 12, European Treaty (ex 6) and banning Member States from imposing transport rates and conditions that favour particular undertakings or industries, except in accordance with authorisation from the Commission.

In an echo of Art. 25, European Treaty (ex 12),[347] Art. 77 is designed to ensure that charges or dues in addition to the transport rates in respect of the crossing of frontiers do not exceed a reasonable level after taking into account the costs thereby actually incurred.

Finally, while Art. 79 provides for the setting-up of an advisory Committee on transport to be attached to the Commission, Art. 80(1) expressly states that the provisions of Arts 70–80 are to apply to transport by rail, road and inland waterway. Article 80(2) provides that it is for the Council, acting by qualified majority, to decide whether, to what extent and by what procedure, appropriate provisions may be laid down for sea and air transport. It should be noted, however, that Art. 80(2) does not prevent the application of the general rules of the European Treaty to transport by sea and by air.[348]

In subsequent chapters of the present study, road and air transport are of the greatest relevance and it is to those areas that we now turn. In each area, there is a wealth of secondary legislation, in the form of regulations and directives. Moreover, the reader should be aware that, in 2001, the Commission published a White Paper making some 60 specific proposals of measures to be taken at Community level under the CTP.[349]

---

343 Notably, in a relatively under-explored area, see Rosa Greaves, *EC Transport Law* (Harlow: Pearson Education, 2000).

344 See para. 4.3.3 above.

345 See para. 12.2.7.4 below.

346 See para. 12.4 below.

347 See para. 12.3.3.2 below.

348 See Greaves, *op. cit.*, p. 22.

349 Commission of the European Communities, *European Transport Policy for 2010: Time to Decide* (Luxembourg: Office for Official Publications of the European Communities, 2001).

Policy preoccupations of the Commission revealed by the White Paper include the promotion of clean urban transport;[350] intermodal transport (that is, integrated transport chains);[351] and transport infrastructure charging policy.[352] We shall return to the last of these in paragraph (2) below.

It is important to stress the same point here, in relation to competition, as was made in relation to energy above. The Community's competition rules apply to transport, just as to other economic sectors; the only aspect of those rules that there will be an opportunity to refer to in what follows is that of state aid.[353]

(2) Road freight transport
Secondary legislation on road transport may be grouped into six areas:[354]

a. market access and pricing relating to goods;
b. market access and pricing in relation to passengers;
c. fiscal harmonisation;
d. social legislation;
e. technology, safety and environment; and
f. transport of dangerous goods.

Whilst noting the broad contours of legislation in each of categories a., b., d. and f., we are concerned most closely here, of course, with categories c. and e. As to market access and pricing, (see a. and b. above), Community law and policy has chiefly been concerned with three areas: with removing restrictions (that is, quotas) on the provision of transport services between Member States; with the right of a non-resident undertaking to provide transport services within a Member State; and with access to the occupation of a transport service operator.[355] The main items of social legislation on road transport (see d. above) are Council Regulation 3820/85/EEC on the Harmonisation of Certain Social Legislation relating to Road Transport[356] and European Parliament and Council Directive 2002/15/EC on the Organisation of the Working Time of Persons Performing Mobile Road Transport Activities.[357] As to f. (that is, the transport of dangerous goods), the main provision is Council Directive 94/55/EC,[358] which transposes international law on the transport of dangerous goods into Community law.

---

[350] See www.europa.eu.int/comm/energy. Also, White Paper, pp. 81–4.
[351] See www.europa.eu.int/comm/transport. Also, White Paper, pp. 41–7 and Greaves, *op. cit.*, p. 125.
[352] See www.europa.eu.int/comm/transport. Also, White Paper, pp. 88ff., and Greaves, *op. cit.*, pp. 125–30.
[353] The reader is referred to the materials cited in para. 12.2.6.4(1)n above; also, in this context, to Greaves, *op. cit.*, ch 7.
[354] These are the six divisions used in the *ABC of the Road Transport* Acquis, available from www.europa.eu.int/comm/transport.
[355] See Greaves, *op. cit.*, ch 3.
[356] (1985) OJ L370 1.
[357] (2002) OJ L80 35.
[358] (1994) OJ L319 7.

Secondary legislation on technology and safety in relation to road transport (see e. above) covers matters such as information on road accident statistics;[359] driver training,[360] driving licences;[361] vehicle speed limitation devices;[362] and vehicle recording equipment (that is, tachographs).[363]

Meanwhile, certain environmental aspects of road transport have been covered in a series of measures relating to sound and air pollution from vehicles.[364] Prominent among these measures is a 1970 directive, which continues to be amended regularly,[365] and which establishes mandatory technical standards for emissions of carbon monoxide, unburnt hydrocarbons, nitrogen oxides and particulates from both petrol- and diesel-engined cars.[366] The Council, which is committed to implementing a research and development programme for the marketing of clean vehicles and fuels,[367] has been authorised to adopt legislation for the stabilisation and reduction of emissions of carbon dioxide and other GHGs from motor cars and to introduce tax incentives for certain types of vehicle.[368] Additionally, the Commission's Auto/ Oil II Programme is aiming for considerable improvements by 2010 in urban air quality,[369] the Commission also having entered into environmental agreements with associations such as the Korean Automobile Manufacturers Association ('KAMA'), the Japanese Automobile Manufacturers Association ('JAMA') and the European Automobile Manufacturers Association ('ACEA')[370] for the reduction of carbon dioxide emissions from cars.[371] Emissions from diesel engines propelling road vehicles are dealt with in Council Directives 72/306/EEC[372] and 88/77/EEC,[373] while polluting emissions from engines powered by NG and LPG are covered by Directive 99/96/EC.[374] Directive 99/96/EC introduces the concept of Enhanced Environmentally Friendly Vehicles, while Directive 96/1/EC[375] permits Member States to introduce tax incentives for vehicles that satisfy certain conditions. Finally, Council Directive 70/157/EEC,[376] as amended, approximates the legislation of

---

[359] Council Decision 93/704/EC, (1993) OJ L329 63.

[360] Council Directive 76/914/EEC, (1976) OJ L357 36.

[361] Council Directive 91/439/EEC, (1991) OJ L237 1.

[362] Council Directive 92/6/EEC, (1992) OJ L57 27.

[363] Council Regulation EEC/3821/85, (1985) OJ L370 8, amended by Council Regulation EC/2135/98, (1998) OJ L274 1.

[364] See Sands, *op. cit.*, pp. 758–9.

[365] See the list of amending measures in Sands, *op. cit.*, p. 758n. It includes Council Directive 91/441/EC, (1991) OJ L242 1; Council Directive 93/59/EC, (1993) OJ L186 21; Council Directive 94/12/EC, (1994) OJ L100 42; and Directive 01/1/EC, (2001) OJ L35 34.

[366] Council Directive 70/220/EEC, (1970) OJ L76 1.

[367] Council Directive 91/441/EEC, amending Council Directive 70/220/EEC.

[368] Council Directive 89/458/EEC, (1989) OJ L226 3, amending Council Directive 70/220/ EEC.

[369] COM (00) 626 final (see Prelex link in Documents section of www.europa.eu.int).

[370] See COM (1996) 561 and COM (1998) 495.

[371] See Sands, *op. cit.*, p. 759.

[372] (1972) OJ L190 20.

[373] (1988) OJ L36 33.

[374] (2000) OJ L44 1.

[375] (1996) OJ L40 1.

[376] (1970) OJ L42 16.

Member States covering noise levels from motor vehicles. The implementation of these measures in UK law has already been discussed in Chapter 6 above.[377]

In addition to certain measures discussed below, which relate to the harmonisation of excise duties on fuel,[378] at least one significant measure has been enacted in relation to the fiscal harmonisation of road transport (see c. above). This is European Parliament and Council Directive 1999/62/EC on the Charging of Heavy Goods Vehicles for Use of Certain Infrastructure.[379] The Directive seeks to reduce the differences between systems of road taxes and charges applicable within Member States; to take better account of the principles of fair and efficient road pricing; and to move further towards the principle of territoriality in charging for road use. It therefore covers, not only vehicle excise duties, but also tolls and user charges.[380] By Art. 2(d), the scope of the Directive is restricted to goods vehicles having a maximum permissible gross laden weight of at least 12 tonnes. Although Art. 4 of the Directive allows each Member State to fix its own procedures for levying and collecting the vehicle excise duties[381] to which it applies, Art. 5 provides that, as regards vehicles registered in the Member States, such duties are to be charged only by the Member State of registration. Annex I to the Directive fixes, subject to the derogations, reduced rates and exemptions in Art. 6, the minimum rates of vehicle excise duty to be applied by Member States. Tolls and user charges may be imposed only on users of motorways and similar multi-lane roads, on users of bridges, on users of tunnels and on users of mountain passes.[382] Art. 7(4) prohibits tolls and user charges from discriminating, whether directly or indirectly, on the basis of the haulier's nationality or the origin or the destination of the vehicle. By Art. 7(5), Member States are enjoined to ensure that tolls and user charges are collected in such a way as to cause as little hindrance as possible to 'the free flow of traffic' and to avoid any mandatory checks at the internal borders of the Community. Under Art. 7(7) of the Directive, Member States are to fix user charges at a level not exceeding the maximum rates laid down in Annex II thereto, while user-charge rates are to be in proportion to the duration of the use made of the infrastructure.[383] Articles 7(9) and 7(10) of the Directive set out the rules for determining user charges and for relating weighted average tolls to the costs of constructing, operating and developing the relevant infrastructure, etc.[384] Finally, Art. 9(2) permits the earmarking of tolls and user charges, allowing Member States to attribute to environmental protection and the balanced development of transport networks a percentage of the amount of the user charge or toll, provide that the amount in either case is calculated in accordance with Arts 7(7) and 7(9) of the Directive.

---

[377] See para. 6.4 above.
[378] See para. 12.3.4 below.
[379] (1999) OJ L187 42.
[380] See para. 7.2.3 above.
[381] The duties falling within the scope of the RCD are listed for each Member State in Art. 3, Directive 99/62/EC.
[382] Directive 99/62/EC, Art. 7(2). 'Tolls' are defined, *ibid.*, Art. 20.
[383] *Ibid.*, Art. 7(8).
[384] See para. 7.2.3 above.

The UK's vehicle excise duty regime is discussed in Chapter 22 below; the UK Government currently intends to introduce road-user charging for lorries by 2006;[385] and the Commission has signalled its intention to amend Directive 1999/62/EC so as to align national systems of tolls and user charges for infrastructure use.[386]

(3) Air passenger transport
The secondary Community legislation on air transport may be divided into 10 categories,[387] six of which are not relevant in the present context, that is, the procurement of air traffic-management equipment and systems; air safety; air security; the protection of passengers; working conditions of employees; and a 'sweep-up' category of (mainly administrative) measures.[388] The remaining four are as follows:

a. market access and pricing;
b. state aids;
c. competition rules; and
d. the environment.

For present purposes, the most relevant categories are a. and d. above, although at least a brief indication of the relevant aspects of Community state aid and competition law is necessary to enable the significance of a. and d. fully to be appreciated. State aid issues, which are part of Community competition law, are discussed below;[389] the competition rules referred to at c. above seek to apply the general rules of Community competition law to the specifics of the air transport sector. Thus, although since 1 May 2004, infringement proceedings under Arts 81 and 82, European Treaty (ex Arts 85 and 86) have been subject to new procedures of general application,[390] there is an exemption from the scope of Art. 81(1) in relation to certain agreements, decisions and concerted practices in the transport sphere, in so far as their sole object and effect is to achieve technical improvements or co-operation.[391] Furthermore, by Council Regulation EEC/3976/87,[392] among the types of agreements, decisions and concerted practices to which the Commission has the power to apply Art. 81(3), European Treaty (block exemptions), are those having as their object the planning and co-ordinating of airline schedules and joint operations on new less busy scheduled air services.[393]

---

385 See para. 27.3 below.
386 See COM(03) 488 (see Prelex link in Documents section of www.europa.eu.int).
387 These are the ten divisions used in the 'Legislation' section of the 'Air Transport' part of the Energy and Transport Directorate-General's website (see www.europa.eu.int/comm/transport).
388 See Greaves, *op. cit., passim.*
389 See para. 12.2.7.4 below.
390 See Council Regulation EC/1/2003, (2003) OJ L1 1, Arts 39 and 41, repealing Council Regulation EEC/3975/87, Arts 3–19 and Council Regulation EEC/3976/87, Art. 6.
391 See Council Regulation EEC/3975/87, (1987) OJ L374 1, Art. 2(1).
392 (1987) OJ L374 9.
393 See Council Regulation EEC/3976/87, as amended by Council Regulation EEC/2411/92, Art. 1(2) (1992) OJ L240 19.

In relation to a. above, air transport raises problems which, although different from those raised by road transport, are no less intractable. As noted in an earlier chapter,[394] the international regulatory background to Community policy in the area is a complex of national law, the 1944 Chicago Convention and a network of bilateral conventions on routes, tariffs, etc. At least since 1987, Community policy has been to work towards the establishment of a genuine internal market in civil aviation, one in which such bilateral agreements are abolished and which is subject to general EU competition law rules.[395] Thus, Council Regulation EEC/2408/92 on Access for Community Air Carriers to Intra-Community Air Routes,[396] has afforded full market access to intra-Community air services by Community air carriers. This market-opening process has been facilitated by Council Regulation EC/95/93, on Common Rules for the Allocation of Slots at Community Airports,[397] and by Council Regulation EEC/2409/92 on Fares and Rates for Air Services.[398] Regulation EEC/2408/92 allows Community undertakings to operate as air carriers anywhere in the Community, regardless of nationality; Regulation EC/95/93 requires Member States to decide on the need for allocating slots[399] according to capacity analyses (but not, significantly, by auction);[400] and Regulation EEC/2409/92, in conjunction with Community competition law generally, regulates fares and rates for air transport services.

The environmental effects of air transport have been a pressing concern of the Commission in recent years (see d. above). In its 1999 Communication, *Air Transport and the Environment: Towards Meeting the Challenges of Sustainable Development*,[401] the Commission set out the four main 'pillars' for integrating environmental concerns into air transport policy, that is: improving technical environmental standards on noise and gaseous emissions; strengthening economic and market incentives; assisting airports in their environmental endeavours and advancing long-term technology improvements. Although the contribution of aircraft both to air and atmospheric pollution and to climate change is considerable, to date Community measures have concentrated on aircraft noise. Thus, Council Directive 80/51/EEC placed restrictions on noise emissions from subsonic aircraft;[402] Council Directive 89/629/EEC[403] banned the registration of so-called 'Chapter 2 aircraft';[404] and Council Directive 92/14/EEC[405] provided for the gradual withdrawal of such

---

394 See para. 8.5 above.
395 Greaves, *op. cit.*, p. 67.
396 (1992) OJ L240 8.
397 (1993) OJ L14 1.
398 (1992) OJ L240 15.
399 A 'slot' is defined as 'the scheduled time of arrival or departure available or allocated to an aircraft movement on a specific date at an airport coordinated under the terms of Regulation EEC/2408/92' (see Art. 2(a) thereof).
400 See para. 27.5 below.
401 COM (99) 640.
402 (1980) OJ L18 26.
403 (1989) OJ L363 27.
404 That is, Chapter 2 of Annex 16 to the 1944 Chicago Convention (see para. 8.5 above).
405 (1992) OJ L76 21.

aircraft from operation in the EU by April 2002. Furthermore, European Parliament and Council Directive 2002/30/EC[406] has embodied in Community law the International Civil Aviation Organisation ('ICAO')'s[407] Resolution A33–7 on the use of a 'balanced approach' to the management of noise around airports.[408] Finally, this last Directive takes effect against the background of a more general framework for limiting noise contained in European Parliament and Council Directive 2002/49/EC,[409] which has the aim of defining a common approach that is intended to avoid, prevent or reduce, on a prioritised basis, the harmful effects of exposure to environmental noise. For the UK provisions, see para. 6.5 above.

12.2.7   State aids

12.2.7.1   General

State subsidy, whether or not on professedly environmental grounds, has long been used to manage economies.[410] Subsidy, that is, a cost or loss of revenue to the public authority and a benefit to recipients, is the essence of state aids.[411] Although state aids are not forbidden under Community law, they are subject to its discipline, which means that they must not be applied so as to discriminate on grounds of nationality or so as to lead to unlawful barriers to trade. Community state aid rules form part of the competition law of the Community. Writing in 2000, Ehlermann and Atanasiu summed up the significance of Community state aid law thus:

> The control of state aids is a unique feature of EU competition policy. No similar control system exists in any of the Member States or in any federal state outside the EU. This model has, nonetheless, an increasing influence beyond the borders of the Community: its rules have been 'exported' to the European Economic Area, and, more recently, to the Central and Eastern European countries (CEECs) which are candidates for EU membership.[412] EU state aid rules and oversight practice have also influenced the evolution of the subsidy discipline imposed at the level of the GATT and the WTO.[413]

Articles **88** and **89**, European Treaty (ex 93 and 94) create a procedure whereby Member States must keep the Commission informed of state aids, so that the

---

[406]   (2002) OJ L85 40.
[407]   See para. 8.5 above.
[408]   Directive 02/30/EC also repeals the so-called 'Hushkit' Regulation (that is, Regulation EC/925/99, (1999) OJ L 115 1), as to which, see Greaves, *op. cit.*, p. 117.
[409]   (2002) OJ L189 12.
[410]   See Weatherill and Beaumont, *op. cit.*, pp. 1018–29; A. Leigh Hancher, Tom Ottervanger and Piet Jan Slot, *EC State Aids*, 2nd edn (London: Sweet and Maxwell, 1999); and Andrew Evans, *European Community Law of State Aid* (Oxford: Clarendon Press, 1997).
[411]   See Evans, *op. cit.*, p. 27.
[412]   The passage was obviously written prior to the accessions of May 2004.
[413]   See the Introduction to the *European Competition Law Annual 1999: Selected Issues in the Field of State Aid*, ed. by Claus-Dieter Ehlermann and Michelle Everson (Oxford: Hart Publishing, 2001), p. xxi.

Commission can assess whether they are consistent with Community law.[414] Art. 88(3) imposes a positive duty on Member States to notify the Commission of any plans to grant or alter state aid and Art. 88(1) obliges the Commission to keep all state aid existing in member States under constant review. The Commission is charged with proposing to Member States any appropriate measures required by the development or functioning of the common market. In appropriate circumstances and, having followed the procedure in Art. 88(2), the Commission can require the Member State in question to alter or abolish the aid within a specified time, as well as to recover the aid in question (for example, where aid has been implemented without notification).[415]

Article 87(1), European Treaty (ex 92(1)) contains the basic substantive rule of state aid law and provides that:

> Save as otherwise provided in this Treaty, any aid granted by a Member State or through State resources in any form whatsoever which distorts or threatens to distort competition by favouring certain undertakings or the production of certain goods shall, insofar as it affects trade between Member States, be incompatible with the common market.

Article 87(2) then lists three categories of state aid that *are* compatible with the common market, while Art. 87(3) lists five categories of aid that *may be* compatible with it. The former (obligatory) categories, subject to certain conditions, are; aid with a social character; aid to combat natural disasters; and aid to compensate certain areas of Germany for the economic disadvantages caused by the division of the country. The latter (permissive) categories include the following:

a. aid to promote the economic development of areas where the standard of living is abnormally low or where there is serious underemployment;
b. aid to promote the execution of an important project of common European interest or to remedy a serious disturbance in the economy of a Member State;
c. aid to facilitate the development of certain economic activities or of certain economic areas, where such aid does not adversely affect trading conditions to an extent contrary to the common interest;
d. aid to promote culture and heritage conservation where such aid does not affect trading conditions and competition in the Community to an extent that is contrary to the common interest; and
e. such other categories of aid as may be specified by decision of the Council acting by a qualified majority on a proposal from the Commission.

State aid is conventionally divided into sectoral and horizontal aid; the former relates to particular industries, the latter cuts across individual sectors. Of the permissive categories listed above, the most relevant in the present context are b. and c., since

---

[414] The detailed rules for the application of Art. 88, European Treaty (ex 93) are contained in Council Regulation EC/659/99, (1999) OJ L83 1 (see Hancher, Ottervanger and Slot, *op. cit.*, ch 19).
[415] Only unlawful state aid may be recovered, as to which, see Hancher, Ottervanger and Slot, *op. cit.*, paras 20–003–20–008.

these will usually be the Treaty Articles under which the Commission will consider a state aid clearance application on the grounds discussed in one or other of the next three paras.

12.2.7.2 *State aid for environmental protection*

In a case where the proffered justification for granting state aid is an environmental one, the Commission will follow its published Guidelines in exercising its discretion.[416] The basis of the discretion is generally Art. 87(3)(c) but the discretion may instead be exercised under Art. 87(3)(b) in an appropriate case.[417] The Guidelines demonstrate the Commission's adherence to the injunction in Art. 6, European Treaty (ex 3c), to integrate environmental protection requirements into the implementation of Community policies.[418] The Commission's approach in the Guidelines consists in determining whether, and under what conditions, state aid might be regarded as necessary to ensure environmental protection and sustainable development, without having disproportionate effects on competition and economic growth.[419]

The UK's environmental taxes, including the exemptions and reliefs which they embody, together with its economic instruments for environmental protection, are the subject of a number of Commission decisions on state aid. Most of these have concerned climate change levy[420] and its associated economic instruments, the UK ETS[421] and the RO,[422] but the special Northern Ireland aggregates levy provisions have also been the subject of Commission scrutiny.[423]

12.2.7.3 *State aid in the energy industries*

Following the initial liberalisation of the electricity market by the 1996 Electricity

---

[416] *Community Guidelines on State Aid for Environmental Protection*, (2001) OJ C37 1. See Hancher, Ottervanger and Slot, *op. cit.*, paras 17–011–17–020 and Evans, *op. cit.*, pp. 357–74.
[417] Guidelines, paras 72 and 73.
[418] See para. 12.2.1 above.
[419] Guidelines, para. 5.
[420] See Decision N 123/2000, *Climate Change Levy* (28 March 2001); Decision N 660/A/2000, *Exemption from Climate Change Levy for Natural Gas in Northern Ireland* (18 July 2001); Decision C 18 and C19/2001, *Climate Change Levy (EC and ECSC)* (3 April 2002); Decision N 539/2002, *Climate Change Levy Exemption for Electricity Exports of Good Quality CHP* (5 March 2003); and Decision C 12/2003 (ex N 778/2002), *Climate Change Levy Exemption for Coal Mine Methane* (17 September 2003). Aspects of the levy were also the subject of state aid decisions under the ECSC (not listed).
[421] See Decision N 416/2001, *Emission Trading Scheme* (November 28, 2001); and Decision N 104/B/2002, *Emission Trading Scheme – Modification to Commission Decision State Aid N 416/2001 of 28 November 2001* (12 March 2002).
[422] See Decision N 504/2000, *Renewables Obligation and Capital Grants for Renewable Technologies* (28 November 2001).
[423] See Decision N 863/2001, *Aggregates Levy* (24 April 2002); and Decision N 2/2004, *Aggregates Levy – Northern Ireland Exemption* (7 May 2004). See para. 13.3 below.

Directive,[424] the main area in which state aid issues arise is that of stranded costs.[425] Schemes for the recovery of such costs through compensatory levies are capable of qualifying as state aid[426] and the Commission deals with such schemes under Art. 87(3)(c), European Treaty, in accordance with a 2001 Communication.[427] However, this methodology is not applicable to state aid granted to support renewables generation; there remains considerable scope for tension between the development of a new competitive framework for the granting of state aids to renewables generators and the environmental provisions of the European Treaty.[428]

The question of state aid to the coal industry is dealt with under its own regime,[429] which is designed to ensure security of supply.[430]

State aids to support the environmental objective of energy *conservation* are dealt with in the environmental Guidelines discussed in para. 12.2.7.2 above. The Commission specifically acknowledges that the use of green taxes may offset the adverse economic effects of state aid in the form of tax reliefs and exemptions.[431]

12.2.7.4 State aid in the air passenger and road freight transport sectors[432]

Pursuant to a Council Decision of 1965,[433] Regulation 1191/69/EEC[434] provided that Member States could require road transport operators, to continue to operate

---

[424] See para. 12.2.6.3(1) above.

[425] That is, costs incurred by electricity utilities prior to market liberalisation, in order to meet customer or governmental needs, and which liberalisation has made uncommercial. Synonymous with the term 'stranded assets' (see Cameron, *op. cit.*, p. lvii).

[426] See Cameron, *op. cit.*, paras 7.99–7.115. The Commission approved the fossil fuel levy and non-fossil fuel levy under what is now Art. 87(3)(b) (see Hancher, Ottervanger and Slot, *op. cit.*, paras 3–020 and paras 21.4.4 and 21.5 below).

[427] See *Commission Communication Relating to the Methodology for Analysing State Aid Linked to Stranded Costs* (not referenced), July 2001, available from *www.europa.eu.int.* See also Commission Decision 99/791/EC, (1999) OJ L319 1 (Northern Ireland Electricity plc and Premier Power, discussed at Cameron, *op. cit.*, para. 7.103).

[428] See Cameron, *op. cit.*, para. 7.116, which includes a discussion of an ECJ decision revealing something of the tension referred to in the text, that is, *PreussenElektra AG v. Schleswag AG (Windpark Reussenköge III GmbH and Another, Intervening)*, C–379/98, [2001] 2 CMLR 36. See van Calster, *op. cit.*, para. 8.4n above.

[429] See Council Regulation EC/1407/02, (2002) OJ L205 1.

[430] See Decision N 4/2002, *State aid to coal production for the period 1 January 2002 to 23 July 2002* (21 January 2003).

[431] See *Community Guidelines on State Aid for Environmental Protection*, (2001) OJ C37 1. See also para. 12.3.5 below.

[432] See Greaves, *op. cit.*, pp. 144–5; Hancher, Ottervanger and Slot, *op. cit.*, ch 14; and Evans, *op. cit.*, para. 5.7.

[433] This was Council Decision 65/271, (1965) 88 JO 1500 (no longer in force), which provided for a legislative programme to harmonise national rules affecting competition in the inland transport sector.

[434] (1969) OJ L156 1.

unprofitable services, by way of a PSO,[435] provided that they compensated the undertakings in question for the financial burden thus incurred. Such aids are expressly declared by Art. 73, European Treaty (ex 77), to be compatible with the Treaty.[436]

In considering state aid applications in relation to the air transport sector, under Art. 87(3), European Treaty, the Commission follows its 1994 Guidelines,[437] introduced in the wake of four major grants in favour of national airlines, that is, Aer Lingus, TAP, Air France and Olympic.[438] The two main concerns of the Guidelines are stated to be the completion of the internal market for air transport and the increase of transparency in the notification and decision-making processes.[439]

12.3   Taxation aspects

12.3.1   General

The Commission has long advocated the use of economic instruments (including environmental levies) in environmental protection. This is shown not only, for example, by the advocacy of such instruments in the Community Strategy for Waste Management, already discussed,[440] but in the Commission's 1997 Communication on the use of environmental taxes and charges ('the 1997 Commission Communication').[441] Whether the ECJ shares this enthusiasm is perhaps to be doubted.[442] For instance, in 2001, the ECJ struck down a Belgian flat rate municipal

---

435   Regulation 1191/69, Art. 2(1).
436   Council Regulation EEC/1107/70, (1970) L130 1, much amended, contains the procedures applicable to such aid.
437   Application of Articles 92 and 93 [now 87 and 88] of the European Treaty and Article 61 of the EEA Agreement to State Aids in the Aviation Sector, (1994) OJ 350 7.
438   See Hancher, Ottervanger and Slot, *op. cit.*, paras 14–044–14–068.
439   See 1994 Guidelines, paras 7 and 8. See also Rosa Greaves, 'Judicial Review of Commission State Aid Decisions in Air Transport', in *Judicial Review in European Union Law* (The Hague: Kluwer, 2000), ch. 39.
440   See paras 12.2.5.1 above.
441   See Commission Communication, *Environmental Taxes and Charges in the Single Market*, COM (97) 9 final, (1997) OJ C224 6 (see para. 1.2.1.5(2) above) and, generally, Paul Farmer and Richard Lyal, *EC Tax Law* (Oxford: Clarendon Press, 1994); David Williams, *EC Tax Law* (London: Longman, 1998); Alexander Easson, *Taxation in the European Community* (London: Athlone Press, 1993); B. Terra and P. Wattel, *European Tax Law* (Amsterdam: Kluwer, 1993); and D. Berlin, *Droit Fiscal Communautaire* (Paris: Presse Universitaire Francaise, 1988).
442   A point cogently argued by Amparo Grau Ruiz and Pedro Herrera in an as yet unpublished paper, entitled 'The Polluting Side of Economic Freedoms: is the ECJ against Environmental Taxes?', given at the International Seminar on Energy Taxation and Sustainable Development held in Madrid on 2 and 3 October 2003. The authors are most grateful to Drs Grau Ruiz and Herrera for making available to them the slides from that paper.

tax on all satellite dishes in a particular municipality, on the ground that it infringed Art. 49, European Treaty (ex Art. 59) by restricting the freedom to receive satellite television broadcasts and by conferring an unfair advantage on the internal Belgian broadcasting market.[443]

12.3.2   Attribution of taxation powers

Like environmental policy, taxation policy is not within the exclusive competence of the Community.[444] This means that, again as with environmental law, Member States remain competent in the taxation field.[445]

By Art. 93, European Treaty (ex 99), the Council,[446] acting unanimously on a proposal from the Commission[447] and after consulting the European Parliament[448] and ECOSOC,[449] is mandated to 'adopt provisions for the harmonisation of legislation concerning turnover taxes, excise duties and other forms of indirect taxation to the extent that such harmonisation is necessary to ensure the establishment and the functioning of the internal market'. Whilst this might at first seem to give the green light, among other things, to the creation of Community-wide environmental taxes, the fact that the Council's duty is specifically related to the establishment of the internal market by 31 December 1992,[450] means that it is at least arguable that it is now spent.[451] More importantly, however, Art. 93 embodies the general principle of the fiscal veto,[452] which is reflected in Art. 175(2)(a), European Treaty (ex 130s), relating to environmental provisions primarily of a fiscal nature.[453]

Article 93 relates specifically to the harmonisation of indirect taxation and environmental taxes, generally speaking, are indirect taxes.[454] To the extent that there is a basis for the harmonisation of direct taxation,[455] however, this is to be discerned in Art. 94, European Treaty (ex 100) which again tasks the Council, acting unanimously on a proposal from the Commission and after consulting the European Parliament and ECOSOC, with issuing directives for the 'approximation of such laws, regulations or administrative provisions of the Member States as directly

---

443   See *De Coster v. College des Bourgmestre et Echevins de Watermael-Boitsfort*, C–17/00, [2002] 1 CMLR 12.
444   See para. 12.2.1 above.
445   See Kirsten Borgsmidt, 'Ecotaxes in the Framework of Community Law' [1999] EELR 270–281. The writers would like to acknowledge a particular debt of gratitude to this work in the preparation of para. 12.3.
446   See para. 4.3.1 above.
447   See para. 4.3.2 above.
448   See para. 4.3.3 above.
449   See para. 4.3.4 above.
450   See Art. 14, European Treaty (ex 7a).
451   See Williams, *op. cit.*, p. 34n.
452   See Art. 95(2), European Treaty (ex 100a).
453   See para. 4.3.1 (point 1) above.
454   See para. 1.2.1.2 above. An exception to this is, of course, the differential against environmentally-unfriendly cars in the income tax provisions for the taxation of the provision of company cars (see para. 23.2 below).
455   See para. 1.2.1.2 above.

affect the establishment or functioning of the common market'.[456] The requirement of unanimity which is present here, as in Art. 93, means that Art. 94 is again of somewhat limited significance and, in the absence of unanimity, in no way detracts from the competence of Member States to legislate in the taxation field.

The overall effect of Arts 93 and 94, are, of course, twofold, that is: (1) that any Community-wide environmental tax would require the unanimous support of Member States; and (2) that Member States are free to create their own environmental taxes, provided that they do not conflict with other provisions of the Treaty.[457] It is now necessary to turn to these other Treaty provisions.

12.3.3   Fiscal barriers to trade

12.3.3.1   Discriminatory internal taxation

(1) Generally
Barriers to trade of a fiscal, as distinct from a general, nature are covered by Art. 90, European Treaty (ex 95).[458] Art. 90 reads as follows:

> No Member State shall impose, directly or indirectly, on the products of other Member States any internal taxation of any kind in excess of that imposed directly or indirectly on similar domestic products.
>
> Furthermore, no Member State shall impose on the products of other Member States any internal taxation of such a nature as to afford indirect protection to other products.

The Article is, of course, a national treatment obligation, and, as such, it is the close equivalent, in relation to intra-EU trade, of Art. III(2), GATT 1994,[459] which is applicable to trade between EU Member States and third countries. Like Art. III(2), Art. 90 applies to internal taxation, rather than to customs duties.[460] It will be noted also that, unlike in relation to non-fiscal barriers to trade,[461] the national treatment obligation in Art. 90 admits of no exceptions, not even (or, perhaps, not surprisingly!) environmental ones.

---

456 See *Asscher v. Staatsecretaris van Financien*, C–107/64, [1996] STC 1025, 1033 (para. 53), '[Article 93] of the Treaty explicitly gives the Council powers of harmonisation in the field of indirect taxation alone. Laws relating to direct taxation may be harmonised ... under [Art 94] ... of the Treaty by the Member States acting unanimously, where they directly affect the establishment or functioning of the Common Market ... ' (Advocate General Leger).

457 For example, the regionally-imposed levy of one euro per day between 2002 and 2003 on holiday makers in the Balearic Islands (see B. Arino, 'Sustainable Tourism and Taxes: an Insight into the Balearic Eco-Tax' (2002) 11 EELR 114–19).

458 For a detailed consideration, see Farmer and Lyal, *op. cit.*, pp. 46–77.

459 See para. 8.4.3 above.

460 In *R (on the application of British Aggregates Association and others) v. C & E Commrs*, [2002] EWHC 926 (Admin), [2002] 2 CMLR 51, aggregates levy was held to be part of a system of internal taxation and, as such, fell to be considered under Art. 90, European Treaty, rather than as a charge equivalent to a customs duty under Art. 25 thereof (see para. 12.3.3.2 below).

461 See para. 12.4 below.

(2) The two rules of Article 90, European Treaty (ex 95)
Article 90 contains two rules: a ban on tax discrimination against 'similar [non-domestic] products' (first paragraph) and, in the second paragraph, a ban on tax discrimination that results in indirect protection.[462]

In the context of the first rule,[463] the ECJ has held that the mere fact that two products contain the same raw materials (in the particular case, alcohol) is not enough to make them 'similar products' for the purposes of the first rule. Similar products are those that '... at the same stage of production or marketing, have similar characteristics and meet the same needs from the point of view of consumers'.[464] As Borgsmidt points out, this is a very broad reading of the first rule, since it means that 'similar products' are not those which are identical but those which have a similar and comparable use, due regard being had to consumer habits in the Community as a whole and not merely in the Member State under consideration.[465]

As to the second rule,[466] this may apply where, even though the 'other products' are not similar in the sense of the first rule, they have sufficient characteristics in common that they are an alternative choice for consumers in some circumstances.[467]

(3) Graduated schemes of taxation
Particular problems may arise in relation to Art. 90 where a Member State uses a graduated scheme of taxation, either for goods in general or for a specific product.[468] The ECJ has mapped out the circumstances in which differentiation between products is permitted in a line of cases,[469] two of which are as follows:

a.   In *Chemial Farmaceutici v. DAF SpA*,[470] Italy had imposed a higher tax on synthetic ethyl alcohol than on ethyl alcohol of agricultural origin, even though the two products could be used interchangeably. The purpose of the differential was to favour the agricultural manufacture of ethyl alcohol and to restrain its synthetic production (that is, the processing of ethylene[471] into alcohol), since ethylene could be used for economically more important purposes. The result

---

[462]   See Borgsmidt, *op. cit.*, p. 278.
[463]   See Farmer and Lyal, *op. cit.*, pp. 57–65.
[464]   See *Rewe v. Hauptzollamt Landau/Pfalz*, C–45/75, [1976] ECR 181, 194 (para 12).
[465]   See *Cogis v. Amministrazione delle Finanze dello Stato*, C–216/81, [1982] ECR 2701; *Commission v. Denmark*, C–206/84, [1986] ECR 833; and *Commission v. Italy*, C–184/85, [1987] ECR 2013. There is a rough parallel between this test and the test used for product substitutability in the competition rules of Arts 81 and 82, European Treaty (ex 85 and 86) (see Farmer and Lyal, *op. cit.*, p. 59).
[466]   See Farmer and Lyal, *op. cit.*, pp. 65–76.
[467]   See *Commission v. UK*, C–170/78, [1980] ECR 417 (wine and beer).
[468]   See Farmer and Lyal, *op. cit.*, pp. 69–76.
[469]   See *Amministrazione delle Finanze dello Stato v. Essevi and Salengo*, C–142–143/80, [1981] ECR 1413; *Commission v. Italy*, C–200/85, [1986] ECR 3953; *Commission v. France*, C–196/85, [1987] ECR 1597; and *Bergandi v. Directeur Général des Impôts*, C–252/86, [1988] ECR 1343.
[470]   C–140/79, [1981] ECR 1; see also *Vinal SpA v. Orbat SpA*, C–46/80, [1981] ECR 77.
[471]   That is, a petroleum derivative.

was that only imported synthetic alcohol was subject to the tax, since domestic production of synthetic alcohol was uneconomic. The ECJ held that the tax was not discriminatory since, although imports were hampered by it, so too was domestic production. In the course of its judgment, the Court said:

> ... As the Court has stated on many occasions ... in its present stage of development Community law does not restrict the freedom of each Member State to lay down tax arrangements which differentiate between certain products on the basis of objective criteria, such as the nature of the raw materials used or the production process employed. Such differentiation is compatible with Community law if it pursues economic policy objectives which are themselves compatible with the requirements of the Treaty and its secondary law and if the detailed rules are such as to avoid any form of discrimination, direct or indirect, in regard to imports from other Member States or any form of protection of competing domestic products.[472]

On this basis, it seems that, given that environmental protection is a legitimate objective,[473] a differential in tax rates for environmental reasons should be lawful for the purposes of Art. 90.

b. In *Outokumpu Oy*,[474] the Finnish government had imposed an excise duty on electricity, the rate of which depended on the method of production. Other features of the duty were as follows:

i. the duty applied only to electricity of domestic origin, not to imported electricity;
ii. the rate of duty was less on electricity produced by water power than on electricity produced by nuclear power;
iii. no duty was charged on electricity produced by certain other methods;
iv. although the duty charged on imported electricity was lower than the highest duty on electricity produced in Finland, it was higher than the lowest rate of duty charged on electricity produced in Finland; and
v. duty was charged on imported electricity without regard to the method of its production.

The ECJ held that the duty was incompatible with Community law, although it stressed that, provided that a duty differential was based on objective criteria, it was lawful for Member States to tax the same or similar products differentially. Art. 90 did not prevent differential tax rates based on environmental considerations, provided that the tax in question did not discriminate against imports. However, in the instant case:

> ... [t]he first paragraph of Article [90] of the EC Treaty precludes an excise duty which forms part of a national system of taxation on sources of energy from being levied on electricity of domestic origin at rates which vary according to its method of production

---

472 See [1981] 3 CMLR 350, 361 (para 14).
473 See Borgsmidt, *op. cit.*, p. 278.
474 Case C–213/96, [1998] ECR I–1777.

while being levied on imported electricity, whatever its method of production, at a flat rate which, although lower than the highest rate applicable to electricity of domestic origin, leads, if only in certain cases, to higher taxation being imposed on imported electricity.[475]

In the light of the wording of Arts 93 and 94, the effect of Art. 90 is therefore that any national environmental tax must not discriminate against goods from other Member States whether on environmental or any other grounds.[476] Equally, however, provided that such a tax is non-discriminatory, it is not prohibited by Art. 90.

(4) Parafiscal charges[477]

A second set of problems may arise in relation to Art. 90 when the application of revenues raised by levies has a discriminatory effect.[478] These problems, which arise in relation to certain types of earmarking, are obviously important to environmental levies because of the strong arguments, already discussed, for the hypothecation or earmarking of the proceeds of such levies.[479]

Where all or part of the revenue of a Member State from a particular levy is used to resource one of that state's own industries, then, depending on the circumstances, there may be a breach either of Art. 90 or of Art. 25, European Treaty (ex 12).[480] In *Compagnie Commerciale de l'Ouest and Others v. Receveur Principal des Douanes de la Pallice Port*,[481] a reference under Article 234, European Treaty (ex 177), the ECJ held that Art. 90 applies to the situation where the revenue raised from a particular tax is used for the benefit of domestic products only, in circumstances such that the advantages accruing to the domestic product offset the charge borne by domestic products in part only. Where such advantages fully offset the charge borne by the domestic product, then Art. 25 applies.[482] In the cases under consideration, importers and distributors of petroleum products had challenged the legality of a parafiscal charge levied in France on the putting into circulation of certain petroleum products, irrespective of whether those products were domestic or imported. The charge had been introduced to fund an Energy Savings Agency, which had then applied the money to finance measures purportedly to encourage and achieve energy-savings as well as the use of under-utilised energy resources.

(5) Article 90, European Treaty (ex 95) and environmental taxes

From the point of view of new green taxes, the danger with the first rule in Art. 90 is, of course, that the current interpretation of the term 'similar products' will mean that products are considered to be similar, even where they raise quite different environmental issues. This issue is specifically addressed in the 1997 Commission Communication, where it is stated that those involved in the design of green levies

---

[475] [1998] ECR I–1777 (para 41). See van Calster, *op. cit.*, para. 8.4n above.
[476] See Borgsmidt, *op. cit.*, p. 274.
[477] Broadly, species of 'hypothecated' levies. See para. 11.2.2 above.
[478] See Weatherill and Beaumont, *op. cit.*, pp. 475–6, and Borgsmidt, *op. cit.*, p. 280.
[479] See para. 11.2.2 above.
[480] *Ibid.*
[481] C–78–83/90, [1994] 2 CMLR 425.
[482] See para. 12.4 below.

should consider:

> ... whether goods with the same function but with different environmental properties due to the content or differences in production methods could be regarded as being different goods.[483]

Although the cases in which national taxes' compatibility with Art. 90 has been considered leave open the possibility of a differentiated tax for environmental reasons, it is clear that imported products must not be subject to higher rates of tax than domestic ones.[484] In this connection, as Borgsmidt points out, '[n]ew production methods and control thereof may not be available in the country of origin.[485] In *R (on the application of British Aggregates Association and others) v. C & E Commrs*,[486] the applicants, the quarry operators' trade association,[487] applied for permission to move for judicial review of the aggregates levy legislation.[488] Moses, J. held that there was no breach of Art. 90 merely because the levy had some protective effect, in the sense that, had it not been imposed on imports, it might have encouraged them;[489] its purpose was environmental rather than protectionist. Moreover, the fact that the revenue raised by the levy benefited people living in the UK and provided no benefit to importers conferred no specific benefit at all on domestic products or producers; the mere existence of the Aggregates Levy Sustainability Fund ('ALSF')[490] and the NIC reduction[491] did not partially offset the levy borne by the domestic production of aggregate.[492]

The possibility that the earmarking of the proceeds from a tax might fall foul of Art. 90 in situations where the liability to tax of those benefited is reduced, although not eliminated, has obvious resonances for environmental taxes.[493] The avowed environmental purpose of such levies is clearly not sufficient, however, to prevent Art. 90 from applying in an appropriate case.[494]

(6) Article 92, European Treaty (ex 98)

Finally, a little-noted provision appears in Art. 92, European Treaty (ex 98).[495] This allows the Council to take limited action where export or import distortions arise from levies that are not indirect taxes. In *R (on the application of British Aggregates*

---

483 COM (97) 9 final, para. 21.
484 See, for example, *Schöttle & Söhne OHG v. Finanzamt Freudenstadt*, C–20/76, [1977] 2 CMLR 98.
485 See Borgsmidt, *op. cit.*, p. 279.
486 [2002] EWHC 926 (Admin), [2002] 2 CMLR 51.
487 See para. 2.5 above.
488 See paras 4.2.1.5 above and 13.1 below.
489 [2002] 2 CMLR 51, para. 56. See also para. 8.4.5.1 above.
490 See para. 4.2.1.2(1)n above and 21.3.1(b) below.
491 See para. 21.2 below.
492 [2002] 2 CMLR 51, paras 57 and 58.
493 See para. 11.2.2 above.
494 See Borgsmidt, *op. cit.*, pp. 280–81.
495 See Farmer and Lyall, *op. cit.*, pp. 81–2; Williams, *op. cit.*, p. 33.

Association and others) v. Customs and Excise Commissioners,[496] the applicants argued that the aggregates levy breached Art. 92, on the basis that it was a direct, rather than an indirect tax. Moses, J. rejected this submission: '... the mechanism for achieving the purpose of the levy, namely by passing the levy on to a consumer of the aggregate, does, as it seems to me, provide a powerful indication that the levy is an indirect tax. The fact that in some cases the levy will not be passed on does not turn it into a direct tax any more than in the case of an excise duty on wine'.[497]

12.3.3.2 Prohibition on customs duties and on charges having equivalent effect ('CEEs')

If the national treatment obligation of Art. III(2), GATT 1994 is mirrored in Art. 90, European Treaty (ex 95) so also, subject to one vital qualification, is Art. II(2), GATT 1994 mirrored in Art. 25, European Treaty (ex 12).[498] The vital qualification, of course, is that customs duties are prohibited between Member States, the Community being based on a customs union.[499] Article 25 provides that:

> Customs duties on imports and exports and charges having equivalent effect shall be prohibited between Member States. This prohibition shall also apply to customs duties of a fiscal nature.

The definition of a charge having equivalent effect (a 'CEE') has already been considered in Chapter 7 above.[500]

The key point, from the point of view of environmental taxes, is that, in order for a CEE to escape the scope of Art. 25 on the basis that it is consideration for services supplied by the importing state,[501] it is necessary for it to be shown that the service in question is of benefit to the importer and not to the general public.[502] As Lyons points out, the ECJ 'has consistently denied that something which is done by a public authority for the benefit of the general public as well as the trader concerned, is a service to the trader for which a charge may be made'.[503] This means that, in the case of an environmental levy on imports which was not part of a general system of taxation,[504] it would be necessary to rely on the only other exception to the concept of CEEs and to show that it was an administrative charge,[505] that is, that it was imposed to cover the costs of services required by Community law or by international agreement.[506]

---

[496] [2002] EWHC 926 (Admin), [2002] 2 CMLR 51.
[497] [2002] 2 CMLR 51, para. 75.
[498] Generally, see Timothy Lyons, *EC Customs Law* (Oxford: Oxford University Press, 2001), pp. 60–72 and ch 13.
[499] See Art. 23, European Treaty (ex 9).
[500] See para. 7.2.2.3 above.
[501] *Ibid.*
[502] See, for example, *Cadsky v. Instituto Nazionale per il Commercio Estero,* C–63/74, [1975] ECR 281, although more recent cases are reviewed at Lyons, *op.cit.,* pp. 70–71.
[503] See Lyons, *op. cit.,* p. 70.
[504] See para. 7.2.2.3 above.
[505] *Ibid.*
[506] See Borgsmidt, *op. cit.,* p. 278.

In *Compagnie Commerciale de l'Ouest and Others v. Receveur Principal des Douanes de la Pallice Port*, discussed above,[507] the ECJ considered the particular problem of the circumstances in which a parafiscal charge[508] would fall foul of Art. 25:

A parafiscal charge applied under the same conditions as regards its collection to both domestic and imported products, the revenue from which is used for the benefit of domestic products only, so that the advantages accruing from it fully offsets [*sic*] the charge borne by those products, constitutes a charge having an effect equivalent to customs duties prohibited by Article [25] ...

In so defining the scope of Art. 25 in relation to such charges, the Court was distinguishing the scope of the Art. from that of Art. 90, the latter applying where the offset is only a partial one.[509]

12.3.4   Harmonisation of excise duties on fuels

Differentiated rates of excise duty were introduced for leaded and unleaded petrol by Council Directive 92/81/EEC[510] on the Harmonisation of the Structures of Excise Duties on Mineral Oils and Council Directive 92/82/EEC[511] on the Approximation of Excise Duty Rates on Mineral Oils. The former was designed to harmonise the structures, exemptions and rate reductions applicable to excise duties on mineral oils; the latter was designed to specify minimum rates or rate bands for each category of oil. Controversially, Directive 92/81/EC exempted from duty oils that are used as fuels for the purpose of air navigation, as well as navigation within Community waters, other than for private pleasure-flying and sailing. Both Directives have now been repealed by the 2003 Energy Products Directive.

The Energy Products Directive ('the EPD')[512] repeals the two 1992 Directives as from 31 December 2003.[513] Unlike that of its predecessors, its scope is not limited to mineral oils but covers most energy products, including electricity, natural gas and coal. The recitals of the EPD specifically refer to the taxation of energy products including, where appropriate, electricity, as an instrument for achieving the Kyoto Protocol objectives.[514] They also acknowledge the possibility that CHP generation and renewables might qualify for preferential taxation treatment.[515] The main substantive provision is Art. 4(1), which provides that Member States must not apply levels of taxation below the prescribed minimum levels to the products covered by the EPD. Article 14 provides that Member States must exempt: (1) 'energy products and electricity used to produce electricity and electricity used to maintain the ability to

---

507   See para. 12.3.3.1(4) above.
508   *Ibid.*, n above.
509   See para. 12.3.3.1 above.
510   (1992) OJ L316 12 (repealed).
511   *Ibid.*
512   Council Directive 03/96/EC, (2003) OJ L283 51.
513   *Ibid.*, Art. 30.
514   *Ibid.*, recitals 7, 12 and 13.
515   *Ibid.*, recital 25.

produce electricity';[516] (2) 'energy products supplied for use as fuel for the purpose of air navigation other than in private pleasure-flying';[517] and (3) 'energy products supplied for use as fuel for the purposes of navigation within Community Waters', other than private pleasure craft and electricity produced on board.[518] Of these three mandatory exemptions, (1) does not, however, prevent Member States from subjecting products to taxation for reasons of environmental policy without having to respect the specified minimum levels.[519] The other two preserve the exemptions in the 1992 Directives. By Article 15, EPD, Member States may exempt, either wholly or partially, *inter alia*, electricity from renewables; electricity produced from 'environmentally-friendly' CHP generation;[520] and energy products and electricity used for CHP generation.[521] The implementation of the EPD in UK law has been discussed in Chapter 6,[522] together with the derogations therefrom which the UK has succeeded in negotiating.

The Biofuels Directive,[523] a second major development in 2003, aims to promote the use of biofuels and other renewable fuels[524] to replace diesel or petrol for transport purposes in each Member State. Climate change, environmentally-friendly security of supply and the promotion of renewables are given as express justifications for the measure.[525] 'Biofuels' are defined as liquid or gaseous fuel for transport produced from biomass.[526] The heart of the Directive is Art. 3(1)(a), which provides that Member States should ensure that a minimum proportion of biofuels and other renewable fuels is placed on their markets and, to that end, must set national targets therefore. This is backed up by the specific monitoring and reporting requirements in Arts 3(3) and 4(1) of the Directive respectively. Again, implementation of the Biofuels Directive in the UK was discussed in Chapter 6 above.[527]

12.3.5   State aids and taxation

Where revenue is paid into Member States' national treasuries and producers are then supported out of general funds, the support may constitute state aid under Art. 87, European Treaty.[528] However, reliefs and exemptions within particular tax codes may also, in certain circumstances, count as state aid.[529] State aid in the form of

---

[516] *Ibid.*, Art. 14(1)(a).
[517] *Ibid.*, Art. 14(1)(b).
[518] *Ibid.*, Art. 14(1)(c).
[519] *Ibid.*, Art. 14(1)(a).
[520] Council Directive 03/96/EC, Art. 15(1)(d).
[521] *Ibid.*, Art. 15(1)(c).
[522] See para. 6.4 above.
[523] Directive 03/30/EC, (2003) OJ L123 42.
[524] Defined, *ibid.*, Art. 2(1)(c).
[525] Directive 03/30/EC, Art. 1.
[526] *Ibid.*, Art. 2(1)(a). Biomass is defined, *ibid.*, Art. 2(1)(b).
[527] See para. 6.4 above.
[528] See para. 12.2.7.1 above. Where the arrangement is also discriminatory, then it may be caught by Art. 90, European Treaty (ex 95) as well as by Art. 87.
[529] See the (unsuccessful) attempt to argue that certain exemptions within the aggregates levy code constituted unlawful state aid (in *R (on the application of British Aggregates Associates and Others) v. C & E Commrs*, [2002] EWHC 926 (Admin), [2002] 2 CMLR 51, paras 79–115).

such exemptions and reliefs is referred to as 'fiscal aid' or 'fiscal state aid'.[530] In considering an application for state aid clearance under Art. 87(3), the Commission follows its 2001 Guidelines,[531] as well as, where appropriate, its 1998 guidelines on the application of the state aid rules to measures relating to direct business taxation.[532]

In 2001, the Commission decided not raise any objections to the introduction of enhanced capital allowances for energy efficient investments, on the basis that the measure did not constitute aid.[533]

In 2003, the Commission decided to close down the Art. 88(2) procedure, with a conditional decision, in relation to the Finance Act 2001, ss.92A and 92B exemption from stamp duty for non-residential properties in disadvantaged areas.[534] For details of the Commission consents granted in relation to the UK's environmental taxes and other economic instruments, the reader is referred to para. 12.2.7.2 above.

In 2004, the Commission decided not to raise any objections to the introduction of a reduced rate of excise duty on bioethanol used for road transport, on the basis that the aid was compatible with the European Treaty.[535]

12.4   Rules on free movement of goods

Part Three, Title I, Ch. 2, European Treaty, as is well-known, prohibits quantitative restrictions between Member States. Article 28, European Treaty (ex 30) deals with imports, while Art. 29, European Treaty (ex 34), deals with exports. By Art. 28, quantitative restrictions on imports and all measures having equivalent effect, are prohibited between Member States. Art. 29 lays down the same rule, *mutatis mutandis*, for exports. To each rule, there are the exceptions in Art. 30, European Treaty (ex Art. 36), including prohibitions or restrictions justified on grounds of the protection of health and life of humans, animals or plants. Any such prohibition or restriction must not, however, constitute a means of arbitrary discrimination or a disguised restriction on trade between Member States.

The question of what constitutes a measure having equivalent effect has for long been governed by the first rule articulated in *Procureur du Roi v. Dassonville*,[536] which reads as follows:

---

[530] See Raymond Luja, 'WTO Agreements versus the EC Fiscal Aid Regime: Impact on Direct Taxation' (1999) 27 *Intertax* 207–25, esp. pp. 216–23.

[531] See para. 12.2.7.2 above.

[532] See *Commission Notice on the Application of the State Aid Rules to Measures Relating to Direct Business Taxation*, (1998) OJ C384 3.

[533] See Decision N 797/2000, *Enhanced Capital Allowances for energy efficient investments* (13 March 2001). See para. 21.3.1 below.

[534] See para. 24.5.3 below.

[535] See Decision N 407/2003, *Reduced Rate of Excise Duty on Bioethanol used for Road Transport* (3 February 2004). See para. 22.2.2 below.

[536] C–8/74, [1974] ECR 837.

All trading rules enacted by member States which are capable of hindering, directly or indirectly, actually or potentially, intra-Community trade are to be considered as measures having an effect equivalent to quantitative restrictions.

This rule was subsequently developed in the ECJ's decision in the so-called '*Cassis de Dijon*' case,[537] which held that Art. 28 was capable of applying to national rules which inhibited trade, not because they discriminated against imported products, but because they were different from the trade rules applicable in the products' country of origin.[538] However:

> Obstacles to movement within the Community resulting from disparities between the national laws relating to the marketing of the products in question must be accepted in so far as those provisions may be recognised as being necessary in order to satisfy mandatory requirements relating in particular to the effectiveness of fiscal supervision, the protection of public health, the fairness of commercial transactions and the defence of the consumer.[539]

The rule thus enunciated has been developed in a line of more recent cases.[540] Of interest in the present context is the *Danish Bottles* case, that is, *Commission v. Denmark*,[541] the ECJ decision that gave rise to the Packaging and Packaging Waste Directive.[542] There, the Danish Government argued that a national rule which stated that a government agency had to approve the specification of beer containers, and that the empty ones had to be returned under a deposit-and-return scheme to be set up by distributors, was justified by a mandatory requirement of environmental protection.[543] To mitigate the difficulties for foreign producers, Danish producers were entitled to market beer in unapproved containers to a maximum of 3,000 hectolitres per annum. Whilst accepting that environmental protection was a mandatory requirement,[544] the Court held that to require the use only of approved containers was disproportionate, and that even the limit of 3,000 hectolitres did not prevent Denmark from being in breach of its Art. 28 obligations.

The *Danish Bottles* case demonstrates that, in applying Arts 28-30, the proportionality principle[545] requires the court to weigh a given restriction on free movement against its environmental objectives. On this basis, it might be argued that an environmental levy, that is, a disincentive to the import or export of a product rather than an outright

---

[537] *Rewe-Zentrale AG v. Bundesmonopolverwaltung für Branntwein*, C–120/78, [1979] ECR 649.

[538] The case therefore embodies a mutual recognition principle.

[539] [1979] ECR 649, para. 8.

[540] See the discussion in Craig and de Búrca, *op. cit.*, ch. 15.

[541] C–302/86, [1989] 1 CMLR 619. Also (confusingly) known as the *Disposable Beer Cans* case. See para. 12.2.5.1(2)(a) above. See also *Commission v. Belgium*, C–2/90, [1992] ECR I–4431 ('the *Walloon Waste* case').

[542] See para. 12.2.5.1(2)(a) above.

[543] That is, in the absence of Community harmonisation measures.

[544] [1989] 1 CMLR 619, 630 (para 8).

[545] See Craig and de Búrca, *op. cit.*, pp. 371–9.

ban, might meet the proportionality concern.[546] This case left open, however, whether the proportionality principle could be applied in this way.[547]

But this issue did arise in the *German mineral water bottles* case, that is, *Commission v. Germany*.[548] The German transposition of the Packaging and Waste Directive[549] was based on the concept of the return of used empty sales packaging by final consumers free of charge, so that it could be re-used. The ECJ upheld a Commission objection that this constituted a burden on mineral water producers in other Member States, and was contrary to Article 28 (ex 30) of the European Treaty.[550] The German Government's defence that this was justifiable as being for the protection of the environment was rejected on grounds of proportionality.[551] Although the German legislation specified a six month transitional period, that was an insufficient safeguard for importers because the real transitional period was indeterminate.[552]

The Single Market means that 'green' electricity can be sourced from France (with the attendant conceptual problems and difficulties of verification presented by this) and the UK road system has, for some years, been subjected to additional wear from 44-tonne lorries coming over from the Continent on the ferries and through the Channel Tunnel. Of more concern to the UK Government is the undermining of its excise duties through 'shopping at Calais'.[553]

12.5 Concluding comments

The authors are reluctant to extend an already long chapter still further. However, as a preliminary to the practical discussion of the UK's green levies and subsidies over subsequent chapters, it might be useful to identify those features of Community law which seem to be of particular significance in shaping the design of those instruments.

Most important, perhaps, is the division of the material between the regulatory and taxation aspects. As to the former, it is important to emphasise that, although Community environmental legislation has not *required* the creation of any of the UK's environmental levies and subsidies, its existence nonetheless helps to provide at least technical justifications for them. How meritorious in fact such justifications are is a different matter and this a major theme of subsequent chapters. For example, the 'landfill tax escalator', which is discussed in Chapter 15 below, is expressly justified by reference to the UK's obligation to meet the reduction targets set out in

---

[546] See Borgsmidt, *op. cit.*, p. 273.
[547] *Ibid.*, *op. cit.*, referring to Jan H. Jans, *European Environmental Law* (The Hague: Kluwer, 1995), pp. 203–4.
[548] C–463/01, decided 14 December 2004. See para. 27.6 below and Christine Trüe, 'The German Drinks Can Deposit: Complete Harmonisation or a Trade Barrier Justified by Environmental Protection?' (2005) 2 JEEPL 142–49.
[549] 'European Parliament and Council Directive' 94/62/EC. See para. 12.2.5.1(2)(a) above.
[550] Para. 84 of the judgement.
[551] *Ibid.*, paras. 75, 78.
[552] *Ibid.*, paras. 79–82.
[553] See para. 22.1 below.

the Landfill Directive. It is nonetheless apparent that – absent special factors – the legitimacy that this may be seen to lend to the tax may be lost in the fact that any increase in landfill tax will, for reasons to be discussed, be passed on to council tax payers. Equally, however, it may be the case that an apparently greater alignment between landfill tax and the Landfill Directive may help to sharpen the focus of the tax and prevent it from in future merely becoming a device for raising revenue. As mentioned above, a recent decision of the Court of Appeal on the concept of 'waste', for landfill tax purposes,[554] shows that at least the courts may take an approach to the tax which is more deeply rooted in its European objectives than has hitherto been the case. Moreover, as regards the control of air and atmospheric pollution, it is difficult to avoid the conclusion that, whatever the professedly environmental dimension of the Community law and policy on energy and transport, their objectives often sit ill with those of the law Community law on air and atmospheric pollution, as exemplified by, for example, the IPPC regime. Market liberalisation and environmental protection, whatever environmental safeguards the former policy may contain, are not obvious bedfellows. It may be that a combination of state aid law and other exemptions from 'single market' measures will offer at least the pragmatic possibility of reconciling these at first not easily reconcilable policy objectives. This is a matter to which we return in Chapter 21 below.

The Community regulatory framework for environmental levies and subsidies, internally inconsistent though it may sometimes be, at least offers an ideological basis for such instruments. The same cannot be said for Community taxation law. Like GATT 1994, these provisions originate with a Treaty the framers of which did not share the modern preoccupation with environmental matters. In consequence, the process of fitting national environmental levies and subsidies into the framework of the European Treaty is to a considerable extent a matter of eluding the various obstacles that the Treaty presents. That this can be done successfully seems to be borne out by the recent decision of the High Court of England and Wales, discussed above, on the validity of the design of aggregates levy. Nonetheless, the reader is left with a very similar impression to that which may be gained from a close reading of the ECJ decisions on the various national attempts to create eco-taxes, that is, that their validity or otherwise, under the taxation rules of the Community, is by no means a foregone conclusion.

---

[554] See para. 15.3 below.

Division 2
National Taxes

Chapter 13

Aggregates Levy

13.1   Introduction

This tax had a chequered history before it even got off the ground.[1] Even the attempt to pretend that it might not be a tax failed, the general statement that it was a *levy* in Finance Act 2001, s.16(1), having to be changed into the explicit one that it was a *tax* by Finance Act 2002, Sched. 38, para. 2.[2]

In a Consultation Paper issued on 15 June 1998, it was proposed that a tax of this nature might be introduced.[3] It would be similar to landfill tax, with no input complications and assessed on site operators, but with the addition of liability for imports and a rebate for exports.[4] The underlying objective would be to promote recycling. In the event, the tax had a long gestation period, the decision to proceed not being announced until Budget 2000, and the legislation being passed with no debate on anything other than the underlying principles and economic effects of the tax just before the 2001 General Election. The start date was 1 April 2002.[5]

In the months leading up to the introduction of the tax, the government machine gave out seemingly contradictory messages. In January 2002, in *Aggregates Levy – at a glance*,[6] Customs said:

> The Government believes that it is essential that there continues to be an adequate supply of aggregates.

But, come Budget Day, the Treasury went on record that the introduction of the tax was 'to reflect the environmental costs imposed by aggregates quarrying'.[7] The answer to this apparent inconsistency is to be found in the framework enshrined in Art. 6, European Treaty (ex 3c), that is, that environmental protection requirements must be integrated into Community policies and activities, in particular, the concept of sustainable development.[8]

---

1   See para. 11.3.2 above.
2   See paras 1.2 and 1.2.1.1 above.
3   See HM Customs and Excise, *Consultation on a Proposed Aggregates Tax*, 15 June 1998.
4   By way of a credit (see para. 16.9 below).
5   See para. 11.3.2 above.
6   Available from www.hmce.gov.uk.
7   HMT 1 (17 April 2002).
8   See para. 12.2.1 above. See also John F. McEldowney and Sharron McEldowney, *Environmental Law and Regulation* (London: Blackstone Press, 2001), p. 42.

13.2   Operation of the tax

The levy applies at the rate of £1.60 per tonne.[9] It is collected by site operators and normally accounted for quarterly.[10] The tax point is ascertained under Finance Act 2001, s.19(1), and takes place, in the course of business,[11] as the first event[12] in the following sets of circumstances:

1   by removal from the site at which it was won[13] other than for an exempt purpose[14] or, if removed from that site without a tax point arising, removal from a site under the same registration[15] or at which an exempt process had been intended to be supplied but had not;[16]
2   when it becomes subject to an agreement to supply it to somebody else (which will be paraphrased hereafter as 'sold');
3   when it is used for construction purposes; or
4   when it is mixed, other than in permitted circumstances, with any substance other than water.

However, whereas landfill tax was able to take advantage of the requirement for there to be a waste disposal licence,[17] the only requirement for the extraction of aggregate is the need for planning permission, and the applicant for that would not necessarily be the subsequent extractor. Customs' approach to this problem has been fourfold:

1   the adoption of a sledgehammer to identify all possible 'business'[18] candidates for the actual controller of the site,[19] making them jointly and severally liable,[20] and requiring them to notify registrability.[21] From 1 April 2003, certain changes were made for discrete commercial exploiters. As explained in *Business Brief* 34/02,[22] this involved:

---

9   VAT is charged on the tax-inclusive figure (see para. 1.4.3.1 above).
10   See Chapter 16 below.
11   Finance Act 2001, s.19(3)(a),(3A).
12   *Ibid.*, s.17(2)(b),(c),(5).
13   *Ibid.*, s.20(1)(a),(b),(d),(2).
14   *Ibid.*, s.17(2)(a),(3),(4).
15   *Ibid.*, s.19(2)(b),(3)(b).
16   *Ibid.*, s.19(2)(c),(3)(c).
17   And was designed around the waste management regime already in force (see para. 6.3.2 above).
18   Finance Act 2001, s.22(2), charities and local authorities being confirmed as capable of coming within this by amendment in 2002: in this context, it needs to be remembered that, under various authorities relating to VAT, it is clear that individuals and trustees who act as landlords are, *prima facie*, carrying on business.
19   Finance Act 2001, ss.21, 22(1).
20   *Ibid.*, s.22(3).
21   *Ibid.*, s.24(2)(a), 24(6)(ca), Sched. 4, para. 1(1)(a).
22   24 December 2002: see Aggregates Levy (Registration and Miscellaneous Provisions) (Amendment) Regulations 2003, S.I. 2003 No. 465.

a. exemption from all obligations was to be given for those who exploited only:

i. soil, vegetable matter or other organic matter; or
ii spoil, waste or other by-products of any industrial combustion process, or the smelting or refining of metal; or
iii. drill-cuttings from licensed oil exploration; or
iv. arising from roads when utilities' work is carried out; whereas

b. exemption from registration and subsequent compliance, but not notification, is be available when the exploitation is only of:

i. coal, lignite, slate or shale; or
ii. spoil, waste or other by-products resulting from the separation of coal, lignite, slate or shale after extraction; or
iii. spoil, waste or other by-products resulting from the separation of specified minerals after extraction; or
iv. china clay and ball clay and spoil or waste resulting from its extraction; or
v. any other clay.

2 reserving the ultimate right to define the extent of the site;[23] and
3 selecting only one person for registration (which is hardly surprising since controlling the weighbridge is of the essence to compliance), that person being 'the one best placed to do so having regard to existing commercial practice, record-keeping, access to a weighbridge etc'[24] and, in general, not using their powers to amalgamate adjoining sites; while
4 making only very limited use of their power to exclude registrables from registration,[25] initially confining[26] this to the three remaining situations for which exemption was provided under section 17(3)[27] –

a. by subsection (3)(b), removal from the site of a building,[28] to the extent that

---

[23] Finance Act 2001, s.24(7), the intent being to be able to counter avoidance: see Financial Secretary to the Treasury (Mr Stephen Timms, MP), Committee of the Whole House, 23 April 2001, col. 84. The issue is considered in more detail in para. 16.8 below.
[24] This phrase was used to several representative body correspondents in April 2002.
[25] Finance Act 2001, s.24(3).
[26] Aggregates Levy (Registration and Miscellaneous Provisions) Regulations 2001, S.I. 2001 No. 4027, reg. 3(3), repealed, with effect from 1 April 2003, by Aggregates Levy (Registration and Miscellaneous Provisions) Amendment Regulations 2003, S.I. 2003 No. 465.
[27] That is, (b)–(d).
[28] Finance Act 2001, s.17(3)(b). In *East Midlands Aggregates Ltd v. C & E Commrs*, (2003) A1, affirmed by Rimer, J., [2004] STC 1582, this relief was extended to the lorry park adjoining a warehouse and not confined to the footprint of the building and the route of its service pipes. But it was not given for the removal of soil further away in order to provide

      it is in connection with its modification or erection,[29] and exclusively for the purpose of laying foundations, cables or pipes;[30]

  b. by subsection (3)(c), dredging (or other removal from a watercourse);[31] and

  c. by subsection (3)(d), removal from the route of a proposed highway, but only to the extent that for the improvement of that highway.[32]

As indicated in 1 above, with effect from 1 April 2003, the regime was changed, with the (remaining) additional heads of exemption being added on the basis that Customs should be notified instead.[33]

Somewhat surprisingly, the regulations laying down the compliance obligations of registered persons did not exclude the registrables not selected for registration,[34] even though it is difficult to see how they could conceivably check up on the authenticity of figures ascertainable only by the person with control of the weighbridge. It is understood, however, that in practice Customs will not seek to enforce these obligations against other than the registered person.

13.3   Problems caused by the legal structure

Two problems arise from the scatter-gun approach to registrability:

1   where more than one party to a commercial arrangement appears to be within the scope of the statutory criteria, it would be wise for the contractual documentation to reflect which of them, *inter partes*, is intended to bear the financial burdens of both:

  a. the tax itself; and

  b. if an unexpected person is selected as the accountable party, the resulting administrative costs.

    Clearly a specific contractual indemnity will be required in almost all cases. In addition, the tax in question being one on turnover, the question of security (perhaps by way of bank guarantee) will need to be considered in many cases. The credit of components of the construction industry is frequently not of the highest, and the provision of security might well not be practicable in many cases. There will therefore be cases where the existence of the tax will cause some people, and

---

a safe bank for the lorry park. In *Pat Munro-(Alness) Ltd v. C & E Commrs*, (2004) A2, it was confirmed that an all-weather football pitch was not a building.

29   Finance Act 2001, s.17(3)(b)(i).

30   *Ibid.*, s.17(3)(b)(ii).

31   *Ibid.*, s.17(3)(c).

32   *Ibid.*, s.17(3)(d).

33   *Ibid.*, ss.17(3)(e),(f) and (4)(a),(c),(d),(e) or (f): in the substituted Aggregates Levy (Registration and Miscellaneous Provisions) Regulations 2001, S.I. 2001 No. 4027, reg. 3(2)–(3).

34   Aggregates Levy (General) Regulations 2002, S.I. 2002 No. 761, regs 5–11.

especially landowners, to decide against entering into arrangements which might bring with it the prospect of what is, in reality, a guarantor's liability.

With this observation in mind, it is necessary to examine the situations which may lead to registrability. These are set out in Finance Act 2001, s.22(1), which defines[35] who is *responsible for subjecting a quantity of aggregate to exploitation*:[36]

a. where the tax point is ascertained by the removal[37] of aggregate from either its originating site[38] or one in the same registration,[39] this is the *operator* of the site.[40] This concept is a creation of s.21, Finance Act 2001, and includes *each* of:

  i.   *the person who occupies the site*[41] (which Customs have defined in correspondence as to be interpreted 'in its literal sense as physical occupation of the site');[42] and
  ii. another person who exercises any right[43] to win aggregate from land at that site,[44] carry out any exempt process at it[45] or store aggregate at it.[46]

b. where the tax point arises by virtue of removal from a site from which it had originally been expected that an exempt process[47] would be applied to the aggregate[48] *both*:

  i   the *operator* (as in a. above); *and*
  ii. the owner of the aggregate at the time of removal.

c. where it is sold *other than* on its originating site or one in the same registration *or* used for construction purposes, the respective vendor or user.

---

[35] Subject, of course, to the *course or furtherance of a business* requirement in Finance Act 2001, s.22(2).
[36] From this starting point, one arrives at registrability through the following chain: Finance Act 2001, s.19(1), s.24(2)(a), Sched. 4, para. 1(1)(a) and Aggregates Levy (Registration and Miscellaneous Provisions) Regulations 2001, S.I. 2001 No. 4027, reg. 2(1).
[37] Finance Act 2001, s.19(2)(a).
[38] Defined *ibid.*, s.20(1) as (a) won from the seabed, (b) at which an exempt process (under s.18) was applied or (d) the site from which the aggregate was most recently won. Under s.20(2), this can include mixture on that site.
[39] Finance Act 2001, ss.19(2)(b), 22(4).
[40] The concept of *site* is critical to the administration of the tax (*ibid.*, s.24(6)–(8)).
[41] *Ibid.*, s.21(1)(a).
[42] See para. 21.8 below (other possible interpretations).
[43] Finance Act 2001, s.21(1)(b).
[44] *Ibid.*, s.22(2)(a).
[45] *Ibid.*, s.22(2)(c), by reference to s.18.
[46] *Ibid.*, s.22(2)(d).
[47] See para. 13.5 below.
[48] Finance Act 2001, s.19(2)(c).

    d.   Where it is sold *while on* its originating site or one in the same registration, *both*:

        i.   the respective vendor or user; *and*

        ii. the *operator* (as in (a) above).

    e.   Where the aggregate is mixed at premises *other than* its originating site or that in the same registration, *both*:

        i.   the owner of the aggregate at that time; *and*

        ii. the occupier[49] of the premises where the mixing takes place.

    f.   Where the aggregate is mixed *at* the originating site or one in the same registration, *both*:

        i.   the owner of the aggregate at that time; *and*

        ii. the *operator* (as in (a) above).

2   The one major concession given to the industry after the June 1998 Consultation was the ability for a company to move aggregate between its extraction and mixing sites,[50] including those within the same group registration,[51] without incurring a tax point.[52] It will readily be apparent that this commercial advantage could be lost by Customs exercising their discretion as to which of alternative candidates to register adversely to the interests of such a person. This possibility will need to be taken into account:

    a.   in negotiating the terms upon which premises are taken for relevant occupation;[53] but also

    b.   in negotiating the circumstances in which others are permitted to carry out activities at those premises,[54] or are likely to do so at adjoining premises.[55]

Under Finance Act 2001, s.30A, which has been replaced by Finance Act 2004, s.291,[56] a five-year transitional rebate scheme for Northern Ireland was authorised

---

49   Presumably the same meaning applies here as in Finance Act 2001, s.21(1)(a), as to which see above.

50   This was to mitigate the 'backload' empty problem (that is, the return journey without any goods to carry).

51   *General Guide to Aggregates Levy* (January 2002) (current text available from www.hmce. gov.uk), para. 10.3. Note, however, that those who have opted for divisional registration will be limited to inter-divisional transfers (*ibid.*, para. 10.5).

52   Finance Act 2001, s.19(3)(b).

53   *Ibid.*, s.21(1)(a).

54   *Ibid.*, s.21(2).

55   *Ibid.*, s.24(7).

56   As a result of EU authorisation on 7 May 2004, and, following Royal Assent, supplemented by Aggregates Levy (Northern Ireland Tax Credit) Regulations 2004, S.I. 2004 No. 1959.

by the European Commission.[57] Subject to the revised scheme receiving state aid approval, operators registered with Customs who have agreed to abide by the Northern Ireland Code of Practice Compliance Scheme ('COPCS') will retain the original abatement of 80 per cent for processed aggregate and be able (in addition) to obtain this for virgin aggregate used in its raw state, until 31 March 2012. Aggregate transferred to Great Britain will not qualify for relief, unless for export, when it will be possible to claim a 100 per cent rebate instead. Imported aggregate will not qualify for relief. Within a year of joining the COPCS, the operator will be obliged to commission an environmental audit to record the base line for the site. The Department of the Environment of Northern Ireland[58] will set improvement targets and carry out verification audits to monitor performance. In the event of EU consent not being obtainable, the UK Government will seek the Commission's agreement to the continuation of the original 80 per cent rebate level. Under the original scheme, this was destined to be reduced to 60 per cent on 1 April 2004.[59]

13.4   Meaning of 'aggregate'

Between the time of the June 1998 Consultation and the Budget 2002 amendments to Finance Act 2001, the subject-matter of the tax underwent considerable revision.[60]

Aggregate is any rock, gravel or sand, with any substances incorporated or naturally mixed with it,[61] including spoil, waste, off-cuts and other by-products resulting from the application of any exempt process[62] to it,[63] but not anything else resulting from the application of that process.[64]

But aggregate is exempt as such if:

1   it is wholly or mainly, or part of, coal, lignite, slate, shale,[65] spoil, waste or other by-products from an industrial combustion process or the smelting or refining of metal,[66] Continental Shelf drill cuttings,[67] anything resulting from works carried out under the New Roads and Street Works Act 1991 (and equivalent Northern Irish provisions),[68] and clay, soil, vegetable or other organic matter;[69] or

---

[57]   *Business Brief* 13/04 (10 May 2004).
[58]   See para. 4.2.2 above.
[59]   CE11 (17 March 2004) (see also para. 21.8 below).
[60]   To avoid confusion in this chapter, an examination of the genesis of the final list history has been deferred until para. 21.8 below.
[61]   Finance Act 2001, s.17(1).
[62]   See para. 13.5 below.
[63]   Finance Act 2001, s.18(1)(a).
[64]   *Ibid.*, s.18(1)(b).
[65]   *Ibid.*, s.17(4)(a).
[66]   *Ibid.*, s.17(4)(c).
[67]   *Ibid.*, s.17(4)(d).
[68]   *Ibid.*, s.17(4)(e).
[69]   *Ibid.*, s.17(4)(f).

2   it comes *wholly* from the excavation of a building site in connection with the
    modification or erection of a building or the laying or foundations or services,[70]
    or from the bed of a river, canal or harbour approach in the course of dredging
    undertaking for improvement or maintenance,[71] or in the course of highway
    construction,[72] or from the spoil or waste of china clay or ball clay (other than
    overburden),[73] or the spoil of coal, lignite, slate, shale, anhydrite, ball clay,
    barytes, china clay, feldspar, fireclay, fluorspar, fuller's earth, gems and semi-
    precious stones, gypsum, any metal or its ore, muscovite, perlite, potash, pumice,
    rock phosphates, sodium chloride, talc or vermiculite.[74]

13.5   The concept of 'exempt processes'

An *exempt process* consists of one of the following:[75]

1   the cutting or any rock to produce stone with one or more flat surfaces –
    originally this was called dimension stone, the idea being to exempt stone used
    for monumental masonry or the facing of historic buildings, but not off-cuts;
2   any process by which one of the following is extracted or separated from any
    aggregate: anhydrite, ball clay, barytes, china clay, feldspar, fireclay, fluorspar,
    fuller's earth, gems and semi-precious stones, gypsum, any metal or its ore,
    muscovite, perlite, potash, pumice, rock phosphates, sodium chloride, talc or
    vermiculite;[76] or
3   any process for the production of lime or cement either from limestone alone or
    limestone and anything else.

Carrying on an exempt process means that tax is not charged by reference to what
which is being done on the site.

13.6   Administration of the levy

The administrative provisions for aggregates levy, climate change levy and landfill
tax share certain common features. Please refer, therefore, to Chapter 16, where they
are discussed together.

---

[70]   Finance Act 2001, s.17(3)(b).
[71]   *Ibid.*, s.17(3)(c).
[72]   *Ibid.*, s.17(3)(d), unless aggregate extraction is the main purpose.
[73]   *Ibid.*, s.17(3)(e).
[74]   *Ibid.*, s.17(3)(f), together with future substances added by Treasury order under s.18(4).
[75]   *Ibid.*, s.18(2).
[76]   *Ibid.*, s.18(3), together with future substances added by Treasury order under s.18(4).

Chapter 14

Climate Change Levy

14.1   Introduction

Climate change levy was the second environmental tax to be placed under the jurisdiction of HM Customs and Excise.[1] It was introduced to assist the government in meeting its emissions reduction target under the Kyoto Protocol.[2]

Despite its being preceded by a specially-commissioned report,[3] the tax's introduction may, however, be considered premature, because it preceded the formation of a proper energy policy.[4] Also, despite the fact that transport accounts for 34 per cent and domestic consumption for 29 per cent of final energy use,[5] in part for political reasons, the scope of the tax is confined to business users.[6]

Moreover, in the context of the post-nationalisation structure of the electricity supply industry,[7] this has meant that the effect of the concessions in favour of fuel sources which are for some reason desirable[8] has been diluted by a complicated apportionment process under which:[9] there are three month balancing periods; and averaging periods which normally run for two years.

---

[1]   The first such tax so to have been entrusted to Customs (see para. 4.2.1.2(2) above) having been landfill tax in 1996 (see para. 15.1 below).

[2]   See para. 8.3.1.4 above.

[3]   Lord Marshall, *Economic Instruments and the Business Use of Energy* (HM Treasury, 1998) (see para. 11.3.1 above).

[4]   See House of Commons Environmental Audit Committee, *8th Report. Energy White Paper – Empowering Change?* (House of Commons Papers, Session 2002–2003, HC 618) (London: Stationery Office, 2003), published on 22 July 2003. The Committee said, in para. 10: 'The Energy White Paper represents a major shift in the approach to UK energy strategy'. See para. 21.6.2 below.

[5]   The former is understandable, in the sense that transport taxation has always been segregated, albeit in some respects, for favoured treatment. The latter is a serious lacuna, especially since domestic energy consumption has increased by 25 per cent since the early 1970s (see Benjamin J. Richardson and Kiri L. Chanwai, 'The UK's Climate Change Levy: Is It Working?' [2003] JEL 39–58).

[6]   See para. 21.6.2 below.

[7]   See paras. 2.4.1 and 6.4.3.1 above and 21.7 below.

[8]   See para. 14.4 below.

[9]   Finance Act 2002, Sched. 6, para. 20.

14.2   Tax base and rates

Where the full rate of levy applies, it is collected from the supplier and payable at different poundages,[10] depending on the type of energy supplied, that is:[11]

electricity at	£0.0043 per kilowatt hour;
gas supplied by a utility at	£0.0015 per kilowatt hour;
liquid natural gas	£0.0096 per kilogram; and
any other substance	£0.0117 per kilogram.

The tax applies to supplies of taxable commodities, that is, electricity, gas, liquid natural gas, coal, lignite and coke, but not to hydrocarbon oil or road fuel gas or waste within the scope of Environmental Protection Act 1990.[12] Within this general ambit, the levy is confined to supplies of:

a.   electricity by a utility to a non-utility,[13] although the position of combined heat and power ('CHP') stations is complex;[14]
b.   gas by a utility to a non-utility;[15] and
c.   other commodities, provided that this is in the course or furtherance of a business.[16]

There are also concepts of:

a.   self-supply,[17] so that electricity utilities and CHP operators are normally brought into charge in respect of their head office complexes, but non-CHP producers of electricity largely for their own consumption are not;

---

[10]   *Government Response to the Environment, Transport and Regional Affairs Committee's Report: UK Climate Change Programme*, 8 August 2000, para. 98, said: 'The initial rates of climate change levy will be based on the *energy* content of the different energy products. Current electricity pool arrangements means that is [*sic*] only possible to determine *carbon* content of electricity as broad average with a "downstream" tax. However, the government has said that it will keep under review the basis for setting climate change levy rates, in light [*sic*] of developments in electricity trading arrangements'. See para. 6.3.2.1 above.
[11]   VAT is payable, in addition, on the levy-inclusive price (see para. 1.4.3.1 above).
[12]   Finance Act 2000, Sched. 6, para. 3. See para. 6.3.2.1 above.
[13]   *Ibid.*, Sched. 6, paras. 5, 150(2); Climate Change Levy (Electricity and Gas) Regulations 2001, S.I. 2001 No. 1136: for the underlying regulatory regime, see paras. 2.4 and 6.4.3 above and 21.7 below.
[14]   See para. 14.5 below.
[15]   Finance Act 2000, Sched. 6, para. 6; Climate Change Levy (Electricity and Gas) Regulations 2001, S.I. 2001 No. 1136.
[16]   Finance Act 2000, Sched. 6, para. 7.
[17]   *Ibid.*, Sched. 6, paras. 17, 23(3), 152; Climate Change Levy (General) Regulations 2001, S.I. 2001 No. 838, reg. 41.

b. deemed supply,[18] that is, self-use,[19] a tag which can be used to allow landlords supplied by utilities some but not all of whose tenants are exempt to achieve the intended end result for their tenants;[20]
c. special utility schemes, which can amalgamate electricity and gas supplies where Customs agree that this is reasonable;[21] and
d. recipient accountability where energy is received from a non-resident non-utility.[22]

14.3   Reduced rates

Under transitional arrangements made with the European Commission:[23]

1. horticultural producers will pay half rate for a period of five years;[24] and
2. installations within the IPPC[25] concept were able to get reductions of 80 per cent for ten years,[26] provided that they entered into climate change agreements;[27] and
3. natural gas for burning in Northern Ireland is exempted on a temporary basis,[28] a concession which was later mirrored for aggregates levy.[29]

14.4   Exemptions

The following are exempt:

1   gas supplies for burning outside the UK[30] – this is logical because fuel oil exports

---

[18]   Finance Act 2000, Sched. 6, para. 23(1),(2); Climate Change Levy (General) Regulations 2001, S.I. 2001 No. 838, reg. 52.
[19]   Except for CHP where for own use only: Financial Secretary to the Treasury (Mr Stephen Timms, MP), Committee of the Whole House, 2 May 2000, col. 89. Some such businesses may be able to claim auto-generator exemption: see Charles Yorke, [2002] IELTR 210.
[20]   *Technical Briefing* No. 8 (March 2001) (available from www.hmce.gov.uk): the alternatives suggested officially are arranging dedicated supplies direct from the utility or (if the relief is small) making a liability sharing agreement with the tenants: but see para. 21.7 below.
[21]   Finance Act 2000, Sched. 6, para. 29; Climate Change Levy (General) Regulations 2001, S.I. 2001 No. 838, reg. 54.
[22]   See para. 16.7 below.
[23]   See para. 21.10 below.
[24]   Finance Act 2000, Sched. 6, para. 43.
[25]   See the Integrated Pollution Prevention and Control Directive (Council Directive 96/61/EC, (1996) OJ L257 26). See paras. 6.2.3 and 12.2.2 above.
[26]   Finance Act 2000, Sched. 6, paras. 44–51.
[27]   See para. 14.6 below.
[28]   Finance Act 2000, Sched. 6, para. 11A.
[29]   Finance Act 2001, s.30A: see paras. 13.3 above and 21.8 below.
[30]   Finance Act 2000, Sched. 6, para. 11.

are not within the scope of the tax, it may also be seen as the equivalent of the (controversial)[31] export rebate given for aggregates levy;

2   supplies for use in public transport, on the railways, or international shipping[32] – public transport is to be encouraged in order to get cars off the road;

3   direct[33] supplies for domestic or non-business charitable use[34] – the latter is consistent with government support of the charitable sector;[35]

4   supplies to:

   a.   producers of taxable commodities other than electricity,[36] except for headquarters operations;[37] or

   b.   other than for use as fuel,[38] being the electrolytic processes, steam reformation, dual use functions and non-heating uses specified in regulations,[39] and which the European Commission confirmed on 4 April 2002, could be extended (retrospectively) to recycling;[40]

   c.   with effect from 29 January 2003,[41] there were included additional dual use functions and non-heating uses connected with non-ferrous blast furnaces, waste combustion, the re-carburising of iron and steel, carbon black production, the manufacture of titanium oxide and the production of lower olefins; and

   d.   following the enactment of Finance Act 2003, seemingly in order to facilitate (c), the supply of a commodity is exempt if the person supplied intends it to be used as fuel in a recycling process for which there is a competing process which is not a recycling process using taxable commodities other than as fuel and producing a product of the same kind as that produced by the recycling process with a greater amount of energy involving a lower levy charge.[42]

Paragraphs 4a. to 4d. above thus confine the tax to its intended objective.

---

31   See para. 21.8 below.

32   Finance Act 2000, Sched. 6, para. 12, as to which note the similarity to Value Added Tax Act 1994, Sched. 8, Group 8, item 4(a).

33   This seems to be the position following the VAT case concerning a 1994 pre-payment scheme subsidiary for Bristol University, *Oval (717) Ltd v. C & E Commrs*, LON/01/1070.

34   Finance Act 2000, Sched. 6, paras. 8–9, as to which again note the similarity to Value Added Tax Act 1994, Sched. 7A, Group 1, Notes 3–5.

35   The former being political (see para. 21.6.2 below).

36   Finance Act 2000, Sched. 6, para. 13. This includes exploration, but so that energy used at exploration sites is not relieved: *Technical Briefing* No. 12 (October 2001) (available from www.hmce.gov.uk).

37   Climate Change Levy (General) Regulations 2001, S.I. 2001 No. 838, reg. 40.

38   Finance Act 2000, Sched. 6, para. 18.

39   Climate Change Levy (Use as Fuel) Regulations 2001, S.I. 2001 No. 1138.

40   *Technical Briefing* No.16 (16 June 2002) (available from www.hmce.gov.uk); *Business Brief* 18/02 (8 July 2002).

41   By ESC – see *Business Brief* 02/03 (6 February 2003) – until 13 March 2003, when Climate Change Levy (Use as Fuel) (Amendment) Regulations 2003, S.I. 2003 No. 665, came into effect.

42   Finance Act 2000, Sched. 6, para. 18A. See Troup, [2003] BTR 416.

5   Although wholesale supplies of taxable commodities are not normally within the scope of climate change levy,[43] such a situation can arise in the following circumstances:[44]

   a. where a gas shipper without a supply licence sells natural gas to a 'burner';
   b. where a gas producer burns it;
   c. where an electricity producer consumes it;
   d. where a gas utility burns it; or
   e. where an electricity utility consumes it.

6   The creation of new biofuels, for example, biodiesel and bioblend, has taken effect from Royal Assent to Finance Bill 2004.[45] The supply of products used in the creation of bioethanol will take effect from 1 January 2005, when the new rate of excise duty is introduced.[46]

7   Renewable source and 'good quality' CHP[47] electricity is also exempted: see para. 14.5 below.

The requirement for certification[48] can, however, give rise to administrative problems where the person to whom energy is supplied by the utility is the landlord of the end user, since it is only the landlord who can provide the utility with the relevant certificate.[49] Following the enactment of Finance Act 2003, provision can be made for situations in which there has been a change of circumstances.[50]

14.5   Combined heat and power and renewable source electricity

Not all actually renewable[51] sources of energy qualify for relief. Large hydro-electric schemes have been excluded because all major UK sources have already been exploited.[52] Nuclear energy is also excluded, presumably because this has been a political hot potato for decades and, at the time the tax was being introduced, no decision had been taken to proceed with the erection of third generation capacity. It

---

43   *Ibid.*, Sched. 6, para. 14.
44   *Technical Briefing* No.11 (March 2001) (available from www.hmce.gov.uk).
45   22 July 2004.
46   Finance Act 2000, Sched. 6, paras. 13 and 13A: CE20 (17 March 2004). For the excise situation, see para. 22.2 below.
47   Climate Change Levy (Combined Heat and Power Stations) Prescribed Conditions and Efficiency Percentages Regulations 2001, S.I. 2001 No. 1140.
48   Climate Change Levy (General) Regulations 2001, S.I. 2001 No. 838, reg. 34. But, by ESC, from 1 May 2002, that which may have been claimed during 2001/02 could be claimed within 2002/03. Amendments to Finance Act 2000, Sched. 6, paras. 15 and 148 enable changes of methodology to be prescribed following the enactment of Finance Act 2003.
49   *Technical Briefing* No. 8 (March 2001) (available from www.hmce.gov.uk).
50   Under the substituted Finance Act 2000, Sched. 6, para. 24.
51   As defined by Utilities Act 2000, s.50.
52   Notice CCL 1/4 (June 2004), para. 2.2n (available from www.hmce.gov.uk).

is, however, in line with the EU approach of excluding nuclear from the definition of renewables.[53] Fossil fuels are obviously excluded, but waste from them can sometimes qualify if the fossil content is below a threshold.[54]

As has been indicated above, the generator/REC[55] method of supplying electricity under the post-nationalisation structure[56] means that the effect of the exemption for renewables or other 'desirable' sources, for example, methane from coal mines,[57] is lost in part:

1   because it is only direct supply contracts to commercial users (and normally only very large ones) which will be capable of full exemption; and[58]
2   where electricity goes through the REC to a mixture of business and residential addresses, the effect of (in effect) exemption at the previous level is diluted.[59]

This gives rise to:

1   the need to ascertain whether the supplier is fully exempt (especially in relation to CHP);[60] and
2   if this is not the case, to calculate the degree of exemption by reference to three monthly balancing and two yearly averaging periods,[61] with Customs obtaining additional powers in 2002 to enable collection to be made if a CHP generator

---

53   See para. 12.2.6.3(2) above.
54   See CCL 1/4 (June 2004), para. 2.1, by reference to Climate Change Levy (General) Regulations 2001, S.I. 2001 No. 838, regs 46–51.
55   Regional Electricity Companies, into which the old area boards were injected at the time of privatisation into which the old area boards were injected at the time of privatisation (see para. 2.4.1 above).
56   See paras. 2.4.1 and 6.4.3.1 above.
57   Announced in Budget 2002, subject to European Commission confirmation: HMT 2 (17 April 2002); this is *treated as if* a renewable source under Finance Act 2000, Sched. 6, para. 19(4A) – by the autumn, however, the fall in electricity prices under NETA had been such that such projects were not deemed to be viable. On 9 April 2003, European Commission permission had still not been obtained, but the government were 'optimistic' that it would be [see Treasury Budget 2003, para. 7.13]. Eventually, this addition was brought in with effect from 1 November 2003: see Climate Change Levy (General) (Amendment) (No. 2) Regulations 2003, S.I. 2003 No. 2633.
58   CCL 1/4 (June 2004), para. 5.1, indicates that the prices in such contracts are normally expressed on a free-of-levy basis, moving the risk of non-compliance onto the supplier.
59   Nonetheless, until 31 March 2003, CHP supplies to a utility or for domestic or non-business charity use were disregarded and the electrical equivalent of any mechanical output for non-generation purposes taken into account: Climate Change Levy (Combined Heat and Power Stations) Prescribed Conditions and Efficiency Percentages Regulations 2001, S.I 2001 No. 1140, reg. 5(3),(4), repealed by Climate Change Levy (Combined Heat and Power Stations) Prescribed Conditions and Efficiency Percentages (Amendment) Regulations 2003, S.I. 2003 No. 861. For consequent subsequent amendments: see CE 4 and CE 5 (9 April 2003).
60   As to which Customs took additional powers to those of disclose in Finance Act 2000, Sched. 6, paras. 19(8) and 20A(8), to direct certification through the Gas and Electricity Markets Authority by the inserted Finance Act 2000, Sched. 6, para. 149A, in 2002.
61   Finance Act 2000, Sched. 6, paras. 20, 20B.

ceases to make such supplies rather than ceases to be a supplier of any taxable commodity;[62] but

3   where self-supply is concerned, CHP is exempt in proportion to qualifying input.[63] Where the supplier produces largely for himself, the normal exemption is, however, replaced in the case of partially exempt CHP where the threshold set by regulations has been exceeded;[64]

4   In the case of a partially exempt CHP producer, the output is nonetheless exempt if it is below the threshold set in regulations;[65]

5   Where supplies other than self-supplies are made to CHP stations a proportion of the supply is exempt, the fraction being determined by Treasury regulations.[66] Under the pre-2002 system, exemption was confined to *good quality* CHP.[67]

Where exemption is to be obtained on the basis that it has been supplied by a renewable source, or, from 1 April 2003,[68] a CHP source, a declaration to that effect must be made by the supplier,[69] the electricity of any other supplier used under contract must confirm to Customs that the relevant conditions have been applied,[70] and the electricity must be generated, as the case may be, from a renewable source in a manner prescribed by Customs,[71] or by CHP from a non-renewable source.[72]

The essence of the scheme applicable to CHP stations which took effect from 1 April 2003, is that:[73]

1   the power station must be certified by Defra (one of the pre-conditions being that it may not receive state aid in excess of the level specified in the guidelines issued the European Commission on 3 February 2003);

2   the station makes monthly returns to its regulator (Ofgem in England and Wales), which issues certificates (CHP levy exemption certificates ('LECs')) two monthly in arrears; and

---

62   Budget Day Press Release, CE21 (17 April 2002).
63   Finance Act 2000, Sched. 6, para. 17(3),(4).
64   *Ibid.*, Sched. 6, para. 17(2), 152.
65   *Ibid.*, Sched. 6, para. 16(2).
66   *Ibid.*, Sched. 6, para. 15.
67   CCL 1/2 (February 2002) (available from www.hmce.gov.uk); *Technical Briefing* No. 15 (reissued June 2002) (available from www.hmce.gov.uk).
68   Finance Act 2002, s.123, (Appointed Day) Order 2003, S.I. 2003 No. 603.
69   Finance Act 2000, Sched. 6, paras. 19(1)(b),(2), 20A(1)(b),(3): certain conditions also have to be fulfilled for a period of time, see *ibid.*, paras. 19(1)(c),(5)–(7), 20A(1)(c),(4)–(7), revised by Finance Act 2003 to permit averaging from 1 April 2003.
70   Finance Act 2000, Sched. 6, paras. 19(1)(d), 20A(1)(d).
71   *Ibid.*, Sched. 6, para. 19(3),(4).
72   *Ibid.*, Sched. 6, para. 20A(4); Climate Change Levy (General) Regulations 2001, S.I. 2001 No. 838, regs 51A–51M, Sched. 2 (inserted by Climate Change Levy (General) (Amendment) Regulations 2003, S.I. 2003 No. 604. As most CHP is used by its generator, it has been questioned how much effect this measure will have on the UK's progress towards its Kyoto targets: see Charles Yorke, [2002] IELTR 210.
73   *Technical Briefing* No. 18 (March 2003, revised March 2004) (available from www.hmce. gov.uk).

3   the station's combined heat and power quality assurance ('CHPQA') certificate
    is issued by Defra annually in arrears;[74]
4   it is permissible to 'trade' CHP LECs independently of the physical route taken
    by the physical supply, or, as Customs put it originally: '[t]his means LECs may
    be issued for suppliers to consolidators or into the imbalance market'.[75] Levy
    exemption certificates have, however, to be allocated in accordance with the
    outputs record.[76]

14.6   Climate change agreements and the UK Emissions Trading Scheme

The 80 per cent levy reduction arrangements are confined to those industries within
the Integrated Pollution Prevention and Control ('IPPC') concept.[77] This was,
initially, only discernable by a process of deduction,[78] but the list in the statute[79] was
replaced (by statutory instrument) before the tax came into operation by one which
incorporated the (amended) statutory instrument incorporating the IPPC concept into
UK law following the gradual replacement of Part I of the Environmental Protection
Act 1990 by the Pollution Prevention and Control Act 1999.[80]

In order to benefit from the 80 per cent levy reduction, an electricity consumer
has first to provide his supplier with Customs' certificate that he has entered into a
climate change agreement.[81] For the first agreement, it was of the essence that this
certificate be served before the reduced rate could be applied.[82] However, where
occupation of the site changes, the new occupier's accession can be backdated.[83]

---

[74] See para. 4.2.1.2(1) above.
[75] Technical Briefing No. 18 (March 2003, revised March 2004) (available from www.
      hmce.gov.uk), para. 2.10 (2003 version): as to the legality of this activity and the
      distinction between it and trading in Green Certificates ('or ROCs'), see para. 20.6
      below.
[76] Ibid., para. 2.9 (2004 version). Although not acknowledged, the reason for this is likely
      to have been the Cogeneration Directive (European Parliament and Council Directive
      04/8/EC on the Promotion of Cogeneration based on a Useful Heat Demand in the
      Internal Energy Market and Amending Directive 92/42/EEC, (2004) OJ L52 50). See
      para. 12.2.6.3(2) above.
[77] See para. 6.2.3 above. This was for state aid reasons (see para. 12.2.7 above) and consent
      is to be sought, during 2004, to the extension of the scheme to other sectors (see Business
      Brief 27/2003 (10 December 2003), and para. 21.10 below).
[78] The debate initiated by the government concerned heavy users of electricity, which is not
      necessarily the same thing as heavy polluters.
[79] Finance Act 2000, Sched. 6, para. 51.
[80] Pollution Prevention and Control (England and Wales) Regulations 2000, S.I. 2000 No.
      1973, am. S.I. 2001 No. 503.
[81] Finance Act 2000, Sched. 6, para. 44, variation notices come within para. 45.
[82] Ibid., Sched. 6, para. 49(1).
[83] Ibid., Sched. 6, para. 49(2). Although this does not appear to require adherence to the same
      umbrella agreement, clearly difficulties could arise if the successor was not a member of
      the trade association, for example, a landlord taking over on an insolvency (see para. 16.6
      below).

There are few direct agreements.[84] Most participants are signatories to an 'underlying agreement' supplemental to an 'umbrella agreement' made between the (then) DETR[85] and their trade association.[86] Moreover, because of the delay in completing negotiations with the trade associations,[87] it will usually be found to be the case that the underlying agreements pre-date the umbrella agreements and provide for the trade association to divide out the reduction target for that particular industrial group subsequently.

The umbrella agreements contain obligations on underlying agreement signatories to comply with certain procedures. These may seem odd in some cases, for instance solo intensive pig or battery hen enterprises. They will include, however, monitoring compliance against an agreed plan.[88] That plan will, in turn, in reality be governed by the signatory's obligations under the Pollution Prevention and Control (England and Wales) Regulations 2000,[89] through:

1 the best available techniques ('BAT')[90] being phased in for existing installations over the period 2001 to 2007,[91] as against the previous standard of best available techniques not entailing excessive cost ('BATNEEC');[92] but

2 the changes required will, to some extent, be site-specific after taking account of:

   a. the technical considerations of the installation;
   b. the local geography; and
   c. the environmental considerations for the area; together with

3 due consideration of energy efficiency, for which guidance material has been published.

Although the targets in the agreements run for ten years, the scheme operates on the basis of two-year certification periods,[93] running to 30 September.[94] But compliance is only required to be verified in the second year of each period.[95] Verification has to

---

84 Under Finance Act 2000, Sched. 6, para. 47. Some of them are joint ones involving groups of retailers.
85 See para. 4.2.1.2(1) above.
86 Under Finance Act 2000, Sched. 6, para. 48.
87 Called 'sector associations' for this purpose.
88 Account also needs to be taken of Climate Change Levy (General) Regulations 2001, S.I. 2001 No. 838, reg. 44, with regard to the delivery of taxable commodities.
89 S.I. 2000 No. 1973, am. S.I. 2001 No. 503. See para. 6.2.3 above.
90 That is, the IPPC concept (see paras. 6.2.3 and 12.2.2 above).
91 Under the Pollution Prevention and Control Act 1999. See para. 6.2.3 above.
92 Under the Environmental Protection Act 1990, Part I. See para. 6.2.2 above.
93 Finance Act 2000, Sched. 6, para. 49.
94 Which has to be compared to direct participants under the UK Emissions Trading Scheme ('UK ETS') participants operating by reference to calendar years (see para. 20.5 below).
95 The first ran from 1 October 2001 to 30 September 2002, even though the start date was 1 April 2001.

be provided, through the trade association, by 31 January following.[96] (It is worthy of note, in this context, that the Regional Development Agencies,[97] who were asked, for the first time, to become involved in the Budget process in 2003, recommended that the target-setting and verification processes need to be made more cost-effective and less burdensome.)[98]

The sanction for non-compliance is exclusion from relief for the subsequent two-year period, but:

1   Participants who do especially well as against their targets may sell their excess pollution capacity on the market set up pursuant to the UK Emissions Trading Scheme ('UK ETS').[99]
2   An incentive to do this is the extraordinary arrangement under the Umbrella Agreement system under which the undershoot by one or more signatories to the Underlying Agreement is first offset against overshoots by other such signatories within the scope of that particular sector, that is, individual Umbrella Agreement.[100]
3   Those who have overshot their targets may avoid disqualification for the subsequent two-year period by buying-in capacity on the trading market.[101]
4   It appears that, if an 'over-shooter' has in fact sold in the market, the position of the sector as a whole, that is, all those covered by the same Umbrella Agreement, may be affected adversely.
5   If the sector as a whole meets its target, the individual participant may carry forward his personal surplus to 2007,[102] when the initial scheme had been expected to be replaced by at least a European one.[103]

In Budget 2004, the government made known its intention to seek Community state aid approval[104] to extend the scope of climate change umbrella agreements outside

---

[96]   Which has to be compared to 31 March for direct participants (see para. 20.5 below).
[97]   These are bodies, such as the South East Regional Development Agency, which promote economic activity, for example, by taking small roles in promoting the development of 'brown' and 'green' field sites which might be difficult to develop without some form of official support.
[98]   Budget Day Press Release, PN 02 (9 April 2003).
[99]   See para. 20.4 below, and also the government's intentions as to the possibility of switching to the EU Emissions Trading Scheme with a similar climate change levy discount in 2005: *Business Brief* 27/2003 (10 December 2003).
[100] Note, in this context, that a particular trade association may be a party to more than one umbrella agreement, for example, the National Farmers Union.
[101] See para. 20.4 below.
[102] With the arrival of the EU Emissions Trading Scheme ('the EU ETS') in January 2005, the opportunity is likely to be given to climate change agreement holders to switch to the EU scheme: *Business Brief* 27/2003 (10 December 2003). In the same breath, however, it was made known that consideration was being given to extending the climate change agreement scheme to industries not within IPPC. It was not altogether clear how these two changes were considered to be consistent (see Jeremy de Souza, *The Report* (the journal of the City of Westminster and Holborn Law Society), March 2004, p. 5).
[103] See para. 20.3 and Chapter 28 below.
[104] See para. 12.2.7 above and 21.10 below.

those industries within the IPPC criteria.[105] To qualify, businesses will have to be above a threshold of energy intensity,[106] measured for the purposes of the Energy Products Directive[107] as one whose 'purchases of energy products and electricity amounted to at least 3 per cent of the production value', under a one-off four-year test (the lowest year being discounted).[108]

All businesses meeting or exceeding a 12 per cent threshold would be eligible to enter into climate change agreements. But where the 3 per cent level had been exceeded, but the 12 per cent level not breached, qualification would only be possible if one of two additional criteria were also present:[109]

1   a percentage of imports to home demand[110] of 50 per cent; or
2   a ratio of exports to total manufacturers' sales of 30 per cent.[111]

14.7   Administration of the levy

Since the model on which climate change levy is administered shares a number of features with those applicable to aggregates levy and landfill tax, the administrative provisions for all three taxes are discussed together, in Chapter 16 below.

---

[105]   Budget Day Press Release, CE19 (17 March 2004). In *Business Brief* 19/04 (23 July 2004), applications for new entrants were requested on a special spreadsheet by 31 August 2004. It was hoped that new members would be able to participate from 1 January 2005 (when the EU ETS starts), but this was subject to Defra being able to obtain the necessary state aid approval by then (see para. 12.2.7.2 above).

[106]   This was the criterion the UK Government would have liked to use when the rebate scheme was first devised, before being obliged by the European Commission to fall back on the IPPC test, thus enabling certain businesses which would not normally be associated with heavy energy use (for example, intensive farming) to qualify.

[107]   Council Directive 03/96/EC, (2003) OJ L283 51. See para. 12.3.4 above.

[108]   Budget Day Press Release, CE19, para. 4.

[109]   *Ibid.*, CE19, para. 6.

[110]   Total manufacturers' sales plus imports minus exports.

[111]   These complications were the subject of criticism from the Engineering Employers' Federation: Houlder, *Financial Times*, 18 March 2004, p. 12.

Chapter 15

Landfill Tax

15.1 Introduction

The decision to introduce a tax on landfill was announced in the 1994 Budget.[1] The extent of the UK's waste disposal problem at that date was made know in a Treasury Press Release. The annual production of controlled waste[2] was 140m tonnes, of which 70 per cent (a higher proportion than in other countries) was being disposed of to landfill. Of this, 15m tonnes was of commercial origin and 35m from the construction industry. The major problem was therefore domestic, a service undertaken by local authorities, and paid for through the local taxation system currently in force.[3]

The tax[4] was introduced from 1 October 1996, pursuant to Finance Act 1996. It took a form similar to VAT, albeit with credits being limited to reuse, recycling or incineration.[5] This was the first of the environmental taxes placed under the management of HM Customs and Excise.[6]

15.2 Tax base and rates

Taxation is by weight,[7] determined in accordance with regulations, the matter being complicated by water content.[8] For this purpose, it is permitted to discount:[9]

1 non-natural water added:

    a. for extraction; or
    b. for transportation purposes; or
    c. if it amounts to 25 per cent, it has been added in the course of an industrial process; or

2 which constitutes the residue, not exceeding the original water content, of a water treatment works unless the water contains pollutants liable to leach from the landfill site.

---

[1] 29 November 1994.
[2] See para. 6.3.2.1 above.
[3] See paras. 2.3 above and 21.4.1 below.
[4] On which VAT is charged on the inclusive price (see para. 1.4.3.1 above).
[5] See para. 16.9 below.
[6] See para. 4.2.1.2(2) above.
[7] Finance Act 1996, s.68.
[8] Landfill Tax Regulations 1996, S.I. 1996 No. 1527, regs 41–44.
[9] *Ibid.*, reg. 44(5)–(8).

The rate of tax was set at £7 per tonne for general waste. This was increased to £10 per tonne on 1 April 1999 and by £1 each year until £15 is reached in 2004.[10] It was announced on 27 November 2002, that the rate was to be set to reach £35, by instalments of at least £3 per annum, starting in 2005.

Inert waste is taxed at £2 per tonne.[11] Where mixed loads are presented at a landfill site, Customs are empowered to give directions[12] (which power has been exercised)[13] as to how those loads are to be treated where the quantity of non-qualifying material is 'small'.[14] This has been held to mean up to 5 per cent, weight being only one factor to be taken into account in assessing whether this level has been exceeded.[15]

The intention behind the extensive definition of *inert waste* given in the current Treasury Order[16] is that this is intended to be confined to *waste that does not physically or chemically react, biodegrade or adversely affect other matter with which it comes into contact in a way likely to give rise to environmental pollution.*[17] The substances entitled to the lower rate are:

1   naturally occurring rocks and soils, which Customs regard as including clay, sand, gravel, sandstone, limestone, crushed stone, china clay, clean building or demolition stone (for example, sandstone, limestone or slate), topsoil, peat, silt and dredgings;
2   ceramic or concrete materials, which are treated as comprising glass, ceramics, bricks, tiles, clay ware, pottery, china, bricks and mortar, reinforced concrete, concrete blocks, breeze blocks and thermalite blocks;
3   process or prepared minerals which have not been used, that is, moulting sands and clays, clay and other absorbents, fuller's earth, bentonite, glass and other man-made fibre, silica, mica and abrasives which have been extended[18] to include foundry sand;
4   furnace slags;
5   bottom ash and fly as from wood or coal combustion (including pulverised fuel ash from the latter);
6   low activity inorganic compounds, that is, titanium dioxide, calcium carbonate, magnesium carbonate, magnesium oxide, magnesium hydroxide, ferric oxide, iron hydroxide, aluminium oxide and hydroxide and zirconium dioxide; and
7   provided that they are disposed of in a site licensed to take only inactive or inert waste,[19] gypsum and plaster (but not plasterboard).

---

10   As to the inadequacy of these rates, see para. 21.4.1 below.
11   Finance Act 1996, s.42(2),(3),(4). This was, however, adjudged to have been a success in House of Commons Environment, Transport and Regional Affairs Select Committee, *13th Report. The Operation of the Landfill Tax* (House of Commons Papers, Session 1998–1999, HC 150–I) (London: Stationery Office, 1999), para. 21.
12   Finance Act 1996, s.63.
13   Currently in Notice LFT 1 (available from www.hmce.gov.uk).
14   Finance Act 1996, s.63(2).
15   *Cleanaway Ltd v. C & E Commrs*, (2003) L.17.
16   Landfill Tax (Qualifying Material) Order 1996, S.I. 1996 No. 1528.
17   Customs Press Release 60/95 (15 December 1995).
18   Albeit by News Release 48/96 (18 September 1996).
19   The power to impose conditions is given by Finance Act 1996, s.63.

15.3   Securing the tax base

Tax is charged on a taxable disposal,[20] which takes place when material is disposed of as waste,[21] by way of landfill,[22] at a landfill site[23] at the time the land is question is within this category.[24]

The accountable party is the landfill site operator of that land at that time.[25] That person is normally the holder of the waste disposal licence for that particular area of land.[26] At the time the tax was introduced, it was thought that the need for such a licence to be held identified with precision the only person who needed to be held to account for the tax. However, it appears that evasion was taking place and Customs therefore thought it necessary to introduce the concept of secondary liability in 2000.[27] A person within the relevant description (called 'the controller') is obliged to identify himself and is, in effect, a guarantor without statutory right of recourse of the 'extractor's' (turnover) tax liability for the site.[28] There is no necessity for there to be a connection between the controller and the licence holder and, indeed, it is possible that some independent landlords could come within the scope of the definition.[29]

A disposal of material as waste takes place when the person initiating[30] the disposal does so with the intention of discarding it as *waste*,[31] it being irrelevant that either he or somebody else could benefit from or make use of the material.[32] The concept of waste has been the subject of litigation. In *ICI Chemicals & Polymers Ltd v. C & E Commrs*,[33] the secondary product of a chemical process, which could be sold to the building industry was used to construct cells to contain noxious substances, in accordance with the terms of the waste disposal licence, it was held that the disposal had not been *of* waste. But, in *NSR Ltd v. C & E Commrs*[34] and *J & S Mackie v. C & E Commrs*,[35] the fact that the acquiring licence holder intended to use imported soil for restoring the site to agricultural use did not avail him. Similarly, in *McIntosh Plant Hire v. C & E Commrs*,[36] the use of waste to metal a road within a landfill site was also held to be taxable. In *C & E Commrs*

---

20   Finance Act 1996, s.40(1).
21   *Ibid.*, s.40(2)(a).
22   *Ibid.*, s.40(2)(b).
23   *Ibid.*, s.40(2)(c).
24   *Ibid.*, s.40(3).
25   *Ibid.*, s.41.
26   *Ibid.*, ss.66(a), 67(a); Environmental Protection Act 1990, s.35. See para. 6.3.2.1 above.
27   Finance Act 1996, Sched. 5, paras 48–61.
28   See para. 16.2 below.
29   See para. 16.16 below.
30   Thus, in *C & E Commrs v. Darfish Ltd*, [2001] Env LR 3, where the licence holder required, and contracted for the delivery of builders' topsoil for site engineering, it was the intentions of the builders, rather than the purchasing licence holder, which were relevant.
31   Finance Act 1996, s.64(1),(3),(4).
32   *Ibid.*, s.64(2).
33   (1998) V&DR 310.
34   (1999) L.7.
35   (2000) L.9.
36   (2001) L.10; see also *Lancashire Waste Services Ltd v. C & E Commrs*, (1999) L.8.

v. Parkwood Landfill Ltd,[37] a local authority disposed of unwanted material, at a fee, to a recycling company ('Recycling') in which it held a minority shareholding. At Recycling's plant, the material was crushed, sorted, mixed with other material and graded before being divided into saleable material. Part of the recycled material was sold to Parkwood, the holder of the waste disposal licence, for use in the restoration of its landfill site. The central issue was that the local authority intended to *discard* the material. It was immaterial either that, as a result of recycling, there was a change in the nature of the material or that, through its shareholding, the local authority stood to benefit from the recycling of some of it.[38]

A further issue in *Parkwood* was whether the use to which it had put the material precluded the recognition of what had happened as a disposal by way of *landfill*,[39] irrespective by whom it was made. It was held, however, that a *disposal* for this purpose could be more than the *legal transaction by which it is effected or property in the material passes*.[40] This is relevant because Finance Act 1996, s.65(1), provides that a disposal by way of landfill takes place if the material is deposited either:

1 on the surface of the land,[41] or a structure set into the surface;[42] or
2 under the surface of the land.[43]

Which head applies can be of significance. In *F A Gamble & Sons Ltd v. C & E Commrs*,[44] when the operator of the site had accepted the waste, for a fee, it was stockpiled for subsequent processing into soil (after intermixture with fertilisers) for use to cover the clay cap and thus achieve site restoration. Section 65(1)(a) being in point, the initial deposit was the relevant event,[45] and, at that stage constituted *waste*.

The Court of Appeal in *Parkwood* overruled the Vice-Chancellor's judgment. All four heads of section 40(2) had to be present before a tax liability arose. In these particular circumstances, head (a) was not because, in determining whether a disposal was *as waste*, it was legitimate to have regard to the purpose of the tax, which was divert waste from landfill into recycling.[46]

---

37 [2002] STC 417.
38 See para. 6.3.2.1 above.
39 For the purposes of the two heads of Finance Act 1996, s.65(1).
40 By reference to Finance Act 1996, s.64(3),(4).
41 Which includes that covered by water above the low water mark of ordinary spring tides: Finance Act 1996, s.65(7).
42 And irrespective of whether placed in a container before being deposited: Finance Act 1996, s.65(2). Where it is intended to cover it with earth subsequently, the disposal takes place when the material is deposited, rather than when it is covered: Finance Act 1996, s.65(4).
43 Irrespective of whether covered by earth (or similar matter) afterwards *or* deposited in a cavity in the first place: Finance Act 1996, s.65(3)(a),(b),(8).
44 (1998) L.4.
45 By reason of Finance Act 1996, s.65(4).
46 [2002] STC 1536. One has to wonder whether it was a pure coincidence that this judgment was given the day after it had been announced in the Pre-Budget Report that the rate of tax was to rise, over a period of years to £35, and in the context of the publication in draft of

It is for consideration whether the Court of Appeal's decision had the effect of overruling earlier decisions, including that of Moses, J. in *Darfish*. In *Bendall v. C & E Commrs*,[47] the intention of the site licensee was taken into account to the extent of withdrawing from charge material taken in for a fee but used to make a hard stand for his scrap and other businesses. In so holding, the tribunal distinguished *NSR, ICI Chemicals & Polymers* and *Gamble*.

15.4   Exemptions

Extensive exclusions from landfill tax are provided for in addition to the reduced rate for inert landfill:

1   dredging operations from the beds of rivers, canals and harbours undertaken in the interests of navigation;[48]
2   the removal of naturally occurring material from the sea in the course of commercial extraction operations;[49]
3   naturally occurring material extracted from the earth in the course of commercial mining or quarrying operations (including opencast), other than such material as has been subjected to or results from either a process forming part of those operations which permanently alters its chemical composition or a process separate from the mining or quarrying;[50]
4   the disposal of inert waste at a landfill site which is or was a quarry, for which there is a planning requirement for total or partial refilling and the waste disposal licence permits only the disposal of inert waste at the site;[51]
5   subject to prior written notice to Customs, the use of inert waste for site restoration (other than capping waste) under the terms of the planning consent or waste disposal licence:[52]

   a.   it had been held in *Ebbcliff Ltd v. C & E Commrs*[53] that this relief was available in circumstances in which a quarry had been filled before 1970 and then restored in accordance with 1950s planning consents, but to a poor standard. It became necessary for it to be recontoured with inert waste deposited under the capping layer. The tribunal held that this constituted continued restoration of the site to modern standards, within Finance Act

---

the Downing Street Strategy Unit's *Waste Not, Want Not*, which did not pull punches on the country's problems in relation to the domestic waste management problem (see para. 6.3.2 above).

[47]   (2003) LON/00/1305.
[48]   Finance Act 1996, s.43(1)–(3).
[49]   *Ibid.*, s.43(4).
[50]   *Ibid.*, s.44.
[51]   *Ibid.*, s.44A (taking effect from 1 October 1999).
[52]   *Ibid.*, s.43C (also taking effect from 1 October 1999).
[53]   (2003) L.16.

1996, s.43C(2). But the tribunal's decision was reversed by Etherton, J.,[54] on the basis of:

   i.   the word *restore* not encompassing the concept of improvement, but restoration: the judge adopted a tribunal case not cited below, *Harley v. C & E Commrs*,[55] where a farmer had imported waste to improve the contours of a valley in order to make it more productive; and

   ii. the words *on completion of* not being apposite to cover a disposal user which had ceased in 1970.

b. The concept of *restoration* was subsequently considered in *Dispit Ltd v. C & E Commrs*.[56] In that case, the Environment Agency had varied the terms of the taxpayer's waste disposal licence permitting tipping in a former railway cutting by requiring steps to be taken to prevent leaching. This was achieved by lining the cutting with inert waste before tipping took place. While this use of waste was not for 'capping', the tribunal distinguished the circumstances in which the works were carried out from those in *Harley*. By looking at the planning consent in its totality, and not just the restoration condition, it was clear that there was a distinction between tipping under it and restoration after the tipping had been completed. The taxpayer was not, in consequence, entitled to relief under this head;

6 Exemption is also available[57] for certain[58] materials removed in the course of the reclamation of contaminated land, which has been certified by Customs after at least 30 days' prior written application, if it involves cleaning the land (or part of it) of some or all of the pollutants causing or having the potential to cause harm, after polluting activity has ceased either:

---

54 [2004] STC 391. Etherton, J.'s judgment was upheld in the Court of Appeal, [2004] STC 1496, on the basis that s.43C(2) was drafted in a manner which presupposed a knowledge of ordinary terms in use in the landfill business. It was generally recognised in the industry that the restoration stage commenced after infilling had been completed. It was therefore proper, when construing the subsection to have regard to industry practice and usage. Parliament's express exemption of capping was hard to explain if the waste below was to be included. There was no overlap with s.44A. Where there was no capping layer, s.43C(2) could still apply, but it would be a question of fact at what point the infilling ended and the restoration began.

55 (2001) L.13.

56 (2004) L.19, which was approved by the Court of Appeal in *Ebbcliff*, [2004] STC 1496.

57 Under Finance Act 1996, s.43A(1) provided that the work is completed before the removal of the source of pollution or, where construction work is involved, before the commencement of that work (*ibid.*, s.43A(2)).

58 This treatment is not, however, available where the removal is required under notices served under the statutory powers specified in Finance Act 1996, s.43A(4), unless the removal is carried out by an official body within the descriptions contained in *ibid.*, s.43A(5),(6).

a. where those pollutants would prevent this being attained, with the object of facilitating development, conservation, the provision of a public park or other amenity, or the use of the land for agriculture or forestry; or

b. to remove or reduce the potential of pollutants to cause harm;[59] and

7 pet cemeteries operating under a waste disposal licence and in which no (other) waste is deposited.[60]

15.5 Temporary disposals

Where material is held temporarily, in an area designated for the purpose,[61] pending one or more of incineration or recycling, removal for use elsewhere, the use of inert waste for restoration of the site at which disposal takes place, or sorting pending removal elsewhere or eventual disposal,[62] the disposal may be treated as wholly[63] or partially[64] exempt under regulations.

Under the text of regulation 38[65] in force from 1 February 2002, where Customs have designated an area in which material is held temporarily under such conditions as Customs impose while the designation lasts and, within a year, material stored within the area is either:

1 recycled[66] or incinerated, or

2 (being inert waste and provided that prior written notice is given to Customs) used other than by way of taxable disposal either for site restoration at the same site or at another place, or

3 sorted pending either application.

15.6 Fly-tipping

It seems to have been envisaged in 1994 that, in addition to their tax collection duties, Customs would carry out an enforcement role, the suspicion being that there were

---

59 Finance Act 1996, s.43B.

60 *Ibid.*, s.45.

61 *Ibid.*, s.62(7)(b).

62 *Ibid.*, s.62(7)(a), from 28 July 2000.

63 Under Landfill Tax Regulations 1996, S.I. 1996 No. 1527, reg. 39.

64 By reason of shifting the disposal date under Landfill Tax Regulations 1996, S.I. 1996 No. 1527, reg. 40.

65 That is, of Landfill Tax Regulations 1996, S.I. 1996 No. 1527.

66 On 19 March 2004, in *Business Brief* 10/04, Customs announced a change in policy relating to recycling. Although the original producer's intention is not relevant, previously Customs regarded it as of the essence that a chemical change was a prerequisite to recycling. Under the revised policy all that is required is processing into a useable material.

quite a lot of unauthorised[67] landfill sites around.[68] However, not only were no such duties assigned to Customs under Finance Act 1996, the position of licensees who have been the victims of fly-tipping has been most unsatisfactory, with Customs able to issue an estimated assessment[69] against which (in the absence of the material not having been weighed) the taxpayer does not realistically have much prospect of success on appeal.

The statutory code includes a provision which makes it clear that the site operator is liable for fly-tipping.[70] It does not, however, include any right of recovery by the taxpayer against any fly-tipper who might be apprehended. The minister in charge of this part of the Bill[71] had this problem drawn to his attention in correspondence, and indicated (no doubt correctly) that, as and when fly-tippers were caught, the authorities would prosecute. But he did not feel able to introduce an amendment to ameliorate the risk to site operators because the general understanding was that local authorities ought to be requiring all licensees to install boundary security,[72] seemingly of a type suitable for that appropriate to the storage of dangerous waste.

It needs to be borne in mind in this context that, in 1996, farmers had not been brought within the waste management licensing regime.[73] It is understood that undertakings were received at that time that, as and when it was extended to farmers,[74] this unsatisfactory situation would be revisited. But when the required general reorganisation of waste disposal site regulation occurred, in July 2002,[75] no such action was taken. The stricter general regime, coupled with the increased obligations on site operators (for which there was no five year transition for new sites), makes it unlikely that farmers will undertake responsibility for their own waste management, thus increasing the costs of an industry in a very poor economic state indeed. Furthermore, the advent of legal responsibility for removal upon the landowner has given rise to serious concern, not least in the context of prospective changes in the regime for the disposal of hazardous waste.[76] In April 2004, the UK

---

67 The relevant licence or permit being required (see paras 6.3.2.1 and 6.3.2.4 above).

68 Nonetheless, in October 1999, the government indicated that the majority of fly-tipping was by householders 'not directly affected' by the tax (see HM Government, *The Government's Response to the Environment, Transport and Regional Affairs Committee's Report: the Operation of the Landfill Tax*, 1999 (Cm 4461, 1999)).

69 Under Finance Act 1996, s.50(5).

70 Finance Act 1996, s.69(2)(b).

71 The Paymaster-General (Mr David Heathcoat-Amory, MP).

72 Letter of 22 February 1996, to Chairman of the Holborn Law Society Revenue Committee.

73 Environmental Protection Act 1990, s.75(7)(c). See para. 6.3.2.1 above.

74 It being known at that date that the European Commission were disputing whether the UK had complied with its obligations under the relevant waste disposal directive in this respect. The UK Government did not capitulate to threats of proceedings until 9 December 2004, when a Written Statement from the Environment Minister (Mr Elliot Morley, MP) revealed that draft regulations had been published, for promulgation in early 2005: HC 9 December 2004, cols 106–107WS.

75 Landfill (England and Wales) Regulations 2002, S.I. 2002 No. 1559. See para. 6.3.2.4.

76 Such clean-up costs have been estimated at £100m, with none of the 12 commercial landfill sites authorised to handle such waste serving London or Wales (see Houlder, *Financial*

Government set up a *flycapture* database[77] following the inclusion in Anti-Social Behaviour Act 2003 of powers for local authorities to take preventive action against regular offenders.[78]

15.7 Administration of the tax

In common with aggregates levy and climate change levy, the administrative machinery applicable to landfill tax is discussed overleaf, in Chapter 16.

Times, 25 March 2004, p. 6). Furthermore, with property developers speeding up their clearance of contaminated sites in anticipation of price rises, the available capacity in the 12 sites is only half the annual demand (see Houlder, *Financial Times*, 15 July 2004, p. 5).

[77] See www.environment-agency.gov.uk (an email address and password is required).

[78] See ENDS Report 350, March 2004, p. 48. A consultation was also then in progress on an amended version of Environmental Protection Act 1990, s.59. A Select Committee commented on developing policy in this field in July 2004 (see the House of Commons' Environmental Audit Committee, *9th Report. Environmental Crime: Fly-tipping, Fly-posting, Litter, Graffiti and Noise* (House of Commons Papers, Session 2003–2004, HC 445) (London: Stationery Office, 2004)).

Chapter 16

Customs' Administrative Model

16.1   Introduction

Although of considerably greater longevity,[1] Customs took on their principal current tax collection role in 1973, when value added tax ('VAT') was introduced upon the entry of the UK into the (then) European Economic Community ('the EEC').[2]

As Customs have been the tax collection agency selected for the collection of national environmental taxes, it was decided to model the collection machinery and the administrative arrangements and powers upon those extant for the purposes of VAT. These operate at three levels – statute ('primary legislation'), statutory instrument ('secondary legislation') and legally binding notices ('tertiary legislation'). It should be appreciated, in this connection, that the primary function of the notices (which can take the form of pamphlets or leaflets and appear under various designations) is to inform taxpayers. Where paragraphs are intended to have the force of law, this is indicated.

Where notices do not have the force of law, but are intended to be explanatory of it and of Customs' practice in interpreting it, the legal system will not regard Customs as estopped by them as regards issues of interpretation. Erroneous or misleading statements in them will, however, be taken into account in deciding whether the imposition of penalties is appropriate.

16.2   Registration

Unlike Inland Revenue taxes before the late 1990s, the collection of indirect taxes[3] outside ports of entry into the UK has always been on a self-assessment basis. In addition, it was an essential feature of VAT that traders purchasing goods or services were (generally) able to credit their liability (called 'input tax') against their sales (or 'outputs') and only accounted to Customs for the net amount. The crediting feature is not normally a feature of environmental taxes, but, where it is not, has been retained for some purposes.

A crediting facility makes it necessary that assessable parties be registered, if only so as to be able to give those who they are dealing with a number to cite on their own documentation.

This means, in turn, that, while registrability may be of importance to the collection of tax, actual registration is of the essence to its administration.

---

[1]   See Graham Smith, *Something to Declare: 1000 Years of Customs and Excise* (London: Harrap, 1980).
[2]   The circumstances are outlined in, for example, Stephen Weatherill and Paul Beaumont, *EU Law*, 3rd edn (London: Penguin Books, 1999), pp. 6 and 36.
[3]   See para. 1.2.1.2 above.

Where, as in the case of VAT or climate change levy, the taxable event is the supply of something, it will be readily apparent that registrability is easy to determine, subject to anti-avoidance measures[4] and small trader thresholds.[5]

However, where what is taxed, although extracted or delivered, is ascertained by reference to weight delivered to or from a site, different criteria apply. Where control of the site requires a licence, as in the case of landfill,[6] there should be no problem in identifying the accountable party.[7] This has not, however, circumvented the need (in the eyes of Customs) for the addition of anti-avoidance legislation, in the form of secondary liability.[8]

Where this identifying factor is not present, the legislature is faced with grave problems. In the case of aggregates levy, it was decided to solve these by providing for joint and several liability between members of a class,[9] the membership varying according to the type of exploitation envisaged.[10] Nonetheless, as Customs have conceded in correspondence, control of a weighbridge cannot be shared and in practice only one candidate can be selected for registration.

Although Customs have power, by regulation, to exempt those not selected from both the requirement to register[11] and administrative responsibility where another has been registered,[12] they initially chose not to do so[13] in relation to aggregates levy.

A different solution was therefore adopted over a period of a year for two similar problems. In 2000, landfill tax evasion was countered by the introduction of the concept of secondary liability, that is, that of a guarantor without right of indemnity,[14] whereas, in 2001, fears of a similar situation arising in relation to aggregates levy were countered by invoking the joint and several liability concept.[15]

---

4 For example, Value Added Tax Act 1994, Sched. 1, para. 1A, and Sched. 9A.

5 As, for VAT, in Value Added Tax Act 1994, Sched. 1, para. 1.

6 See Environmental Protection Act 1990, s.35 (see para. 6.3.2.1 above).

7 Although there is still a requirement to notify registrability and prospective registrability (see Finance Act 1996, s.47(2),(3)).

8 Finance Act 1996, Sched. 5, paras 48–60 (added by Finance Act 2000, Sched. 37).

9 Finance Act 2001, s.22(3).

10 *Ibid.*, ss.21 and 22(1),(2).

11 *Ibid.*, s.24(4)(a), which they chose to exercise by requiring notification of situations in which exemption was conferred by ss.17(3)(b)–(d) (see Aggregates Levy (Registration and Miscellaneous Provisions) Regulations 2001, S.I. 2001 No. 4027, reg. 3(2), until its repeal on 15 August 2002, by Aggregates Levy (Registration and Miscellaneous Provisions) (Amendment) Regulations 2002, S.I. 2002 No. 1929).

12 Finance Act 2001, s.24(4)(b).

13 As to administrative responsibilities: see Aggregates Levy (General) Regulations 2002, S.I. 2002 No. 761, regs 4–11, and Aggregates Levy (Northern Ireland Tax Credit) Regulations, S.I. 2002 No. 1927.

14 Although described as involving joint and several liability (see Finance Act 1996, Sched. 5, para. 57(1)).

15 These powers are, however, as nothing when compared to the new Value Added Tax Act 1994, s.77A and the amended Sched. 11, para. 4(2), inserted by Finance Act 2003, ss.18 and 19, to combat *carousel* and *missing trader* frauds which exploited the European single market by moving computer chips and mobile phones between Member States with the

A complication arose in relation to climate change levy by virtue of the regulatory structure governing the supply of electricity. Since it was not desired to tax domestic supplies, it was not possible to distinguish at the point of taxation between polluting and renewable or cleaner sources. The accountable party had, therefore, to be a person with a licence from the regulator to supply electricity or gas.[16] This is the person making the supply.[17] Where, however, the supply is made by a person who is neither a utility nor resident in the UK, it is the recipient of the supply who has to register.[18]

It was therefore necessary for the statute to provide for people to notify Customs of their registrability.[19] As a result both climate change levy and aggregates levy have regulations which require candidates for registration to identify themselves to Customs.[20] (Aggregates levy had an additional complication because, for revenue protection reasons, Customs have been given the right to determine the boundaries of a site.[21]) The concept of self-identification[22] came in, needless to say, in 2000,[23] and in conjunction with the introduction of secondary liability for landfill tax.[24]

16.3   Deregistration

Although it might not seem necessary for a taxpayer who has ceased to be accountable for a tax to deregister – rather than just make a nil return, Parliament has seen fit to prescribe penalties in this situation for aggregates levy[25] but not landfill tax[26] or climate change levy.[27]

---

importer defaulting on the reverse VAT charge. Such operations were costing billions of pounds. Nonetheless, for the Customs to obtain power to recover the missing tax off innocent third parties not necessarily the immediate purchaser from the fraudster must be rated extraordinary.

[16]   To get round certain administrative difficulties, Customs can designate people as deemed utilities, under Finance Act 2000, Sched. 6, para. 151, but these are purely fiscal fictions: see *Technical Briefing* No. 8 (March 2001) (available from www.hmce.gov.uk).

[17]   Finance Act 2000, Sched. 6, para. 40(1).

[18]   *Ibid.*, Sched. 6, para. 40(2).

[19]   *Ibid.*, Sched. 6, para. 55(1), which should be compared to aggregates levy's Finance Act 2001, s.24(2) and Sched. 4, para. 1(1).

[20]   See Climate Change Levy (Registration and Miscellaneous Provisions) Regulations 2001, S.I. 2001 No. 7, reg. 2(1), and Aggregates Levy (Registration and Miscellaneous Provisions) Regulations 2001, S.I. 2001 No. 4027, reg. 2(1) and (3).

[21]   Finance Act 2001, s.24(7).

[22]   In the circumstances, one might almost say 'self-incrimination'.

[23]   When all the other fiscal collection agencies went over to it.

[24]   See Finance Act 1996, Sched. 5, para. 60(3)(a).

[25]   Finance Act 2001, Sched. 4, para. 3.

[26]   Finance Act 1996, s.47(4) merely provides for an obligation to notify. Landfill Tax Regulations 1996, S.I. 1996 No. 1527, reg. 6 sets out a timescale for this.

[27]   Climate Change Levy (Registration and Miscellaneous Provisions) Regulations 2001, S.I. 2001 No. 7, reg. 4, sets out a timescale for notification.

Needless to say, where circumstances have changed, Customs must be informed.[28]

16.4   The form of registration

It goes without saying that Customs are empowered to specify the format of an application for registration or notification of registrability. However, the way in which they use their powers can affect the way in which the tax is collected.

To start with the VAT position, there are just two forms – single and joint (and several) registration. The former deals with both sole traders and companies. The latter encompasses partnerships, trusts and co-ownership situations.[29]

The VAT position in relation to partnerships is strange, with limited partners being ignored[30] and businesses with the same composition of general partners being amalgamated.[31] This oddity comes from the fact that, for the purposes of that tax, persons, rather than businesses, are registered, with the result that several businesses with turnovers below the registration threshold can come within the charge to tax where, taken separately, they would not have done.

For landfill tax, climate change levy and aggregates levy, partnerships and unincorporated associations[32] are treated in the same way.[33] While there has been no litigation on the point, there seems to be no reason why businesses should not be treated as segregated for the purposes of these taxes. The point is not, however, an idle one so far as aggregates levy is concerned, because the one major concession obtained by the industry under the first consultation was the ability to move aggregate from the extraction site to another site in the same registration without triggering a tax charge.[34]

---

28   Landfill Tax Regulations 1996, S.I. 1996 No. 1527, reg. 5; Climate Change Levy (Registration and Miscellaneous Provisions) Regulations 2001, S.I. 2001 No. 7, reg. 3; Aggregates Levy (Registration and Miscellaneous Provisions) Regulations 2001, S.I. 2001 No. 4027, reg. 4.

29   Value Added Tax Act 1994, s.45 is, however, confined to partnerships.

30   *Saunders and Sorrell v. C & E Commrs*, (1980) VATTR 53.

31   *C & E Commrs v. Glassborow*, [1974] STC 142.

32   For VAT, clubs are dealt with separately by Value Added Tax Act 1994, s.46(2)–(3), being regarded as deemed traders by s.94(2).

33   By, respectively, Finance Act 1996, s.58(1)–(3), with Landfill Tax Regulations 1996, S.I. 1996 No. 1527, reg. 8, Finance Act 2000, Sched. 6, para. 117, with Climate Change Levy (Registration and Miscellaneous Provisions) Regulations 2001, S.I. 2001 No. 7, regs 12 and 13, and Finance Act 2001, Sched. 4, para. 2(4), with Aggregates Levy (Registration and Miscellaneous Provisions) Regulations 2001, S.I. 2001 No. 4027, regs 12 and 13: the provisions relating to unincorporated association liability are based on Value Added Tax Regulations 1995, S.I. 1995 No. 2518, reg. 8.

34   Finance Act 2001, s.19(3)(b): a dispute could reach the litigation system under s.40(1)(c), by reason of a request for a review by a registrable party not selected for registration. See para. 4.2.1.5 above.

16.5   The problem of trusts

Although the difference between a partnership and a trust fund[35] is fundamental, with one variant, Customs treat the two situations in the same way for VAT. The difference is that each settlement[36] is treated as a discrete taxpayer. Where a bank is a sole trustee or executor, the trust's activities are not dealt with on the bank's own (and, no doubt, partially exempt), registration number.

Despite VAT having been extant since 1973, proper arrangements for land held upon either statutory or substantive trust have never been put in place.[37] There are problems which have never been sorted out satisfactorily, the underlying problem being that some of the people Customs would prefer to have included in the registration do not have control of the activities of the fund and, not unreasonably, feel that they should not be included in a system of taxation which involves personal liability.

Where there is a tenant for life under the Settled Land Act 1925, the legal estate is vested in him alone, and the function of the trustees is to give a good receipt for capital monies. In practice, transactions are therefore treated as carried out by the tenant for life alone, and the VAT paid to him other than as part of the price. Where, however, there is a single income beneficiary under a trust of land, the rental income, technically, belongs to that person as it arises,[38] and Customs' preference is for the beneficiary and the trustees to be registered jointly. This is not generally acceptable to the income beneficiary's solicitor, and in practice registration by the trustees alone is agreed by Customs. Where a unit trust is involved, Customs' theory is that the trustee is registered 'on behalf of the unitholders', but this is not generally accepted as having taken place. Where a co-ownership is involved, the official view is that no waiver of exemption[39] can take place unless all join in, and that dealings in beneficial interests are outside the scope. Although this has not been challenged in the Courts, it is very doubtful whether this analysis is correct. Such an issue could not, however, arise in connection with environmental taxes, because it is not possible to be a 'semi-operator'.

In addition, for VAT, at present, some conveyancing situations have to be dealt with by interposed registrations.[40] As a general rule, this complication does not apply to environmental taxes, because there is no concept of input tax in relation to them.[41]

---

[35] For general guides to the English law of trusts, see, for example, *Hanbury and Martin: Modern Equity*, ed. by Jill E. Martin, 16th edn (London: Sweet and Maxwell, 2001) and Gary Watt, *Trusts and Equity* (Oxford: Oxford University Press, 2003).

[36] That said, the distinctions discussed in, inter alia, *Roome v. Edwards*, [1981] STC 96, are not matters with which Customs should be expected to be conversant, even at Head Office level.

[37] Value Added Tax 1994, s.51A was added in 1995, but has never been brought into effect and, it is understood, is not now intended to be.

[38] *Williams v. Singer (No. 2)*, (1920) 7 TC 387.

[39] Under Value Added Tax Act 1994, Sched. 10, paras 2–3.

[40] Under Value Added Tax Act 1994, s.47.

[41] But see Climate Change Levy *Technical Briefing* No. 8 (March 2002) (available from www.hmce.gov.uk) in relation to interposed landlords and *deemed utility* directions under Finance Act 2000, Sched. 6, para. 151, and the list released on 6 May 2003, which may reflect this.

There is no recognition of trusts as such in the environmental taxes legislation.[42] Plainly some form of joint and several registration would be required, but whether as partnerships or as unincorporated associations is unclear.[43] While, as a general rule, only those engaged in some form of business activity can be involved in the registration process,[44] the aggregates tax legislation was amended in 2002 to make it clear that Government departments, local authorities and charities could come within this concept.[45] It also appears to be the case that Customs are under the impression that landlords can be registrable as waste disposal operators,[46] and in consequence become primarily liable for landfill tax.[47] Quite apart from this, it is clear that letting property can, and usually does, constitute *economic activity* for the purposes of VAT,[48] and personal representatives (and sometimes trustees) do sometimes carry on trading, especially that of farming.[49]

It follows that the treatment of trusts for the purposes of environmental taxes is by no means an academic issue, and it is far from clear that Customs' practice in relation to VAT would necessarily be adopted by the Courts.

16.6   Groups, divisions and going concerns

Provision is made, however, for both group and divisional registration, and for transfers of going concerns:

1   The first can be requested for Companies Act groups and have a substantive effect because it has the effect of removing boundaries between members, who assume joint and several liability for tax owed. Customs have therefore had to introduce controls over the use of this medium,[50] and may have resort to an extremely wide set of anti-avoidance provisions.[51] The reason for this is that exempt transactions

---

[42]   This is particularly noticeable in relation to the definition of non-resident for the purposes of climate change levy (see Finance Act 2000, Sched. 6, para. 156).

[43]   The point might be of some significance for climate change levy and aggregates levy, where Climate Change Levy (Registration and Miscellaneous Provisions) Regulations 2001, S.I. 2001 No. 7, reg. 2(2) and Aggregates Levy (Registration and Miscellaneous Provisions) Regulations 2001, S.I. 2001 No. 4027, reg. 2(3) require a partner to fill in Forms CCL 2 and AL 2 as well as Forms CCL 1 and AL 1. By way of contrast, it should be noted that, for landfill tax, only the partnership files particulars under Landfill Tax Regulations 1996, S.I. 1996 No. 1527, reg. 4(3).

[44]   See Finance Act 2001, s.22(2), in relation to aggregates levy.

[45]   See the addition to Finance Act 2001, s.22(2).

[46]   Although Environmental Protection Act 1990, s.35(2)(a) is, in terms, clearly inconsistent with this interpretation. See para. 6.3.2.1 above.

[47]   See *Landfill Tax Briefing* (7 August 2000).

[48]   *Commission of the European Communities v. France*, C–50/87, [1989] 1 CMLR 505.

[49]   In relation to which climate change agreements may be relevant: see Finance Act 2000, Sched. 6, para. 51.

[50]   Value Added Tax Act 1994, ss.43A, 43AA and 43B.

[51]   *Ibid.*, s.43C, Sched. 9A.

generate breaks in the transparency chain which is of the essence to the structure of this European tax up until the stage of supply to the final consumer. Such considerations do not affect environmental taxes, for which the facility has been retained for purely administrative reasons,[52] Customs having no power, under regulations made since the issue was raised in Parliament, to add companies to a group of their own motion.[53]

2   Bodies corporate may have separate divisional registrations. The existence of these has no effect on liability, however. Such registrations are also permissible for landfill tax[54] and aggregates levy.[55]

3   For VAT, clubs, insolvencies and estates in the course of administration are dealt with by regulations made under the same statutory provision.[56] For landfill tax, climate change levy and aggregates levy, death and incapacity[57] and insolvency[58] are dealt with separately.

4   For VAT, a transfer of going concern is treated as a non-event, and such treatment can be compulsory. It is, furthermore, sometimes extremely difficult to ascertain whether such a state of affairs exists at the time of the tax point. Where such a state of affairs exists, it is neither obligatory nor desirable[59] for the transferee to take over the transferor's registration number. The same pertains for environmental taxes.[60]

---

[52]   By, respectively, Finance Act 1996, s.59, Finance Act 2000, Sched. 6, para. 116, with Climate Change Levy (Registration and Miscellaneous Provisions) Regulations 2001, S.I. 2001 No. 7, regs 8–11, and Finance Act 2001, s.35, Sched. 4, para. 2(2), and Sched. 9, with Aggregates Levy (Registration and Miscellaneous Provisions) Regulations 2001, S.I. 2001 No. 4027, regs 8–11.

[53]   Climate Change Levy (Registration and Miscellaneous Provisions) Regulations 2001, S.I. 2001 No. 7, regs 8–9 and Aggregates Levy (Registration and Miscellaneous Provisions) Regulations 2001, S.I. 2001 No. 4027, regs 8–9.

[54]   Finance Act 1996, s.58(3).

[55]   Finance Act 2001, Sched. 4, para. 2(3).

[56]   Value Added Tax Act 1994, s.46.

[57]   By, respectively, Landfill Tax Regulations 1996, S.I. 1996 No. 1527, reg. 9, Finance Act 2000, Sched. 6, para. 118 with Climate Change Levy (General) Regulations 2001, S.I. 2001 No. 838, regs 55 and 57, and Finance Act 2001, s.38, with Aggregates Levy (General) Regulations, S.I. 2002 No. 761, regs 34 and 36.

[58]   By, respectively, Landfill Tax Regulations 1996, S.I. 1996 No. 1527, reg. 9, Finance Act 2000, Sched. 6, para. 120 with Climate Change Levy (General) Regulations 2001, S.I. 2001 No. 838, regs 56 and 57, and Finance Act 2001, s.37, with Aggregates Levy (General) Regulations, S.I. 2002 No. 761, reg. 35.

[59]   Because liability for penalties etc is transferred with the registration (see *Ponsonby v. C & E Commrs*, [1988] STC 28).

[60]   By, respectively, Finance Act 1996, s.58(5), with Landfill Tax Regulations 1996, S.I. 1996 No. 1527, reg. 7, Finance Act 2000, Sched. 6, para. 119, with Climate Change Levy (General) Regulations 2001, S.I. 2001 No. 838, reg. 59, and Finance Act 2001, s.39, with Aggregates Levy (General) Regulations, S.I. 2002 No. 761, reg. 37.

16.7   Non-UK residents

Where an overseas taxpayer is involved, the position is far fairer for VAT than for environmental taxes, where a formula devised for the purposes of air passenger duty[61] and insurance premium tax[62] and adapted for the purposes of liability on offshore life policies[63] has been followed.

Although, for VAT, while a person who accepts the position of tax representative is personally liable for all the VAT owed by his principal,[64] irrespective of whether the relevant transactions were under his control, for both climate change levy[65] and aggregates levy,[66] it is (theoretically) possible for one to be nominated as an agent by Customs without having any connection with the business, let alone the ability to comply with the administrative and financial obligations of registration. When pressed on this extraordinary situation during the Committee Stage of Finance Bill 2000,[67] the Financial Secretary to the Treasury said that the taxpayer had two safeguards – the review procedures[68] and judicial review.[69]

For climate change levy, a non-resident must notify Customs within 30 days.[70] Customs may then either 'require or permit' the non-resident to appoint a resident tax representative and oblige him the non-resident to request them to approve the person nominated,[71] or a replacement.[72] They may also require such a replacement.[73] Where the non-resident fails to comply, Customs may direct the nomination of a person of their own choice,[74] who is both 'eligible to act'[75] (which means being resident in the United Kingdom)[76] *and* 'suitable in all the circumstances to be the tax representative for the relevant non-resident taxpayer'.[77] Such a person is not permitted to withdraw from such appointment.[78]

---

[61]   Finance Act 1994, s.35(1)(a), albeit not in such outrageous terms!

[62]   It should be noted, however, that Finance Act 1994, s. 57(11) specifies that the person in question should have been the agent of the insurer!

[63]   In the inserted Income and Corporation Taxes Act 1988, s.552A(5)(g).

[64]   Value Added Tax Act 1994, s.48.

[65]   Finance Act 2000, Sched. 6, paras 114 and 115, see also Climate Change Levy (Registration and Miscellaneous Provisions) Regulations 2001, S.I. 2001 No. 7, regs 14–19.

[66]   Finance Act 2001, ss.33–34, see also Aggregates Levy (Registration and Miscellaneous Provisions) Regulations 2001, S.I. 2001, No. 4027, regs 14–19.

[67]   Standing Committee H, 18 May 2000 (Mr Stephen Timms, MP).

[68]   Which normally precede an appeal to a tribunal, the latter not being available in this case.

[69]   See para. 4.2.1.5 above.

[70]   Climate Change Levy (Registration and Miscellaneous Provisions) Regulations 2001, S.I. 2001 No. 7, reg. 14(2),(3).

[71]   *Ibid.*, reg. 14(4),(6).

[72]   *Ibid.*, reg. 15(1),(3).

[73]   *Ibid.*, reg. 16.

[74]   *Ibid.*, reg. 17(1),(2),(3).

[75]   *Ibid.*, S.I. 2001 No. 7, reg. 17(4)(a).

[76]   *Ibid.*, reg. 14(4)(a).

[77]   Climate Change Levy (Registration and Miscellaneous Provisions) Regulations 2001, S.I. 2001 No. 7, reg. 17(4)(b).

[78]   *Ibid.*, reg. 18(2)(b).

There are similar provisions for aggregates levy. The non-resident must notify Customs within 30 days.[79] Customs require or permit the appointment of a resident tax representative and oblige the non-resident to request their approval of the nominee,[80] or his replacement.[81] They may also request his replacement.[82] If the non-resident does not comply, they may select a tax representative,[83] who is both 'eligible'[84] and 'suitable in all the circumstances',[85] and who may not resign.[86]

Both formulae are, to put it mildly, 'comprehensive' from the point of view of HM Government, but fail to provide either a statutory right of indemnity or a statutory right of disclosure to the Customs-nominated 'representative'.[87]

There are additional complications where climate change levy is concerned. While the supplier of energy is normally the registrable party, where that person is both non-resident[88] and not a utility[89] the recipient has to register.[90] It seems therefore that the very wide powers of selection given to Customs in relation to that tax can only apply where either:

a. the utility is non-resident; or
b. the supplier is not a utility and is outside the stretched definition of residence *and* the recipient is non-resident under the same definition.

16.8   Site registration

For the purposes of aggregates levy, Customs are able to fix the boundaries of a site.[91] When registering, applicants have, therefore, to submit site details.[92] Although Customs

---

[79] Aggregates Levy (Registration and Miscellaneous Provisions) Regulations 2001, S.I. 2001, No. 4027, reg. 14(2),(3).
[80] *Ibid.*, reg. 14(4),(6).
[81] *Ibid.*, reg. 15(1).
[82] *Ibid.*, reg. 16.
[83] *Ibid.*, reg. 17(1),(2),(3).
[84] *Ibid.*, reg. 17(4)(a), that is, resident in the UK (see reg. 14(4)(a)).
[85] *Ibid.*, reg. 17(4)(b).
[86] *Ibid.*, reg. 18(2)(b).
[87] Neither follows from the fact that the person in question is stated to be 'entitled to act' under Finance Act 2000, Sched. 6, para. 115(1) or Finance Act 2001, s.34(1).
[88] Albeit restricted by reference to Finance Act 2000, Sched. 6, para. 156, which treats a person as resident where it has an established place of business in the UK, *or* has a usual place of residence in the UK, *or* is a firm or unincorporated body which has amongst its partners or members at least one individual with a usual place of residence in the UK. (It should be noted that the third criterion does not apply to trustees, giving rise to considerable difficulty to purchasers of energy.)
[89] As defined in Finance Act 2000, Sched. 6, para. 150.
[90] Finance Act 2000, Sched. 6, para. 40(2). Value Added Tax Act 1994, s.9A introduces a reverse supply mechanism for VAT where a VAT-registered trader imports electricity or natural gas from a non-UK person.
[91] Finance Act 2001, s.24(7).
[92] On Form AL 1A: see Aggregates Levy (Registration and Miscellaneous Provisions) Regulations 2001, S.I. 2001, No. 4027, reg. 2(6).

will normally accept this description, they may override it. The Financial Secretary to the Treasury justified this to the Committee of the whole House as follows:

> Customs & Excise is given powers to decide the boundaries of any premises for registration purposes to safeguard against avoidance of the levy by businesses that might otherwise locate taxable activity outside previously agreed boundaries.[93]

Presumably an adjoining landowner aggrieved by such a decision could have resort to judicial review, even if the review procedures were inoperable,[94] because registrability could be in dispute. The issue could be relevant in connection with construction activity, not least in the context of the concept of *occupation* being a qualification for joint and several liability.[95] Concern on this front has, however, been reduced following the receipt of confirmation from Customs to two representative bodies in April 2001 that the concept of *occupation* in this context must involve physical presence.[96]

Site registration is also required for landfill tax, but only where a registered taxpayer has more than one site.[97]

16.9   The credit concept

It is of the essence to VAT that it does not matter how many traders contribute inputs to the make-up of the product or service sold to the final consumer. The underlying theory is known as the *neutrality principle*. In order to achieve this, the price each trader pays to his own suppliers is credited against the price he receives for the product he sells on, and on which he has to account for VAT. For the system to work, each trader receives from and delivers to other traders an invoice specifying the VAT element of the transaction. As a general rule, only if a trader receives such an invoice can he claim a credit for the VAT paid to his supplier.

It follows that invoicing is of the essence to VAT.[98] This is especially important because the VAT system is not fully transparent, breaks in the recovery chain are created by both non-business user and also business outputs to which VAT is not charged (known as *exempt supplies*). This complication can create difficulties in the calculation of input recoveries, especially where overheads are concerned.

---

93   HC, 23 April 2001, col 84.
94   In theory Finance Act 2001, s.40(1)(c) enables a review, and s.41(1)(a) an appeal to a tribunal, to take place in relation to the registration of *premises*.
95   Either under Finance Act 2001, s.22(1)(e) or through s.22(1)(a),(b),(d) or (f), by reason of s.21(1)(a).
96   It seems that they may also take that view in relation to the interpretation of Environmental Protection Act 1990, s.35(2)(a), and the position of landlords (who do not have legal *occupation*), and in relation to which they are generally believed to be wrong.
97   Landfill Tax Regulations 1996, S.I. 1996 No. 1527, reg. 4(2).
98   Where an invoice is required, clearly it has to contain a minimum amount of information: see Landfill Tax Regulations 1996, S.I. 1996 No. 1527, reg. 37, and (for the climate change levy equivalent) Finance Act 2000, Sched. 6, paras 27(5), 28(4) and 143(2).

The factor just discussed is not present in environmental taxes. There are, however, circumstances in which credits can be generated. One of these, bad debt relief, is dealt with in the next para. As a general rule, credits can only be generated where an invoice is issued.[99] But, unless credits are in point, the issue of invoices is not obligatory.[100] This gives rise to potential problems in relation to supplies received from non-utilities for the purposes of climate change levy,[101] since the recipient[102] will be responsible for accounting for the tax if his supplier is non-resident and, unless a climate change levy accounting document (which is what an invoice is called in the context) is issued, he will be put on enquiry as to the residential status of the supplier, a task not assisted by the extra-ordinary definition of *non-resident* which applies for the purposes of that tax.[103]

Credits can arise for landfill tax,[104] but generally only in relation to bad debts[105] or sums paid to qualifying environmental bodies.[106]

The purpose of *accounting documents* and *invoices* in climate change levy is to fix the tax point.[107] Separate provision is made, however, for the timeous[108] (or early)[109] and late[110] issue of such documents.[111] The accounting document system does not apply, however, where a special utility scheme is in force.[112]

Where a relevant commodity other than electricity or gas is made by a non-resident, the supply is treated as taking place on its delivery (or earlier making available) to the UK recipient unless that recipient elects for it to be deferred to the date of an election made within 14 days.[113]

Credits and repayments[114] are made in the following circumstances:[115]

---

[99] It has, however, been found possible to dispense with them for the purposes of aggregates levy, see Aggregates Levy (General) Regulations 2002, S.I. 2002 No. 761, regs 10(i) and (j), 12, 13.

[100] Just to make things complicated, Finance Act 2000, Sched. 6, paras 31 and 32 use the word *invoice* for supplies of other than electricity and gas, for which the equivalent is called a *climate change levy accounting document*!

[101] Where an accounting document only has to be issued in the circumstances specified in regulations (which powers have not been exercised in this respect) (Finance Act 2000, Sched. 6, para. 143(1)).

[102] Finance Act 2000, Sched. 6, para. 40(2).

[103] *Ibid.*, Sched. 6, para. 156.

[104] Finance Act 1996, s.51; Landfill Tax Regulations 1996, S.I. 1996 No. 1527, regs 17–20.

[105] Finance Act 1996, s.52.

[106] *Ibid.*, s.53.

[107] Finance Act 2000, Sched. 6, paras 26(2),30(2).

[108] *Ibid.*, Sched. 6, para. 27.

[109] *Ibid.*, Sched. 6, para. 31.

[110] *Ibid.*, Sched. 6, paras 28, 32.

[111] For this a system of *supplier certificates* is prescribed by Climate Change Levy (General) Regulations 2001, S.I. 2001 No. 838, regs 34–40 and Schedule 1.

[112] Finance Act 2000, Sched. 6, paras 26(4),29; Climate Change Levy (General) Regulations 2001, S.I. 2001 No. 838, reg. 54.

[113] Finance Act 2000, Sched. 6, para. 33(2),(4).

[114] *Ibid.*, Sched. 6, para. 102.

[115] *Ibid.*, Sched. 6, para. 62(1).

1   a subsequent change in circumstances which, had they existed at the time of
     supply, would have meant that it was not taxable;
2   after a taxable supply has been made on that basis, it is determined that it was to
     no extent such a supply;
3   after a taxable supply has been made on the basis that it is neither a half-rate nor
     a reduced-rate supply, it was determined that it was to *any* extent such a supply;
4   levy is accounted for on a half-rate supply as if it were neither a half-rate supply
     nor a reduced-rate supply;
5   after a charge has arisen for the supply of a taxable commodity to a person who
     uses it in producing taxable commodities primarily or for his own consumption,
     that recipient makes supplies of any of the resulting commodities;
6   the bad debt provisions apply; or
7   the making of a taxable supply gives rise to a double charge.[116]

For aggregates levy, credits can be generated where aggregate is:[117]

1   exported;[118] or
2   has an exempt process applied to it; or
3   is used in a prescribed industrial or agricultural process;[119] or
4   is disposed of in a manner not constituting use for construction; or
5   where the bad debt provisions apply.

To justify such a credit, documentation must be retained.[120]
     Where a taxpayer issues an invoice showing tax which is not in fact chargeable, that
tax is nonetheless payable to the Crown.[121]

16.10   Bad debt relief

Relief for bad debts under VAT has taken some time to read the situation under which
there is an add-back if the invoice remains unpaid after six months. Relief is limited
to the VAT proportion of what is unpaid, rather than a VAT-LIFO basis operated.
This has given rise to problems for some taxpayers. On the other hand, the scheme
has not been without its problems for Customs. At one stage the use of interposed
registrations[122] (which are, more often than not, created at the request of Customs)
gave rise at one stage to double deductions being sought, frequently as a result of
an administrative muddle, but sometimes by design, before the rules were changed.

---

116   *Ibid.*, Sched. 6, para. 21.
117   Finance Act 2001, s.30(1).
118   This is the export side of the BTA (see paras 1.2.1.2 and 8.4.5.1 above).
119   Aggregates Levy (General) Regulations 2002, S.I. 2002 No. 761, Sched.: reformulated
       with effect from 1 April 2003 – see *Business Brief* 29/02 (November 14, 2002).
120   Aggregates Levy (General) Regulations 2002, S.I. 2002 No. 761, regs 10(i) and (j).
121   Value Added Tax Act 1994, Sched. 11, para. 5(2),(3), Finance Act 1996, Sched. 5, para.
       44(1)(b).
122   Under Value Added Tax Act 1994, s.47.

More recently there has been the problem of how to enforce an input tax add-back by defaulting business customers.

All of the above seems to have made Customs wary of debt arrangements for environmental taxes, even though Parliament has provided for them in all cases.[123]

For landfill tax, the requirements include payment of the tax, invoicing within 14 days, the issue (and retention of a copy) of an invoice, the writing off of the debt, a year having elapsed and the recording of the claim.[124] A claim may not be made, however, where the transaction is with a connected person.[125]

For climate change levy, the conditions for relief include the issue of a climate change levy accounting document, the writing off of the debt and six months having elapsed.[126] In the case of this particular tax, provision is made for this relief to apply to transactions for non-monetary consideration.[127] Here again, a claim may not be made for a transaction with a connected person.[128]

For aggregates levy, relief is limited to situations in which a sale or a removal from a site are involved.[129] No relief is available where the chargeable event is mixture or use for construction purposes. Relief is also denied where the counterparty is a connected person.[130] It is also restricted to circumstances in which the customer has become insolvent or gone into liquidation.[131]

16.11   Transitional provisions

For landfill tax, adjustments may fall to be made in three sets of circumstances:

1   where, after a contract has been entered into, there is a change in the rate of tax, unless the contract provides for no adjustment to be made;[132]
2   where material undergoes a landfill disposal under a construction contract entered into before 30 November 1994, unless the contract provides for no adjustment to be made;[133] and
3   where a turnover rent arises under a contract entered into before 30 November

---

[123]   Finance Act 1996, s.52, Finance Act 2000, Sched. 6, para. 62(1)(f), Finance Act 2001, s.30(1)(e).
[124]   Landfill Tax Regulations 1996, S.I. 1996 No. 1527, regs 23(b), reg. 23(e)(i), reg. 23(d), reg. 23(c), reg. 25(a) and reg. 26 respectively.
[125]   *Ibid.*, reg. 23(a).
[126]   Climate Change Levy (General) Regulations 2001, S.I. 2001 No. 838, reg. 10(1)(c), reg. 10(1)(d) and reg. 10(1)(e) respectively.
[127]   *Ibid.*, reg. 10(7).
[128]   *Ibid.*, reg. 10(1)(b),(2) specifying Income and Corporation Taxes Act 1988, s.839 as relevant in this context.
[129]   Aggregates Levy (General) Regulations 2002, S.I. 2002 No. 761, reg. 12(1)(a).
[130]   *Ibid.*, reg. 12(1)(c), (2) specifying Income and Corporation Taxes Act 1988, s.839 as relevant in this context.
[131]   Aggregates Levy (General) Regulations 2002, S.I. 2002 No. 761, reg. 12(1)(e).
[132]   Finance Act 1996, Sched. 5, para. 45.
[133]   *Ibid.*, Sched. 5, para. 46.

1994 and it can reasonably be expected that landfill tax would have been ignored in calculating such turnover.[134]

For climate change levy, where a contract has been entered into before 1 April 2001, the supplier may change the price to include the levy.[135] Thereafter changes in the amount of the levy are taken into account, whatever the terms of the contract.[136]

For aggregates levy, only two circumstances are provided for:

1   where a contract has been entered into before 1 April 2002, for the supply of a quantity of aggregate;[137] and
2   where a turnover rent is paid under an agreement made before 1 April 2002, and the circumstances are such that it can reasonably be expected that aggregates levy would have been ignored in calculating such turnover.[138]

16.12   Record keeping and inspection

As with VAT, environmental taxes are normally accounted for quarterly.[139] Where a climate change levy payer has an annual liability of under £2,000, he is given the option of annual accounting.[140]

In order to achieve compliance, the taxpayer is placed under a duty to make and retain records and produce them for inspection,[141] and Customs are given power to enter and search for, copy and remove documents, and take samples.[142]

16.13   Disputes

Where Customs and the taxpayer do not agree on the sum due, Customs issue either an assessment or a ruling and the taxpayer has the right to ask for this to be reviewed.[143] In

---

134   *Ibid.*, Sched. 5, para. 47.
135   Finance Act 2000, Sched. 6, para. 142(1),(2).
136   *Ibid.*, Sched. 6, para. 142(3)–(5).
137   Finance Act 2001, s.43(1).
138   *Ibid.*, s.43(2).
139   Landfill Tax Regulations 1996, S.I. 1996 No. 1527, reg. 2(1); Climate Change Levy (General) Regulations 2001, S.I. 2001 No. 838, reg. 3; Aggregates Levy (General) Regulations 2002, S.I. 2002 No. 761, reg. 5.
140   Climate Change Levy (General) Regulations 2001, S.I. 2001 No. 838, regs 6A–6G.
141   Finance Act 1996, Sched. 5, paras 2–3; Finance Act 2000, Sched. 6, paras 125–7; Finance Act 2001, Sched. 7, paras 2–4.
142   Finance Act 1996, Sched. 5, paras 4–5,7–10; Finance Act 2000, Sched. 6, paras 5–10; Finance Act 2001, Sched. 7, paras 5–10.
143   Finance Act 1996, s.54, Sched. 5, para. 59; Finance Act 2000, Sched. 6, para. 121; Finance Act 2001, s.40: but it should be noted that value added tax reviews are non-statutory and that in relation to environmental taxes this means that the costs of any successful appeal to a tribunal from a review only run from after the completion of the review process (see *C & E Commrs v. Dave*, [2002] EWHC 969 (Ch), [2002] STC 900).

most cases, an appeal may then be lodged with a tribunal.[144] Appeals from tribunals lie to the Courts on points of law (see para. 4.2.1.5 above).

In some cases, refunds from Customs carry interest.[145] In the case of landfill tax[146] and aggregates levy,[147] however, refunds are restricted by the doctrine of unjust enrichment.

16.14 Irregularities

Where tax is overdue, it carries interest.[148]

Where tax is overdue regularly or returns are made incorrectly, Customs are able to levy civil penalties (which carry interest on a different basis), subject to control by the tribunals.[149]

In the case of fraudulent evasion, criminal proceedings can be taken.[150]

16.15 Enforcement

Customs are given wide powers, including arrest,[151] distress and diligence,[152] and requiring the provision of security.[153]

These must, furthermore, be read against the historical background of the Commissioners' original functions being directed substantially against smuggling, leaving them with greater powers of entry and detention without warrant than the police![154]

Where customs and excise duties are concerned, they also have the power to seize vehicles and other objects within which contraband is carried or concealed.[155] An

---

[144] Finance Act 1996, s.55; Finance Act 2000, Sched. 6, para. 122; Finance Act 2001, s.41.
[145] Finance Act 1996, Sched. 5, para. 29; Finance Act 2000, Sched. 6, para. 66; Finance Act 2001, Sched. 8, para. 2.
[146] Finance Act 1997, Sched. 5, paras 1–4.
[147] Finance Act 2001, s.32(2)–(4).
[148] Finance Act 1996, Sched. 5, paras 26–7; Finance Act 2000, Sched. 6, paras 70–71, 81–9; Finance Act 2001, Sched. 8, paras 6–7.
[149] Finance Act 1996, Sched. 5, paras 18–23; Finance Act 2000, Sched. 6, paras 98–113; Finance Act 2001, s.46, Sched. 6, paras 7–9.
[150] Finance Act 1996, Sched. 5, paras 15–17; Finance Act 2000, Sched. 6, paras 92–5; Finance Act 2001, Sched. 6, paras 1–4.
[151] Finance Act 2000, Sched. 6, para. 97; Finance Act 1996, Sched. 5, para. 6; Finance Act 2001, Sched. 6, para. 6.
[152] inance Act 1997, ss.51–2; Finance Act 2000, Sched. 6, para. 90; Finance Act 2001, Sched. 5, paras 14–16.
[153] Finance Act 1996, Sched. 5, para. 31; Finance Act 2000, Sched. 6, para. 139; Finance Act 2001, s.26.
[154] Customs and Excise Management Act 1979, ss. 138, 139: their attitude has also come in for criticism recently, see para. 22.1 below.
[155] Customs and Excise Management Act 1979, s.141(1).

appeal can be made to the Courts in condemnation proceedings,[156] or, if the seizure is not contested, application made to the Commissioners for the exercise of their discretion to restore on terms.[157] A decision not to do so may be reviewed internally, upon application. If the taxpayer is dissatisfied with that, there is the right of appeal to the VAT and Duties Tribunal, but only on a supervisory[158] basis and with the best outcome only the ordering of a new review.[159]

16.16   Problems for landowners

Finance Acts 2000 and 2001 contained a number of ill-thought out provisions imposing considerable potential burdens on landowners, especially on those who let property some years ago. Professional advisers would have needed a crystal ball to counter legislation which could not then have been within anyone's contemplation.[160]

The secondary liability provisions applicable to landfill tax provide a striking example. These were added to Part VIII of Sched. 2 to the Finance Act 1995 by Finance Act 2000. In the Committee Stage of the Finance Bill 2000, the provisions were not even debated. Even so, the 'controller' who can come within this concept (and thus become, in effect, an unsecured guarantor of somebody else's turnover tax liability) is somebody who *determines, or is entitled to determine, what disposals of material, if any, may be made.*[161] While most post-1990 leases will not give the landlord overriding control covenants,[162] leases drafted before that date would be likely to do so, in order to preclude the dumping of dangerous waste. Such landlords could, therefore, have become personally liable following this change in the law. Their position has not been made easier by the example given by Customs to the House of Commons in the Explanatory Notes to this part of the Finance Bill,[163] and, astonishingly in view of the reaction to those, repeated in a *Landfill Tax Briefing* issued on 7 August 2000.[164] This example of the situation giving rise to concern was one in which a farmer had let a site to an operator and the farmer held the waste disposal licence. Bearing in mind that a tenant has legal occupation of the land, this situation was not one which should have arisen under Environmental Protection Act 1990, s.35(2)(a).[165]

The problem with landfill tax persisted in relation to the aggregates levy legislation, which had been issued in draft at the time of Finance Bill 2000, and which was placed

---

156   *Ibid.*, Sched. 3. See para. 4.2.1.5 above.
157   *Ibid.*, s.152(b).
158   That is, that the reviewer's decision was unreasonable in the sense of having either applied the law wrongly or taken irrelevant material into account.
159   Finance Act 1994, s.16(4).
160   See de Souza, *Taxation*, 21 March 2002, pp. 609–10.
161   Finance Act 1996, Sched. 5, para. 48(1).
162   Because of the fear of overlapping with the responsibilities of what is now the Environment Agency (see para. 4.2.1.3 above).
163   Explanatory Notes, para. 24.
164   *Ibid.*, para. 1. See de Souza, *Taxation*, 7 September 2000, p. 600.
165   As enacted. See para. 6.3.2.1 above.

before Parliament under a shortened scrutiny procedure in order to enable the 2001 General Election to be held in May.[166] No points of detail were debated, the House of Commons' available time being, rather understandably, directed to the highly dubious economic and environmental justifications put forward for the very existence of the tax. The problem is, yet again, that of *occupation*,[167] the point being of considerable significance because building sites are involved and, for obvious reasons, developers who are to obtain long leases are let into possession under licences, the lease being granted only when the landlords' surveyor is satisfied that they have been erected in accordance with the agreed specification. In such circumstances, legal *occupation* is retained by the landlord.

The issue could also arise, however, in relation to quarries where the taxpayer's accountant has sought to avoid liability to income tax under Income and Corporation Taxes Act 1988, s.34(1) by instructing the solicitor to draw the agreement up as a licence, rather than a lease.[168]

Concern had also been expressed at the prospect of landowner liability under Finance Act 2001, s.22(1)(e) or (f), where a mixing process carried out by another was not such as to require the exclusive possession which the grant of a lease would create and resulted in the production of other products. The only one identified at the outset was coated limestone, clearly not one to be created in such circumstances, but developments in technology could easily result in others which might.[169]

No definition of 'occupation' was inserted in the statute. The question therefore arose as to whether the rating concept of *occupier* was to apply instead – under this the person actually in possession is liable.[170] Uncertainty on this persisted until after the registration period for the tax had expired, Customs telling a representative body[171]

---

[166] The consultation process in relation to the draft had, however, been extremely drawn out. The first of these appears to have been produced at a time when a value (rather than weight) basis of taxation was Customs' preferred option. Had such a basis been practicable (and it was found not to be) it is possible that a multiple accounting party arrangement would have been somewhat less unreasonable in principle. In the events which have happened, Customs take the view that the wide range of potential registrables is merely there for the protection of the revenue, and that in practice the one with real control over the operation has to be picked as a sole registered taxpayer, with the duties attributed to registrables confined to him alone. While the former must be regarded as envisaged by the statute, a 'guarantor' liability situation for the other can clearly arise under it.

[167] Finance Act 2001, ss.21(1)(a), 22(1)(e).

[168] It also needs to be noted, in this context, that in an old rating case, *Andrews v. Hereford RDC*, (1964) 10 RRC 1, a farmer was held to be the rateable occupier where the extractor appears to have been granted some form of non-exclusive licence. But this case was, however, doubted in *Re Briant Colour Printing Co Ltd*, [1977] 1 WLR 942, and (in any event) subsequent health and safety legislation would have made it extremely unlikely that those particular circumstances would be repeated.

[169] Jeremy de Souza, (1999) 63 Conv 7.

[170] *Re Briant Colour Printing Co Ltd*, [1977] 1 WLR 942, *John Laing & Son Ltd v. Kingswood Area Assessment Committee*, [1949] 1 KB 344, at 350, *London County Council v. Williams*, [1957] AC 362, *Ratford v. Northavon District Council*, [1987] 1 QB 357, at 377H, *Verrall v. Hackney LBC*, [1983] 1 QB 445, at 462E.

[171] The Country Land and Business Association on 15 April 2002.

that the word *should be interpreted in its literal sense as physical occupation of the site*.[172] Nevertheless, this issue remains open, unless and until a Court has a chance to adjudicate on the matter. The seriousness of the position is that if a landowner comes within the concept, he becomes, in effect, a guarantor of the operator's turnover[173] tax, that is, in this case, landfill tax.[174]

Unfortunately, these are not the only areas in which the hazards to landlords have not been properly thought out. Climate change levy, and the attendant UK ETS, contains provisions which could create problems for them:

1   where a tenant becomes insolvent, and had been a party to a climate change agreement,[175] the landlord does not automatically succeed to it. Indeed, as a non-member of the former tenant's trade association (which will have entered into the umbrella agreement),[176] it is likely to be difficult to accede because the main agreement will have been made by that association. Negotiating a stand alone agreement is likely to be difficult. Power supplies could therefore become liable to the full rate of levy.[177]

2   Where a former tenant has been receiving energy supplies from a non-resident non-utility, it will have had to register and account for the levy as the recipient of supplies liable to the levy.[178] This is unlikely to be known to the landlord and, there being no obligation upon suppliers accounting for the levy to deliver an invoice to their customers, may not immediately be apparent.

3   With regard to the interface between climate change levy and the UK ETS, when the UK ETS's rules were finalised on 1 February 2002, the crucial concept of *change of operation* was defined by reference to the concept of *management control*, as set out in the proposals published on 14 August 2001. Under these it seemed that, in some cases, landlords could find themselves within this concept, for instance the following formulation:

Alternatively, a Direct Participant could exercise dominant influence over the emissions from a source by virtue of the terms and conditions contained in the contract governing the operation of the source.[179]

This could well prove embarrassing for all parties if this was first discovered (as indeed seems more likely than not) during due diligence for a take-over.

---

172   The subsequent decision of a tribunal on the basis of VAT liability under Value Added Tax Act 1994, Sched. 10, para. 3A(7)(a), in *Brambletye School Trust Ltd v. C & E Commrs*, LON/00/458, might, however, be cited in favour of this view.

173   Rather than one based on profits, and therefore a much more onerous obligation.

174   Finance Act 2001, s.22(3). Furthermore, the transitional provisions take no account of the possibility that he might, indeed, even be selected for registration.

175   Finance Act 2000, Sched. 6, para. 50. See para. 14.6 above.

176   *Ibid.*, para. 48. See para. 14.6 above.

177   *Ibid.*, para. 49(2).

178   *Ibid.*, Sched. 6, para. 40(2).

179   *Ibid.*, Annex A, para. A4.

4   There are understood to be some situations in which the terms of the pre-2000 lease may not permit the landlord to pass on the levy element of his disbursements to the tenants.[180] In such circumstances, legal advice must clearly be taken as to the situation and the options available to the landlord to overcome it.

5   There is, furthermore, a problem that Customs acknowledge[181] in relation to the operation of climate change levy. The tax is applied to supplies from suppliers to businesses.[182] If the supply is to the landlord of premises in multiple occupation, the following situations could arise:

a.   none of the occupiers is entitled to exemption, in which case no problem arises, because the tax is due and has just been paid one step up the line;

b.   all of the occupiers are entitled to exemption, in which case the landlord has to certify this to the supplier;[183]

c.   there is a mixed situation, in which case one of the following courses of action must be adopted:

i.   arrangements are made for the supplier to make direct supplies to the 'odd ones out'; or

ii.   Customs are requested to register the landlord as a deemed utility,[184] so that sub-certification can take place[185] – Customs have, however, very fairly made the point that landlords will incur not only administrative costs, but also recovery delays if they agree to do this;[186] or

iii. landlord and tenants come to an agreement as to how the cost of the levy is borne between them; or

iv.   things will have to be left as they are and the appropriate proportion of the levy collected from each tenant, irrespective of his personal 'direct' exemption position.

---

[180]   Kennedy and McClenaghan, *The Tax Journal*, 21 May 2001, p. 6.

[181]   *Technical Briefing* No. 8 (March 2002) (available from www.hmce.gov.uk).

[182]   See para. 14.2 above.

[183]   Although he cannot where direct certification to the supplier by the end-user has been specified in Finance Act 2000, Sched. 6, that is, in relation to non-fuel or horticultural use.

[184]   Seemingly under Finance Act 2000, Sched. 6, para. 27(3), in which case it is not possible to claim renewables or CHP relief (see paras 19(1)(a),20A(1)(a)). No such persons appeared on the list of *deemed* utilities published by Customs in May 2002 on www.hmce.gov.uk.

[185]   Although this is not normally done retrospectively: *Technical Briefing* No. 9 (March 2002) (available from www.hmce.gov.uk).

[186]   *Technical Briefing* No. 8 (March 2002) (available from www.hmce.gov.uk): as to the subsequent uptake of which, see the revised of *deemed utilities* list issued by Customs on 6 May 2003 (the list for the time being being available from www.hmce.gov.uk).

Division 3
Local Levies

Chapter 17

Workplace Parking Levies

17.1   Introduction

The concept of workplace parking levy was introduced in the legislation setting up the new Mayor and Assembly for Greater London.[1] Similar powers were subsequently given to other local authorities in England and Wales,[2] and this legislation also amended the earlier provisions relating to London.[3]

However, whereas in London, it was road user charging which hit the headlines, the government seems to have intended workplace parking charges to be pioneered outside London. During the passage of the legislation, the Department of the Environment, Transport and the Regions ('the DETR')[4] appears first to have tried, through an approach to Hampshire County Council, to obtain the consent of Winchester City Council to act as guinea pig, based apparently on the fact that the Council controlled most of the inner city parking and that a park-and-ride scheme (albeit one under which most of the traffic had to pass through the City to get to it) was in place.[5] This did not prove to be acceptable to the City Council.[6]

The government next turned its attention to the West Midlands,[7] but ran into opposition from Coventry, Solihull and Walsall.[8] Nottingham City Council seems to have been quite enthusiastic, however, proposing a daily charge of 70p, but ran into heavy opposition from local employers, even though Boots plc (and others)[9] intended to pass the cost on to the employee.[10]

Two years after the passage of the relevant legislation no such scheme was in operation. Nottingham still had the possibility of a charge of £150 per annum on the agenda, however, to start in April 2005.

---

1   Greater London Authority Act 1999, s.296 and Sched. 24.
2   Transport Act 2000, ss.178–90.
3   *Ibid.*, Sched. 13, paras 19–43.
4   See para. 4.2.1.2(1) above.
5   *South Hants Weekly News*, 11 March 1999, p. 6.
6   *Ibid.*, 28 March 1999, p. 12.
7   Jowit, *Financial Times*, 20 May 1999, p. 10.
8   Guthrie, *Financial Times*, 25 May 2000, p. 7.
9   That is, the landowners, from whom the charge would be collected.
10   Ward, *Evening Post*, 22 January 2000, pp. 1–2. By 2004, the proposal had become one for £150 p.a. per parking space over and above an exempt maximum of ten, expected to rise to £300 over a decade. It appears that Councillors thought that this would be politically acceptable because the vast majority of commuters lived outside that local authority district's boundaries (see Guthrie, *Financial Times*, 14–15 February 2004, p. 3).

17.2   Changes in planning practice

It is indicative of the previous culture (of encouraging the provision of office parking) that, from 1988, car parking facilities were provided for employees at their employer's expense should have been free of income tax.[11] (Somewhat surprisingly, only since 1999, following the point being taken on a PAYE inspection of a FTSE 100 company's main premises, has this been extended to the parking of motor cycles and bicycles.)[12]

In 1997, the incoming Labour Government had changed the general planning guidance from one under which developers were expected to provide adequate off-street parking to one in which this was discouraged in order to reduce private car commuting in favour of the use of public transport.

Indeed in 1999, while the relevant legislation was still before Parliament, it was reported that a charity carrying out a mixed residential and retail development in central Winchester with no on-site parking (and for which it had been made clear that residential parking permits would not be made available by the City Council) had been asked by Hampshire County Council to pay for 19 car parking spaces in the out-of-town park-and-ride at a cost of £3,300 per space.[13]

17.3   Double payment

Transport Act 2000 contained no provision for the reduction of the rateable value of premises subject to workplace parking levy, despite the fact that there was recent authority making it clear that the availability of parking had to be taken into account in arriving at rateable value.[14]

Also absent was a provision permitting tenants who had to pay this levy to apply to the court for a downward revision of their rents. Three factors should be noted in this context:

1   rateable value was based on the open market rental, so that the identification of the attribution of the former to car parking as such was clear evidence of its availability being reflected in rental values;
2   until the late 1990s, it was common practice for new office developments to be let on the basis of a 25 year lease with upwards-only rent reviews; and
3   between the mid-1970s and 1997, local planning authorities normally only gave permission to developers of such accommodation.

---

11   Income and Corporation Taxes Act 1988, s.197A; Income Tax (Earnings and Pensions) Act 2003, s.237.
12   Finance Act 1999, s.49.
13   *South Hants and Winchester Weekly News*, 16 December 1999, p. 1.
14   *Benjamin v. Austin Properties Ltd*, [1998] 19 EG 163.

17.4   Metropolitan schemes

In London, either Transport for London or the Common Council (of the City) or any London Borough Council may establish a scheme to licence persons to provide workplace parking.[15] The scheme is made by the relevant authority and confirmed by the Mayor[16] acting on behalf of the Greater London Authority[17] and has to conform to the Mayor's transport strategy.[18] Each scheme has to designate the areas and times at which it applies and specify rates of charge,[19] which may be differential.[20] Although joint licensing schemes are permissible,[21] the same premises cannot be subject to more than one scheme.[22] Licences may not be granted for a period of longer than a year,[23] and the Secretary of State is empowered to prescribe exemptions for numbers of spaces, classes of vehicle and reductions in and limits on rates of charge.[24] The scheme may also provide for the net proceeds of charging to be divided between the Greater London Authority, Transport for London and certain Metropolitan Boroughs for relevant transport purposes[25] under both a ten-year and a four-year plan.[26]

17.5   Other schemes

These schemes operate in relation to a maximum number of parking spaces in premises covered by them (as specified in the licence application),[27] and are normally the responsibility of the occupier of the premises in question.[28] They are made by non-metropolitan traffic authorities, or a combination of them and London traffic authorities,[29] over the whole or part of their area.[30] In the case of a joint scheme, provision has to be made for the net recoveries to be apportioned between the relevant authorities.[31] The order is made by the relevant licensing authority,[32] which has to specify the areas and hours to which it relates and specify the rates of charge.[33] It

---

[15]   Greater London Authority Act 1999, s.296. See para. 4.2.3 above.
[16]   *Ibid.*, Sched. 24, para. 2.
[17]   *Ibid.*, Sched. 24, para. 7.
[18]   *Ibid.*, Sched. 24, para. 8.
[19]   *Ibid.*, Sched. 24, para. 11.
[20]   *Ibid.*, Sched. 24, para. 13.
[21]   *Ibid.*, Sched. 24, para. 8.
[22]   *Ibid.*, Sched. 24, para. 12.
[23]   *Ibid.*, Sched. 24, para. 15.
[24]   *Ibid.*, Sched. 24, para. 17.
[25]   *Ibid.*, Sched. 24, para. 24.
[26]   *Ibid.*, Sched. 24, paras 25–28.
[27]   Transport Act 2000, ss.178(1),(4), 188.
[28]   *Ibid.*, s.178(2)(a).
[29]   *Ibid.*, s.178(5).
[30]   *Ibid.*, ss.179(1), 180(1), 181(1).
[31]   *Ibid.*, Sched. 12, para. 3.
[32]   *Ibid.*, s.183.
[33]   *Ibid.*, s.186.

has to be confirmed by the Secretary of State in England and the National Assembly for Wales in that Principality,[34] both of which may restrict the scope of schemes by regulation in relation to exemptions and reductions in or limits on charges.[35]

17.6   Those subject to schemes

Licensing schemes, when made, apply (within the relevant area) where the occupier of premises provides a parking space occupied by a motor vehicle of his (or another person with whom he has made an arrangement or, in the case of a company, a member of the same group)[36] for attending at a place (or in its vicinity)[37] where that person carried on a trade, profession, the functions of office holder, government department, local authority or other statutory body[38] being the vehicle of that person, his employees, agents, suppliers, business customers or business visitors or by a pupil or student if the business is the provision of education or training or a local councillor in the case of the local authority.[39] It should be noted that this formulation could include the making available by a householder of parking, by arrangement, to a neighbouring businessman.

---

[34]   Transport Act 2000, ss.184, 198(1).
[35]   *Ibid.*, s.187.
[36]   Greater London Authority Act 1999, Sched. 24, para. 3(2),(3); Transport Act 2000, s.182(2),(3).
[37]   Greater London Authority Act 1999, Sched. 24, para. 3(1); Transport Act 2000, s.182(1).
[38]   Greater London Authority Act 1999, Sched. 24, para. 3(4); Transport Act 2000, s.182(4).
[39]   Greater London Authority Act 1999, Sched. 24, para. 3(1),(4); Transport Act 2000, s.182(1),(4).

Chapter 18

Road User Charging Schemes

18.1   Introduction

Although first advocated in the Smeed Report in 1964,[1] road user charging was not invented in the UK, with Singapore, Oslo, Bergen, Trondheim and Stavanger having them up and running before any British scheme was authorised.[2] In advocating an increase in this,[3] a Report published on 20 July 2004, with the White Paper *The Future of Transport*, entitled a *Feasibility Study of Road Pricing in the UK*, emphasised the need for the establishment of public trust.[4] The Report revealed that one-fifth of the UK's total carbon emissions came from road traffic and that, with engines operating less efficiently, these emissions were higher in congested areas.[5]

18.2   Central London

The concept of road user charging was introduced in the legislation setting up the new Mayor and Assembly for Greater London.[6] Most of the publicity generated by this concept has been related to these powers, in the context of Mr Ken Livingstone's plan to charge[7] £5 per day from 17 February 2003, for each vehicle[8] entering the

---

[1]   *Road Pricing: The Economic and Technical Possibilities*: HMSO 1964, publication of which is understood to have been delayed until after the General Election of that year by the Conservative Minister of Transport, Mr Ernest Marples.

[2]   Jowit, *Financial Times*, 22 August 2002, p. 21. See, generally, Stephen Ison, *Road User Charging: Issues and Policies* (Aldershot: Ashgate, 2004), esp. ch. 3. The Norwegian schemes were, however, of limited duration and to fund infrastructural development: House of Commons Transport Committee, *1st Report. Urban Charging Schemes* (House of Commons Papers, Session 2002–03, 390–I) (London: Stationery Office, 2002). In this context, it is worthy of note that Nottingham's espousal of the workplace parking levy – see para. 17.1 above – was intended to achieve a combination of specific objectives, namely the subsidisation of local bus and train services and the funding of a new tramway system (House of Commons Transport Committee, *op. cit.*, para. 37).

[3]   And extolling, inter alia, the establishment of 'HOT' commuter freeway lanes in California (House of Commons Transport Committee, *op. cit.*, para. 2.17).

[4]   House of Commons Transport Committee, *op. cit.*, para. 2.21.

[5]   *Ibid.*, para. 3.7.

[6]   Greater London Authority Act 1999, s.295 and Sched. 23.

[7]   Payment has to be made by 10 pm on the day in question and may be achieved by telephone, email, at machines in car parks (for example, the National Theatre) and some retailers.

[8]   After Mr Livingstone's re-election as Mayor of London in June 2004, consideration was given to doubling the congestion charge for 4×4 vehicles, which an interim report from the Liberal Democrat group in the London Assembly were purchased by 3 per cent more

central London 'box' between 7 am and 6.30 pm Monday to Friday (with exemption for certain specified categories):[9]

1   vehicles for the disabled displaying a blue badge or exempt from vehicle excise duty;
2   emergency service vehicles, the armed forces, the Coastguard, Port of London authority and lifeboat service and certain vehicles of the London Fire Brigade, the eight local authorities whose areas are covered and the Royal Parks Agency;
3   certain National Health Service staff and patient journeys, including those by on-call emergency services, and NHS vehicles exempt from vehicle excise duty;
4   buses, coaches, community minibuses, 'black' taxis and London licensed minicabs;
5   motorcycles;[10]
6   electrically propelled vehicles and the 'cleanest' alternative fuel vehicles; and
7   special recovery and breakdown vehicles.

The reliefs are:

1   90 per cent discounts for residents;[11] and
2   reductions for vehicles using environmentally-friendly fuels.[12]

The authorities will no doubt have been gratified to learn of employers' general refusal to meet the bill for their employees' cars.[13] The effect of the scheme on retailers within the charging zone was not positive. It is, however, a matter of

---

Londoners than in UK as a whole (see Blitz, *Financial Times*, 14 July 2004, p. 6). This news came out shortly after the French Government revealed plans to impose a tax on 3,200 Euro on (*inter alia*) new cars of this type from January 2005 (see Johnson, *Financial Times*, 3–4 July 2004, p. 6).

9   A challenge to the Mayor of London's proposal in July 2002 by Westminster City Council and some residents of Lambeth, based largely on the absence of an environmental impact assessment was rejected by the High Court: *R (on account of The Mayor, Citizens of Westminster and others) v. The Mayor of London*, [2002] EWHC 2440 (Admin), and see G. Sector, 'Sleepless Nights at City Hall', NLJ, 24 January 2003, pp. 82–96.

10   Which seems to have generated a purchasing boom, giving rise to consideration of the possibility of the rules being changed before the scheme started (see Jowit and Griffiths, *Financial Times*, 20 January 2003, p. 4).

11   Of 90 per cent, although the saving of £1,000 p.a. would go nowhere towards funding the premium payable for residential property within the zone (see Brun-Rovet, *Financial Times*, 6 January 2003, p. 4).

12   For those powered by hydrogen, electricity or liquified petroleum gas, this is 100 per cent. This relief was welcomed by the House of Commons Transport Committee, *op. cit.*, para. 78. It should be noted, in this context, that the World Business Council for Sustainable Development's *Mobility 2030* Report, which was backed by eight international car makers, had concluded that fiscal reform, rather than regulation, was required to induce consumers to buy environmentally-friendly cars (see Mason, *Financial Times*, 6 July 2004, p. 9).

13   Buck, *Financial Times*, 6 January 2003, p. 4.

dispute how much of the reduced trade in Oxford Street was caused by this, rather than the effect of terrorism and the world economic slowdown on trade from tourists.[14]

The Mayor's objectives were:

1 to cut traffic by 10 to 15 per cent;
2 to reduce congestion by double that;[15] and
3 to generate a net operating profit of £121m.[16]

During the first fortnight, traffic in central London appears to have been cut by 20 per cent.[17] It was anticipated that this would have the following consequences:

1 the Mayor of London would extend the scheme to cover other central areas, initially the rest of Westminster, Kensington and Chelsea;[18] and
2 other UK cities were likely to proceed with further consideration and/or implementation of schemes of their own.[19]

---

[14] Wight, *Financial Times*, 5 September 2003, p. 3. Another factor, not canvassed in that article, might be the increasing difficulty of getting into London on the railway network. There was, however, a direct effect on some businesses (see Blitz, *Financial Times*, 27 November 2003, p. 6; Blitz, *Financial Times*, 14–15 February 2004, p. 3; Fifield, *Financial Times*, 16 February 2004, p. 4). In April 2004, a study by Imperial College, London for the John Lewis Partnership attributed a decline of 8.2 per cent in the takings of its Oxford Street store to the introduction of the congestion charge (Blitz, *Financial Times*, 22 April 2004, p. 6).

[15] In a Report published on 20 July 2004, with the White Paper *The Future of Transport*, Cm 6234, entitled a *Feasibility Study of Road Pricing in the UK*, para. 3.12, it was said that a 15 per cent fall in traffic had led to an average 30 per cent reduction in congestion. It was also suggested, in para. 3.8, that there had been a proportionately greater decrease in accidents involving personal injury within the charging zone.

[16] Newman, *Financial Times*, 8–9 February 2003, p. 1.

[17] Jowit and Blitz, *Financial Times*, 1–2 March 2003, p. 4. It later settled down at between 17 and 18 per cent: Jowit, *Financial Times*, 15 April 2003, p. 11. There was subsequently a reverter to 20 per cent, coupled with a realisation that there had been no material extra congestion on outer London routes which it had been thought would be affected adversely by the creation of the central charging zone, leading to puzzlement as to where the traffic had gone (see Blitz, *Financial Times*, 17–18 May 2003, p. 5).

[18] Possibly at half the central zone rate: although a poll commissioned by the London Assembly showed 53 per cent of the residents of Westminster and Kensington and Chelsea as favouring an extension of the existing zone: Blitz, *Financial Times*, July 25, 2003, p.4. From the point of view of efficiency, an extension of the boundaries of the initial zone at the same daily charge would be controversial because it would create a large number of 90 per cent discounted motorists. The Mayor, however, seems to be thinking in terms of a pilot scheme for the whole country in 2010 on roads such as the North Circular (see Blitz, *Financial Times*, 24 October 2003, p. 3).

[19] See para. 18.3 below.

By the anniversary, the reduction in traffic had stabilised at 20 per cent, with 17,000 travellers seeming to have switched to the buses.[20] There were two unexpected fiscal consequences of the London scheme:

1   Employers did not normally agree to reimburse the congestion charge to staff bringing their cars to work, but where they did so and company cars had been used, there was no addition to the benefit scale for which the employee was assessed.[21]
2   Where subsidised child care facilities for which no benefit charge was assessed on the employee[22] were available, these were likely to be close to the place of work, rather than the employee's home. Employees of businesses within central London would almost certainly have no alternative means of transporting young children than by car, and were not exempted from liability for the congestion charge. It is worthy of note that this underlying problem was not addressed in the Inland Revenue's proposals, issued in late February 2003, for the revision of child care benefit relief.

18.3   Other cities

Similar powers were subsequently given to other local authorities in England and Wales,[23] and this legislation also amended the earlier provisions relating to London.[24] In August 2002, Ministerial approval was given to a £2 per diem charge for entering the centre of Durham between 10 am and 4 pm Monday to Saturday. The scheme started on 1 October 2002. Bristol is understood to be considering a charge of £5 per day and Cardiff and Chester are also giving consideration to the possibility of congestion charging.

Were a similar scheme to London's to be implemented throughout England, it has been estimated that it might raise £16bn per annum, increasing motoring costs for urban motorists from 12p to 20p per kilometre.[25]

---

20   Blitz, *Financial Times*, 14–15 February 2004, p. 3.
21   Deloitte and Touche Press Release, 26 February 2003. It is interesting to note, in this context, that the House of Commons Transport Committee was concerned at the effect of the charging zone on employees with antisocial hours on low wages (see House of Commons Transport Committee, *op. cit.*, para. 73).
22   Under Income and Corporation Taxes Act 1988, s.155A; Income Tax (Earnings and Pensions) Act 2003, s.318.
23   Transport Act 2000, ss.163–77.
24   *Ibid.*, Sched. 13, paras 1–18.
25   Blitz, *Financial Times*, 14 October 2003, p. 8. The House of Commons Transport Committee drew attention to the fact that the underlying policy was to *reduce congestion*, without causing problems elsewhere, and that the current model of measuring congestion was unsuitable (see House of Commons Transport Committee, *op. cit.*, paras 15, 20 and 26 respectively). It was also observed that the 'by default' adoption of a twin track policy of workplace parking and congestion charging contributed to a Departmental policy muddle (*ibid.*, para. 132). In this context, it is worthy of note that the main objective of the City Councils in Bristol and Durham, the protection of the City centre, was at variance with the Government's preferred economic criteria (*ibid.*, paras 42, 43).

In Scotland, Edinburgh is considering introducing a charge of £2 per day from 2006. For this the City Council will have to hold a public enquiry, followed by a referendum.[26]

18.4   Metropolitan schemes

Transport for London, a London Borough Council or the Common Council of the City of London may establish schemes for 'the keeping or use of motor vehicles on roads' in their areas.[27] The scheme is made by the relevant authority and confirmed by the Greater London Authority,[28] acting through the Mayor,[29] (and which may require London Boroughs to implement a joint scheme)[30] and must be in conformity with the Mayor's transport strategy.[31] The scheme has to designate the area to which it applied, specify the classes of motor vehicle to which it applies, designate the roads in respect of which a charge is imposed, and specify the charges.[32] Modifications to the boundaries and the roads may, however, be required by the Mayor on behalf of the Greater London Authority.[33] The Secretary of State has power to specify exemptions and reductions in or limits on charges.[34] The person liable, who may be the registered keeper (that is, registered at Swansea) unless he can show that some other person was the driver, is to be provided for in regulations.[35] The net proceeds of any scheme for the first ten years of potential operation may only be applied for relevant transport purposes.[36]

18.5   Other schemes

Charges can be imposed on the registered keeper of a vehicle (that is, the person registered at Swansea)[37] under a charging scheme 'in respect of the use or keeping of motor vehicles on roads'[38] made by a non-metropolitan traffic authority alone or jointly with another such or a London traffic authority, in respect of its own roads,[39] or (in relation to a trunk road)[40] by the Secretary of State in England or the National Assembly for Wales in that Principality.[41]

---

26   Nicholson, *Financial Times*, 1 October 2002, p. 2; 24 February 2004, p. 6.
27   Greater London Authority Act 1999, s.295(1).
28   *Ibid.*, Sched. 23, para. 4.
29   *Ibid.*, Sched. 23, para. 2.
30   *Ibid.*, Sched. 23, para. 7.
31   *Ibid.*, Sched. 23, para. 5.
32   *Ibid.*, Sched. 23, paras 8, 10.
33   *Ibid.*, Sched. 23, para. 9. See para. 4.2.3 above.
34   *Ibid.*, Sched. 23, para. 11.
35   *Ibid.*, Sched. 23, para. 12. See para. 4.2.1.2n above.
36   *Ibid.*, Sched. 23, paras 16–22.
37   Transport Act 2000, s.163(2).
38   *Ibid.*, s.163(1).
39   *Ibid.*, ss.164(1), 165(1), 166(1). See para. 4.2.3 above.
40   *Ibid.*, s.167.
41   *Ibid.*, s.163(3). See para. 4.2.2 above.

Orders are made by the relevant authority[42] and confirmed by the Secretary of State or National Assembly for Wales (as appropriate).[43] The scheme has to designate the roads, classes of motor vehicle, chargeable events, rates of charge and the period for which the scheme is to continue in force,[44] but is subject to the relevant confirming authority's power to specify, by regulation, exemption and reduced rates and limits on charges.[45] In the case of a joint scheme, the net charges are to be apportioned.[46]

One way of removing cars from the roads of conurbations is to provide additional rail capacity. Central to this strategy as far as London is concerned is the long-mooted Crossrail link involving a tunnel between Paddington and Liverpool Street. A special tax on businesses benefited is under consideration in this context.[47]

---

[42] *Ibid.*, s.168.
[43] *Ibid.*, ss.169, 170, 198(1). See para. 4.2.2 above.
[44] Transport Act 2000, s.171.
[45] *Ibid.*, s.172.
[46] *Ibid.*, Sched. 12, para. 3.
[47] Blitz, *Financial Times*, 15 July 2003, p. 3. Light rail schemes elsewhere, including (significantly, Nottingham) may require taxpayer subsidy (see Wright, *Financial Times*, 13–14 December 2003, p. 4).

Division 4
Other Economic Instruments

Chapter 19

The Packaging Regime Route

19.1   Introduction

The Packaging and Packaging Waste Directive[1] was brought into effect in the UK
on 6 March 1997, by the Producer Responsibility Obligations (Packaging Waste)
Regulations 1997,[2] made under powers conferred by Environment Act 1995, ss.93–5.
   The aim of the Directive is to reduce the environmental effect of packaging and the
waste from it by a combination of:[3]

1   packaging waste prevention;
2   increased reuse through recycling and recovery; and
3   reductions in the quantity of packaging discarded.

The Directive set targets for recovery and recycling and Member States were required
to implement it by 30 June 1996, with the new regime coming into force on 1 January
1997.[4] The UK was not the only Member State to fail to meet the deadline.
   The Directive sets out a general framework. Within that, each Member State can
adopt a discrete approach. Germany and Denmark were amongst those which had
schemes already in existence. Although one of the stated purposes of the Directive
was to harmonise national systems, companies trading across national frontiers are
faced with having to comply with a number of separate regimes.
   The Environment Agency[5] has placed on record[6] that, in 2001, 9,000 companies in
England and Wales came within the original legislation and 3.5m tonnes of packing
waste had been recycled in the UK. Seeing its function as kick-starting sustainability,
Mr Howard Thorp was quoted as saying:

> Producer Responsibility could be the key to making that shift in industrial production and
> consumer behaviour. The concept is simple. It tackles the generation of waste at source.
> It encourages producers to prevent pollution, reduce use of energy and other resources
> throughout the life cycle of a product by better design and production. It is prevention
> rather than cure.
>   Producer responsibility makes producers take some responsibility for the environmental
> impacts of their products, including the raw materials and what happens after the product
> is used.[7]

---

[1]   European Parliament and Council Directive 94/62/EC, (1994) OJ L365 10.
[2]   S.I. 1997 No. 648. See para. 1.4.2.1(2) above.
[3]   Council Directive 94/62/EC, Art. 4(2).
[4]   See para. 12.2.5.1(2) above.
[5]   See para. 4.2.1.3 above.
[6]   Press Release of 29 May 2002.
[7]   *Ibid.*

19.2 Packaging

This concept is widely defined,[8] being divided into three categories – sales packaging, group packaging and transport packaging. Containers are, however, excluded from the last of these. There is, however, a lack of uniformity between Member States as to what is covered. It was revealed in the test case about flower pots, *Davies v. Hillier Nurseries Ltd,*[9] that nine Member States had concluded that plastic pots were covered, France had not and the position was unclear elsewhere. In that case, the conclusion of the stipendiary magistrate, that the items under consideration were not 'packaging', was overruled by the Divisional Court.[10]

The basis of that decision is instructive. Hilliers are well-known nurserymen. They grew plants and shrubs from cuttings with a view to selling them both to trade customers and the public. The cuttings were not sold, but nurtured to a size at which they could be sold for replanting. Once rooted, a cutting was placed in a liner. It was subsequently transferred to a plastic pot, the size being determined by the needs of the plant, to continue its growth to the size at which it was offered for sale. After sale, it was expected that a high proportion of purchasers would remove the plant from the pot.[11] The question at issue was whether the plastic pot was 'packaging' as defined by the Producer Responsibility Obligations (Packaging Waste) Regulations 1997, reg. 2(1)(a):

> sales packaging or primary packaging, that is to say packaging *conceived* so as to constitute a sales unit to the final user or consumer at the point of purchase.[12]

At the time the plants were potted, the purpose was to nurture them to a saleable size. The pot therefore performed the dual function of keeping the plant in good condition until replanted and containing the plant for the purchaser's ease and convenience in handling and transporting it.

The magistrate concluded that the pot was not within the definition of 'packaging' because its primary purpose was the growing and nurturing of the plant, and the Divisional Court concluded that he was entitled to come to that conclusion of fact. The question that should have been addressed, however, was the meaning of the work 'conceived'. The Divisional Court concluded that this required consideration of all the circumstances present at the time of the potting of the plant, in order to determine whether its use *at that time* (rather than the time of sale) included its (contemplated) use in the sales process, even though the growing and nurturing function of the pot could not be regarded as incidental to the sales process. Although the primary purpose

---

8   Producer Responsibility Obligations (Packaging Waste) Regulations 1997, Sched. 3.
9   (2001) Env LR 726.
10 See para. 4.2.1.5 above.
11 Although not gone into by the Divisional Court, the underlying concept in the Directive is that packaging becomes waste when the owner discards it or intends to discard it (see para. 12.2.5.2 above). This can be for a variety of reasons, including becoming obsolete or inappropriate.
12 Emphasis added.

of potting was growing and nurturing, the plants were 'conceived so as to constitute a sales unit and, at the time of sale, constitute a sales unit' and were 'packaging' within the relevant definition. Hilliers should therefore have registered as a producer under the Regulations.[13]

19.3   Choice

Back in 1997, packaging users had until 31 August to decide how to meet their recycling an recovery obligations. The choice was either to register with the official agencies or to join a collective scheme that took on their obligations.[14]

19.4   Registration

Where a producer is liable to be registered and is not a member of a collective scheme,[15] it must do so every year[16] with the appropriate agency[17] if it comes within the relevant criteria set out in Schedule 1 to the 1997 Regulations – basically having an annual turnover of packaging or packaging materials in excess of 50 tonnes.

In England, the appropriate agency is the Environment Agency,[18] whose practice is to prosecute those who fail to register.[19] A fee is charged for the annual application for registration.[20] The Environment Agency may refuse or cancel a registration.[21] It also has monitoring obligations.[22]

19.5   Obligations once registered

A registered person is obliged to:[23]

1   take reasonable steps to recover and recycle packaging waste in relation to each relevant class in accordance with Sched. 2 to the 1997 Regulations;[24] and

---

13   See para. 19.4 below.
14   For example, the officially promoted Valpack, Wastelink's Wastepak or the Dairy Industries Federation's Difpak.
15   Producer Responsibility Obligations (Packaging Waste) Regulations 1997, reg. 4.
16   *Ibid.*, reg. 5.
17   *Ibid.*, reg. 6.
18   See para. 4.2.1.3 above.
19   The *Hillier* case was, technically, that, albeit arranged openly as a test case on registrability.
20   Producer Responsibility Obligations (Packaging Waste) Regulations 1997, reg. 9(2).
21   *Ibid.*, reg. 10 and 11(1).
22   *Ibid.*, reg. 25.
23   Producer Responsibility Obligations (Packaging Waste) Regulations 1997, reg. 3(5)(b).
24   This provides for different percentages for different functions, with targets increasing year-by-year, in accordance with the UK targets set out in *ibid.*, Sched. 10.

2   furnish a certificate of compliance in accordance with regulation 23.
3   If its main activity is that of seller, there is also an obligation to provide the information to consumers specified in regulation 3(5)(c).
4   Where its turnover exceeded £5m, a recovery and recycling plan has to be lodged with the annual application for registration.[25]
5   Businesses have the option of meeting their recovery obligations themselves or by joining a registered compliance scheme, which will meet the obligations of others.[26]
6   It is, however, possible to comply with this obligation by purchasing packaging waste recovery notes ('PRNs') issued by accredited reprocessors.[27] This market has operated since January 1998, initially on a non-statutory basis,[28] but, from 1 January 2004, on an official one,[29] as an adjunct to the statutory command and control regime enshrined in the 1997 Regulations. Indeed, the submission of PRNs seems to be the main way in which obligated businesses, including compliance schemes, demonstrate their compliance. They acquire the PRNs either on their issue by accredited packaging waste reprocessors or by purchase on the open market from other organisations. Concern has, however, been expressed on a number of fronts:

a.   at the way in which the scheme operates in practice, not least in connection with the development of a secondary market in which non-obligated parties trade in PRNs for their own profit;[30] and
b.   at the fact that PRNs were not equally available to all obligated parties;[31] coupled with

i.   what was sometimes a significant price being set without any indication of the factors or policy behind it;[32] leading to

---

[25]   *Ibid.*, reg. 6(4)(dd).
[26]   See para. 19.3 above and also Defra's Consultation Paper, *Review of the Producer Responsibility Obligations (Packaging Waste) Regulations 1997* (7 August 1998), para. 7.3: see www.defra.gov.uk.
[27]   *Ibid.*, para. 7.2.
[28]   Which was before any accreditations had even been applied for: *ibid.*, para. 7.4. A similar issue also arose in relation to Green Certificates (see para. 21.5 below).
[29]   Producer Responsibility Obligations (Packaging Waste) (Amendment) (England) Regulations, S.I. 2003 No. 3294. This also created the concept of the PERN (packaging export recovery note): see para. 1.4.2.1(2) above.
[30]   See Defra Consultation Paper, *Review of the Producer Responsibility Obligations (Packaging Waste) Regulations 1997* (7 August 1998), para. 7.4, point 4 This concern was stated to be that of the obligated parties, although the underlying problems were not identified: see, however, para. 19.7 below. It runs counter to what subsequently became general government policy, namely the support of environmental taxation by trading mechanisms (see Chapter 20 below).
[31]   *Ibid.*, points 2 and 5.
[32]   *Ibid.*, point 1. Although between May and August 2003, prices fell from £8 to below £4 per tonne: ENDS Report 346, November 2003, pp. 19–20.

ii. the thought that resources might not have been, as intended, flowing towards increasing collection and reprocessing capacity and developing markets for recyclate.[33]

Subsequent investigations into wood and plastic recyclers led to the suspension or cancellation of the accreditation of 14 firms between September 2003 and June 2004.[34]

19.6   Waste disposal obligations

Official experience in relation to packaging[35] can be expected to be put to use in the fields of motor vehicle,[36] electronic and electrical equipment.[37] It appears, however, that the motor industry is not to have to contribute in relation to the first of these until 1 January 2007[38] (the date from which the owner of the vehicle may not be charged for delivery to an authorised treatment facility), and even this prospective obligation is giving rise to concern in the trade because the prospective charge of £50 per car far exceeds the £5–£10 which some members of the public with old cars are currently reluctant to pay scrap dealers,[39] with 350,000 cars being abandoned in the UK each year. However, although the prior requirement that, by 1 January 2006, at least 80 per cent of such vehicles have to be reused or recycled, generated less publicity initially, the combination of the falling price of steel and the rising cost of recycling has since created doubt as to the economic viability of that part of the proposal.[40]

Moreover, the manufacturers and retailers of electrical equipment cannot agree over a proposal by the former,[41] based on schemes extant in Belgium, the Netherlands and

---

33   *Ibid.*, point 1. This problem was solved by the insertion of regs 5(5B) and 21B and the reformulation of reg. 7, from 1 January 2004.

34   ENDS Report 353, June 2004, p. 19.

35   Considered by the European Packaging Industry Association to have resulted in 'a massive transfer of costs from the public sector to the packaging and packaged goods industry' (see Houlder, *Financial Times*, 16 October 2003 (special report Sustainable Business), p. 6). The cost of recycling continued to cause concern, in the context of the 2008 EU targets (see Houlder, *Financial Times*, 28–29 February 2004, p. 5).

36   End-of-life Vehicles Directive, 2000/53/EC, OJ L 269, partially implemented by the End-of-Life Vehicles Regulations 2003, S.I. 2003 No. 2635.

37   Albeit, initially, subject to the expectation of the Better Regulation Task Force that the dumping of electrical equipment will increase in the short term (see Houlder, *Financial Times*, 28 July 2003, p. 3). Concern has, furthermore, been expressed as to the prospect of hazardous waste being stockpiled and illegally dumped after the introduction of new EU regulations in the summer of 2004 restricting the number of landfill sites able to take such waste (and especially in the south of England), reducing the number of sites able to take it from 200 to ten (see Houlder, *Financial Times*, 8 December 2003, p. 6).

38   Eaglesham, *Financial Times*, 21 June 2002, p. 3.

39   Guthrie, *Financial Times*, 15 October 2002, p. 4.

40   Dowen, *The Times*, 18 March 2003, p. 10 (law).

41   Made before the passing of the new EU rules under which, by 2006, the industry would have to recycle or reuse over half the old equipment, with less than 30 per cent going to landfill or incineration (see Parker and Houlder, *Financial Times*, 19 December 2002, p. 6).

Norway, that a £10 charge be imposed at the point of sale to finance the recycling both of goods made by companies which have gone out of business ('orphan' products); and 'historic' products made ten to 20 years ago.[42] It may be possible for this problem to be ameliorated by following the example of the computer industry and espousing leasing, which gives manufacturers the ability to refurbish equipment, where practicable.[43]

The European Commission has estimated that recycling costs between 5 and 8 billion euros per annum, saving an equivalent amount in waste disposal cost.[44] It is now considering the possibility of introducing similar legislation for spent batteries.[45]

Concern has, however, been expressed by the House of Commons Environment, Food and Rural Affairs Committee not only on the progress made by the UK in implementing the relevant Directives but also as to the adequacy of the existing bureaucracies (especially the Environment Agency and Defra)[46] to do so effectively within the relevant timescales.[47]

19.7   The future of packaging waste recovery notes

The decision made by the UK Government in 2003 to 'legalise' PRNs and weed its base of accreditation holders was not the only option available to it.

It might easily have decided to supplement the recycling obligation summarised in paragraph 19.3 above by new taxes or schemes of the types considered in paragraph 27.6. It is perhaps not surprising that none of these routes was adopted because packaging is degradable waste and with the (compulsory, albeit belated) introduction of special sites for noxious waste in 2004, those solutions would not have contributed to the main objective of taking waste away from landfill.

A more logical solution would have been to concentrate on the objective referred to in paragraph 19.3 and close down the diversion into tradable obligations, not least because the latter route does not seem to offer a solution to the more serious problem on the horizon referred to in paragraph 20.7, namely both the recapture of vehicular and electronic goods and the subsequent recycling of the parts. The former could hardly be dealt with satisfactorily on an EU-wide basis[48] because of the greenhouse

---

42   Pickard, *Financial Times*, 17–18 August 2002, p. 4; Houlder, *Financial Times*, 28–29 December 2002, p. 4.

43   Houlder, *Financial Times*, 27 May 2003, p. 11.

44   In relation to cars alone, see Bream, *Financial Times*, 12 January 2004, p. 4.

45   Houlder, *Financial Times*, 27 May 2003, p. 11.

46   See para. 4.2.1.2(1) above.

47   *End of Life Vehicles Directive and Waste Electrical and Electronic Equipment Directive: Fourth Report of Session 2003–2004*, HC 103, 11 February 2004.

48   If this were to be attempted, it would undoubtedly come as an unexpected lifeline to the Channel Ferry operators, who had been reducing their services as a result of a combination of competition from the Channel Tunnel and 'no frills' airlines and a reduction in the number of those travelling to Calais to buy alcohol and tobacco at prices which did not include UK excise duties.

gas consequences of long-distance transportation to disposal sites. This has already been the cause of concern in relation to post-July 2004 noxious waste disposal. But it is possible that the invention, under the new statutory regime which took effect for PRNs on 1 January 2004, of packaging waste export recovery notes ('PERNs'),[49] may have been in part with a view to a similar regime being introduced for vehicle and electronic goods manufacturers.

It is, furthermore, appropriate to analyse both the differences between the PRN 'product' (that is, the provision of services) and that of the official[50] and unofficial[51] markets in allowances:

1   There is a very significant difference between a market in which allowances or permits are to be traded, when the offeror merely needs to buy-in (if necessary at a loss) in order to deliver, and one in the provision of services, where the offeror may either not have any capacity to supply the relevant services[52] or be offering what the authorities may or should consider to be a substandard product.

2   Although Defra's August 1998 Consultation Paper[53] highlighted as an area of concern the presence of non-obligated[54] parties in the unofficial market, this has not been a factor which has been voiced in public in relation to the other unofficial markets, both of which deal in allowances (rather than services to be supplied).

3   Most puzzling from an economic point of view has been the record of price movements. The Defra Consultation noted both lack of universal availability; and a pricing structure which was not readily capable of analysis. However, the prospective removal of some accreditations, as part of the solution, would not normally have expected to result in the halving of prices in the course of four months.[55] For this reason it would therefore be premature to assume that the January 2004 legitimation of the market has produced a viable end product in the form of properly performed services.

In conclusion, while it must be accepted that the PRN route chosen by the government is consistent with both its 'traded market' approach and the apparent preferences of

---

49   See para. 1.4.2.1(2) above.
50   See Chapter 20 below.
51   See para. 20.6 and 21.5 below in relation to Green Certificates, the deficiencies of which it has been possible to fix by introducing tougher non-compliance sanctions, and the 'grey' market in EU ETS, in which some major Continental banks are participating without (it seems) their respective Central Banks being overtly concerned (for the latter, see Chapter 28 below).
52   That is, what is required to achieve the objective set out in para. 19.3 above.
53   See para. 19.5 above.
54   By which is presumably meant speculators, rather than financial institutions acting as formal market makers. As to the former, it may be recalled that, in the February 1974 Election, Mr Harold Wilson (the Labour Leader of the Opposition) made great play with 'land speculation', generating a fuss about the difference between this and what he later described as 'land reclamation', ending up in the Millhench affair.
55   See para. 19.5n above.

the 'industry', it is too early to say whether this major departure from the original objective[56] is not a dead end in relation to the packaging recycling problem, let alone the basis upon which the recapture and recycling obligations to be imposed on vehicle and electronic goods suppliers can be resolved.[57]

---

[56]   That is, that in para. 19.3 above.
[57]   See para. 19.6 above. As to which, it should be noted that the answer to the disposal of electrical goods problem which was favoured in the autumn of 2004 was the introduction of a levy payable by purchasers of between 2 and 5 per cent of the retail price: Marsh, *Financial Times*, 6–7 November 2004, p. 5.

Chapter 20

Emissions and Waste Trading Schemes

20.1   Introduction

The concept of transferable pollution permits was first developed in the US, in relation to power station emissions into the atmosphere. It first entered the UK political arena with the publication of *Economic Instruments for Water Pollution*.[1] In the event, however, the concept did not develop in that direction in the UK.

In March 2000, the UK Emissions Trading Group[2] made some proposals to the government, and, in an official Press Release on 6 November 2000, the government let it be known that it had been decided to put these into effect, albeit subject to further consultation. On the basis that the necessary initial research would have been completed by early 2001, and targets set by March 2001, and the bidding process under way by the autumn of 2001, £30m was to be made available from public funds in 2003/04 to kick-start the process.

The process was expected to go hand in hand with the negotiations between the then DETR[3] and trade associations representing IPPC users of energy in relation to the terms of climate change agreements.[4] This was because the 6 November 2000 Paper had identified the interrelationship between the climate change levy 80 per cent rebates available to parties to climate change agreements and the subsidisation of successful bidders under the initial auction of tradable permits as a key issue. Consequently considerable slippage in the timetable for negotiating the terms of the umbrella climate change agreements[5] led, inevitably, to a deferment in the timetable for tradeable permits. The recommendations received by the government in January 2001 were only published on 3 May 2001. It was still envisaged that trading would start in April 2002, but the publication date for the draft rules had been put back to July 2001.

A pilot project had, however, been set up in February 2001, operated on a non-monetary basis, between 11 oil companies flaring gas from 58 offshore installations.[6]

20.2   Design and implementation of the UK Emissions Trading Scheme

In the event, the government's proposals, entitled the *UK Emissions Trading Scheme* ('UK ETS'), were not published[7] until 14 August 2001. Under this document, it was

---

[1]   DETR, 20 January 1998.
[2]   See para. 2.4.5 above.
[3]   See para. 4.2.1.2(1) above.
[4]   Under Finance Act 2000, Sched. 6, para. 51.
[5]   See para. 14.6 above.
[6]   Buchan, *Financial Times*, 6 February 2001, p. 8.
[7]   By the DTI. See para. 4.2.1.2(1) above.

envisaged that the draft rules would be issued in November and finalised in December, when the draft Protocol. would become final. That was to remain open for comment until 1 October 2001, by which date prospective entrants were urged to contact the government, which needed to receive their source lists for consideration by 31 October 2001. Bidding was scheduled to start in January 2002, on the basis of unverified baseline figures, but independent verification would be required before any allocations were made in April 2002, in relation to targets running from January 2002.

Certain areas had, however, to be left open in the August 2001 document. It was hoped, at that time, that additional clarity would have been possible in relation to the international ramifications at the Marrakesh Conference in November 2001. In the event, nothing of great significance came out of that event, mainly because before it started President Bush had announced that the Kyoto Protocol had been withdrawn from treaty consideration before the Senate.

This was not, however, the only problem. Kyoto had been constructed on the basis that Russia would be able to sell pollution capacity to the West and Japan, and in particular the USA. Indeed, without Russian ratification it cannot come into effect. The Kremlin's failure to submit the Treaty for approval by the Duma had, therefore, given rise to puzzlement.[8] Indeed it seemed that the ratification procedures were unlikely to be completed during 2004 unless Russia was guaranteed substantial financial benefits from ratification.[9] The EU's outgoing Energy Commissioner, Mrs Loyola de Palacio, had expressed the view, however, that Russia's accession is of the essence to the EU continuing to espouse the Kyoto mechanisms,[10] not least because (following the withdrawal of US) Russian ratification is a pre-requisite to the Treaty coming into force at all.[11] In April 2004, however, there were indications that the re-elected Russian President, Mr Putin, might be willing to ratify if the EU were to drop their demands for energy price liberalisation[12] as far as Russian natural gas was concerned.[13] But, at the end of the following month, there were indications that the Russian Government was considering an alternative strategy, generating economic growth with the assistance of a general absence of emissions controls. This uncertainty lasted until the end of September 2004, when the Russian Government sent the Treaty to the Duma with a recommendation for ratification.[14]

---

[8]   Grubb and Safonov, *Financial Times*, 15 July 2003, p. 19.

[9]   Jack, *Financial Times*, 27–28 September 2003, p. 8.

[10] *Financial Times*, 2 March 2004, p. 18. Her letter generated a considerable reaction: Grubb and Bunzl in *Financial Times*, 18 March 2004, p. 18, and Okonski in *Financial Times*, 8 March 2004, p. 18.

[11] A point made when such ratification did not take place when talks with the EU took place in May 2004: Houlder, *Financial Times*, 20 May 2004, p. 17; Jack, *Financial Times*, 21 May 2004, p. 8.

[12] See para. 12.2.6.3(2) above.

[13] Ostrovsky, *Financial Times*, 23–25 April 2004, p. 7.

[14] Shlaes, *Financial Times*, 24 May 2004, p. 17. The problem of Russia and its relationship with the EU in this respect was also addressed by Jacqueline Karas in *Russia and the Kyoto Protocol: Political Challenges* (London: RIIA, Sustainable Development Programme, March 2004). But, by the end of September, the Ministerial power struggle had been won by the ratification camp (see Ostrovsky and Harvey, *Financial Times*, 30 September

The development in the previous paragraph was, needless to say, not immediately predictable in 2001. There had, however, been further slippage in the implementation timetable for UK ETS. The draft rules (with considerable gaps in them) were only made available for consultation on 25 January 2002, and observations invited by 'close of play' on 31 January, the day before the bidding process was to start.

One of the results of this chaotic situation was that the concept of *management control*, which was critical to that of *change of operation*, ended up by being defined by reference to the original proposal document, *UK Emissions Trading Scheme*, Annex A. This was almost certainly not intended to act as (rather than to form the basis of) a legal document and it is possible that problems may arise,[15] not least in the context of takeovers and mergers.[16]

The auction[17] produced 34 successful participants (since reduced to 32), who were to be entitled to £53.37 per tonne of carbon dioxide equivalent cut under a five-year programme, in return for reductions of 1.1m tonnes. The first credits, to these 'winners' were made in the register on 2 April 2002, to the extent of approximately 4m allowances. Nearly 600 allowances had been purchased.[18]

20.3   Subsequent development of the UK Emissions Trading Scheme

The initial scheme was designed as a support for the climate change agreement programme, under which 6,000 signatories to 43 climate change agreements had to meet energy usage reduction targets (usually on an industry-apportioned basis)[19] in alternate years, during the five year period for which those agreements ran. The scheme was negotiated with the European Commission on the basis of being temporary state aid for the period of ten years and, officially at any rate, was unlikely to be renewed. The opening of the register had not, in fact, been the commencement of trading, applicants having hedged their positions since the previous autumn. Trading in depth was not expected, however, to develop

---

2004, p. 12; Harvey and Ostrovsky, *Financial Times*, 1 October 2004, p. 8). Although the Governing Party had a majority in the Duma, ratification could not, however, be guaranteed (see editorial, *Financial Times*, 1 October 2004, p. 20). The President's proposal did nonetheless, subsequently, obtain the approval of both Houses of Parliament.

15   Especially in relation to para. A.4.

16   This being the time at which the legal small print is likely to be examined by outside lawyers – where premises are being acquired rather than the operating entity, a vendor well within his prospective target is likely to sell units in the market as a means of increasing his effective sale price for the going concern.

17   Seemingly completed on 11–12 March 2002.

18   HM Treasury's Budget 2003 (9 April 2003), para. 7.15 (see also Brian Jones and Peter Hawkes, [2001/2002] 3 ULR 39).

19   This has the extraordinary effect of permitted a delinquent to benefit from an overall industry undershoot if he has not covered himself in the tradable permits market, whilst encouraging a likely overshooter to sell his prospective surplus in order to prevent delinquents benefiting from this.

before climate change agreement participants[20] were in a position to estimate the amount of their individual divergences from their first year targets in the autumn of 2002.[21]

On 1 December 2004, the Environment Minister, at a speech launching the London Climate Change Services Providers Group, let it be known that, in its first two years, UK ETS had achieved emissions reductions of 9.8m tonnes of CO_2 equivalent and that six of the leading participants had pledged additional reductions of 8.9m tonnes.[22]

The ultimate intention was to run the trading arrangements on until they could be merged with an international trading scheme into the first Kyoto Protocol[23] target saving period, 2008–2012.[24] Whether this will ever happen is, however, far from certain. The US is the largest emitter of greenhouse gases[25] and the Kyoto Protocol had been designed on the basis that it could mitigate its obligations under the reduction programme by buying permits from Russia, which had been allocated a very generous quota,[26] thus helping to subsidise the democratic regime which had succeeded the collapse of the Soviet Union less overtly than hitherto. The withdrawal of the draft treaty from consideration by the Senate (where it was heavily bogged down) by President Bush in the summer of 2001 put into question whether any international permit trading scheme would be viable, except possibly within the EU, which, in late 2002, was preoccupied with the more pressing problems of the enlargement process.[27]

In the spring of 2002, it was anticipated that a voluntary European scheme would start in January 2005[28] and that the discrete British trading scheme would come to

---

[20] In relation to whom, it needs to be remembered that the criterion for membership was not that of heavy power usage, but having come within the IPPC heavy polluter categorisation (see para. 6.2.3 above).

[21] Buchan, *Financial Times*, 2 April 2002, p. 4.

[22] http://www.defra.gov.uk/corporate/ministers/speeches/em041201.htm: Mr Elliot Morley, MP.

[23] See para. 8.3.1.4 above.

[24] The first such sale was by the Slovak Republic to a Japanese group on 6 December 2002, and a decision in principle on an EU-wide system was expected to be agreed on 9 December 2002 (see Dombey and Houlder, *Financial Times*, 6 December 2002, p. 14).

[25] Following the dramatic blackout in Ontario, Ohio, Pennsylvania, Michigan and New York in August 2003, doubts have been expressed as to the credibility of US environmental policy (see Coldwell, *Financial Times*, 16–17 August 2003, p. 11).

[26] A. Moe and K. Tangen, *The Kyoto Mechanisms and Russian Climate Politics* (London: RIIA, 2000).

[27] Relations between USA and many of the 'old' EU states deteriorated as a result of the former's initiation of the Iraq War. At the Nairobi conference convened under the auspices of the 1987 Montreal Protocol, US negotiators sought exemption for methyl bromide, a fumigant used in farming, and were opposed by the EU delegation (see Houlder, *Financial Times*, 12 November 2003, p. 14).

[28] It is worthy of note in this context that a wholesale price surge in electricity is considered likely to accompany this event, despite 'windmills' replacing retiring Magnox (nuclear) and coal power stations (see *Financial Times*, Lex Column, 22 August 2003, p. 18).

an end and merge into it by December 2006.[29] It was also envisaged that the British Government would set the penalty level for overshooting at about the same level, of approximately twice the open market rate.[30]

In August 2001, it had, however, also been envisaged that a number of non-domestic greenhouse emission projects would be set up and that participation in them might earn credits which could be traded within the UK ETS.[31]

It was also envisaged in August 2001 that one possible future development of the scheme might be the creation of a facility for non-governmental organisations to buy in capacity for cancellation and/or companies to buy-and-cancel in order to offset greenhouse emissions voluntarily.

In the Pre-Budget Report on 27 November 2001, mention was made of the possibility of UK ETS being extended to biodegradable municipal waste.[32]

It is, however, possible that the scope of the UK ETS may, instead, be narrowed. In February 2002, the Cabinet Office Performance and Innovation Unit[33] evaluated two possible methods of providing a fiscal incentive for carbon abatement,[34] a carbon tax (which was ruled out by the Treasury more or less immediately)[35] and changing UK ETS[36] to confine it to carbon emissions permits.[37] Membership of the revised scheme would be mandatory for all participants in the fossil fuel market.[38]

It was therefore rather surprising that, when the Department for Transport published *Powering Future Vehicles* on 30 July 2002,[39] it was revealed that work was under way on a *projects* entry route, and that transport was to be one of the priority sectors for this.[40] This is especially so when account is taken of the fact that the compulsory scheme on which the European Commission was understood to be working at the

---

[29] In the event, the scheme approved by EU Ministers on 9 December 2002, although starting in 2005 and relating to carbons, involved a compromise which the Commission feared might impair its efficiency. Both companies and sectors would be permitted to opt out until 2008, and after that date Member States are to be able to include more sectors or gases and to auction off up to 10 per cent of the carbon quotas (see Dombey, *Financial Times*, 10 December 2002, p. 13). Come November 2003, however, there was no internal consistency of approach between UK Government departments (see Eaglesham and Taylor, *Financial Times*, 16 November 2003, p. 2). For the EU Emissions Trading Directive, see Chapter 28 below.

[30] These were authorised by Waste and Emissions Trading Act 2003, ss.38, 39.

[31] *UK Emissions Trading Scheme* (14 August 2001), section 6.

[32] Para. 7.82.

[33] See para. 4.2.1.2(1) above.

[34] *The Energy Review* (February 2002), paras 3.57, 3.59 and Box. 3.1.

[35] Written Answer, 25 February 2002, vol. 380, No. 106, col. 1031.

[36] That is, The UK Greenhouse Emissions Trading Scheme 2002, the rules of which spell out in detail the obligations of direct participants.

[37] Entrants to the 2002 auction had the option of basing their targets on either carbon dioxide or on all six Kyoto gases (see para. 8.3.1.4 above).

[38] See also Taylor, *Financial Times*, 8 August 2002, p. 4.

[39] Containing a foreword by the Prime Minister, The Rt Hon. Tony Blair, MP.

[40] Para. 29.1.

time covered a multitude of areas, but not transport.[41] However, in *The Future of Air Transport*,[42] the British Government indicated that it proposed to use its 2005 Presidency of the EU to bring aviation within the EU Emissions Trading Scheme ('EU ETS')[43] from 2008.

In the context of the Cabinet Office Performance and Innovation Unit proposals, it should be borne in mind that the following limitations were set upon those who could apply to join the 2002 auction scheme:

1   a renewable obligations certificate scheme being under consideration, electricity and heat generators could only participate in relation to their own headquarters and generation site where all the energy was consumed on site;

2   sites covered by climate change agreements, which provided a 20 per cent reduction in levy as an incentive, and in relation to which the proprietors were participants in the *relative* sector of the trading mechanism, a *gateway* preventing net transfers from this to the *absolute* sector in which those successful in the auction were to have their credits;[44]

3   those whose emissions were from land or water transport;

4   where the emissions were from households, in relation to which the Energy Efficiency Commitment was to come into force in April 2002;

5   the emission of methane from a landfill site covered by the Landfill Directive;[45] and

6   where the flaring of natural gas was concerned, allowances purchased would not be capable of substitution for consents under Energy Act 1976.

In the Energy White Paper published in February 2003, *Our Energy Future – Creating a Low Carbon Economy*, emissions trading was said to be 'central' (to the White Paper's thesis),[46] and indications given that it might be extended to projects involving CHP[47] and coal mine methane.[48]

---

[41] Including energy, the production and processing of ferrous metals, the mineral industry and industrial plants from the production of pulp from timber or paper and board with a daily capacity of 20 tonnes.

[42] Cm 6046 (December 2003), Annex B: see para. 27.5 above.

[43] See Chapter 28 below.

[44] How these arrangements are to be fitted into the 2005 European emissions trading arrangements could be an issue of some difficulty, since such participants do not have a true 'cap' on their emissions and their climate change agreements do not represent an allocation from the UK national total for the purposes of the EU ETS Directive (see Fiona Mullins and Jacqueline Karas, *EU Emissions Trading: Challenges and Implications of National Implementation* (London: RIIA, 2003), p. 61). The solution favoured by the UK Government is the creation of a level playing field through the introduction of an equivalent climate change levy reduction for agreement participants who prefer their direct emissions to be covered by the EU scheme (see *Business Brief* 27/2003 (10 December 2003)).

[45] See para. 12.2.5.1(2) above.

[46] Para. 2.27.

[47] *Ibid.*, para. 4.18.

[48] *Ibid.*, para. 6.67.

Furthermore, in June 2004, in a joint Defra-Treasury Consultation, *Developing Measures to Promote Catchment-Sensitive Farming*, the possibility was mooted of a trading element in an economic instrument solution option.[49] It appeared, however, that Dutch experience with manure trading[50] and the inability of transferability to generate any trading in relation to local reservoir schemes in Colorado[51] would militate against the adoption of such a scheme.[52] The responses to the Consultation were published on 2 December 2004 and revealed that the majority had opposed the use of economic instruments.[53] On the same day, the minister issued a written Parliamentary Statement containing no reference to such instruments in his assessment of the way forward.[54]

However, a more fundamental problem to the future development of the current system was becoming apparent at the beginning of August 2003. The EU ETS,[55] to be introduced in January 2005, affects 2,000 industrial sites in the UK. How the UK's national allocation of allowances would be divided out was becoming highly controversial and a consultation was therefore put in hand. Industries with high abatement costs are likely to have to buy allowances to the extent that they have been unsuccessful in obtaining enough of the 95 per cent expected to be allocated free of charge. According to the research accompanying the consultation document, an increase in electricity prices would be likely to result if trading took place at £10 per tonne of carbon dioxide. There was, subsequently, a considerable debate over the prospective allocation of allowances between energy suppliers and industry in general.[56]

In the event, although it had produced a draft national allocation plan ('NAP') before the end of 2003,[57] the UK Government was not able to meet the 31 March

---

49 See para. 21.9 below.

50 Para. E9.

51 *Ibid.*, para. E10.

52 This is, perhaps, not altogether surprising in the light of the review of international precedents in Annex 3 to *Economic Instruments for Water Pollution*, published by the DETR on 20 January 1998.

53 www.defra.gov.uk.

54 Mr Elliot Morley, MP: HC 2 December 2004, col. 51WS.

55 See Chapter 28 below.

56 Houlder, *Financial Times*, 14 August 2003, p. 2; Taylor, *Financial Times*, 9 January 2004, p. 3; Taylor, *Financial Times*, 12 January 2004, p. 4; Miller (letter), *Financial Times*, 13 January 2004; Whitelegg and Nicholson (letters), *Financial Times*, 15 January 2004; *Financial Times* (leader), 20 January 2004; Taylor, *Financial Times*, 20 January 2004, p. 2; Taylor, *Financial Times*, 24–25 January 2004, p. 4; Taylor, *Financial Times*, 27 January 2004, p. 4. The EU Energy Commissioner at that time, Mrs Loyola de Palacio, was subsequently understood to regard the reduction of emissions as a preferable option for maintaining energy industry competitiveness (see Buck, *Financial Times*, 26 February 2004, p. 10). However, she made clear in correspondence on that article that her concerns related to the state of uncertainty resulting from Russia's (then) failure to ratify the Kyoto Protocol (see *Financial Times*, 2 March 2004, p. 18).

57 Providing for a reserve of 5.7 per cent for new entrants: ENDS Report 348 (January 2004), p. 18. As to this, it is worthy of note that the Commission's acceptance, on 7 July 2004, of the UK's late filing was conditional on two aspects: the need to include Gibraltar;

2004[58] deadline for announcing its allocation of allowances between generators and industry.[59] The main reason was thought to be the lack of base line emissions data,[60] especially in the energy sector.[61] It was indicated in a written answer to a House of Commons Question, however, that the installation-level allocation would be published in June.[62] In the event, the UK filed on 30 April 2004,[63] scaling back the previous target for savings between 1990 and 2010 from 16.3 per cent of carbon emissions to 15.2 per cent.[64] The official prediction was that electricity prices would rise by 6 per cent, but this was disputed by the Energy Intensive Users Group,[65] who predicted an increase of between 25 and 30 per cent by 2007.[66]

The government published a Consultation entitled *Towards a UK Strategy for Biofuels* on 26 April 2004. One of the questions posed was the possibility of introducing a

---

and the provision of insufficient information on the way in which new entrants would be able to begin participating in EU ETS (see IP/04/862).

[58] Making the criticism of other Member States in the joint statement issued by Mrs Beckett and Sir Digby Jones, the Director-General of the Confederation of British Industry on 9 June 2004 somewhat surprising (see www.defra.gov.uk).

[59] It was, needless to say, not the only Member State to miss the deadline! It seemed, however, that informal trading in EU ETS was not attaining volume because of the absence of certainty as to national allocation plans (see Morrison, *Financial Times*, 18 March 2004, p. 63). But the international market in carbon emissions did seem to be increasing in size (see Houlder, *Financial Times*, 10 June 2004, p. 11).

[60] It should be noted, in this context, that Art. 3(6) of the Electricity Acceleration Directive, 2003/54/EC, (2003) OJ L176 37 (see para. 12.2.6.3(2) above) obliges Member States to require suppliers to notify customers of their source mix. Although the principle behind NETA does not facilitate this to be done, the UK has accepted that this will have to be incorporated into the UK regime, albeit by increasing the administrative obligations of suppliers (see DTI Consultation, *Implementation of EU Directive 2003/54 Concerning Common Rules for the Internal Market in Electricity*, paras 3.3.11–17. See paras 6.4.3.1(4) and 12.2.6.3(2) above).

[61] See 350 ENDS Report, March 2004, pp. 47–8.

[62] HC PQ 160232, 29 March 2004, cols 1196-1197W.

[63] After which the Commission had been considering legal action against the six Member States who had still not filed. Only three of the new joiners had filed. In addition, the Commission was in correspondence with some of those who had filed, including Germany (see Minder, *Financial Times*, 19 May 2004, p. 8). The German Government has therefore found itself assailed from both sides, with EnBW considering taking legal action to obtain a higher share of the allowances for its lignite and gas plants to take account of its nuclear plants being phased out under that government's decision to shut them down (see *Reuters*, reported in *Financial Times*, 3 June 2004, p. 8).

[64] Houlder, *Financial Times*, 7 May 2004, p. 2. The final allocation may, however, be delayed until February 2005 (see ENDS Report 355, August 2004, pp. 55–6).

[65] See para. 2.4.5 above.

[66] Houlder, *Financial Times*, 7 May 2004, p. 2. Although not the concern of that particular organisation, this issue is particularly important in the context of *UK Fuel Poverty Strategy*, where it seems to be of the essence to the effect on the poorer sections of the domestic consumer market that NETA and other government structural changes bring down prices (see also para. 21.6.2 below). See also Taylor, *Financial Times*, 15 September 2004, p. 2.

Renewable Transport Fuels Obligation. There was, however, a consensus amongst the responders that regulation would be extremely difficult. Indeed some oil companies took the view that such arrangements would be inconsistent with the European Single Market. The 'most popular' option was said, however, to be a system of tradable certificates.[67] The government then announced that it would undertake a feasibility study and consultative process.[68]

The government is also consulting on the possibility of introducing trading in *white certificates*.[69] This concept involved the concept of energy saved (or the use of which had been avoided) rather that the UK and EU ETS concept of an allowance or credit of emissions (called a *black certificate*). It was accepted, however, that monitoring and verification problems would arise, as well as the issue of interaction with UK and EU ETS, were such a system to be introduced.

20.4 Operation of the UK Emissions Trading Scheme

Direct participants are those who were successful in the February 2002 auction.[70] If they meet their five-year targets, they will obtain a subsidy. To the extent that they have done so, the surplus may either be sold[71] or 'banked' for carry forward to future years up to, but not later than 2007.[72] It is not envisaged that the subsidy element will be present in successor regimes, which will operate by reference to compulsory emission control levels.[73]

---

[67] A summary of the replies was published on 2 December 2004: www.dft.gov.uk.

[68] Consultation issued 2 December 2004, by Defra, *Review of the UK Climate Change Programme*, para. 8.17.

[69] *Ibid.*, para. 7.25.

[70] This was an electronic exercise, with a series of rounds.

[71] The National Audit Office's investigation of the initial operating period for UK ETS indicated that it had helped to develop a small core of trading expertise in the City of London, with five brokers participating. Those brokers had, however, expressed disappointment with the low prices and small transaction volumes (see Houlder, *Financial Times*, 21 April 2004, p. 6). In the context of the pending merger with EU ETS, for which Continental trading was taking place in an unregulated over-the-counter market, the news that the London-based International Petroleum Exchange was planning a link up with the Chicago Climate Exchange (which had established the first trading market for US greenhouse gas emissions in 2003) should therefore be seen as a significant development (see Tait and Morrison, *Financial Times*, 22 April 2004, p. 29).

[72] But with the prospect of being able to transfer a maximum of the overall target saving into the subsequent first Kyoto Protocol period.

[73] The UK was originally to be permitted to opt out of the pilot phase of the EU ETS, which is to run from 2005 to 2007. Under the EU ETS, Member States have to draw up their own Kyoto commitment allocation plans by March 2004 and seem likely to auction 5 per cent of their carbon dioxide emissions licences. It is expected that higher wholesale electricity prices will result (see Houlder, *Financial Times*, 26 June 2003, p. 10).

If direct participants look like falling short of their targets,[74] they may cover their shortfall for the calendar year by the following 31 March. If they do not, they suffer not only the loss of the financial incentive (alias subsidy), but also a reduction in the following year's target (that is, a requirement to reduce emissions by more) multiplied, in the short term by a factor of between 1.1 and 2.[75]

Direct participants may buy and sell credits[76] on the basis of spot trading carried out though the auspices of the Emissions Trading Authority,[77] the ownership of credits being maintained by the Registry. Trading may either be direct or through a broker and takes place on a seller-liability basis, that is, if the target was not met, the consequent penalties would be visited on the seller rather than the buyer.

It is not necessary, however, to be either a direct participant or the signatory to a climate change agreement (known as 'agreement participants') to take part in trading. The Financial Services Authority is not, however, involved in the operation of this market.

20.5   The regime applicable to direct participants

Before being accepted as qualifying, direct participants had to establish a *baseline* figure and have it verified by an independent firm selected from a panel whose members have been accepted as professionally competent by the government. The baseline figure will usually have been the emissions of the three years. The subsequent task of the verifier will be to monitor divergence from the target based on this by 31 March following the calendar year end.[78] Divergences are treated as material if they exceed 5 per cent.

---

[74] The problem about the scheme seems, however, to be that, out of the £215m committed to 31 companies, the National Audit Office (see para. 4.2.1.3 above) discovered that the DTI had felt that an even handed approach had meant that bids had had to be accepted from four chemical and oil companies (BP, Invista UK, Ineous Fluor and Rhodia Organique Fine) who managed to obtain between half the first year's commitment for attaining targets (by a very wide margin) which, in the NAO's view, they would have achieved without the aid of the scheme (see Houlder, *Financial Times*, 21 April 2004, p. 6). The NAO Report was heavily criticised for ignoring BATNEEC: ENDS Report 351 (April 2004), p. 27.

[75] The intention was to replace this by a financial penalty once the necessary legislation could be put through Parliament, which turned out to be in the Waste Emissions Trading Bill, a piece of legislation the government's Parliamentary business managers did not think it necessary to organise expedited passage on the floor of the House of Commons before the 2003 Summer Recess!

[76] Each equivalent to one tonne of carbon dioxide equivalent.

[77] The purpose of this entity is to police the system. It alone has the authority to transfer credits out of the compliance accounts of direct and agreement participants into the national retirement account, out of which no transfers can be made.

[78] By way of contrast, agreement participants operate to years from 1 October to 30 September have to notify by 31 January and only have to comply in alternate years, that is, 2001/02, 2003/04, 2005/06, 2007/08 and 2009/10. What happens in the 'fallow' years is ignored.

Candidates for the 2002 auction were able to select five-yearly targets *either* in carbon dioxide alone *or* in all six Kyoto gases.[79]

Candidates also had to select a sector of their business and provide a *source* list of their plants in that sector. Cherry picking was not permissible. All sources within the same *management control* had to be included.

The concept of *management control* was clearly central to this part of the scheme. This was intended to mean the ability to direct the financial and operating policies governing the omissions from the source. Clearly it includes, but is not confined to, voting control and the ability to nominate the composition of the board of directors. It can also, however, include contractual control.[80]

It follows from this that it becomes necessary to be able to identify when a *change of operation* has taken place and the resulting situations, differentiating between:

1 transfers between participants, when the target for the source is switched unless either:

   a. the change threshold of the lower of 25,000 tonnes of carbon dioxide equivalent or 2.5 per cent of the verified original base line has not been breached; or
   b. the parties contract otherwise;

2 the transferor is a participant, but the transferee is not and does not join the scheme, and the change threshold is breached;
3 a participant obtains control from a non-participant, when nothing happens; or
4 the acquisition of a substitute source by a participant, when a substitution takes place.

In the event of withdrawal from the scheme, the whole of the incentive money received is repayable, with interest.

20.6 Green Certificate trading

The concept of a tradable Green Certificate arose out of the introduction of the RO in the electricity sold into the climate change levy system from 1 April 2002.[81] A generating station holding such certificates may convert them into UK ETS allowances at the rate of 0.43 *allowances* per *certificate*. It is not, however, permissible for *allowances* to be converted into *certificates*.

Customs have made it clear that this form of trading has to be distinguished from, and should not be compared to, trading in (climate change levy) levy exemption certificates ('LECs') after 31 March 2003, which they have been advised is lawful

---

[79] Carbon dioxide, methane, nitrous oxide, hydrofluorocarbons, perfluorocarbons and sulphur hexafluoride. See para. 8.3.1.4.
[80] *UK Emissions Trading Scheme* (14 August 2001), Annex A, para. A.4.
[81] See para. 21.5 below. As with PRNs (as to which, see para. 19.5 above), trading started initially on a non-statutory basis.

provided that the balancing requirement is maintained and that an audit trail exists which leads back to appropriately-generated electricity.[82]

20.7   Landfill Allowances Trading Scheme ('the LATS')

Under the Waste and Emissions Trading Bill introduced in the House of Lords on 14 November 2002,[83] tradeable allowances from local authorities were to be created. This was by way of implementation of the commitment contained in the White Paper *Waste Strategy 2000: England and Wales*. This was part of a strategy to restrict the amount of municipal biodegradable waste[84] sent to landfill[85] in order to comply with the targets for 2010, 2013 and 2020 contained in the Landfill Directive.[86]

Provision is made for the setting of a maximum amount of degradable municipal waste (to be specified by weight)[87] that may be sent to landfill on a county-by-county basis, by reference to years beginning on 17 July.[88] There will then be an allocation of landfill allowances to waste disposal authorities by the county councils.[89] These allowances will be tradeable. The details of the scheme, which will make provision for borrowing and 'banking', as well as trading, are to be contained in secondary legislation.[90] In preparation for the scheme, each county is to be required to prepare a strategy for reducing the amount of such waste going to landfill.[91]

---

[82]   *CCL Information Sheet* 01/03 (May 2003) (see para. 14.14 above).

[83]   But stalled in the House of Commons before the 2003 Summer Recess, since the completion of its Committee Stage there on 29 April 2003. The reason seems to have been the need to formulate a policy following the departure of Mr Meacher from the government in June 2003. The government's approach came under very heavy criticism from both national Opposition parties in the Report Stage and Third Reading debate held on 28 October 2003.

[84]   Waste Emissions and Trading Act 2003, s.21(2) defines this as a combination of *biodegradable waste* as defined in s.21(1) and *municipal waste* as defined in s.21(3).

[85]   Defined in Waste Emissions and Trading Act 2003, s.22(1) as either a waste disposal site or a site used for the storage of waste, as defined in s.37, subject to the matters referred to in s.22(2) and the activities set out in s.25(1) being left out of account.

[86]   See para. 12.2.5.1 (2) above and the Landfill Directive, Arts 5(1),(2) (a four-year derogation has been obtained under the latter).

[87]   Waste Emissions and Trading Act 2003, ss.1, 2: depending upon whether the year is a *target* year, for which the Secretary of State has to agree a target with the county in question in advance. If agreement cannot be reached, s.3 specifies default rules. Section 23 empowers the Secretary of State to change the target years.

[88]   Scheme years run from 2004 to 2019, target years are those starting in 2009, 2012 and 2019. These are the years by which compliance with the Landfill Directive is to be judged by the European Commission.

[89]   Waste Emissions and Trading Act 2003, s.4. Under s.24, the devolved authorities are responsible in Scotland, Wales and Northern Ireland. Under s.5, this allocation must be before the beginning of the relevant year. The ban on exceeding the target is contained in s.9.

[90]   Waste Emissions and Trading Act 2003, ss.6, 7.

[91]   It is thought that an authority with substantial incineration capacity, such as Birmingham, will be at a substantial advantage under this system.

On 11 May 2004, Mr Elliot Morley, MP (the Minister of State) announced that landfill allowance trading ('the LATS') would commence on 1 April 2005.[92] This followed the outcome[93] of a consultation[94] on the draft regulations.[95] As a result of this, the government decided that scheme years should run to 31 March.[96] The allocation formula would not have a population growth formula built into it,[97] and it was felt that it would be impractical to run a scheme without penalties.[98] The first review was to take place in 2007.[99] Allowances were to be divided out on the basis of convergence by 2010,[100] with the potential for some increases before then (rather than capping).[101] The reductions during the intervening years were to be on a back-end loaded (rather than straight line) basis,[102] with a reduction in the first year[103] and a trajectory of 10/15/20/25/30.[104] A mass balance approach was to be used in calculating the rate,[105] with 2003/04 as a base year.[106]

It was not considered appropriate for a specified percentage of the funds generated to be reinvested in waste management.[107] There was to be a borrowing limit of 5 per cent of the following year's allocation,[108] but with no restriction on the banking of unused allowances.[109]

There was to be no requirement to use a broker,[110] as to whom the possibility of FSA authorisation was under consideration.[111] Defra would issue guidance,[112] predict the outturn and make a preliminary reconciliation available,[113] and monitoring would

---

92 HC, 11 May 2004, col. 58WS.
93 *Landfill Allowance Trading Scheme Consultation Outcome*, DEFRA, April 2004.
94 By internet, opened 29 August 2003, closing 21 November 2003, with comments on the draft regulations published on 14 November 2003, accepted up to 31 January 2004.
95 To be entitled: the Landfill Allowances and Trading Scheme (England) Regulations 2004 and Landfill Allowances Scheme (Scheme Years and Landfill Targets) Regulations 2004.
96 *Consultation Outcome*, para. 4.11. The start is being deferred until 2005/06 (see paras 4.5, 4.17).
97 *Ibid.*, para. 4.18.
98 *Ibid.*, paras 4.6, 10.7, 10.8, 10.9, 10.15, 10.20, 10.23, 10.27.
99 *Ibid.*, para. 4.34. It would focus on operational arrangements (*ibid.*, para. 4.35).
100 *Ibid.*, para. 5.4.
101 *Ibid.*, para. 5.15.
102 *Ibid.*, para. 5.20.
103 *Ibid.*, paras 5.21, 11.3, 12.4, 12.8, 12.10, 12.12, 12.14, 12.16, 12.20.
104 *Ibid.*, para. 5.22.
105 *Ibid.*, para. 5.28.
106 *Ibid.*, para. 5.29.
107 *Ibid.*, para. 5.17.
108 *Ibid.*, para. 6.9.
109 *Ibid.*, para. 6.10.
110 *Ibid.*, para. 7.4. It would be up to each local authority to decide how many people should be authorised to deal and their level of seniority (*ibid.*, para. 8.6).
111 *Ibid.*, paras 7.5, 7.15. It will have been noted above that UK ETS brokers were not within the FSA's jurisdiction.
112 *Ibid.*, para. 7.10. There was some controversy over the cost estimate in the Partial RIA which had been issued (*ibid.*, para. 13.2).
113 *Ibid.*, para. 9.2.

be on a mass balance approach with reconciliation by the Environment Agency six months after the scheme year end.[114] Defra would formulate a communications strategy.[115] Non-monetary trading would be permitted (albeit with no decision made on how this would be recorded.[116] The public would have 'read only' access to the register and to the average prices paid (but not to individual prices),[117] forward trades would have to be registered,[118] but only in the most exceptional circumstances would trading be suspended.[119]

Coupled with this development, albeit less heralded by the UK Government, was the placing of PRN trading on a statutory basis.[120] In relation to these, the government's view in 1998 had been that the market was a source of instability.[121] What effect the LATS will actually have on the bringing forward of capital expenditure on recycling facilities is uncertain, not least because the existence of central government council tax capping powers[122] may deter local authorities who would otherwise have undertaken such expenditure from commissioning it.

---

114 *Ibid.*, paras 9.10, 9.15, 9.16, 9.20, 9.21.
115 *Ibid.*, para. 7.24.
116 *Ibid.*, para. 7.20.
117 *Ibid.*, para. 4.18.
118 *Ibid.*, para. 8.4.
119 *Ibid.*, para. 7.9.
120 See para. 19.5 above.
121 *Ibid.*
122 Which were exercised in 2004 (see para. 21.4.1 below).

Division 5
The Instruments in Operation

Chapter 21

Policies in Practice (1)

21.1 Introduction

It was stated in Chapter 1 above that the overriding aim of the book has been to give a comprehensive, contextual and critical account of environmental taxation law in the UK.[1]

In this division of Part III, we attempt to relate the theory of environmental taxation, as discussed in Chapter 5 above, to the events that have occurred since the introduction of each of the post-1997 environmental taxes, respectively in 2001 and 2002, as well as the ongoing modifications to landfill tax, and the absence of taxes on agricultural activity. We begin with an examination of the vexed question of the so-called 'employment double dividend'.[2] As we have already mentioned, this notion is much contested by economists and, since we are not economists, we set out simply to show what has happened to rates of employers' NICs in the years since the introduction of climate change levy in 2001. The facts are rather striking. From there, we look at another legitimating factor of environmental taxes, that is, the extent to which the revenue raised by them is 'recycled' or 'hypothecated'.[3] Again, the way in which the hypothecation debate has 'played out' in practice discloses some surprising problems with what may seem at first a self-evidently sensible idea.

The main theme of this government, however, is the extent to which the energy policy which underlies the biggest of the environmental taxes, climate change levy, is internally inconsistent in itself. The failure of the government to create the conditions for a viable renewables sector, while structuring the tax so as to prevent further damage to what remains of the UK coal industry, is only one of the inconsistencies disclosed by this examination. Similar contradictions are disclosed by the relationship between the landfill tax escalator and local government taxation and by the imposition of aggregates levy on a product for which the government, via its various agencies, is the largest customer.

Throughout, the main source of the criticisms made are the reports of a number of Parliamentary Select Committees who have chosen in the last couple of years to examine the UK's energy policy and, with it, its environmental regulation, including the operation of climate change levy.

21.2 The employment 'double dividend'

When landfill tax was introduced at a principal rate of £7 per tonne, the anticipated revenue was such that a reduction of 0.2 per cent could be made in employers' national

---

[1]   See para. 1.2.2 above.
[2]   See para. 5.3 above, especially the materials footnoted at para. 5.3n.
[3]   See para. 11.2.2 above.

insurance contributions ('NICs'). Although the Labour government increased the rate of tax to £10 per tonne in 1999 and by £5 in five annual increments from 2000 to 2004, no variation in the NIC rate has been announced because the landfill tax has become a net non-producer.[4] When climate change levy was introduced in 2001, the rebate was increased, but by only 0.3 per cent. Again, when aggregates levy was brought in, in 2002, there was a further rebate, but by only 0.1 per cent.

The last two reductions were so low as to make little difference even (in the case of climate change levy) to the few employers unaffected by the new tax. Such changes were, furthermore, expected to be short-lived by reason of the general increase in life expectancy's knock-on effect on the future levels of national insurance contributions, which, historically, have been increased in line with the underlying 'insurance' element.

Three points need to be made about the underlying policy decision:

1   Employers' payroll taxes in the UK do compare very favourably with those of its immediate European neighbours, with some French businesses moving to the southeast in order to obtain relief on this score. This climate is, furthermore, an important one in the context of attracting overseas inward investment. The extent to which the UK Government's decision to stay out of the euro has undermined this advantage, and therefore policy, however, has not yet been quantified.

2   The German Government's 1999 Law Initiating Ecological Tax Reform is based on a similar premise (higher energy taxes being offset by reduced social security contributions).

3   Whereas the effect of climate change levy has been to increase industrial energy costs by approximately 15 per cent, politically the importance of the NIC offset does not seem to have been got across to many small and medium sized businesses.[5]

In the event, the Chancellor's announcement of a general surcharge of 1 per cent in Budget 2002, to take effect in April 2003, has put an end to whatever the process was intended to be.[6] In consequence, it is difficult to avoid the conclusion that environmental taxes are just that, taxes, and presumably ones which, in due course, will become just as much available for general use as the 'infamous' road fund tax of the 1920s.[7]

---

4   The figures given by the Financial Secretary to the Treasury (Mr Stephen Timms, MP) to Standing Committee A in the afternoon of 8 May 2001, were a net take, after the rebating scheme, of £452m, as against NIC rebates of £690m.

5   See the survey of businesses in the northwest of England in Benjamin J. Richardson and Kiri L. Chanwai, 'The UK's Climate Change Levy: Is It Working?' [2003] JEL 39–58.

6   Indeed in 2003 Pre-Budget Report, para. 7.50, it was said: 'The Government does not think there is a strong case for recycling the increases in landfill tax revenue through any further tax cuts'.

7   See Martin Daunton, *Just Taxes: The Politics of Taxation in Britain, 1914–1979* (Cambridge: Cambridge University Press, 2002), pp. 129–32. See also Roy Jenkins, *The Chancellors* (London: Macmillan, 1998), pp. 167–68, 317 and 322.

21.3 Hypothecation, 'recycling' of revenue and tax subsidies

21.3.1 Post-1997 taxes

The government has linked both climate change levy and aggregates levy with the setting up of specified fixed funding to deliver environmental improvements:

1 as part of the measures to meet the UK's Kyoto targets,[8] of which the main component was the introduction of climate change levy, there were set up:

   a. an Energy Efficiency Fund of £50m, part of which was earmarked to assist with the five year transition for horticulture;[9] and
   b. the Carbon Trust,[10] to which day to day responsibility for administering, again in connection with the Kyoto targets, the enhanced capital allowances scheme for the adoption of specified environmentally-friendly energy equipment passed in July 2002.[11]

2 Furthermore, the following direct tax reliefs were given:

   a. 100 per cent first year capital allowances for the installation of environmentally-friendly energy equipment.[12] This started off, in 2001, covering good quality CHP, certain motors, variable speed drives, boilers,[13] refrigeration equipment and thermal screens, lighting and pipe insulation. These were extended in 2002 to oil-fired condensing boilers, heat pumps for space heating, radiant and warm air heaters, compressed air equipment, refrigeration display cabinets and compressors. They were further extended in 2003 to certain water meters, monitoring equipment, flow controllers, leakage detection equipment, lavatories and taps.[14] The reliefs were then yet

---

8   See para. 8.3.1.4 above.
9   This was part of the amelioration package contained in Budget 2002: see Press Release REVCEC 4/00 (21 March 2000). The Carbon Trust is a different body from the Energy Saving Trust, which had been set up by the previous Conservative Government after the 1992 Rio Conference (see para. 4.2.1.2(1) above).
10 See para. 4.2.1.2(1) above.
11 The Carbon Trust also makes loans to small and medium-sized enterprises. However, the setting-up of both it and the Energy Saving Trust ('the EST') were seen as political gestures by the House of Commons Science and Technology Committee in its Report issued on 2 April 2003, entitled *Towards a Non-Carbon Fuel Economy: Research, Development and Demonstration*, paras 41 and 44. In the 2004 Spending Review, issued on 12 July 2004, para. 17.3, it was revealed that, by 2007/08, at least £40m per annum would be used to expand the Carbon Trust's Programmes. One of the projects being backed was Ceres Power, a developer of fuel cells (see Cookson, *Financial Times*, 13 July 2004, p. 6).
12 Capital Allowances Act 2001, s.45A.
13 A by-product of this seems to have been a retail boom in green household boilers (which do not qualify for capital allowances) (see Marsh, *Financial Times*, 7 January 2003, p. 4).
14 Press Release REV BN 26 (9 April 2003); Capital Allowances Act 2001, ss.45H–J.

further extended in 2004 to air-to-air energy recovery equipment, compact heat exchanges and heating, ventilation and air conditioning zone controls on the 'energy saving' side and rainwater harvesting equipment on the 'water' side; and

b. the 5 per cent VAT rate was made available for the grant-aided installation in dwellings of energy saving materials[15] and heating equipment.[16]

3   In conjunction with the start of aggregates levy, the Aggregates Levy Sustainability Fund ('the ALSF') of £35m was set up 'to deliver local environmental improvements'.[17]

21.3.2   Landfill tax

Landfill tax was, of course, already in existence when the present government came to power in 1997 and had built into it the possibility of an operator making direct contributions to one or more environmental trusts, the only conditions being that:[18]

1   the landfill operations of the contributor could not benefit from the activities of that particular trust;[19]
2   only 90 per cent of the contributions would qualify for relief from landfill tax – but contributions to research and development, educational or disseminational[20] trusts apart, it is unclear whether (and if so why) the remaining 10 per cent is deductible from profits under Case I of Schedule D;[21] and
3   there was an overall relief cap of 20 per cent of an operator's total liability.[22] Thus, for instance, say Dump-it PLC has a total landfill tax liability of £5m. If no contributions had been made, this would have cost it £3.5m after corporation tax at 30 per cent. However, if it had contributed £1m to environmental trusts, it would

---

15   Value Added Tax 1994, Sched. 7A, Group 2. Insofar as these cover construction materials, such as 'Wallform' blocks, the relief may be ineffective because the supply may be *incidental* to the *principal* supply of construction (see *Beco Products Ltd and BAG Contractors v. C & E Commrs*, MAN/01/4).
16   Value Added Tax 1994, Sched. 7A, Group 3. From 1 June 2004, this was extended to *ground source heat pumps*: via Group 2, Note 1(h). Microchip units are to be added in 2005.
17   See para. 4.2.1.2(1) above.
18   Finance Act 1996, s.51; Landfill Tax Regulations 1996, S.I. 1996 No. 1527, regs 17–21, 30–32.
19   Except in the case of trusts set up for research and development and the dissemination of its results.
20   That is, ones which collate evidence of best possible practice and distribute it back to the industry.
21   See paras 1.4.3.1 above and 24.8 below. The solution to this problem has not been helped by the article in *Tax Bulletin* No. 23 (June 1996) indicating that it was for the taxpayer to justify deducting the 90 per cent under Income and Corporation Taxes Act 1988, s.74(1)(a)! (See paras 1.4.3.1 above and 24.1 below.)
22   However, from 1 April 2003, this has been reduced to 6.5 per cent. See Landfill Tax (Amendment) Regulations 2003, S.I. 2003 No. 605, reg. 3(a).

have reduced its landfill tax bill by £900,000, but have increased the (probable) cost after corporation tax by only £70,000. The company's management might well have seen this as a viable alternative to advertising as a means of improving the company's public image.

Such entities were set up under the auspices of Entrust,[23] and were extremely popular,[24] both with site operators, who fill the *pro bono* parts of their annual reports with extolling their contribution to the public good through these entities, and with local churches[25] and other land-based charities, who have been quick on the uptake when waste disposal operations have been set up in the *vicinity*. Entrust are prepared to regard projects within a radius of ten miles as qualifying under that head, and Customs are prepared to treat *environmental protection* as including both the protection of buildings as well as the creation of wildlife habitats, conservation areas, 'urban forestry' and the promotion of positive land management and community involvement.

Eligible bodies may be corporate or unincorporated, and include trusts and partnerships.[26] They must not be controlled by, *inter alia*:

1   local authorities, or
2   those registered for landfill tax, or
3   people connected with them,

and are precluded from distributing their profits or applying any of their funds for the benefit of contributors. Any profit has, furthermore, to be applied in furtherance of their objects, which must include one or more of the following:[27]

1   unless this benefits the polluter or is required by a statutory notice,[28] the reclamation, remediation or restoration of land or any operation intended to facilitate the economic social or environmental use of land for which such use has ceased because of the carrying out of some activity over the land;
2   subject to the same restrictions, where the use of land has ceased because of pollution, the prevention, reduction, remediation or mitigation of pollution;
3   research, development, education or the collection and dissemination of information about waste management practices generally – in relation to this object and the following one, contributors can obtain general (but not specific) benefit;[29]

---

23   Finance Act 1996, s.53. As a result of controversy as to the effectiveness of the degree of supervision attained, improvements in the scheme's operation were put into effect in the autumn of 2003. Less information was required, common systems adopted, better information recorded on project funding and audit processes improved (see *Business Brief* 19/03 (1 October 2003)).
24   And have to comply with Landfill Tax Regulations 1996, S.I. 1996 No. 1527, reg. 33A.
25   See Churches Main Committee circulars 1996/6, 1997/2 and 1997/8.
26   Landfill Tax Regulations 1996, S.I. 1996 No. 1527, reg. 33(1)–(1C).
27   *Ibid.*, reg. 33(2).
28   *Ibid.*, reg. 33(3).
29   Discontinued with effect from 1 April 2003 (see Landfill Tax (Amendment) Regulations 2003, S.I. 2003 No. 605, reg. 4).

4   the same in relation to recycling;[30]
5   provided that this is not required under a statutory notice and is not to be operated
     with a view to profit,[31] the provision, maintenance or improvement of a public
     park or other public amenity in the vicinity of the landfill site where this is for the
     protection of the environment;
6   following the Treasury's reassessment of the efficiency of this type of support,
     with effect from 1 October 2003,[32] the conservation or promotion of biological
     diversity[33] where this was for the protection of the environment and neither carried
     out compulsorily[34] or for profit,[35] through either the provision, conservation,
     restoration or enhancement[36] or the maintenance or recovery of a species in its[37]
     natural habitat;
7   subject to the same restrictions as in 5, the maintenance, repair or improvement
     of a place of worship or of historic or architectural interest which is open to
     the public and in the vicinity of the landfill site where this is required for the
     protection of the environment; and
8   the provision of financial and administrative services to (only) such bodies – a
     function which, it should be noted, cannot be charitable, so that an inheritance
     tax issue will arise if close companies and unincorporated businesses contribute
     to this class of body.[38]

The system, as set up in 1996, was, however, exploited by the industry, with Customs
confirming that it was in order, under the original scheme, for an arrangement to be
made for a third party to put up the 'missing' 10 per cent on the basis that he would be
reimbursed. However, when suggestions began being made to local authorities that
they should join in, the rules were changed, with effect from 1 January 2000.[39]

The Labour government let it be known, in Budget 2001, that it was attracted to
the idea of replacing Environmental Trusts with a public spending programme on
sustainable waste management.[40] In the Pre-Budget Report on 27 November 2001,
these thoughts were refined into:

---

[30]   Discontinued with effect from 1 April 2003 (see Landfill Tax (Amendment) Regulations
       2003, S.I. 2003 No. 605, reg. 4).
[31]   Landfill Tax Regulations 1996, S.I. 1996 No. 1527, reg. 33(6).
[32]   Under Landfill Tax (Amendment) (No.2) Regulations 2003, S.I. 2003 No. 2313.
[33]   As defined in the UN's 1992 Framework Convention on Climate Change (see para. 8.3.1.3
       above) and Landfill Tax Regulations 1996, S.I. 2003 No. 1527, reg. 33(2A).
[34]   Landfill Tax Regulations 1996, S.I. 1996 No. 1527, reg. 33(3A)(a)–(e).
[35]   *Ibid.*, reg. 33(3A)(f).
[36]   *Ibid.*, reg. 33(2)(da)(i).
[37]   *Ibid.*, reg. 33(2)(da)(ii).
[38]   This is by reason of the requirement under Inheritance Tax Act 1984, s.10(1) that *no
       gratuitous benefit* should be intended to accrue to any person.
[39]   Malcolm Gammie, QC, and Jeremy de Souza, *Land Taxation*, release 53 (London: Sweet
       and Maxwell looseleaf), para. E3.008A.
[40]   In commenting on the 1999 Thirteenth Report of the Environment, Transport and Regional
       Affairs Select Committee, *The Operation of the Landfill Tax*'s criticisms of the uneven
       distribution of funds [para. 48], the point had been made that a 'key strength' of the scheme
       was that it encouraged private sector involvement in environmental protection schemes:

1   consideration of the possibility of replacing environmental trusts by discretionary expenditure;[41]

2   the inclusion of a biodegradable municipal waste permit within the UK ETS;[42] and

3   the setting up of a New Opportunities Fund to encourage recycling.[43]

This process was given encouragement, in summer 2002, by the deliberations of the House of Commons Public Accounts Committee, which concluded that inadequate public benefit had resulted from the subsidy of £400m given to environmental trusts since 1996.[44]

In the Pre-Budget Report, made to the House of Commons on 27 November 2002, it was made known that the level of funding would be capped at its 2002/03 level, with one third (approximately £47m) continuing to be available from 1 April 2003, through a reformed tax credit scheme for spending on local community environmental projects. The remainder (£100m in 2003/04, rising to £110m in 2004/05 and 2005/06) was to be allocated to public spending to encourage sustainable waste management,[45] in a manner yet to be determined[46] after due consideration of the Downing Street Strategy Unit's Report, *Waste Not, Want Not*,[47] with an emphasis on recycling in partnership with local government.[48]

In a written statement to the House of Commons on 3 February 2003,[49] the Economic Secretary to the Treasury (Mr John Healey, MP) announced certain changes in the scheme, to take effect on 1 April 2003:

1   the maximum percentage of the operator's total liability to be available within the scheme was to be reduced from 20 to 6.5 – somewhat unexpectedly increased to 6.8 per cent from 1 April 2004;[50] and

---

    [2000] *JPL* 18, at 23. The change in policy plainly risks dissipating the advantages of this approach and well as giving rise to the likelihood in the longer term that ring fencing will be lost and the money be diverted to general Budgetary purposes.

41   *Ibid.*, para. 7.81.

42   *Ibid.*, para. 7.82 (see Chapter 20 above).

43   *Ibid.*, para. 7.84.

44   Eaglesham, *Financial Times*, 27 July 2002, p. 5.

45   *Ibid.*, para. 7.53.

46   The House of Commons' Environmental Audit Committee, in its Fourth Report of 2002–03 on Pre-Budget Report 2002, HC 167, published on 1 April 2003, seems to have suspected, however, that this presaged diversion to strategic central objectives.

47   Published in draft the same day and, finally, on 3 December 2002 (see para. 6.3.2 above).

48   *Pre-Budget Report*, para. 7.56. It is noteworthy in this context that, in 2004, West Oxfordshire District Council (a traditionally low spending council) had to consider a 33 per cent Council Tax increase in order to cover unfunded expenditure (largely on this score) and came under pressure from the Labour Government to reduce the increase (see *BBC News* website, 25 February 2004).

49   Col. 5.

50   By Landfill Tax (Amendment) Regulations 2004, S.I. 2004 No. 769, reg. 3. See also *Business Brief* 09/04 (18 March 2004).

2   categories 3 and 4 above to be discontinued, although
3   transitional relief was to be available, until 31 March 2004, only, for the funding
     of projects for which a written contractual obligation had been entered into before
     3 February 2003.

As part of this process, the contribution years were converted into periods ending on
31 March.[51]
  The only developments announced in the 2003 Budget were, however:[52]

1   the announcement of continuing discussions on the LTCS, with a view to the
     making of new regulations in the summer of 2003 which would include habitat
     creation;
2   attempts, in conjunction with Entrust to simplify the administration of the scheme
     and provide better information from it; and
3   the transmogrification of the Waste Minimisation and Recycling Fund into
     a Waste Management Performance Fund for English local authorities, to take
     effect on a date to be announced, after consultation with the local authorities.[53]

At the same time, decisions would be made on how the scheduled landfill tax increases
would be made revenue neutral to local government. A package of measures was
scheduled to be announced in the 2003 Pre-Budget Report. In the event, however, no
substantive measures were announced in the 2003 Pre-Budget Report. In February
2004, however, HM Treasury published[54] a Final Report commissioned by it from
Integrated Skills Ltd entitled *An Assessment of Options for Recycling Landfill Tax
Revenues*. The results, in relation to how the 2005/06 landfill tax surplus should be
applied, were distinctly tentative.[55]
  However, the Integrated Skills Ltd report is extremely important for the way in
which it approaches the problem, leading the authors to wonder how the £35 revised
landfill tax cap had been fixed:

1   This report identifies three alternatives to disposal to landfill:

     a.   Energy from waste, which, it transpires, will only be competitive from an
          economic point of view when the £35 level has been reached, the typical cost
          levels being between £35 and £55 per tonne;
     b.   composting, which has a restricted application to business waste, but will
          become competitive with landfill within a few years, at a price of £20

---

51   Landfill Tax Regulations 1996, S.I. 1996 No. 1527, substituted regs 31(4),(5),(6),(6A);
     Landfill Tax (Amendment) Regulations 2003, S.I. 2003 No. 605, reg. 5.
52   Press Release PN 04 (9 April 2003).
53   See para. 6.3.3.5 above.
54   See www.hm-treasury.gov.uk.
55   Because of the possibility of infringing Community state aid law, as to which Appendix
     B contains the observation: 'EU policy on state aid for environmental protection faces the
     difficulty that it must fulfill two objectives which are sometimes seen as contradictory'
     (see paras 12.2.7 above and 21.10 below).

per tonne for open window schemes and £25–45 per tonne for in-vessel schemes;[56] and

c. materials recycling, the lower end cost equivalents[57] for all Defra-surveyed materials being lower[58] than the £35 landfill tax rate cap, but the upper end ranges for steel[59] and wood[60] being slightly above this level and that for plastics (at £207) very significantly in excess of it.

2   Even though incineration appears to be the alternative to landfill most under consideration by county councils,[61] the report does not give any consideration to this as an alternative to landfill. Energy from waste could, no doubt, be seen as a version of incineration, and was indeed, in its CHP form, seen as having such advantages by the Official Opposition spokesman,[62] its immediate economic (and therefore political) consequences[63] have been laid bare by the Report.

3   Most materially, and worryingly, the Report reveals that businesses do not look at their waste disposal costs in a manner that draws attention to the taxation element in the use of landfill. It is clear, therefore, that the government will have an educational task, as well as the anticipated one of provider of subsidies until the rate of tax reaches much higher levels.[64]

In a report issued in April 2004,[65] the Commission on Sustainable Development[66] expressed concern at progress on the waste disposal, and especially recycling, front,[67] making particular mention of the position of local authorities,[68] whose 'Catch 22' situation in relation to council tax capping was ignored.

---

56   In May 2004, TEG Environmental was to begin the roll-out of a new type of large scale composting plant (see Hall, *Financial Times*, 27 April 2004, p. 24). By way of contrast, the extraordinary size of the legal fees involved in the cross offers for the UK landfill sites of Shanks Groups should be noted (see Taylor, *Financial Times*, 28 May 2004, p. 26).

57   Para. 2.5.2.

58   Albeit, in the case of plastics, at £31, not significantly lower.

59   £42.

60   £45.

61   See para. 27.7 below. See also para. 4.2.3 above.

62   *Ibid.*

63   At a time when county councils in the south have come off badly from a reorganisation of the central government grant regime and are under political pressure to economise both from their electorates and from central government.

64   As indicated in para. 21.10 below, the Report concludes that additional tax relief for research and development or enhanced capital allowances will not produce results in this field. The implication of this, as indicated in para. 2.5.2 of the Report is that, until the rates of landfill tax reach much higher levels, subsidies will be required because there are 'significant cost barriers to overcome'.

65   *Shows Promise. But Must Try Harder.*

66   See para. 4.2.1.4 above.

67   *Ibid.*, paras 75, 120–22.

68   *Ibid.*, paras 123, 157. See also para. 4.2.3 above.

21.3.3 Land remediation reliefs

The pollution legacy of the industrial revolution affects most industrial countries.[69] Under the general income tax law, remediation expenditure is likely to be classified as capital,[70] and therefore not relieved before sale (when the expenditure is likely to be deductible under the enhancement heading).[71] An excellent example of this is the US case, *Northwest Corp v. Commissioner of Internal Revenue*,[72] where the removal of asbestos in the course of remodelling a building was held to be capital. Following this decision, the Taxpayer Relief Act 1997 added section 198 to the Federal Income Tax Code giving relief (by election) for the abatement of hazardous substances up to 31 December 2000. In Australia, expenditure on environmental impact assessments is relieved by equal instalments over ten years, and clean-up expenditure over a single year.[73]

In the UK a very limited relief was enacted in 2001, following the publication of Lord Rodgers' Urban Task Force report, *Towards an Urban Renaissance*.[74] The clean-up expenditure relief takes the form of a 150 per cent corporation tax[75] deduction for newly purchased contaminated land for which there was:

1 no right of recovery from the vendor;
2 no entitlement to a capital allowance;
3 no grant aid;
4 no subsidy from a third party; and
5 where the deduction would not be a trading expense.[76]

On the face of it, this relief is confined to capital expenditure, but the Inland Revenue have confirmed that it can be taken against stock-in-trade.[77] Where available, the relief can be claimed either by an acquiring trader or by an acquiring landlord.

21.4   Political and industrial aspects

21.4.1   The waste management industry

A good environmental and economic case can be made for landfill tax. The purpose behind making the dumping of waste more expensive to the end user is to encourage

---

[69] See para. 24.5.1 below.
[70] Income and Corporation Taxes Act 1988, s.74(1)(f).
[71] Taxation of Chargeable Gains Act 1992, s.38(1)(b).
[72] 108 TC 265 (1997).
[73] Income Tax Assessment Act 1997, subdivisions 400–A and 400–B.
[74] The package of measures also included the very restricted flats-above-shops 100 per cent first year allowances (see Capital Allowances Act 2001, ss.393A–393O).
[75] That is, there is no deduction for income tax purposes.
[76] Finance Act 2001, Sched. 22.
[77] *Taxation*, 7 March 2002, p. 546.

recycling.[78] Yet it is quite clear that, even though the rate of £15 per tonne has been reached in 2004, the hoped-for end result will not be achievable.[79]

The decision to introduce the £10 to £15 escalator was examined in the Thirteenth Report of the House of Commons Environment, Transport and Regional Affairs Select Committee, *The Operation of Landfill Tax*, published on 14 July 1999. This had heard evidence to the effect that this level was far too low to encourage investment in recycling technologies.[80] It recommended that further price rises up to £30 per tonne should be considered[81] and that the increased revenue should be used to deter fly-tipping and reduce the additional burden on local authorities.[82] On the other hand, the evidence before the Committee did support the retention of the £2 rate for inert waste, this being a sufficient incentive to encourage alternative means of disposal for this type of waste.[83]

The Committee also took evidence as to the undesirable consequences of the introduction of the tax, which were said to include fly-tipping, increased waste management costs for local authorities, the diversion of industrial waste into the household disposal stream, a negative effect on the recycling of metals and white goods, a downturn in the use of pulverised fuel ash (and therefore increased disposal of it) and the use of virgin aggregate and mineral for landfill engineering (because of the lack of suitable waste).[84] The planning controls on developments (such as golf courses) involving landfilling and landscaping should be reviewed,[85] and that inert material used for site engineering should be exempted from the tax.[86] It also found that the credit scheme was very complicated and difficult to monitor as to quality.[87]

In the light of the Committee's recommendations, it is appropriate to consider why landfill tax rates were originally set so low why they have only gradually been increased. Although it has not been admitted, the reality is that the bulk of the waste committed to landfill is collected from householders by contractors on behalf of local authorities. The cost of the tax has, therefore, to be found out of the council tax.[88]

---

[78]   Environmental Protection Act 1900, s.52, encourages recycling through the making of payments by central government to local authorities. The net saving of expenditure is determined under Environmental Protection (Waste Recycling Payments) (England) Regulations 2004, S.I. 2004 No. 639. See also para. 1.4.2.1(3) above.

[79]   Houlder, *Financial Times*, 22 October 2002, p. 6; Eaglesham, *Financial Times*, 24 October 2002, p. 4.

[80]   *Ibid.*, paras 15, 16.

[81]   *Ibid.*, para. 18.

[82]   *Ibid.*, para. 20.

[83]   *Ibid.*, para. 21.

[84]   *Ibid.*, para. 22.

[85]   *Ibid.*, para. 27. As to a follow-up on this, see RICS Education Trust's *Can the Waste Planning System Deliver?* (Royal Institute of Chartered Surveyors, 2004).

[86]   Select Committee Report, para. 28.

[87]   *Ibid.*, para. 76.

[88]   The Minister for the Environment (Mr Elliot Morley, MP) has said that amendments to Environmental Protection Act 1990, ss.45 and 51 would be require before local authorities could charge householders for collecting household waste or delivering it to Civil Amenity sites (see HC WA 158231, 10 March 2004, col. 1544W). See also para. 4.2.3 above.

The level of local taxes is politically a very sensitive issue indeed. It will be recalled that the main reason for the political demise of Mrs Thatcher was the previous (and wildly unpopular) attempt at a replacement for domestic rates, the community charge (colloquially referred to as 'the poll tax'). It is, therefore, just not politically possible for a government of any complexion to increase landfill tax suddenly to a level at which a major increase in council tax levels round the country will result. Indeed, it is reasonable to presume that this was behind the report that the Performance and Innovation Unit[89] was thinking of recommending the imposition of collection charges of either a £5 per month rubbish collection charge or one of £1 per bag.[90] Some support for this approach can be had from British Columbia's User Pay Waste Management Initiative, which combined massive increases in tipping fees with an annual collection charge for a single bag per household, and additional fees for each extra bag (paid for by stickers purchased in advance), and the approach was supported by the Environment Agency. It was, however, opposed by the Treasury.[91]

On the other hand, there is a school of thought (which appears to have some ministerial support)[92] to the effect that the best way to encourage recycling would be to induce local authorities to enforce the separation of materials by householders,[93] as in Switzerland. Indeed the contrast between the waste disposal profile between the two countries is remarkable. In 2000, Switzerland recycled 45 per cent and incinerated 48 per cent, leaving only 7 per cent to go to landfill. For the UK, the figures were 11 per cent, 8 per cent and 81 per cent. With the UK facing the prospect of fines of up to £180m per annum if it failed to meet the targets set for 2010, 2013 and 2020,[94] and appearing to have no prospect of doing so on the basis of trends at the end of 2002, drastic action had to be, and was, recommended in the Prime Minister's Strategy Unit's[95] report, *Waste Not, Want Not*. Essential to this was the dissemination of the political message that the public perception of the extent to which local council tax bills were dependent upon waste disposal costs (including landfill tax) was too pessimistic. Whether it will be possible to get this message across remains to be seen, especially in the southeast where county councils were to come out extremely badly, in relative terms, from the reallocation of local government subsidy arrangements to be announced the following week.[96] Indeed, by September

---

89   See para. 4.2.1.2(1) above.
90   Newman, *Financial Times*, 12 July 2002, p. 2.
91   Guha, *Financial Times*, 13 August 2002, p. 2.
92   See Houlder, *Financial Times*, 7 May 2003, p. 6.
93   See Houlder, *Financial Times*, 22 July 2002, p. 4.
94   Under the Landfill Directive (see para. 12.2.5.1(2) above).
95   See para. 4.2.1.2(1) above.
96   For instance, in Hampshire, the increase of 3.7 per cent was thought likely to provoke an increase in the County Council's Council Tax precept of 15 per cent in 2003/04 (see Guthrie, *Financial Times*, 8 December 2002, p. 4). Ironically, *Waste Not, Want Not* (see above) had complimented that particular authority on its innovatory policies with regard to waste disposal. Indeed, a green garden material composting business fed by the county's 26 Household Waste Recycling Centres has been so successful that it has made a planning application to expand its composting site near Basingstoke (*The Business Magazine*, Solent and South Central, December 2003/January 2004, p. 3).

2003,[97] a considerable degree of opposition from low income groups, and especially pensioners, had been organised against the continuation of the council tax system following the implementation of the changed central government subsidy arrangements.

The changes, which were announced at a time when the increased regulatory obligations were already encouraging consolidation in the waste disposal industry,[98] were likely to take the form of:

1   an increase in the rate of landfill tax, which was very low compared to its equivalent in other Member States,[99] to £35 per tonne, starting with an uplift of £3 in 2005/06[100] and increasing by at least the same rate thereafter[101] – pursuant to the Strategy Unit's finding that a two year lead in time was required to encourage a switch to other methods of disposal;[102]
2   the reduction of two thirds in the credits available to Entrust-registered entities in favour of publicly funded purposes, such as the development of new technologies, perhaps supported by the Challenge Fund announced in the 2000 Spending Review;
3   the use of the increased revenue stream to lighten the burden on industry – measures which would require consultation;[103]
4   putting pressure on local authorities,[104] which may be difficult to achieve in the absence of the provision of additional funds by central government, since long term capital spending is difficult to attain under a system in which the government has retained the ability to 'cap' the expenditure of individual councils and can

---

[97]   Not least in the context of the Liberal Democrats seeking support for their long-espoused policy of the replacement of the council tax by a local income tax. Although some American cities have such a tax, how practicable it would be in the UK, even if collected through central government machinery, must be debatable. With council tax and its two predecessors, the community charge (or 'poll tax') and generally applicable rates, it was possible for the local authority to estimate its receipts within a very small margin of error. This would not be possible with an income tax supplement.
[98]   Felsted, *Financial Times*, 18–19 January 2003, p. 13.
[99]   *Waste Not, Want Not* focused on the Netherlands's £45 and Denmark's £34.
[100]   Announced in the Pre-Budget Report on 27 November 2002.
[101]   The House of Commons' Environmental Audit Committee, in its Fourth Report of 2002–03, HC 167, published on 1 April 2003, paras 19 and 21, considered that the projected rate of increase would take nine years to reach a level at which recycling would be induced, and that local authorities would be likely to go for the cheaper alternative of incineration in the meantime.
[102]   However, it does not seem to have been realised that the costs of implementing the Landfill Directive have been such (at least to one major operator) that major customers' willingness to fund them by price increases has been breached (see Rafferty, *Financial Times*, 6 November 2003, p. 26).
[103]   Press Release HMT 1 (27 November 2002).
[104]   Eaglesham, *Financial Times*, 30 November/1 December 2002, p. 5. The major problem seems, however, to be winning over doubtful financiers (see Cowe, *Financial Times*, 25 August 2004, p. 10).

(and, in 2003, did) change the formula under which central government subsidies are allocated;

5   the possibility of local authorities being able to offer householders discounts for undertaking composting;

6   the possibility of an incineration tax being imposed – albeit, it seems, without this being expected to operate as an incentive to move to recycling;[105]

7   the realisation that, as 25 per cent of the country's methane discharge comes from landfill sites,[106] a reduction in the amount of waste so disposed of would help meet climate change targets;[107]

8   shortage of land for landfill in the southeast and northwest, with the prospect of transporting waste to landfill elsewhere, thereby increasing the emission of GHGs; and

9   setting up a system of tradable landfill allowances,[108] and, possibly, in the final analysis, a ban on biodegradable material.

21.4.2   The electricity supply industry

Politics has also a great deal to do with the problems behind climate change levy because a policy decision[109] has been taken to exempt domestic households from it.[110] This was inevitable in the light of the government's reduction, shortly after taking office in 1997, in the rate of VAT on domestic power supplies from 8 per cent

---

[105]   At the same time as advertising the fact that the planning system did not make it easy for local authorities to set up new incineration sites. See also House of Commons Environmental Audit Committee, *Waste – An Audit*, Fifth Report of Session 2002–03, HC 99, published 23 April 2003, para. 63, and para. 27.7 below.

[106]   By way of comparison, the ministerially-described 'very comprehensive' *Review of Environmental and Health Effects of Waste Management: Municipal Solid Waste and Similar Wastes*, by Enviros and the University of Birmingham, published by Defra on 6 May 2004, indicated that the incineration of municipal solid waste accounts for less than 1 per cent of dioxin emissions.

[107]   Subsequently the Household Waste Recycling Act 2003 introduced new provisions into Environmental Protection Act 1990: (1) s.45A, obliging English local authorities to make arrangements for the collection of recyclable household waste unless alternative arrangements were available or the cost would be unreasonably high; (2) s.45B, enabling the National Assembly for Wales to extend this obligation to the local authorities in that Principality; and (3) s.47A, obliging the Secretary of State to report to Parliament on progress in relation to English local authorities by 31 October 2004.

[108]   Under Waste and Emissions Trading Bill published 15 November 2002 (see para. 20.7 above).

[109]   This must be contrasted with the position in the Netherlands and Sweden, where lower rates of energy tax apply to industrial, as against domestic consumers (see Benjamin J. Richardson and Kiri L. Chanwai, 'The UK's Climate Change Levy: Is It Working?' [2003] JEL 39–58). In Norway, Sweden, Finland, the Netherlands and Denmark regard is had to the carbon content of fuels taxed. New Zealand is to introduce a carbon-based tax. Germany and Italy have also introduced energy taxes (*ibid.*).

[110]   See also the Warm Homes and Energy Conservation Act 2000.

to 5 per cent.[111] Indeed, it has actually been suggested that this reduction in VAT stimulated the demand for energy.[112]

Nonetheless, the design of the tax means that, in the context of the structure through which electricity is supplied, the incentives being introduced into the structure of that tax to favour renewable or 'clean' sources of energy cannot have the maximum effect because the distribution system lessens the effect of those incentives. For instance, say an electricity supply company has a 75 per cent domestic and 25 per cent commercial customer split. If it draws all its electricity from traditional sources through the market mechanism, 25 per cent of its turnover will be taxable. However if, for instance to secure price stability, its management decided to contract to purchase 20 per cent of its normal load from renewable sources outside the market mechanism, the effect would not be to reduce its tax by 20 per cent, but by 5 per cent.[113] In order to improve upon this it would have to seek regulatory consent to restructure itself into two companies with separate electricity supply licences. Only if the tax point were to be shifted from that of distribution to final consumer, to that at which power was fed by the generator into the network, could this problem be overcome. However, unless there is the political will to subject domestic consumers to the full amount of the tax, this change could not be made. The political will does not, however, seem to be there. This has been demonstrated clearly in *Economic Instruments to Improve Household Energy Efficiency: Consultation Document on Specific Measures*,[114] which makes it clear that, while the use of economic instruments (such as taxation) is more effective than regulation,[115] any increase in the rate of VAT on domestic power or subjecting it to climate change levy has been ruled out.[116]

However, the Commission on Sustainable Development,[117] in a report issued in April 2004,[118] advocates very strongly the extension of environmental taxation to domestic consumers and, indeed, goes as far as advocating the replacement of climate change levy by a carbon tax in order to achieve this. Faced with a statutory requirement to report on the energy efficiency of residential accommodation in England,[119] Mrs Margaret Beckett, the Secretary of State for the Environment, Food and Rural Affairs, published *Energy Efficiency. The Government's Plan for Action* on 27 April 2004.[120] The previous day, her Department, in conjunction with the DTI, had issued a *Consultation on the Methodology used for Calculating the Number of*

---

[111] Value Added Tax 1994, Sched. 7A, Group 1.

[112] *Daily Telegraph*, 21 March 2000, p. 29.

[113] This is the case even though, from 1 July 2004, Community tracing rules oblige UK suppliers to disclose source proportions on electricity bills (see para. 12.2.6.3 above). Only a small minority are likely to be influenced by this as to their choice of supplier.

[114] HM Treasury and Defra, August 2003.

[115] *Ibid.*, paras 16–21. (In the sense of 'command and control'.)

[116] *Ibid.*, para. 28, leaving fiscal amelioration to reducing the price of products which used or saved energy: as to the latter (see para. 24.4 below).

[117] See para. 4.2.1.4 above.

[118] *Shows Promise. But Must Try Harder*, paras 57, 72 and 73.

[119] See para. 21.6.2 below.

[120] According to *Hansard*. According to Defra's website, the document was published the previous day (see www.defra.gov.uk).

Households in Fuel Poverty for England.[121] As to the latter, it is to be wondered whether the former may not be of the greatest value if there was such a degree of lacuna in the underlying statistics.[122]

In the 2004 Spending Review, issued on 12 July 2004, it was revealed that spending on the *Warm Front Programme* would be £95m higher in 2007/08 than in 2004/05, in order to 'ensure further progress towards the 2010 target'. The programme's resources were also to be 'better targeted on the genuinely fuel poor' by reviewing the eligibility criteria.[123] The rush to print had been followed, in May 2004, by the initiation by Defra of a consultation[124] on the terms of the draft of the statutory instrument[125] to be promulgated in October 2004 to govern RECs for the period 1 April 2005 to 31 March 2008. The individual targets were to be apportioned out by Ofgem proportionate to the respective numbers of their domestic consumers, with smaller companies having a lower relative target.[126] For the purposes of this exercise, CHP is *assumed* to be a form of energy efficiency.[127] Illustrations of possible measures of energy efficiency were:[128]

1   cavity wall insulation;
2   loft insulation;
3   A-rated boilers;
4   fuel switching;
5   heating controls;
6   compact fluorescent lights;
7   fridgesaver-type schemes;
8   A-rated appliances;
9   tank insulation; and
10 draughtproofing.

However, it had to be admitted that the *energy efficiency* objective applied to all domestic[129] users,[130] the change was expected to provide particular help to low-income consumers[131] and therefore contribute to the alleviation of fuel poverty.[132]

---

121   See www.dti.gov.uk.
122   As to which, the government's *The UK Fuel Poverty Strategy 2nd Annual Progress Report: 2004* also published on 26 April 2004, Chapter 2, is indicative of a distinctly Rooseveltian New Deal hotchpotch of measures rather than a joined-up strategy.
123   *Ibid.*, para. 17.6.
124   *The Energy Efficiency Commitment from April 2005*.
125   To be entitled the Electricity and Gas (Energy Efficiency Obligations) Order 2004.
126   Consultation proposals, paras 3.18, 3.20.
127   *Ibid.*, para. 3.28.
128   *Ibid.*, Annex 1, para. 8.
129   And, although the document was directed at the residential sector, presumably also other users of buildings.
130   Annex 2 (Partial RIA), para. 10 stated: 'Energy efficiency has a key role in meeting the Government's climate change targets by reducing emissions of carbon dioxide'. Para. 23 went on to quantify the proposed environmental benefit at 0.7MtC in 2010.
131   *Ibid.*, Annex 2, para. 8.
132   *Ibid.*, Annex 2, para. 11.

The government's conclusion was that the proposals would not only benefit the environment,[133] but also provide social benefits through the reduction of fuel bills and improvement of comfort.[134]

On 30 November 2004, Defra published *Fuel Poverty in England: The Government's Plan for Action*.[135] The objective proclaimed in this was to eliminate fuel poverty by 2010.[136] It appeared, however, that the government's efforts to date, which had been targeted at social housing,[137] amounted to a mere 20 per cent of the households involved, as against 65 per cent of owner occupiers.[138] The Warm Front Scheme is, therefore, to be amended from June 2005 to make 'the greatest possible impact' on the private sector,[139] with up to 25 per cent of the vulnerable fuel poor not eligible under the current scheme,[140] 60 per cent of them single elderly householders.[141]

21.4.3   The coal industry

Historically, the National Union of Mineworkers had been a major force behind the Labour Party. By the time Labour returned to office in 1997, however, the coal industry had been denationalised and no longer played the role that it had done in the national economy between the Wars. Nonetheless, when world prices fell, the government chose to carry out a U-turn in its energy policy and, in 1999, vetoed the building of gas-powered stations in order to prop up the coal industry.[142] The view has been expressed that, without this, the Kyoto targets could have been met without resorting to climate change levy.[143]

---

[133] *Ibid.*, Annex 2, para. 23.
[134] *Ibid.*, Annex 2, para. 24. This said, para. 25 then states that the cost of this exercise would be £4 per customer per fuel at 2004 prices. No recognition is given to the racing certainty that the introduction of EU ETS is likely (on the official figures) to increase electricity prices by 6 per cent (see para. 28.3 below).
[135] www.defra.gov.uk.
[136] Lord Whitty's Foreword. However, Annex A, para. A4.2 suggests that it is only social housing which will be brought into 'a decent condition' by 2010, which the targets for the private sector being fixed at 65 per cent by 2006 and 70 per cent by 2010.
[137] That is, subsidised local authority or housing association rented accommodation. See para. 5.3.4.
[138] Para. 3.3.1. It should also be noted from Annex B, para. B1.9 that the National Audit Office review of the Warm Front programme had concluded that the current programme provided very few effective options for homes seen as traditionally hard to heat.
[139] Para. 3.3.8. This cost of this was not possible to estimate: para. 5.3.6.
[140] Para. 4.2.1.
[141] Para. 4.2.2.
[142] This attitude persisted into 2003 when £60m was put up with a view to preventing further deep mine closures for five years (see Eaglesham and Taylor, *Financial Times*, 12 February 2003, p. 1). Yet at the end of November 2003, UK Coal was saying that the award of state aid of £36m for 2004, as against the requested £79m, was insufficient to prevent further mining job losses (see Bream, *Financial Times*, 29–30 November 2003, p. 4).
[143] See *Daily Telegraph*, 30 March 1999, p. 35; Brown, *Financial Times*, 23 February 2000, p. 3. Dieter Helm in *Energy, the State and the Market: British Energy Policy since 1979*, revised edition (Oxford: Oxford University Press, 2004), pp. 301–4, sees this as the point where New Labour's energy policy went off the rails.

However, it appears that the New Electricity Trading Arrangements ('NETA') put in place in 2001,[144] under the Utilities Act 2000, reduced the price for coal-generated electricity by 40 per cent over the course of the first year,[145] resulting in the Fifoots Point power station in Wales being placed in receivership in 2002 and the huge Drax power station being in technical default on part of its debt.[146] Furthermore, the acquisition of the Eggborough station turned out to have been an additional problem for British Energy plc when that company ran into a financial crisis in August 2002.[147] The role played by NETA in British Energy's problems was acknowledged, subsequently, by the Minister of State for Energy (Mr Brian Wilson, MP) before the House of Commons' Trade and Industry Select Committee.[148]

The crisis in the autumn of 2002 was not, however, confined to coal and nuclear generators. On 9 October Powergen announced the shut down of 25 per cent of its electricity capacity, including the oil-fired Isle of Grain and the gas-filed Killinghome plants.[149] The fall in the wholesale market price was nowhere near matched by a modest decline in retail prices. One of British Energy's problems was the lack of a retail base but, on 14 October, it had become apparent that the US TXU group was in difficulties,[150] said to be caused not by its imbalance in generating capacity (three coal-fired stations to service 4m customers) but by having entered into five long-term contracts at prices significantly above the market price then current.[151] Following the demise of TXU Europe, a number of power stations with exclusive supply agreements with that company halted output, doubling wholesale prices in the 'balancing' market.[152] Nonetheless, it was thought likely that five US owned powers stations would be put up for sale.[153] In January 2003, following its acquisition of the coal-fired High Marnham and Drakelow stations from TXU, the German-owned

---

[144] Which emphasised the problem considered in the last part of para. 21.4.4 below.

[145] The effect on domestic consumers was, however, less marked, with those who moved supplier gaining as against those who stayed put, leading the National Audit Office to question whether the lower pricing benefits had been worth the cost (as estimated by Ofgem) of introducing the new scheme (see Taylor, *Financial Times*, 9 May 2003, p. 5).

[146] Jones, *Financial Times*, 22 August 2002, p. 25. Manouevring as to how to restructure this entity was still continuing in December 2003 (see Bream, *Financial Times*, 29–30 November 2003, p. M2, Taylor, *Financial Times*, 2 December 2003, p. 27, and 11 December 2003, p. 28).

[147] Roberts and Taylor, *Financial Times*, 7–8 September 2002, p. 4; Marsh, *Financial Times*, 30 September 2002, p. 6.

[148] Buck, *Financial Times*, 11 December 2002, p. 26.

[149] Taylor, *Financial Times*, 10 October 2002, p. 3. By August 2003, however, there was talk of the former being taken out of mothballs (see *Financial Times*, Lex column, 22 August 2003, p. 18).

[150] TXU Europe later went into administration (see Taylor and van Duyn, *Financial Times*, 20 November 2002, p. 1).

[151] *Financial Times*, 15 October 2002, p. 20.

[152] Taylor, *Financial Times*, 22 November 2002, p. 2.

[153] Taylor, *Financial Times*, 21 November 2002, p. 26. With regard to AES's problems with Drax, leading to it walking away from the station on 5 August 2003 (see further Taylor, *Financial Times*, 24 July 2003, p. 19; 6 August 2003, p. 19; 24 February 2004, p. 25).

Powergen closed them.[154] Moreover, by July 2003, the British banking industry had become sufficiently concerned as to the financial viability of the traditional sector to commission a study into the restructuring possibilities for the UK independent power industry.[155]

The reality was that, by August 2003 (a year after British Energy admitted to having difficulties), only those generators with retail capacity were not in financial difficulty. It had only been avoided in the case of them by failing to reduce retail prices in line with wholesale prices. This development represented not only a failure of the government's substitution of NETA for the original arrangements[156] but also the undermining of the original denationalisation structure, which separated out generation from distribution.[157]

---

[154] Taylor, *Financial Times*, 10 January 2003, p. 2. It appears that, before this acquisition, it had been planning to mothball both its oil-fired station on the Isle of Grain in Kent and the rest of its gas-fired Killingholme plant in Lincolnshire (see Taylor and Eaglesham, *Financial Times*, 7 January 2003, p. 7). The US group, NRG Energy, was also experiencing debt problems with its Killingholme plant (see Taylor, *Financial Times*, 8 January 2003, p. 6). In addition, the regulator was preparing to ring-fence the REC, Midlands Electricity, whose US parent had been unsuccessful in attracting adequate offers for it (see Taylor: *Financial Times*, 9 January 2003, p. 4). In February 2004, the bankers' consortium company, CGE Power, made an indicative bid for Drax stated to be conditional upon the simultaneous acquisition of an appropriate mix of other plants (see Taylor, *Financial Times*, 24 February 2004, p. 25). In July 2004, the Fifoots power station in south Wales was sold by its administrators to the Rutland Fund (see Taylor, *Financial Times*, 7 July 2004, p. 3). Scottish and Southern later bought Ferrybridge and Fiddler's Ferry from AEP (see Taylor, *Financial Times*, 31 July–1 August 2004, p. M2).

[155] Kipphoff, *Financial Times*, 7 July 2003, p. 20. However, the plan put forward in early September, for the lenders to all the insolvent power stations other than Drax to be transferred into a single entity called 'Joe', but dubbed 'BustCo' by the *Financial Times*, has given rise to concern (see *Financial Times*, Lex column, 6–7 September 2003, p. 16). In December 2003, four British and two German banks formed a company, CGE Power, to buy financially distressed UK power stations (see Taylor, *Financial Times*, 12 December 2003, p. 27). One large US power group was, nonetheless, continuing to seek purchasers for its UK plants (see Taylor, *Financial Times*, 10–11 January 2004, p. M2 and 23 January 2004, p. 26). However, in May 2004, CGE Power withdrew all its offers in the face of stiff price competition from electricity utilities (see Taylor, *Financial Times*, 25 May 2004, p. 21). Subsequently, Scottish Power purchased the Damhead Creek power station in Kent and Centrica (originally a gas supplier, which had expanded into electricity) was understood to be likely to acquire Killingholme in Lincolnshire (see Taylor, *Financial Times*, 3 June 2004, p. 23).

[156] Despite the optimism of the Director-General of Ofgem in a speech in November 2002 (see [2001/2002] 6 ULR 170). The New Electricity Trading Arrangements favoured gas-powered stations. In September 2004, the House of Commons Trade and Industry Committee decided to launch an investigation into why wholesale gas prices had risen in the UK to a level much higher than in Continental Europe (see Taylor, *Financial Times*, 17 September 2004, p. 2). The effect on industrial users was also becoming serious (see Workman (letter), *Financial Times*, 17 September 2004, p. 18).

[157] See paras 2.4.1 and 6.4.3.1 above.

As to the former of the foregoing points, although at the end of August 2003, there had been an increase of 30 per cent in the base prices for the following winter and the Isle of Grain power station was being recommissioned.[158] The blackout in New York, Ohio, Michigan and Ontario gave rise to initial concerns, but the conclusion was reached that the genesis of the US's deep-seated problem was political.[159] Ironically, the blackout in London, albeit seemingly for a different reason, followed within a matter of days, reopening the initial concerns about whether the UK's national grid was also vulnerable to this type of problem; moreover, although reconnection was relatively quick, the knock-on effect on the transport system was severe and more prolonged.[160] This (and a similar occurrence in Birmingham) was, however, discovered to be due to incorrectly installed protection equipment, rather than a lack of investment.[161] Shortly after there were significant blackouts in Sweden, Copenhagen and, in the early hours of 28 September, the whole of mainland Italy and Sicily (an admitted case of massive underinvestment nationally) as a result of a series of disconnections in the power lines from Switzerland, Austria and France.[162] In these circumstances, the government ordered the preparation of emergency plans to avoid blackouts in the UK over the 2003/04 winter.[163] Since privatisation, the generation cushion has fallen from 30 per cent to 17 per cent.[164] The National Grid had, furthermore, identified 16 weeks between November 2003 and March 2004 when the reserve margin might fall below that 17.7 per cent cushion.[165] In consequence it sought new powers to cope with the potential problem.[166] The Secretary of State for Trade and Industry let it be known, however, that she was satisfied that, with some capacity reopening, generators would be able to meet winter demand.[167]

Furthermore, a survey by London Electricity published in 2002 suggested that the imposition of climate change levy had not resulted in any saving of energy. Not only would the Kyoto obligation of 87.5 per cent of 1990 level by 2008/12 fail to be met; neither would the government's more ambitious 2010 target of 10 per cent from renewable sources.[168]

By September 2003, concern was being expressed that, with the introduction of the EU measures intended to accompany the start of the EU ETS in 2005, coal-fired generation might become uneconomic[169] and that consideration was being given to

---

[158]   *Financial Times*, Lex column, 22 August 2003, p. 18.
[159]   Dizard, *Financial Times*, 25 August 2003, p. 17.
[160]   *Financial Times*, 29 August 2003, pp. 1, 4.
[161]   Taylor, *Financial Times*, 1 October 2003, p. 2.
[162]   See *Financial Times*, 29 September 2003, Kapner and Graham (p. 1) and Kapner, Buck, Graham, Hoyos and Taylor (p. 6).
[163]   Eaglesham, *Financial Times*, 2 October 2003, p. 1.
[164]   BBC *Breakfast*, 14 October 2003.
[165]   Taylor, *Financial Times*, 11–12 October 2003, p. 6.
[166]   Taylor, *Financial Times*, 15 October 2003, p. 3.
[167]   Taylor, *Financial Times*, 17 October 2003, p. 6.
[168]   Jones, *Financial Times*, 24 June 2002, p. 4.
[169]   However, in November 2003, UK Coal landed a large order from EdF Energy, one of the biggest suppliers (see Taylor, *Financial Times*, 13 November 2003, p. 28). And it was estimated, subsequently, that coal-fired stations increased their generation by 11 per cent in

British Energy, which had by then managed to meet the UK Government's terms for temporary state aid,[170] might seek ways of extending the life of its nuclear power stations by up to five years in order to avoid 20 per cent of the UK's generating capacity going off stream at a critical time.[171]

Furthermore, the release, on 25 March 2004, of the carbon dioxide emissions statistics for 2003, showing an increase of 1.5 per cent, although described by the Environment Minister as a 'blip' and largely due to electricity imports from the Continent,[172] came at a time when, worldwide, concern was being expressed as the 'comeback' coal was making as a generation fuel.[173] The demand for natural gas was, nonetheless, still growing faster than that for competing fuels.[174] In addition, two-thirds of City of London investors believed that the UK would miss its renewables generation targets,[175] and business leaders were understood to have appealed to ministers to reconsider the tough emissions targets thought likely to cause UK electricity prices to rise further than those in other EU Member States.[176]

2003 (see Taylor, *Financial Times*, 27 February 2004, p. 4). Scottish and Southern Energy saw its future, however, in gas-fired stations (see Taylor, *Financial Times*, 7 November 2003, p. 28). At the same time it was recorded that developers had been shelving plans to construct new conventional power plants (see Taylor, *Financial Times*, 13 November 2003, p. 9).

[170] Taylor, *Financial Times*, 13–14 September 2003, p. 2. But by the beginning of August 2004, with the necessary EU consent not yet through, shareholders were complaining that they were to be short changed (see Boswell, *Financial Times*, 2 August 2004, p. 20, and Taylor, *Financial Times*, 5 August 2004, p. 20). The Decision, *Rescue Aid to British Energy plc*, NN 101/2002, came through on 27 November 2004. When the company's results for the half year to 30 September 2004, were published, however, it became apparent that the company had not been in a position to benefit from the upturn in market rates because the bulk of its power had been pre-sold under long-term fixed-price contracts: Taylor, *Financial Times*, 11–12 December 2004, p. M4.

[171] See Taylor, *Financial Times*, 22 September 2003, p. 1.

[172] Defra News Release, 25 March 2004.

[173] Morrison, *Financial Times*, 12 March 2004, p. 10.

[174] Chung, *Financial Times Special Report, Gas Industry*, 19 March 2004, p. 2. Concern has also been expressed as to the adequacy of storage facilities for imported natural gas (see Taylor, *Financial Times*, 10–11 April 2004, p. 2). Helm, *op. cit.*, p. 422, also expresses concern as to the security of supply aspects. This should be contrasted with the setting up of the British National Oil Company in 1976 with the express objective of achieving this for North Sea oil (see para. 2.4.3 above) and the idea underlining the pool (see para. 2.4.1n above) before its replacement by NETA (see, respectively, Helm, *op. cit.*, pp. 40 and 131–7).

[175] Taylor, *Financial Times*, 24 March 2004, p. 5.

[176] Taylor, *Financial Times*, 12 March 2004, p. 6. Defra was subsequently understood to have reduced the 2005–07 target in response to this problem (see Houlder, *Financial Times*, 10–11 April 2004, p. 1). The electricity trading documentation was later revamped in order to allow futures and options to be written (see Skorecki, *Financial Times*, 17–18 July 2004, p. M2). In 1985, these had been the prerequisite to increasing gilt turnover after 'Big Bang' on the London Stock Exchange.

21.4.4   The renewables sector

Reliance on King Coal for electricity generation has been a problem which successive UK governments have for decades tried to address. The UK was one of the first countries to espouse nuclear generation, in a programme starting with the commissioning of the Calder Hall Magnox reactor in 1957. In 1965, the government of the day took the view that the future lay with advanced gas-cooled reactors, but the programme took a long time to implement.[177] By the 1990s, nuclear power had become a controversial source of generation.[178] Indeed, the only perceived advantage of nuclear generation is that it does not cause carbon emissions. With the second generation of reactors coming to the end of their working lives,[179] this source of energy was excluded in terms from the statutory definition of 'renewable sources'.[180] Early closure would, however, create unforeseen difficulty for the British Government in meeting its Kyoto obligations.[181] The following problems are acting as a deterrent to their replacement:[182]

1   they are more expensive to operate than traditional sources, having had to be subsidised by both a price levy[183] and by quotas when denationalisation occurred.[184] They gave rise to a crisis in early September 2002 in the listed utility (which had 20 per cent of the country's generating capacity) when, following the introduction of NETA in March 2001, the 25 per cent planned overcapacity cushion provided for under the 1989 Pool produced a collapse in wholesale prices of between 19 per cent (baseload) and 27 per cent (peak) to below the break even operating level.[185]

2   Decommissioning is an extremely costly business (having, in relation to the older stations, to be underwritten fully by the taxpayer).[186]

---

[177]   Kay, *Financial Times*, 19 September 2002, p. 19.
[178]   As to the debate about continuing to use nuclear power, see P. Beck, *Prospects and Strategies for Nuclear Power* (Earthscan, 1994), P. Beck and M. Grimston, *Double or Quits? The Global Future of Civil Nuclear Energy* (RIIA Briefing Paper, April 2002).
[179]   Only Sizewell B (near Saxmundham in Suffolk) will still be in commission after 2023. This is the UK's only pressurised water reactor: the circumstances leading up to its commissioning are considered in Helm, *op. cit.*, pp. 101–6. The other British Energy reactors are of the advanced gas cooled type.
[180]   Electricity Act 1989, s.32(8), as substituted by Utilities Act 2000, s.62.
[181]   Turner, *Financial Times*, 7–8 September 2002, p. 4. It seems, furthermore, that the crisis faced by British Energy in late August 2002 was triggered by technical problems causing the closure of two generating plants (see Plender, *Financial Times*, 9 September 2002, p. 28).
[182]   Despite which a public consultation involving 4,500 people found opinion evenly split (see Jones, *Financial Times*, 6 December 2002, p. 4).
[183]   Called fossil fuel levy, under Electricity Act 1989, s.33 (abolished by Utilities Act 2000, s.66). See para. 6.4.3.1 above.
[184]   See para. 21.7 below.
[185]   Buchan, *Financial Times*, 26 August 2002, p. 3.
[186]   When the newer stations were privatised in 1996, as British Energy plc, the older Magnox plants had to be retained under the banner of British Nuclear Fuels Ltd. In August 2002, discussions were initiated with a view to the former's financial position being bolstered

3   The disposal of waste from them constitutes a serious, and very political, problem.
4   One particular facility, Sellafield, has been a constant source of safety concerns (irrespective of whether justified) for 20 years or so.[187]
5   The consequences of accidents, whether of a Chernobyl or a Three Mile Island scale, would be, to say the least, extremely serious in a small and densely populated country.
6   after 11 September 2001, the possibility of a nuclear station being the target of terrorist (or, indeed, more formal hostile) activity cannot be ruled out.

When the nuclear industry ran into difficulties in the late summer of 2002, British Energy plc put forward the request that such power be exempted from climate change levy. Making the government's decision not to agree to this known, the Secretary of State for Trade and Industry[188] was reported as having written to the Combined Heat and Power Assocation[189] that climate change levy was not a carbon tax[190] but a downstream energy tax to encourage all sectors of business and the public sector to use electricity efficiently, and that withdrawing a fifth of the supply from the levy would reduce 'its beneficial effects on carbon emissions'.[191] The government subsequently introduced legislation to enable it to give financial assistance, including by way of renationalisation.[192] At the time, it was understood, however, that, while the construction of new nuclear plants was not ruled out, it was not envisaged that this would be at public expense. Subsequently, this possibility was virtually ruled out in the Energy White Paper.[193]

---

by the transfer of the latter, but without the close-down responsibility (see Porter and Bailhache, *The Business*, 25–27 August 2002, pp. 1, 2; Odell, *Financial Times*, 27 August 2002, p. 19). Talks with BNFL, which was also not in a good financial condition, broke down, however, on the issue of reprocessing British Energy's existing waste (see Taylor, *Financial Times*, 7–8 September 2002, p. 4). The government had to offer financial guarantees to prevent the latter going into administration (see Taylor and Eaglesham, *Financial Times*, 9 September 2002, p. 1).

[187]   As to which it is significant that it is anticipated that the government will decide to wind down the Thorp reprocessing centre there one third of the way into its anticipated 30 year life (see Roberts, *Financial Times*, 27 August 2003, p. 2).

[188]   The Rt Hon. Mrs Patricia Hewitt, MP.

[189]   See para. 2.4.1 above.

[190]   The Royal Society subsequently advocated the substitution of a carbon tax to apply to all (including householders) and which would be likely to benefit generation from wind, wave, tidal, nuclear and carbon sequestration sources (see Houlder, *Financial Times*, 18 November 2002, p. 4).

[191]   Taylor and Eaglesham, *Financial Times*, 11 October 2002, p. 2. Various types of assistance can be provided under Electricity (Miscellaneous Provisions) Act 2003, s.1. In this context it is interesting to note that s.4 provides that, where the Secretary of State for Trade and Industry undertakes to make a ground under Electricity Act 1989, Sched. 12, para. 1(1) and, under GAAP, this passes through the profit and loss account, and is to be ignored for tax purposes.

[192]   Taylor, *Financial Times*, 10 January 2003, p. 4.

[193]   See para. 21.6 below. But see Houlder and Burns/Taylor: *Financial Times*, 15 September 2004, p. 4.

Environmental Taxation Law

The proposed financial assistance ran into very heavy opposition from fossil fuel operators, but the European Commission is understood to have felt that it had no choice but to approve the UK Government's request, given the high risk associated with nuclear fuel.[194] Nonetheless, at the end of May 2003, the Commission's approval for the rescue package had not been given and two US coal power station owners were understood to be considering a challenge to it in the ECJ.[195] However, on 23 July 2003, the Commission commenced an investigation into the matter, albeit in the expectation of having, ultimately, to approve at least some of the decommissioning aid.[196] The UK Government set a deadline of 30 September 2003, for British Energy plc to be restructured in the way which was normal in the early twenty-first century, with the shareholders' interests being reduced to 2.5 per cent (originally less than one per cent had been mooted) and bondholders and banks forced to convert part of their debt into shares. Agreement was reached early on 1 October 2003, and involved government assistance with certain future liabilities, such as decommissioning costs and procuring a variation in British Energy's contract with the state-owned British Nuclear Fuels Ltd, in return for 65 per cent of net cash flow.[197] A year later, however, after a substantial rise in wholesale prices and with the EU consent first imminent and then granted, this agreement had become the subject of dispute, with US hedge fund shareholders seeking a renegotiation.[198]

Hydro-electric plants have been used as the source of renewable electricity for some decades, especially in Scotland. However, it is considered that it would not be possible

---

194   Guerrera and van Duyn, *Financial Times*, 11 March 2003, p. 4.

195   Taylor, *Financial Times*, 2 June 2003, p. 21.

196   Dombey, *Financial Times*, 24 July 2003, p. 5. On 17 June 2004, the UK Government announced that it was delaying presenting the information requested by the European Commission until the autumn. It was thought that the concerns behind this delay were not related to a likely rejection but in relation to possible challenges by the anti-nuclear lobby (see Taylor, *Financial Times*, 18 June 2004, p. 23).

197   Taylor, *Financial Times*, 1 October 2003, p. 2; Taylor, Batchelor and Tassell, *Financial Times*, 2 October 2003, p. 3. Nonetheless, British Energy's finances have continued to be the cause of concern (see Bream, *Financial Times*, 31 October 2003, p. 24, Taylor; *Financial Times*, 4 November 2003, p. 24; Taylor, *Financial Times*, 14 November 2003, p. 6; Taylor, *Financial Times*, 26 November 2003, p. 23; Bream, *Financial Times*, 28 November 2003, p. 4; Eaglesham, *Financial Times*, 28 November 2003, p. 22; Taylor, *Financial Times*, 8 December 2003, p. 24; Taylor, *Financial Times*, 27 February 2004, p. 24).

198   Boswell, *Financial Times*, 2 August 2004, p. 20; Taylor, *Financial Times*, 5 August 2004, p. 20; Nicholson, *Financial Times*, 6 August 2004, p. 21; Taylor, *Financial Times*, 1 September 2004, p. 21; Taylor, *Financial Times*, 2 September 2004, p. 8; Taylor and Minder, *Financial Times*, 2 September 2004, p. 19; Taylor and Eaglesham, *Financial Times*, 9 September 2004, p. 4; Taylor, *Financial Times*, 15 September 2004, p. 24; Taylor, *Financial Times*, 17 September 2004, p. 22; Taylor, *Financial Times*, 18–19 September 2004, p. M2; Taylor, *Financial Times*, 20 September 2004, p. 23; Taylor and Buck, *Financial Times*, 23 September 2004, p. 23; Taylor, *Financial Times*, 24 September 2004, p. 21, Dickson, *Financial Times*, 24 September 2004, p. 22; Bream, *Financial Times*, 25–26 September 2004, p. M3; Dickson, *Financial Times*, 1 October 2004, p. 24; Taylor, *Financial Times*, 1 October 2004, p. 25.

to achieve a major increase in capacity from this source. The effect of existing hydro-electric capacity from Scotland coming on stream with the substitution of BETTA for NETA[199] is unclear. The speed of reaction of gas-fired stations gave them a major advantage under NETA. In principle hydro-electric stations should be in a similar position. With most capacity under the control of Scottish and Southern Energy plc, however, one of the companies best placed to satisfy its retail needs from in-group generation capacity, it might well turn out to be the case that the predominance of gas-powered generators will persist on the traded market.

At one stage, the use of tidal power was being considered for the Severn Estuary, following its initiation at the mouth of the Loire, but nothing had been heard of this for some years before January 2003, when a government-commissioned study by Sir Robert McAlpine was published.[200]

Harnessing natural gas burned off from oil rigs and methane from landfill sites[201] may produce a small contribution. Harnessed to this is the generation of power from waste, such as straw, for which a plant was set up near Ely in the mid-1990s[202] and one was commissioned in the West Country in mid-2002.

In December 2002, plans were announced for a power station in south Wales burning hydrogen produced from coal that had been used to generate electricity.[203] Moreover, in August 2003, a proposal was approved for the construction of a 'clean' coal-to-hydrogen power station close to the Hatfield Colliery in Yorkshire.[204]

From 29 November 1994, coppicing has been treated, for tax purposes, as part of the statutory trade of farming.[205] Moreover, farmers who plant up agricultural land can be provided with funding by the Defra Energy Crops Scheme. In 1994 it had been hoped that it would be possible to set up regional biomass generators, based on drawing willow from farmers within a 40 mile radius, on the basis of a three yearly growing cycle. Plants were planned for Cricklade (Buckinghamshire),[206] Eye (Suffolk),[207] Eggborough (Yorkshire),[208] Swindon (Wiltshire) and Basingstoke

---

[199] See para. 6.4.3.1 above.
[200] Taylor, *Financial Times*, 14 January 2003, p. 6: see also letter from Napier, 16 January 2003, p. 18.
[201] Or indeed coal mines.
[202] Abel, *Farmers Weekly*, 3 November 1995, p. 59.
[203] Taylor, *Financial Times*, 20 December 2002, p. 2.
[204] Taylor, *Financial Times*, 6 August 2003, p. 3. Friends of the Earth (see para. 2.2 above) have, however, criticised the Minister for extending a scheme under which conventional power stations were allowed to qualify for green energy incentives by burning imported olive residues and palm kernels, thereby discouraging the construction of new biomass power stations. Such arrangements appear to have been made in relation to the financially-troubled Fiddlers Ferry and Ferrybridge stations (see Taylor, *Financial Times*, 20–21 December 2003, p. 4). These had been put on the market by their US owner, AEP, and a privatised UK company, International Power, appeared to be the front-runner to purchase them (see Taylor, *Financial Times*, 13–14 December 2003, p. M2).
[205] Finance Act 1995, s.154.
[206] Harris and Hill, *Farmers Weekly*, 3 November 1995, p. 60.
[207] A plant based on forestry waste (see Green, *Farmers Weekly*, 30 October 1998).
[208] Barker, *Farmers Weekly*, 3 November 1995, p. 61; 20 November 1995, p. 10.

(Hampshire).[209] In 2002, however, the ARBRE Project in Yorkshire was placed in liquidation and a number of planned projects were placed on hold.[210] There appear to have been two difficulties with this: the problems of storage of wood chips[211] and the need to generate an adequate return for farmers on the prairie value set aside for this use.

Wood-burning is, however, proving a success as a carbon dioxide reducing boiler fuel for large houses and municipal buildings, such as Worcester's County Hall.[212] However, this oddity apart, it has to be recognised (although it does not seem to concern Ministers unduly) that, whilst renewable, biomass is not really an environmentally friendly source of power.[213] Indeed, in January 2003, upon acquiring a controlling interest in Fibrowatt, the owner of three plants in East Anglia burning a combination of poultry litter, meat and bone meal, the chief executive of Energy Power Resources (which already had a poultry-litter station in Fife) described organic products' station as *carbon-neutral*, on the basis that they emitted no more carbon dioxide than was taken in when the straw or wood in question was growing.[214]

At around the same time as coppice-generated stations, onshore wind farms were being set up. These were all, however, small-scale operations and dependent upon the continued receipt of part of the fossil fuel levy by way of subsidy. Furthermore, with their generating capacity dependent upon the current state of the wind, the small operators who ran them found themselves at the risk of severe penalties through causes outside their control if they offered their power for sale through the post-March 2001 NETA market system.[215] Difficulty is also encountered by reason of the cost of linkage to the national grid.[216]

All of the above made it inevitable that the main means of the government actually achieving the desired increase in generation from renewable sources, from 3 per cent in 1999 to 10 per cent in 2010, lay in the field of offshore wind farms. However, although there has been some progress in obtaining permissions,[217] the rate required is making this target extremely hard to meet, especially where defence

---

[209] Dreweatt Neate's *Broadcast*, April 1995.
[210] Lockhart, *Bulletin of the Agricultural Law Association*, winter 2002/03, p. 10.
[211] Blake, *Farmers Weekly*, 10 November 1995, p. 56.
[212] Lockhart, *Bulletin of the Agricultural Law Association*, winter 2002/03, p. 10.
[213] Nonetheless, it has been forecast that biomass could provide 15 per cent of the electricity needs of industrialised countries (see Williams, *Financial Times*, 28 May 2004, p. 12). Furthermore, funders are coming forward to finance biomass plants in the UK (see Taylor, *Financial Times*, 1 June 2004, p. 4).
[214] Taylor, *Financial Times*, 28 January 2003, p. 26.
[215] See para. 21.7.1 below.
[216] Taylor, *Financial Times*, 27 October 2003, p. 3.
[217] Pickard, *Financial Times*, 1 August 2002, p. 4; Taylor, *Financial Times*, 28–29 December 2002, p. 2; Taylor, *Financial Times*, 28–29 February 2004, p. 4. Mr Stephen Timms, MP, Minister of State for Energy, estimated that, in order to generate 10 per cent of electricity from renewable sources by 2010, between 3,500 and 5,000 onshore and offshore wind turbines would be required (see *Hansard*, 23 February 2004, WA 155481, col. 72W). As to the position in Scotland, see Pirie, *Estates Gazette*, 28 February 2004, p. 142.

considerations are involved.[218] In addition, the international legal situation is very far from clear:[219] in territorial waters, problems may arise from two European Directives, on Habitats[220] and Wild Birds,[221] whereas the UK Government's previous failure to declare a full Exclusive Economic Zone, coupled with uncertainty as to the function of the Crown Estates Commissioners in the 200 nautical mile zone[222] seem likely to generate a need for primary (that is, statute) legislation[223] at a time when Parliament has a very crowded programme. Furthermore, the regulator, Ofgem,[224] does not appear to be minded to allow renewable energy and CHP schemes to pass on the cost of connection to the networks unless this would result in greater value for money for consumers.[225] Nonetheless, in July 2003, some progress was being made on this front[226] albeit not on a sufficient scale to satisfy at least one influential commentator.[227] Indeed, concern was still being expressed as to planning delays in May 2004.[228]

Consideration is also being given, and research conducted into, renovating old windmills to produce electricity, in place of the traditional grinding corn or pumping water, possibly on an unmanned basis.[229]

21.5   The Renewables Obligation

The legal mechanism under which the government aims to achieve their target[230] is to be found in Electricity Act 1989, s.32A,[231] under which a qualifying renewable source obligation can be set.[232] Initially, such sources are required to be UK based,

---

218   Taylor and Tait, *Financial Times*, 31 July 2002, p. 4; Trinick (quoted), *The Business Magazine (Solent and South Central)*, November 2002, p. 10; Taylor, *Financial Times*, 1 March 2004, p. 4.
219   See Glen Plant, [2003] *JPL* 939.
220   Council Directive 92/43/EEC, (1992) OJ L206 7.
221   Council Directive 79/409/EEC, (1979) OJ L103 1.
222   Where both the Habitats and Wild Birds Directives, 92/43/EEC and 79/11/EEC, also apply.
223   See Plant, *op. cit.*
224   See para. 4.2.1.3 above.
225   Taylor, *Financial Times*, 16 September 2002, p. 4. The regulator is also opposed to subsidies for remote wind farms (see Taylor, *Financial Times*, 14–15 February 2004, p. 4).
226   Taylor, *Financial Times*, 12–13 July 2003, p. 2 and 15 July 2003, p. 3.
227   15 July 2003, p. 18.
228   Taylor, *Financial Times*, 1 June 2004, p. 4.
229   Taylor, *Financial Times*, 4–5 January 2003, p. 4.
230   A target which, it should be noted, was specified on the basis of what was politically possible, rather than what environmental considerations really demanded (see Richardson and Chanwai, *op. cit.*, pp. 39–58).
231   Inserted by Utilities Act 2000, s.63.
232   See para. 6.4.3.1(2) above. Green Certificates can be issued under Electricity Act 1989, s.32B, and it seems that, in practice, these can be traded. Renewables Obligation Order 2002, S.I. 2002 No. 914, Art. 3(1)((b) envisages compliance through other electricity suppliers. Energy Act 2004, s.116 made amendments to Electricity Act 1989, s.32B.

but, in the course of obtaining state aid clearance from the European Commission, the UK Government indicated that it would be prepared to allow trading between this system and equivalent Community systems on a bilateral agreement basis.

As a general rule, renewable sources comprise all sources other than fossil fuels or nuclear, but not all renewable source generating stations[233] qualify for the issue (in an electronic register run by Ofgem) of Renewable Obligation Certificates, generally known as ROCs or green certificates. Further, hydro electric power is excluded unless the generating station produces less than 20 MW DNC; or was commissioned after April 2002. Waste qualifies only if it is either biomass or manufactured into a fuel using advanced conversion technologies, that is, gasification, pyrolysis or anaerobic digestion. Fossil fuel qualifies up to 2011 if co-fired with biomass, but, from 2006, at least 75 per cent of the biomass must be from energy crops. Finally, all generating stations commissioned before 1990 are excluded unless substantially refurbished since then or co-firing fossil fuels and biomass.

Schedule 1 to the Renewable Obligation Order 2002,[234] made on 31 March 2002, required the following to be drawn down from renewable sources:

Year starting April 1	Percentage of total supplies
2002	3.0
2003	4.3
2004	4.9
2005	5.5
2006	6.7
2007	7.9
2008	9.1
2009	9.7
2010–2026	10.4[235]

There is, however, the option to make payments in lieu,[236] by making a *buy-out payment* to Ofgem.[237] Ofgem then recycles the payment by apportionment between

---

[233] The Green Certificate system is applied to generating stations. There is, however, an overlap with the preceding non-fossil fuel obligation, some supply contracts for which run until 2018 (see para. 6.4.3.1(2) above).

[234] S.I. 2002 No. 914.

[235] A study by the Carbon Trust published on 1 December 2003, indicated that uncertainty was being created amongst investors by Ministers' failure to specify targets after 2010. The Minister for Energy was therefore expected to announce an increase to, perhaps, 15 per cent between 2010 and 2015 (see Taylor, *Financial Times*, 1 December 2003, p. 5).

[236] Electricity Act 1989, s.32C; Renewable Obligation Order 2002, S.I. 2002 No. 914, Art. 7. As to the RO, see Daniel Edmonds, [2001] 2 ULR 46; Mike Nash, [2003] IELTR 45. As to the trading background post-Enron, see Richard Tyler, [2003] IELTR 135. Energy Act 2004, s.115 made amendments to Electricity Act 1989, s.32C.

[237] This option was taken by 13 out of the 66 suppliers during the first year (see Ofgem's first *Renewables Obligation Annual Report* (27 February 2004). For the year from 1 April 2004, the buy-out price was increased from £30.51 per MWh to £31.39 per MWh (see Ofgem Information Note, 11 March 2004).

Green Certificate-holders in proportion to usage. There is therefore a double advantage to tendering Green Certificates.[238] Nonetheless, in the summer of 2002, it was reported that some generators were charging businesses specifying the supply of only 'green' energy a premium.[239]

Green Certificates may be carried forward for up to 25 per cent of a generator's obligation (a practice known as *banking*). 'Borrowing' from future years is, however, not permitted. Whilst there are separate RO systems for England and Wales, Scotland and (prospectively)[240] Northern Ireland, the transfer system is nationwide. It is, however, a matter of some debate whether Green Certificates are, in theory, tradable, since they have to be tendered with electricity. In practice they also seem to be traded independently.[241]

There is, furthermore, the option for the holder of Green Certificates to convert them into UK ETS allowances at the fixed rate of 0.43.[242] The reverse is not, however, allowed.

21.6   The 2003 Energy White Paper

21.6.1   The proposals

The Energy White Paper laid down a further target of reducing carbon dioxide emissions by 60 per cent by 2050[243] but against the background of the likelihood that, by 2020, three quarters of the UK's primary energy would be imported. The government's means of resolving this tension was the maintenance of reliability of supply,[244] very largely an international political (that is, foreign and defence policy) issue;[245] the promotion of competitive markets,[246] through reform within the EU,[247] and the transformation of a 'competitive' UK electricity market into a 'mature'

---

[238] With effect from 1 April 2004, Renewables Obligation (Amendment) Order 2004, S.I. 2004 No.924, replaced Art. 3(4) of the 2002 Order with a provision enabling part fossil fuel and part biomass months in which there was no other fuel intermixed to count for up to 25 per cent between 1 April 2002 and 31 March 2006, 10 per cent between 1 April 2006 and 31 March 2011 and 5 per cent thereafter until 31 March 2016.

[239] Buchan, *Financial Times*, 28 August 2002, p. 4.

[240] The extension to Northern Ireland was made possible by the introduction of Electricity Act 1989, ss.32BA and 32C(5) by Energy Act 2004, ss.117, 118 and 120.

[241] A similar problem arose, initially, for PRNs (see para. 19.5 above).

[242] See para. 20.6 above.

[243] *Our Energy Future – Creating a Low Carbon Economy*, Cm 5761: released 24 February 2003.

[244] *Ibid.*, para. 1.33.

[245] Albeit not recognised as such in terms. But this also applies to the internal distribution network, for which an investment of £3bn is required to bring new wind farms on stream (see *Financial Times*, 23 April 2004, p. 4).

[246] Cm 5761, para. 1.35.

[247] *Ibid.*, para. 6.20, something which, it should be noted, had not been achieved over more than a decade.

one.[248] The government also has the objective, however, that no household would live in fuel poverty by 2016/18, which presumably rules out extending climate change levy to domestic consumers.[249] With emphasis on the renewables element,[250] albeit with no percentage being set for any particular element, and, within this, the 2010 target being doubled by 2020,[251] the strategy was for:

1   changes in the planning system;[252]
2   a role for Scotland, Wales and Northern Ireland and, when and if set up, the English Regions,[253] in the hope of achieving local generation projects;
3   the increasing of grants by £60m up to 2005/06,[254] when there was to be a review;[255]
4   placing considerable hope on the contribution to be made by new technologies, and especially through the support of wave and tidal prototypes,[256] 'a key research area' being solar PV[257] and a 'long-term' role for fuel cells;[258] but

---

[248]   *Ibid.*, para. 7.10.
[249]   In the first report generated by Sustainable Energy Act 2003, s.2(1), Mrs Beckett launched *Energy Efficiency. The Government's Plan for Action*: see HC, 27 April 2004, col. 37WS. This followed the release of *The UK Fuel Poverty Strategy, 2nd Annual Progress Report: 2004* by Defra the previous day. The latter appears to try to link the relief of fuel poverty by the allocation of the various tax credits initiated by the Treasury over the previous years and build upon the effect of NETA and other aspects of the evolution, and regulatory control, of electricity privatisation in reducing prices. This is set, however, against the background of a realisation that the introduction of EU ETS in order to supplement Kyoto is likely to result in increases in such prices, albeit unquantified. Yet, at the same time, Ofgem has chosen to withdraw from the field of price regulation (see Taylor, *Financial Times*, 17 September 2004, p. 3).
[250]   The target clearly relates to generation within the UK. However, in the first year of climate change levy approximately a quarter of the exemptions accrued to French and Belgian renewable energy imported into the UK and, were this to be taken into account, the 2010 target for UK sources would be halved (see ENDS Report, August 2002).
[251]   Cm 5761, para. 4.11.
[252]   *Ibid.*, paras 4.30, 34. See also Taylor, *Financial Times*, 5 November 2003, p. 6. The government approved plans for four wind farms off East Anglia and Lincolnshire (see Taylor, *Financial Times*, 28 October 2003, p. 7). By early 2004, Scottish and Southern (one of the major electricity companies) had invested in three sites in Scotland (see Taylor, *Financial Times*, 17 February 2004, p. 26).
[253]   Cm 5761, para. 6.22.
[254]   *Ibid.*, para. 4.13.
[255]   *Ibid.*, para. 4.12.
[256]   *Ibid.*, para. 4.53. There is no agreed methodology for this, a major problem being securing the device to the seafloor (see Harvey, *Financial Times*, 16 October 2003 (special report Sustainable Business), p. 4).
[257]   *Ibid.*, para. 4.57. The problem here is the high cost of silicon substrates (see Harvey, *Financial Times*, 16 October 2003 (special report on Sustainable Business) p. 4).
[258]   *Ibid.*, paras 4.58, 59. As to the likely timescale here, see Roberts, *Financial Times*, 27–28 September 2003, p. 10. The main problem is the storage and transportation of hydrogen: Harvey, *Financial Times*, 16 October 2003 (special report on Sustainable Business), p. 4.

5   with nuclear seen as unattractive on economic grounds, albeit with the possibility of a future role after the 'fullest' consultation,[259] if the planned reduction in capacity resulting from projected closures was not taken up by a combination of (currently) mothballed plant and the increased contribution from renewables.[260]

In addition to the above, there were three other main proposals. First, a 10 per cent improvement in transport carbon dioxide emissions by 2020 was expected to be achieved from the adoption of advanced technologies, none of them yet proved.[261] Secondly, a major contribution was expected[262] on the home[263] emissions front from a combination of improved boiler technology and beefed up building regulations,[264] which have to be introduced in Member States under the Energy Performance of Buildings Directive[265] by January 2006[266] and under which there will be minimum construction standards for both old and new buildings, energy certificates for all buildings, and efficiency testing of boilers and air conditioning systems in commercial buildings (in due course) leading to reduced consumption.[267] Thirdly, it was anticipated green technology contributions from both increased gas boiler efficiently offsetting CHP demand; and emissions from the coal mines assisting enhanced oil recovery from the North Sea.[268]

Finally, carbon emissions trading would play a 'central' role, prospectively in an EU context,[269] but with the possibility of a Kyoto element,[270] and a possible extension to CHP projects[271] and coal mine methane projects.[272]

---

259   Cm 5761, para. 4.69.
260   *Ibid.*, para. 6.41, in which context it was said that the case for price caps had not been made out: see para. 6.43.
261   *Ibid.*, paras 5.4,16.
262   Albeit downgraded with the publication of *First annual report on implementation of the Energy White Paper*, by Defra in April 2004 (see Eaglesham and Cookson, *Financial Times*, 24 April 2004, p. 4).
263   Which should be contrasted with the criticism accorded to the government subsequently for failing to follow 'green' policies in relation to the way departments ran their affairs (see Houlder, *Financial Times*, 14 November 2003, p. 6).
264   Paras 3.8,28,45 (the last includes a contribution from the introduction of conveyancing 'sellers' packs' in (then) forthcoming Housing Bill, of which it was said: 'This will be necessary for us to comply with the requirements of the EU buildings directive'!).
265   European Parliament and Council Directive 02/91.EC, (2002) OJ L1 65.
266   See Smy, *Financial Times*, 3 November 2003, p. 11. See also para. 21.6.2 below.
267   For the sort of developments which might be expected here, see Peter Sinclair, Cluttons' *businesslines*, October 2003, p. 9. In a parallel development, an initial public offering of shares was made in November 2003 in Romag Holdings PLC, a glass manufacturer whose windows contained solar cells to generate electricity (see Blackwell, *Financial Times*, 22 October 2003, p. 33).
268   Cm 5761, paras 6.60,62.
269   *Ibid.*, para. 2.27.
270   *Ibid.*, para. 2.19. See Grubb, *Financial Times*, 18 November 2003, p. 23.
271   *Ibid.*, para. 4.18. As to which see also *The Government's Strategy for Combined Heat and Power to 2010*, published by DEFRA in May 2004.
272   Cm 5761, para. 6.67. There were, subsequently, hopes for carbon capture and storage (see Houlder, *Financial Times*, 23 January 2004, p. 17).

21.6.2   Select Committee criticism

The White Paper is generally reckoned to have been a disappointment, not least to the House of Commons' Science and Technology Committee, which, in its report published on 3 April 2003, *Towards a Non-Carbon Fuel Economy: Research, Development and Demonstration*,[273] concluded that there was no prospect of meeting the 2010 and 2020 targets.[274] (This observation was disputed by the Treasury on the basis of a reduction in carbon dioxide emissions during 2002 of 3.5 per cent.)[275] Furthermore, the Committee concluded that:

1   the RO did not provide the incentive[276] to stimulate the immediate development of less mature technologies;[277]
2   although, during the last quarter of 2003, considerable advances had been made,[278] renewables were not coming on stream fast enough[279] for further nuclear facilities to be ruled out,[280] and, in particular, it would be unwise to run down investment in fusion research; and

---

273   Fourth Report of 2002–03, HC 55.
274   *Ibid.*, para. 215. Ofgem later reported that the energy companies had missed their targets by 40 per cent in 2002/03: Taylor, *Financial Times*, 26 November 2003, p. 4.
275   Eaglesham, *Financial Times*, 2 April 2003, p. 8.
276   HC 55, para. 188.
277   *Ibid.*, para. 208.
278   See Taylor, *Financial Times*, 13–14 December 2003, p. 5; 18 December 2003, p. 2; 19 December 2003, p. 3. But this led to protests that the process was being rushed to the detriment of 'huge expanses of unspoilt landscape' and that the DTI was being overoptimistic in claiming that 15 per cent of British homes could be supplied by the offshore wind programme (see respectively) letters to *Financial Times*, 20–21 December 2003, p. 10 (Cameron Beattie) replied to on 27–28 December 2003, p. 10 (T.R.H. Kimber), and 22 December 2003, p. 18 (John Bower)).
279   It appeared that offshore wind farms has still not been attracting sufficient financial backers because of concerns as to the government's commitment (see Taylor, *Financial Times*, 26 September 2003, p. 3). However, an initial public offering of shares in Novera Energy Europe is planned to fund the construction of a Welsh wind farm of 17 turbines (see Blackwell, *Financial Times*, 22 December 2003, p. 23). Furthermore, Centrica plc, the household gas and electricity supplier, agreed to pay £9m for two wind power sites off the east coast of England as part of a £500m green electricity expansion programme (see Taylor, *Financial Times*, 23 December 2003, p. 20). Scottish and Southern Energy was also given permission to start a £90m onshore windfarm in South Ayrshire (see *Financial Times*, 24 December 2003, p. 4). The Spanish renewable energy specialist, EHN, plans to invest £352m in the UK over five years (see Taylor, *Financial Times*, 30 December 2003, p. 3). It does appear, however, that Ministers' concerns over wind farm refinancing are less than they were (see Taylor, *Financial Times*, 20 January 2004, p. 8). Ministers are also hopeful that renewable energy will create jobs (see Bream, *Financial Times*, 16–17 January 2004, p. 2).
280   Making the construction of more gas-powered stations inevitable, thus advancing the date upon which importation has to start, improving neither security of supply nor the atmosphere (see Buchan (book review), *Financial Times*, 1 April 2003, p. 12).

3 climate change levy should be replaced with a carbon and renewable energy tax, the proceeds of which should not be used to reduce employers' NICs, but be given to a new Renewable Energy Authority set up to fund research, design and development [RD&D].[281]

A more general criticism has been made – that the aim of reducing carbon emissions is inconsistent with the goal of reducing fuel poverty. The former will result in higher prices, while the latter is dependent upon keeping them low.[282] This has, furthermore, to be considered against the background of a position, at the beginning of June 2003, in which energy market competition at the retail level was feared to be in danger of failing.[283]

By the end of April 2003, concern was being expressed generally that the projected 20 per cent of Britain's power would in fact come from renewables by 2020, at any rate without a long-term subsidy regime.[284] However, in early 2004, the British Wind Energy Association[285] expressed confidence in the increased capacity from wind farms as a result of extended incentives announced at the end of 2003.[286]

In the spring of 2003, ministers were beginning to concede that electricity prices might have to rise by between 5 and 10 per cent by 2010.[287]

With the retirement of the Minister of State at the Department of Trade and Industry with responsibility for energy, Mr Brian Wilson, MP in June 2003, no full time replacement was appointed, responsibility being given to Mr Stephen Timms, MP, who was already responsible for overseas postal services, e-commerce and corporate social responsibility. Although the government insisted that this did not denote that energy strategy had fallen by the wayside, it was the occasion of criticism from both the main Opposition parties.[288] Yet when Mr Timms moved back to the Treasury in September 2004, his successor, Mr Mike O'Brien, MP retained the same diverse portfolio of responsibilities.

The House of Commons' Environmental Audit Committee, published its Eighth Report of Session 2002–03, *Energy White Paper – Empowering Change?*,[289] after the beginning of the summer recess, on 22 July 2003. Its observations included:

---

281 HC 55, para. 216.
282 Buchan (book review), *Financial Times*, 1 April 2003, p. 12.
283 Taylor, *Financial Times*, 16 June 2003, p. 1.
284 Within the UK, it is expected that Northern Ireland will have a surplus of renewably generated electricity to sell to mainland generators to abate the shortage of, especially, wind farm capacity (see Murray Brown, *Financial Times*, 2 July 2003, p. 6). Since then, the position appears to have worsened, with the revelation that CHP targets had not been met for 2002 (see Taylor, *Financial Times*, 1 August 2003, p. 4).
285 See para. 2.4.1 above.
286 Taylor, *Financial Times*, 24 February 2004, p. 4.
287 Eaglesham, *Financial Times*, 28 April 2003, p. 3.
288 Eaglesham, *Financial Times*, 20 June 2003, p. 4. That criticism was echoed in the recommendations in the 4th Report of the House of Lords' Science and Technology Committee for the Session 2003–04, HL 126, *Renewable Energy: Practicalities*, published on 15 July 2004, paras 2.18 and 10.6.
289 HC 618.

'The Energy White Paper represents a major shift in the approach to UK energy strategy'. [290] In the context of this book, this begs the question,[291] as to why climate change levy was introduced in advance of the policy which it purports to implement. Furthermore, the Report goes on to say that: 'Our fears about implementation have proved largely justified. The Energy White Paper is weak on specific measures and contains little that is new ... It remains largely an act of faith on the government's part that present policies, with their reliance on market mechanisms, will in fact deliver'.[292] Moreover, in the context of renewables policy, reference was made to the bankruptcy of ARBRE, the flagship biomass project,[293] the paucity of the current solar PV programme,[294] and the absence of an adequate strategy for other renewables.[295] Concern was also expressed at the absence of targets for energy efficiency,[296] and, in the context of *fuel poverty*, the need to ensure that 'the domestic sector bears its proper share of the costs of reducing greenhouse emissions'.[297] The latter goes, of course, to the heart of the structural deficiency in-built into climate change levy.[298] Perhaps the most significant message of all from the Committee, however, was the unmasking of a major inconsistency in the government's thinking. Combined heat and power can only make a contribution to the problem if energy prices increase, yet that 'flies in the face' of the wish to reduce such prices.[299] Moreover, the dramatic fall in energy prices over recent years had been due to an excess of capacity, which the introduction of NETA had exposed.[300] NETA itself was 'a system for big players' which resulted in the renewable obligation premium being withheld from small renewable generators.[301] The end result was that:

> We highlighted last year our conviction that a transition to an environmentally benign energy system cold not be achieved on the basis of unsustainably 'cheap' energy, as the Prime Minister's foreword to the PIU report indicated was a priority. The Government's approach remains inconsistent, and the price of energy is likely to rise.[302]

---

[290] *Ibid.*, para. 10.
[291] See para. 14.2 above.
[292] HC 618, para. 18.
[293] *Ibid.*, para. 31, which starts: 'A substantial contribution from biomas, in particular, is widely seen as essential to the achievement of the renewable targets'.
[294] *Ibid.*, para. 33.
[295] *Ibid.*, para. 34.
[296] *Ibid.*, para. 39.
[297] *Ibid.*, para. 53. On 27 April 2004, Mrs Beckett issued *Energy Efficiency. The Government's Plan for Action* pursuant to her statutory duty to report on the energy efficiency of residential accommodation in England under section 2(1) of the Sustainable Energy Act 2003.
[298] See para. 21.4.2 above.
[299] HC 618, para. 44.
[300] *Ibid.*, para. 64.
[301] *Ibid.*, para. 63.
[302] *Ibid.*, para. 80. As to actual rises, see Hall, *Financial Times*, 4 August 2004, p. 3; Taylor and Morrison, *Financial Times*, 9 September 2004, p. 1; Taylor, *Financial Times*, 9 September 2004, p. 4.

The Environmental Audit Committee report was followed, on 15 July 2004, by the report of the House of Lords' Science and Technology Committee, *Renewable Energy: Practicalities*.[303] After confining the ambit of its enquiries to the renewables aspect, the Committee expressed particular concerns as to the following aspects of government policy. First, the RO[304] was operated in a way which was unlikely to guarantee the funding of new projects[305] which could not be implemented by 2005,[306] because it acted as a cap on renewable output.[307] In particular, it was felt that, in itself, the new exemption of coalmine methane from climate change levy[308] would be unlikely to stimulate activity in that sphere without it becoming an eligible source for the purposes of the renewables obligation.[309] Secondly, although the fears that had been expressed as to the degree of cover required to allow for wind turbines' 'intermittency' were felt to have been exaggerated,[310] a total reliance in excess of 10 per cent would necessitate this being taken into account.[311] Moreover, only the larger offshore wind farms could be linked up to a high voltage transmission network.[312] Small generators would have to be connected direct to local distribution networks.[313]

Thirdly, as a source, biomass needed nurturing. Straw-based stations had storage problems and could, in reality, only draw in from farmers in their locality.[314] In addition, transporting biomass from outside that locality created additional carbon dioxide emissions,[315] and the size of plant would never be such that linkage to the National Grid would be feasible.[316] Moreover a plant which utilised chicken waste had been prevented by the regulator from burning feathers at all (on the grounds that they were classified as industrial waste) and the variability of the chicken litter which it did burn (classified as agricultural waste under the Incineration Emissions Directive)[317] was giving rise to difficulty.[318] The Committee therefore recommended that the UK Government and Ofgem should conduct an urgent regulatory review in order to mitigate this sort of problem[319] and (overall) to establish a regime which favours small-scale, locally sourced, biomass plants.[320]

---

[303] HL 126.
[304] See para. 21.5 above.
[305] Especially long-term projects such as one small particular forestry-based estate CHP plant: see HL 126, paras 5.25 and 5.26.
[306] *Ibid.*, para. 5.26.
[307] *Ibid.*, paras 5.6 and 5.8.
[308] See para. 14.4 above.
[309] HL 126, paras 3.12 and 10.7.
[310] *Ibid.*, para. 7.2.
[311] *Ibid.*, para. 7.7.
[312] *Ibid.*, para. 6.2: this was a problem which had restricted the size of biomass stations.
[313] *Ibid.*, para. 6.10.
[314] *Ibid.*, para. 4.16.
[315] *Ibid.*, paras 4.32 and 10.27.
[316] *Ibid.*, para. 6.2.
[317] European Parliament and Council Directive 00/76/EC, (2000) OJ L145 52.
[318] HL 126., para. 4.21.
[319] *Ibid.*, para. 4.22.
[320] *Ibid.*, para. 10.27.

Finally, the Committee found that using energy crops for co-firing under the RO[321] could not be achieved on a meaningful scale before 2009.[322]

21.6.3 The government's response

In a Written Ministerial Statement on 21 July 2004,[323] the Parliamentary Under-Secretary of State at the Office of the Deputy Prime Minister[324] announced the publication of two consultation documents which represented progress on the Environmental Audit Committee's concern over the absence of targets for energy efficiency.[325] The first was on the amendment of the buildings regulations energy efficiency provisions and the implementation of the Energy Performance of Buildings Directive.[326] The second was the government's proposals for amending official guidance on the ventilation of buildings regulations. The Minister said:

> Energy used in buildings is responsible for roughly half the UK's carbon dioxide emissions. Wasting energy costs money whereas measures such as loft insulation, boiler replacements and more effective controls often pay for themselves within a few years.
> One of the cornerstones of saving energy is to make building structures more airtight to minimise heating or cooling energy losses due to air leaking through gaps in the structure.[327]

On 9 August 2004, the ODPM[328] issued a revision of *Planning Policy Statement 22: Renewable Energy*. This laid down the policies to be taken into account by regional planning bodies in the preparation of regional spatial strategies and local planning authorities in the preparation of local development plans. Renewable energy covered such as flowed naturally from the sun, the wind, the fall of water, the movement of the oceans[329] and from biomass. In technology terms, this meant onshore wind farms, hydro electric, photovoltaics, passive solar, biomass, energy crops and energy from waste (other than by incineration), landfill and sewage gas. It advocated regional capacity targets,[330] but not specific policies relating to the impact of wind turbines on aviation or separation distances.[331] There was, moreover, to be no allocation of pre-

---

[321] See para. 21.5 above.
[322] HL 126, para. 10.25. A pilot project by Syngenta, to power 1,000 homes in Yorkshire using rapeseed oil has since been announced (see Taylor, *Financial Times*, 24–25 July 2004, p. 5).
[323] HC, 21 July 2004, col. 29WS.
[324] Mr Phil Hope, MP. See para. 4.2.1.2(1) above.
[325] Col. 29WS.
[326] 2002/91/EC. The following day, the House of Lords passed what later became Sustainable and Secure Buildings Act 2004, amending the Building Act 1984.
[327] Col. 29WS.
[328] See para. 4.2.1.2(1) above.
[329] Although no specific commitment was given to this, its mention may be significant in view of the support being given by the Carbon Trust to a wave energy converter project (see Tighe, *Financial Times*, 10 August 2004, p. 3).
[330] PPS 22, para. 2.
[331] *Ibid.*, para. 25.

determined space in local plans.[332] In the case of biomass, local planning authorities were enjoined to minimise the increase in road traffic.[333]

The foregoing was considered to be a major boost to the wind farm industry,[334] but, despite advocating consideration of the opportunity to incorporate small scale schemes into both new developments and existing buildings,[335] a considerable setback to the solar sector.[336]

21.7 Problems with electricity supply industry structures

On two occasions, attempts have been made to convert climate change levy into a carbon tax – both before the legislation was presented to Parliament by the House of Commons Environment Committee[337] and in the Performance and Innovation Unit Report published in February 2002.[338] Both were rejected by the Treasury.[339]

21.7.1 Denationalisation

The major problem behind the efficacy of climate change levy is the structure resulting from the denationalisation of electricity, under which consumers (other than very large commercial users) acquired their supplies through RECs (the successors of the old regional electricity boards), which in turn bought their supplies from generators, mainly (at that date) proprietors of large coal fired power stations, with more expensive nuclear capacity having to be subsidised through a fossil fuel levy on bills.[340] This end of the supply chain has been changed since it was set up in a number of ways.

First of all, renewable[341] and 'clean'[342] sources have entered the market in quantity, these now being considered desirable from, respectively a conservation and an environmental point of view – indeed the 37 per cent of the country's electricity generated from natural gas in 2001 is expected to rise to between 60 and 70 per cent in 2020, a high proportion of it imported (and therefore needing to be assessed, from

---

332 *Ibid.*, para. 6.
333 *Ibid.*, para. 24.
334 Blitz, *Financial Times*, 10 August 2004, p. 3.
335 PPS 22, para. 18. In this context, a letter from Mr William Orchard published in *Financial Times* on 6 August 2004, p. 16, advocating local CHP networks, can be seen as timely.
336 Whose funding looked like being undermined from March 2005 (see Houlder, *Financial Times*, 10 August 2004, p. 3). See also W Patterson, *Generating Change* and *Networking Change: Keeping the Lights On Working Papers Nod. 2 to 3* (RIIA Sustainable Development Programme, 2003 and 2004).
337 Barker, *Daily Telegraph*, 21 March 2002, and see also Watson and Dobson, *The Tax Journal*, 2 August 1999, p. 9.
338 *The Energy Review*, para. 3.57.
339 The latter by Written Answer within a matter of days: 25 February 2002, vol. 380, No. 106, col. 1031.
340 See para. 6.4.3.1 above.
341 Such as biomass: as to which, see W. Patterson, *Power from Plants* (Earthscan, 1994).
342 For example, wind power.

a security of supply viewpoint, in an EU context).[343] Secondly, the government has become much more interested in the potential of CHP schemes[344] to help meet their target of the generation of 10 per cent of electricity from renewable sources by 2010.[345] Thirdly, the present government reorganised the auction procedures in March 2001 in a way which made it difficult for small producers lacking flexibility in capacity[346] to compete because of the need to provide an assured supply during the contractual period.[347] Finally, and perhaps more significantly than anything else, there has been a world-wide trend in favour of replacing large coal-fired power stations with much smaller natural gas stations, the former being environmentally 'dirtier', slower (in the sense that they take more time to come on stream after being started up) and needing decades for the investment in them to be depreciated. Gas stations are able to fill gaps in the supply chain very much quicker than coal-fired stations.[348]

21.7.2   The end game

Suffice it to say that the ultimate consequences of these developments may not, as yet, be fully apparent. This is especially the case in the light of the fact that global privatisation was taking place at the same time as fundamental changes were occurring in the economic structure of the electricity industry. Factors needing to be considered in this context include the fact that trading electricity does not sit easily beside traditional 'big' coal fired stations with their distribution networks. Secondly, although the maintenance of a national distribution network has been seen as critical, the creation of *private wires* (local monopolies linking local generation capacity with local loads) outside the national or regional generation network, needs to be taken into account.[349] Finally, although fuel powered stations are dependent, economically, on the future (variable) cost of fuel, the dominant economic factor governing hydro, wind and photovoltaics-powered plants is the return on their construction cost, which is determined by their degree of use.

The problem which has not generally been appreciated is, however, that different financial techniques need to be used in order to compare the desirability of the alternative techniques. The conclusion reached by one commentator is that energy policy needs to take account of a wider list of factors than was normal under the traditional *fuel and power* approach.[350]

---

343   See J. Stern, *Security of European Natural Gas Supplies* (London: RIIA, 2002).
344   Which are, of course, neither *renewable* nor particularly *emission friendly* modes of generation.
345   But progress is not encouraging (see Taylor, *Financial Times*, 4 August 2004, p. 3).
346   For example, windpower and CHP (see Taylor, *Financial Times*, 11 July 2002, p. 4).
347   See Fairley and Ng, [2001/2002] 3 ULR 57.
348   See W. Patterson, *Transforming Electricity* (Earthscan, 1999).
349   By December 2004, Ofgem's estimate of the cost of rewiring the UK to connect green energy projects to both the national and local networks had risen to over £1bn: see Taylor, *Financial Times*, 18–19 December 2004, p. 2.
350   W. Patterson, *The Electricity Challenge, Generating Change* and *Networking Change: Keeping the Lights On Working Papers Nod. 1* to *3* (RIIA Sustainable Development Programme, 2003 and 2004).

For all of these reasons, it may be of significance that the government has announced its intention to undertake further (and detailed) consultations on specific measures to encourage household energy efficiency.[351] The government's immediate reaction, however, was to pass the Sustainable Energy Act 2003, under which the Secretary of State[352] is obliged to make an annual report on the progress made in four areas:[353]

1 cutting the UK's carbon emissions;
2 maintaining the reliability of the UK's energy supplies;
3 promoting competitive energy markets in the UK; and
4 reducing the number of people living in fuel poverty in the UK.

The Act also provides that, within a week of s.2 coming into effect, the Secretary of State has to designate at least one energy efficiency aim for residential accommodation.[354] A debate was held in Westminster Hall on 30 March 2004 on this subject. During that debate, concerns were expressed at the lack of 'joined up government' in this respect. But, when replying to the debate, the Under-Secretary of State at Defra[355] revealed that a new sustainable policy network had been set up to deliver this.[356]

By s.4 of the Act, the Secretary of State is empowered to direct energy conservation authorities to take such energy conservation measures as the recipient authority considers likely to achieve the improvement to the energy efficiency of residential accommodation specified, by the date specified, in the direction. Moreover, by 31 December 2003, the Secretary of State was required to specify CHP targets (for government use of electricity) for particular periods up to the calendar year 2010.[357] Finally, the Gas and Electricity Markets Authority[358] is required both to carry out impact assessments for important proposals[359] and to pay up to £60m to the government for the Secretary of State to use to promote the use of energy from renewable sources.[360]

Despite all of the above, concerns at the absence of 'certainty of supply' under NETA[361] continued to be widespread, centring especially on the apparent future dominance of imported natural gas.[362] Parliamentary concerns culminated in a

---

[351] HM Treasury's Budget 2003 (9 April 2003), para. 7.23.
[352] Which can mean *any* Secretary of State since the various holders are, in theory, co-holders of a single office, but, currently, Mrs Beckett, the one in charge of Defra.
[353] Sustainable Energy Act 2003, s.1.
[354] Under s.3, Sustainable Energy Act 2003, the National Assembly for Wales (see para. 4.2.2 above) is obliged to do the same.
[355] Mr Ben Bradshaw, MP. See para. 4.2.1.2(1) above.
[356] HC, 30 March 2004, col. 380WH.
[357] Sustainable Energy Act 2003, s.5.
[358] See para. 4.2.1.3 above.
[359] Sustainable Energy Act 2003, s.6, inserting Utilities Act 2002, s.5A.
[360] *Ibid.*, s.7.
[361] See para. 2.4.1 above.
[362] See BBC2's '*If ... the lights go out*', 10 March 2004, 9 pm; and letters to *Financial Times* on 16 March 2004 (Professor Peter Odell) and 18 March 2004 (Mr Jonathan Stern).

House of Lords' amendment to the government's Energy Bill in March 2004, requiring the Secretary of State to 'ensure the integrity and security of electricity and gas supply'.[363] However, after a certain amount of 'Parliamentary ping-pong' between the two Houses, what emerged was the reformulation[364] of s.3A, Electricity Act 1989, to require the Secretary of State (presumably for Trade and Industry) and the Gas and Electricity Markets Authority to have regard to the achievement of sustainable development[365] while protecting the interests of consumers[366] and securing that 'all reasonable demands for electricity are met'[367] and that licensed suppliers were able to finance their statutory activities.[368] These proceedings also saw the insertion[369] of subsections (1A) to (1C) into s.1 of the Sustainable Energy Act 2003 requiring reports by the Secretary of State to state what had been done during the reporting period: to develop specified renewable sources; to ensure the maintenance of such scientific and engineering expertise with the UK as is necessary for the development of potential energy sources (including nuclear); and to achieve the existing energy efficiency aims specified in ss.2 and 3.

Finally, the House of Lords' amendments to the Energy Bill also saw the imposition of an obligation upon the Secretary of State to make an annual report to Parliament on both the long and short term availability of electricity and gas for meeting the reasonable demands of UK consumers, and covering generating, distribution, infrastructure and gas conveyance capacity within both timescales.[370]

21.8   The case of aggregates levy

It is sometimes very difficult to see where current government policy on energy and the environment is heading. Ideas which started out under the auspices of Mr Michael Meacher, the Minister for the Environment between May 1997[371] and June 2003 (his former sub-department is currently part of Defra) got worked on in the Treasury and sometimes seem to see the light of day as purely fiscal measures. In

---

[363]   Taylor, *Financial Times*, 1 April 2004, p. 4.
[364]   By Energy Act 2004, s.81(2).
[365]   Electricity Act 1989, s.3A(1)(b).
[366]   *Ibid.*, s.3A(2)(a).
[367]   *Ibid.*, s.3A(2)(b)(i).
[368]   *Ibid.*, s.3A(2)(b)(ii).
[369]   By Energy Act 2004, s.82(2).
[370]   *Ibid.*, s.172.
[371]   Mr Meacher was born in 1939 and educated at Berkhampstead and New College, Oxford. Elected to Parliament in 1970 as a member of the (old) far left of the Labour Party, he had held junior office between 1974 and 1979 and was, between 1983 and 1997, a member of Labour's elected Shadow Cabinet. He was generally considered to have been sidelined when given the environment portfolio after Mr Blair attained the leadership, but took the opportunity to make an in-depth study of the problems underlying this portfolio. When Labour won power, he was not given a Cabinet seat, but compensated with a Privy Councillorship. Because of this, the importance of his contribution to the evolution of policy in this field has been acknowledged across the party spectrum.

some cases, it is therefore legitimate to ask to what extent the end product actually achieves its alleged environmental objective.

Aggregates levy is the most obvious case in point, not least because the government, through its various agencies, is the largest user of aggregate. It is understood that originally the main environmental concern was the effect on rural communities of excessive numbers of lorries passing along small roads and such user late at night and on public holidays. The tax has, however, nothing at all to do with concerns of that sort.

Instead, aggregates levy seems to be based on the idea of promoting environmentally friendly quarrying, a concept which one might well consider to be a nonsense in itself. Indeed the fact that exports are rebated[372] and imports brought into charge destroys this hypothesis at the outset. The official Treasury justification was, however, that the purpose of the tax was to protect virgin sources of aggregate, so as to encourage recycling.[373] During the consultation initiated in June 1998, the scope of the tax was confined to sand, gravel and crushed rock, with coal, clay and minerals intended for industrial use excluded. The first draft of the proposed legislation made no reference to 'crushed' and incorporated any intermixed substances. Spoil, waste and other by-products of dimension stone (itself exempt) were included, as were the extraction of certain minerals, including waste derived from the extraction of anhydrite, ball clay, barytes, calcite, china clay, china stone, fireclay, fluorspar, fuller's earth, gems and semi-precious stones, gypsum, metals and metal ores, potash, sodium chloride and talc. The production of lime or cement from limestone was also included.[374]

The second consultation, of April 1999, made Customs realise that a use-based tax would be impractical, because the intended use might change after the tax point. In the second draft of the legislation, therefore, the industry was assisted by aligning the meaning of *aggregate* more closely with the terminology in the DETR's Planning Guidance Note 6 (a document with which the industry was familiar). This resulted in the removal of 'stones' from charge and clay ball waste from exemption. Flint,[375] feldspar,[376] rock phosphates and manganese[377] were excluded in their entirety. Limestone was defined to include dolomite. Furthermore, coal spoil heaps, slag and other waste products of industrial combustion, drill cuttings from Continental Shelf

---

[372] A great deal of hostility was encountered on this in the responses to the June 1998 Consultation, some even suggested that a higher rate of tax should be imposed on exports.

[373] The case for this tax seems to be distinctly weak. It was, therefore, not too surprising to find on p. 65 of *Study into the Environmental Impacts of Increasing the Supply of Housing in the UK*, published by Defra in May 2004, the statement that building at a higher density achieved the greatest reduction in the use of aggregates and the subsequent economic costs by between 25 per cent and 50 per cent. This section of the document then concluded with the statement that the tax 'internalises these external costs'.

[374] The 2002 amendments had, however, to include wording to Finance Act 2001, s.18(2)(c) to make it clear that all input into lime and cement manufacture was to be exempt, not just limestone itself.

[375] Being mainly used for tableware and grit in poultry food.

[376] A wider term than china stone, which was therefore removed from the list.

[377] Used as fertilisers.

operations (which are used to line landfill sites) and spoil disposed of as landfill following its excavation for highway repairs were excluded from taxation. Also excluded were silica sand, limestone used for industrial purposes, china and ball clay and recycled aggregate. Complaints as to the prospective effect of the tax on the competitive position of crushed aggregate led to changes in the substances being covered by the tax between Royal Assent on 11 May 2001 and the start date for the tax on 1 April 2002, including bringing uncrushed rock within its scope, but with the exclusion of dimension stone[378] and building stone production. Flint is subject to the tax when used as an aggregate, exempt when cut and used as building stone and eligible for credit when used in non-aggregate applications, such as agriculture and animal feed. Calcite was removed from the exempt list in order to ensure a level playing field with the market for calcium carbonate, both being eligible for a credit when used in non-industrial applications.[379] Even so, this rearrangement of the boundaries left out clay and slate products, giving rise to further concerns as to unfair competition and triggering judicial review proceedings that, in the event, had not been completed by the start date.[380]

Indeed, the government do not now appear to dispute that what the tax has done is to introduce distortions into the market place,[381] although they have not admitted to all of those that have been identified by others, that is,:

1   small quarries and those far away from the southeast are very considerably disadvantaged, at the rate of £1.60 per tonne;
2   those in Northern Ireland are exposed to untaxed competition from Eire;
3   the creation of an annual market of 4m tonnes of slate waste products (via two new railheads) from Welsh quarries operated by Alfred McAlpine;[382] and even
4   (in the light of the extension in Budget 2002 to certain uncrushed materials) an increase in the cost of repairing harbours in the north of Scotland.

The government ran into almost unanimous criticism from the main Opposition parties on the principles underlying the tax during the debates on both the 2001 and 2002 Finance Bills. During the Report Stage of the latter, the (newly appointed) Economic Secretary to the Treasury, Mr John Healey, MP, said:

---

[378]   The phrase, a critical one in the Consultations, actually disappeared from the revised statutory text.
[379]   Although enacted under both a provisional collection resolution and a Finance Bill introduced after the start date, the revisions were made known in principle in *Business Brief* 17/01 (28 November 2001) and draft clauses published at the end of March 2002.
[380]   See *R (on the Application of British Aggregates Association and Others) v. C & E Commrs*, [2002] EWHC 926 (Admin), [2002] 2 CMLR 51 (see para. 11.3.2 above).
[381]   Customs are almost alone in trying to defend the tax, in *Aggregates Levy Question and Answer Briefing*, on the basis that other countries have similar taxes, including Denmark, France and Sweden within the EU, with the Netherlands considering a tax on minerals: see www.hmce.gov.uk.
[382]   Batchelor, *Financial Times*, 1 October 2001, p. 12.

The levy, in building in a recognition of the environmental costs of extraction, recognises those costs by making the price of aggregates better reflect their true social and economic costs; encouraging more efficient use of aggregates; encouraging the use of alternative materials such as recycled materials and certain waste products in construction; and encouraging the development of a range of other alternatives, including the use of waste glass and tyres in mixed aggregate[383] ... Armourstone, the large blocks of aggregate that are often used in sea walls or coastal protection, was brought into the scope of the levy as a result of the levy's extension to uncrushed rock. This was a necessary change. Because aggregates can be produced from rock without crushing it, consultation with the industry suggested that there was a danger of distortion of competition and the distinct possibility of avoidance.[384]

By way of contrast, it should be noted that the Regulatory Impact Assessment issued by the government when it decided to go for the 'taxation' option[385] revealed that the voluntary package put forward by the Quarry Products Association,[386] albeit conditionally on the government's agreeing to limit its purchase of aggregates to firms which had met certain environmental conditions (which smaller firms would have found difficult and might be open to legal challenge), had offered the following 30 elements:

- industry-wide protection of ISA 14001;
- production of a QPA environmental best practice guide;
- environmental management guidance for smaller operators;
- piloting of ISO 140001;
- universal introduction of community liaison committees;
- a no quibble guarantee of Environmental Impact Assessments;
- production of best practice guidance for Environmental Impact Assessments;
- surrender of dormant quarrying permissions that will not be reactivated, in National Parks;
- agreement not to operate National Park dormant sites on behalf of other owners;
- strict qualifying criteria for new quarrying applications in National Parks;
- financing fundamental research into the impact of quarrying on Sites of Special Scientific Interest;
- introducing key sustainability indicators;
- establishing an index-linked Sustainability Foundation financed by the industry;
- major investment in recycling plant and equipment;
- promotion of recycled materials with construction clients;
- establishing a restoration guarantee scheme for all aggregates;
- funding and joint management of the Aggregates Advisory Service;
- a compulsory code of conduct for Transport;
- mandatory membership of a 'Well Driven' scheme;

---

[383] HC Debates, 4 July 2002, col. 465.
[384] *Ibid.*, col. 466.
[385] See para. 11.3.2 above.
[386] See para. 2.5 above.

- introduction of transport plans for all aggregates supply sources;
- mandatory use of low sulphur fuels in the transport fleet;
- environmental training for all drivers, including subcontractors;
- environmental training for all employees;
- 50 per cent of employees to obtain NVQ by 2004;
- the introduction of planning enforcement fees;
- the extension of local authority air pollution controls to sand and gravel processing;
- mandatory use of low sulphur fuels on internal quarry plant;
- the introduction of energy reduction targets;
- the establishment of a Quality Mark for environmental performance; and
- the promotion of environmental purchasing policies with clients.

Suffice it to say that it is very far from apparent to the authors how many of these objectives are likely to be met at all, let alone more likely to be achieved, as a result of increasing operators' costs through the imposition of aggregates levy.

In October 2003 was published a report to Customs by Symonds Group Ltd entitled *Assessment of the State of the Construction Aggregates Sector in Northern Ireland*.[387] The Province had been given a transitional regime,[388] so that the rates of aggregates levy in force were considerably lower than obtained in the rest of the UK.[389] The findings in the report may therefore have greater significance than may at first be apparent.[390] Before the introduction of aggregates levy, says the report, most quarries had been able to sell their scalpings and low-grade materials. This was no longer the case, resulting in stockpiling which, in some cases, prevented seams of aggregate being exploited. Moreover, the proportion of operators expecting their sales to decline over a three year period exceeded those who expected them to increase. There was, furthermore, a significant unlicensed sector offering low grade materials at low prices. These people were not only evading aggregates levy and VAT as well, but also (it was assumed) not complying with health and safety and other relevant legislation.

The Symonds Group report continues by stating that there had been an increase in exports to Eire, coupled with a very low admitted level of importation in return. Enquiries suggested, however, that the actual level of importation was much higher than returns for the tax suggested. Aggregate prices remained low and, although tests had shown that some types of recycled material would be an effective substitute for

---

387 www.hmce.gov.uk.
388 See para. 21.10 below.
389 The reduction is 80 per cent of the full rate and can last until 31 March 2012. Under the original scheme, this was to expire on 31 March 2007, and abated by 20 per cent in 2003–2004, 40 per cent in 2004–2005, 60 per cent in 2005–2006 and 80 per cent in 2006–2007.
390 Or, perhaps, extrapolated from the security situation in the Province even though the paramilitaries were on ceasefire. The implication that Customs do not seem able to shut down seemingly quite substantial unlicensed operations is a surprising one. One would not have expected as similar situation to have been extant on the mainland.

aggregate, the road construction industry had preferred to use the traditional material. Recycling output was estimated at 280,000 tonnes, as opposed to 21 million tonnes of primary aggregate. Finally, one of the reasons why so little construction and demolition waste was recorded as going to landfill was the widespread availability of unlicensed landfill sites.

The sequel was a decision, announced in the Pre-Budget Report on 10 December 2003,[391] to apply to the European Commission for state aid consent to restructure the temporary differential rate system on the basis of negotiated agreements, coupled with an 80 per cent reduction in the rate of aggregates levy, both on processed products and – something which would be new – on virgin aggregate, from a date in 2004.[392]

21.9 Water and the farmers

During Labour's final period in opposition, the shadow cabinet member responsible for environmental policy, Mr Michael Meacher, undertook a detailed study into the part taxation might play in achieving the next government's environmental objectives.[393] Following the return of the New Labour government in May 1997, Mr Meacher (a left winger) was not given a Cabinet seat, but was created a Privy Councillor and appointed to the position of Minister of State at the mammoth Department of Environment, Transport and the Regions ('the DETR'),[394] responsible for environmental policy.

21.9.1 Economic instruments for water

The first significant environmental act of the New Labour Government in 1997 was the publication of Mr Michael Meacher's 'Grey' Paper.[395] Although not so confined in terms, in reality this directed attention to the problem of the pollution of the water table from ongoing agricultural and industrial processes and considered both the imposition of taxation on farmers' use of fertilisers and pesticides, but also the possibility of tradable permits being created for the industrial sector.[396] In the event, however, water pollution is almost the only area for which tradable permits have not

---

391 Cm 6042.
392 *Business Brief* 27/2003 (10 December 2003). See para. 13.3 above.
393 The previous Conservative government had introduced landfill tax in 1996, in order to try and encourage recycling as a preferred method of waste disposal (see Chapter 15 above).
394 See para. 4.2.1.2(1) above.
395 *Economic Instruments for Water Pollution*, DETR, 20 January 1998. See Simon Conran, 'Water Pollution, Abstraction and Economic Instruments', in *Environmental Policy: Objectives, Instruments and Implementations*, ed. by Dieter Helm (Oxford: Oxford University Press, 2000), pp. 203–15.
396 By reference to experience in the USA, which subsequent analysis showed applied to situations not replicated in the UK.

been prescribed as part of 10 Downing Street's list of solutions.[397] The reason for this may be that no policy document has emanated from that source on the contamination of the water table.[398] That, in turn, may owe something to the facts that research suggested that fertilisers did not contribute to this, and that the pesticides debate had become becalmed in the 'war' between the proponents of GM and organic crops.[399]

21.9.2   The privileged historic position of the UK's farmers

Since the Second World War, the farmers have tended to be treated with kid gloves. They have numerous tax exemptions, for example, the ability to recover the VAT they pay as input tax through monthly recoveries based on the zero-rating of food;[400] the exemption from VED for farm vehicles;[401] and the exemption from fuel duty through the 'red diesel' scheme.[402] Indeed the post-war 'Williams' subsidy regime instituted by Agriculture Act 1947 produced such a plethora of financial assistance for farmers that, in 1950, it was possible for a recently-appointed Parliamentary Secretary to the Ministry of Food[403] to say: 'No nation featherbeds its agriculture like Britain'.[404]

21.9.3   1997 and 2004 compared

However, the problems of the farming industry as perceived in 1997 are totally different from those as seen following the 2001 foot and mouth disease outbreak. The Ministry of Agriculture, Fisheries and Food ('MAFF') has been abolished, being absorbed into the Defra mega-department. The Environment unit of the old DETR has also been included within the scope of Defra. The Prime Minister's policy for the countryside is, furthermore, now directed towards the support and preservation of rural businesses, rather than farming, with a few additional sideshows.

21.9.3.1   Common Agricultural Policy reform

Farming still predominates but its subsidy base is being changed. Under the Mid-Term Review (of CAP) agreement made between the Member States of the 'old' EU

---

[397] The possibility of introducing economic instruments was, however, mentioned in the Pre-Budget Report of 10 December 2003, para. 7.58. Against that has, however, to be set the insistence of the House of Commons Environment, Food and Rural Affairs Select Committee that it should not be water utility customers who pick up the bill (see Taylor, *Financial Times*, 18 December 2003, p. 4). See para. 20.3 above in relation to the rejection of various agricultural possibilities.

[398] See para. 21.9.4 below.

[399] See para. 21.9.6 below.

[400] Value Added Tax Act 1994, Sched. 8, Group 1.

[401] Vehicle Excise and Registration Act 1994, Sched. 2, para. 20A.

[402] Hydrocarbon Oil Duties Act 1979, Sched. 1, para. 2.

[403] Mr Stanley Evans, MP, subsequently the only Labour MP to vote in favour of the Suez operation.

[404] He was dismissed by Prime Minister Attlee the next morning.

on 26 June 2003, the old arable area payments and animal subsidy schemes are to be replaced in 2005 by a single farm payment (alias 'SIP').[405] As brought in in the UK,[406] this may turn out to be far more land management, than agricultural activity, based than the EU norm of the assistance of active farmers.

21.9.3.2 World Trade Organization pressures

In addition, under the Doha Round, in September 2003, a main item on the agenda of the Cancún conference was the possibility of the US and the EU abandoning farm support systems which involved export subsidies. Although the talks broke down, this was primarily on other issues and should agreement for the phasing out of this type of farm support regime be reached eventually,[407] the pre-SIP financial base of the UK farming community, the subsidisation of crops rather than the occupation of land, would have been undermined.

In March 2004, there were, furthermore, indications that a resolution of the Doha Round might be achieved in 2005,[408] and (despite the initial US reaction) this is likely to be enhanced by the success, in June, of Brazil's challenge in the WTO to the system of subsidies for US cotton farmers.[409] Indeed, it was Brazil's Foreign Minister who played an important role in all sides arriving at the interim accord which was struck at Geneva on 1 August 2004, before the mandate of the current US and EU negotiators expired.[410] But this accord really did not amount to more than

---

[405] While a primary Regulation, EC/1782/2003, and a number of secondary Commission regulations, had been promulgated, a great deal depended upon the options given to Member States, which had to be exercised by 1 August 2004. Which said, for reasons which are unclear, none of the UK components had laid statutory instruments under the European Communities Act 1972 by that deadline. The scope for national variations had to be especially wide in order to obtain French concurrence. Article 87 of the primary Regulation contains an 'aid' of 45 Euros per hectare per annum for areas sown with energy crops under certain conditions. For this purpose, energy crops are defined as biofuels listed in the Biofuels Directive (European Parliament and Council Directive 03/30/EC, (2003) OJ L123 42), Art. 2(2), or electric and thermal energy produced from biomass.

[406] The regimes for England, Wales, Scotland and Northern Ireland are to differ.

[407] As to the Brazilian Government delegation's view of which, see Colitt, *Financial Times*, 16 September 2003, p. 9.

[408] *BBC News* website 19 March 2004; de Jonquières. *Financial Times*, 19 March 2003, p. 12. See also Duncan Brack and Thomas Branczik, *Trade and Environment in the WTO after Cancun* (RIIA Briefing Paper No. 9, February 2004). An interim accord was struck at Geneva on 1 August 2004, but really did not amount to the end of the 'talks about talks' stage, in the sense that, although it was agreed that agricultural subsidies would be eliminated, no final date was fixed (see de Jonquières, *Financial Times*, 2 August 2004, p. 5).

[409] See Colitt and Alden, *Financial Times*, 19–20 June 2004, p. 12. This was followed by a successful challenge against the EU sugar regime (see Buck, Colitt and de Jonquières, *Financial Times*, 5 August 2004, p. 1; Williams, *Financial Times*, 2 September 2004, p. 8).

[410] See Williams and de Jonquières, *Financial Times*, 2 August 2004, p. 1.

the end of the 'talks about talks' stage, in the sense that, although it was agreed that agricultural subsidies would be eliminated, no final date was fixed.[411] The agreement was, nonetheless, described as 'a minor miracle' in the *Financial Times'* Leader of 2 August 2004.[412]

It must follow that any continuation of support for the European farmer in the longer term will have to take the form of an environmental management subsidy available to rural businesses generally.

21.9.3.3 The Water Framework Directive

In June 2004, a DEFRA consultation had been launched into the promotion of catchment-sensitive farming[413] in the light of the UK's obligations under the Water Framework Directive.[414] Economic instruments were one of the four ways canvassed in order to take this forward, the others being targeted advice, a voluntary approach and bringing in command and control regulation in advance of the 2009 date specified in the Directive itself.

21.9.4 Fertilisers

In November 1997, it seemed that the first new environmental tax was likely to be one on fertilisers. This would have had a serious effect on the profitability of farming, then already under pressure. It appears, however, that detailed work into the idea was discontinued following the revelation that the 15 per cent reduction in nitrogenous fertiliser use over the preceding decade had not resulted in any reduction in the level of pollution in the water table.[415]

This needs to be read against the background, at that time, of taxing farmers being a political taboo. In the aftermath of the 2001 foot and mouth disease outbreak, with government policy much more geared to the rural community, there may be less reluctance to provoke resistance from the farming community than previously.[416]

The advent of the Water Framework Directive[417] timetable, under which regulation will be required between 2009 and 2015, brought the issue of fertiliser taxation to the fore again. This was in the context of the presence of pesticides and veterinary medicines presence in aquatic ecosystems being capable of resulting in environmental damage.[418] Farming itself also generated phosphorus and, where certain fertilisers

---

[411]   See de Jonquières, *Financial Times*, 2 August 2004, p. 5.
[412]   See *Financial Times*, 2 August 2004, p. 16.
[413]   See *Developing Measures to Promote Catchment-Sensitive Farming* (available from www.defra.gov.uk). See also para. 20.3 above.
[414]   European Parliament and Council Directive 00/60/EC, (2002) OJ L327 1.
[415]   See *Farmers Weekly*, 30 January 1998, p. 14.
[416]   As to which it should be noted that New Zealand is planning a farm aroma reduction tax, likely to cost each of its farmers an average of £110 per annum (*Financial Times*, 6–7 September 2002, p. 12).
[417]   European Parliament and Council Directive 00/60/EC.
[418]   *Partial Regulatory Impact Assessment on Proposals Resulting From the Diffuse Water Pollution from Agriculture Action plan* ['PRIA'] (June 2004), p. 5.

were used, nitrogen.[419] This had, however, to be considered in the context of both the decoupling of subsidies from farming activities, in favour of area payments, under the EU's Mid Term Review;[420] and the Entry Level schemes which have been introduced by the Environment Agency in order to reduce diffuse water pollution. Four possible strategies for the promotion of catchment-sensitive farming are also under consideration:

1 reliance upon, and enhancement of, the Entry Level scheme;
2 earlier regulation than required by the Water Framework Directive;
3 a supportive approach to voluntary action by the farmer; and
4 the use of economic instruments.

The fourth option was, however, considered to be difficult to operate in relation to the catchment-sensitive farming policy as far as concerned trading schemes (seen as difficult to apply in such circumstances, difficult to enforce and unlikely to be cost-effective)[421] or levies (for example, to promote better soil management).[422] The scope of economic instruments was therefore likely to be confined to taxing inputs[423] or nutrient surpluses,[424] but was dependent upon farmers being able to pass the cost on to consumers. This was considered unlikely to take place in practice unless costs increased across the EU with the implementation of the Water Framework Directive.[425]

Whilst no decision had been made,[426] it was noted that France was considering placing a levy on nitrogenous fertilisers,[427] and that Sweden, Austria and Finland already had such taxes. Experience in those other EU Member States suggested that, for each 1 per cent increase in price, there would be a reduction in demand of between 0.1 and 0.5 per cent,[428] with the possibility that CAP decoupling might increase this,[429] at a cost of approximately 1 per cent of revenues.[430]

21.9.5   Pesticides

Customs undertook detailed work, and issued consultations on both aggregates and pesticide taxation, and, at the end of 1999, the indications were that, if a tax on

---

[419] PRIA, p. 15.
[420] Council Regulation, EC/1782/2003 (see para. 29.9.3.1n above).
[421] *Development Measures to Promote Catchment-Sensitive Farming* ['DMPCSF'], a joint Defra-Treasury Consultation (June 2004), paras 4.61, 5.28 and E7–12.
[422] PRIA, p. 10.
[423] DMPCSF, paras 4.64 and E16–17. These could be applied to agrochemicals, veterinary medicines, nitrogen and phosphorus in fertilisers, feeds and manures.
[424] Seemingly based on a model, these might be either national or local (see DMPCSF, paras 4.63 and E14–15).
[425] PRIA, p. 24; DMPCSF, para. 4.65.
[426] DMPCSF, para. 5.29.
[427] Astonishing, politically, as this might seem for a right-wing government in that country (see DMPCSF, para. E18).
[428] *Ibid.*, para. E19.
[429] *Ibid.*, para. E20.
[430] *Ibid.*, para. E21.

the former was not introduced (and the environmental case for it is almost non-existent),[431] one on pesticides would be. At that time, it was thought that the latter would be constructed on a cumulative basis, and result in 2 per cent of farms being put out of business, a reduction of 3 per cent in farm incomes and the loss of between 1,000 and 2,000 jobs.[432]

On the basis of pesticide use, it was estimated that the gross margins of the following crops were very sensitive to a tax on pesticides: winter wheat, winter malting barley, dried peas, main crop potatoes, carrots, dry onion bulbs, dwarf beans, spring greens, parsnips and leeks.[433]

At the time of Budget 2000, it was understood that negotiations were in progress with the British Agrochemicals Association, with a view to revised proposals for a non-fiscal voluntary agreement being adopted to achieve the government's environmental objectives.[434]

A consultation was undertaken in response to the resulting proposals, the responses to which were published after the Pre-Budget Report on 8 November 2000, when the government let it be known that they still thought that a 'charge' might play a role in reducing the use of pesticides, but as one of a number of additional instruments. At that time, it was envisaged that a charge would be calculated on a combination of weight and hazard potential, this element being banded in five categories increasing either arithmetically, on a scale of one to five, or geometrically, on a scale of one to sixteen. In order to catch imports, the tax point was likely to be the first sale or use.

At the times of Budget 2001,[435] the following Pre-Budget Report[436] and Budget 2002,[437] the success of the voluntary scheme was being evaluated and possibility of a pesticides charge being introduced was said to be 'a real option' in the future. Furthermore, such a development was not opposed in the Report of the Policy Commission on the Future of Farming and Food.[438] No mention was made, however, of this subject in the supporting Press Notices to the 2003 Budget, on 9 April 2003.

At the time of the 2002 Pre-Budget Report, the government was still urging the parties to make faster progress, keeping in place the threat of a tax in the event of the failure of the *voluntary package*.[439] Shortly afterwards, the Department of Environment, Food and Rural Affairs announced a series of *whole farm audits* to support a new environmental scheme to be open to all farmers.[440]

---

[431]　See Chapter 13 above.
[432]　ENDS Report, November 1999, p. 22.
[433]　Department for Environment, Food and Rural Affairs, *Design of a Tax or Charge Scheme for Pesticides*, 29 April 2000 (republished).
[434]　ENDS Report, March 2000, p. 26.
[435]　HMT 1, 7 March 2001.
[436]　HMT 2: *Protecting the Environment, Today and for the Future*, 27 November 2001.
[437]　HMT 2, 17 April 2002.
[438]　*Farming & Food – A Sustainable Future* (29 January 2002).
[439]　HMT 2 (27 November 2002): this was supported by the House of Commons' Environmental Audit Committee's Fourth Report of 2002–03, HC 167, issued on 1 April 2003, para. 41. Yet in the Treasury's Budget 2003 (9 April 2003), para. 7.65, it was merely stated that work was being pursued on options.
[440]　Mason, *Financial Times*, 9 December 2002, p. 4.

It is, however, worthy of note that the head of steam which the Customs' environmental tax unit in Salford appeared to have been able to build up before the amalgamation of the environment part of the old DETR with the Ministry of Agriculture, Fisheries and Food, to form Defra, in the aftermath of the 2002 food and mouth disease outbreak, seems to be deflating. While the Treasury is still going through the motions of holding open the possibility of a pesticides levy, it seems possible that the bringing together of both sides of the non-fiscal equation under a single Secretary of State may have damaged this project beyond repair.

21.9.6 The GM diversion

A further factor which needs to be borne in mind is the removal of Mr Michael Meacher, who had been Environment Minister since the Blair Government was elected in May 1997, in the June 2003 reshuffle, and his replacement by Mr Elliot Morley, a teacher by profession and an MP since 1987, whose previous ministerial experience had been in the agricultural and countryside sector.

Mr Meacher's views on the outcome of the genetically modified crop trials in progress when he left office had been thought to be out of step with Downing Street.[441] Nonetheless the subsequent publication of the Strategy Unit's study was not as 'pro' these as had been anticipated. Furthermore, in September 2003, it became public knowledge that the Agriculture and Environmental Biotechnology Commission, which had been set up to advise the UK Government on the GM-organic crop controversy,[442] had been unable to reach a consensus, thus delaying a Ministerial decision on the issue.[443] When the results of the trials into three GM crops, oilseed rape, sugar beet and maize, were published in mid-October – revealing increased threats to the environment in two cases – they were generally thought to be a vindication of Mr Meacher's position.[444] There was also a certain amount of scepticism about the government's 'neutral' position on the issue.[445]

Moreover, the position was not clarified by the publication the following month of the report of the Agriculture and Environment Biotechnology Commission, which had

---

[441] Mason, *Financial Times*, 11 July 2003, p. 5, and 12–13 July 2003, p. 6. It was indicated subsequently, however, that the Prime Minister was by no means as favourable to GM crops as had previously been supposed (see Mason and Houlder, *Financial Times*, 22 August 2003, p. 5).

[442] Mason, *Financial Times*, 11 September 2003, p. 1.

[443] Adams, *Financial Times*, 22 September 2003, p. 2; Mason, *Financial Times*, 26 September 2003, p. 4.

[444] Mason and Cookson, *Financial Times*, 17 October 2003, p. 3, and Mason, *Financial Times*, 18–19 October 2003, p. 13. They did, however, come in for very substantial criticism, subsequently, from both the House of Commons Environmental Audit Committee in its Report, *GM Foods – Evaluating the Farm Scale Trials*, and Friends of the Earth. As to the latter, see *Financial Times*, 8 March 2004, p. 4. In both cases the main cause of concern was a failure to take appropriate account of North American experience.

[445] Eaglesham, *Financial Times*, 17 October 2003, p. 3.

failed to agree on how organic farmers might be protected.[446] The GM companies' position had, however, been deteriorating, politically, in Brussels[447] before then, and had been hit, technically, by the ban on the weed killer atrazine.[448] Although BBC News' website reported that qualified approval for GM maize was imminent on 19 January 2004, a decision by the Cabinet was, however, delayed until 4 March 2004.[449] This was taken, furthermore, in the context of a disagreement as to the efficacy of trial undertaken on the basis of a weed killer which would not be available in future. While scientists appeared to take the view that this did not invalidate the trials,[450] this was not the opinion of the House of Commons Environmental Audit Committee.[451]

On 9 March 2004, Mrs Margaret Beckett (the Secretary of State) made a statement in the House of Commons that the government would be prepared to licence (only) one type GM maize for use as animal feed, 'Chardon LL', subject to both EU confirmation and to relicensing in 2006.[452] The biotech industry was unhappy at being told that it, rather than the government, would be responsible for compensating farmers for contamination damage.[453] The concerns of the anti-GM lobby were, however, assuaged when Bayer, the manufacturer of LL Chardon, announced at the end of the month that the government's relicensing requirements meant that it would not be proceeding with marketing it in the UK.[454]

It is, therefore, worthy of note that, in the 2003 Pre-Budget Report, it was indicated that, while the government was 'continuing' to examine tax and economic instrument options, it believed that the most effective method of reducing the environmental impact of pesticides was for the voluntary initiative to be implemented in full.[455] In the 2004 Budget Report, this option was kept open, albeit on a very contingent basis,[456] and the spotlight switched to a pending consultation on the use of financial

---

[446]   Mason, *Financial Times*, 26 November 2003, p. 7; Mason, *Financial Times*, 14 January 2004, p. 4; Eaglesham and Mason, *Financial Times*, 2 February 2004, p. 11; Firn and Eaglesham, *Financial Times*, 20 February p. 4.

[447]   Mason, *Financial Times*, 14 October 2003, p. 10; 22 January 2004, p. 14.

[448]   Mason, *Financial Times*, 13 October 2003, p. 3.

[449]   Eaglesham and Mason, *Financial Times*, 5 March 2004, p. 6.

[450]   Mason, *Financial Times*, 4 March 2004, p. 4.

[451]   *GM Food –Evaluating the Farm Scale Trials*, Second Report of 2003–04, HC 90 (3 March 2004). The Chairman of the Committee, Mr Peter Ainsworth, MP, said that he was writing to the Secretary of State in the context of leaked Cabinet Committee Minutes. The Royal Society did not, however, agree with the Committee's criticism and it was not expected that the government would defer its announcement because of this development (see Mason, *Financial Times*, 6–7 March 2004, p. 3).

[452]   Cols 1381–3.

[453]   Mason, *Financial Times*, 10 March 2004, p. 5.

[454]   Mason, *Financial Times*, 31 March 2004, p. 1.

[455]   Delivered 10 December 2003, paras 7.59–60.

[456]   Para. 7.69. Nonetheless The Rt Hon. Alun Michael, MP, Minister of State, said in answer to a written Parliamentary Question on 27 April 2004 that the matter was still 'under consideration': col. 863W, No. 168082.

instruments to tackle diverse water pollution,[457] following from the publication of the Defra Discussion Document on that subject in April 2003.[458]

The virtual abandonment of a pesticides tax has clearly been a considerable disappointment to the Commission on Sustainable Development, whose report published in April 2004,[459] recorded that, in the southeast of England and Thames Gateway,[460] water shortages were acute and wildlife and habitats already under strain,[461] and that, on the biodiversity front, there was a strong case for early introduction of a pesticides tax, the economic difficulties of the farming sector notwithstanding.[462] Whilst accepting that the creation of Defra had helped to bring about 'a significant input of sustainability thinking into agriculture and food policy', the Commission placed on record that there was a 'serious downside' in the resulting separation of the lead responsibility for planning and local government from the ODPM and transport responsibility in the DFT.[463]

21.10 The effect of Community law and policy

The effect of Community law and policy on the structure of the UK's environmental taxes[464] should not be underestimated, especially in relation to climate change levy. The availability of 80 per cent reductions available under climate change agreements has been constricted, by reference to Community rules on state aid,[465] in two respects: by confining the processes which qualify to those within the IPPC Directive[466] and by the likelihood of those reductions only being available for ten years.[467]

In addition, it prevented giving any relief to an industrial heavy energy user, who was not an IPPC regime heavy polluter. As a general rule, both manufacturing and small businesses pay much more in climate change levy than they save under the

---

[457] HC 301, para. 7.68.

[458] *Strategic Review of Diffuse Water Pollution from Agriculture*. It has to be said, however, that it is not immediately apparent to the authors what form of economic instrument could be designed to encourage best practice in the context of the non-pesticide issues discussed very fully and fairly in this document. Mention has, however, been made of 'rainwater harvesting'.

[459] *Shows Promise. Must Try Harder*. See para. 4.2.1.4 above.

[460] The south coast of Essex bordering on the north bank of the Thames, an area scheduled for considerable urbanisation.

[461] Para. 56.

[462] Para. 76.

[463] Para. 69.

[464] See Chapter 12 above.

[465] See para. 12.2.7 above.

[466] Finance Act 2000, Sched. 6, para. 51 (albeit reformulated before the effective date by reference to the amended UK regulations): the underlying Community legislation is the Integrated Pollution Prevention and Control Directive (see paras 12.2.2 and 14.5 above).

[467] Both were confirmed by the Financial Secretary to the Treasury (Mr Stephen Timms, MP) to Standing Committee H on 18 May 2002.

NICs rebate.[468] The government announced in the 2003 Pre-Budget Report,[469] however, that it would apply for state aid approval to extend the eligibility criteria during 2004 to sectors meeting a specific energy-intensive threshold (account being taken of any competitive distortions in those sectors).[470]

Moreover, in relation to aggregates levy, although the Courts rejected a general challenge[471] based, *inter alia*, on the concept of unlawful state aid[472] in favour of materials and processes not within the scope of the tax, the concept of unlawful state aid not only prevented the government reaching an agreement with the Quarry Products Association, under which quarries with their proposed quality mark would be given procurement preference,[473] it also meant that the relief in favour of producers in Northern Ireland against untaxed competition from Eire was only obtainable on a transitional basis.[474]

Much of the complexity behind the aggregates levy has been caused by the prohibition on the imposition of import taxes under the European Treaty.[475] Although the Court challenge raised this issue on the structure adopted, the end result has been held to be watertight on this point.[476] Having to skirt round the prohibition has, nevertheless, meant that the tax collection machinery could not be confined to extraction sites. Collection at the place of importation being prohibited, Customs had to fall back on the premises at which processing and/or mixing first took place. This is, in large measure, responsible for many of the liability complications that are amongst this tax's most unattractive features.

State aid challenges to the structures of Member States' national taxes might have been thought a serious threat to the latter's fiscal options. Recent developments, in unrelated areas, indicate that this might not, after all, be the case.[477] In *GIL Insurance*

---

468 Jones, *Financial Times*, 24 June 2002, p. 4; Houlder, *Financial Times*, 17 July 2002, p. 2.

469 Cm 6042.

470 *Business Brief* 27/2003 (10 December 2003).

471 *R (on the application of British Aggregates Association and others) v. C & E Commrs*, [2002] EWHC 926 (Admin), [2002] 2 CMLR 51: primarily on the basis of the official explanation that the purpose of the tax was to protect virgin sources (*ibid.*, paras 108–15, 132).

472 There does not seem to have been a jurisdictional problem. The concept of 'direct effect' has been recognised by the ECJ. The challenge was mounted, and failed, on the basis of community law restrictions alone. Similar obligations binding on the UK Government under the WTO Agreement on Subsidies and Countervailing Measures were presumably not thought to be arguable before the national court at the instance of the subject (see paras 8.4.4 and 8.4.5.2 above).

473 *Financial Times*, 2 May 2001, p. 6.

474 Finance Act 2001, s.30A: Press Release HMT 2 – *Protecting The Environment, Today and for the Future* (17 April 2002). See also Press Release CE11 (17 March 2004) and para. 13.3 above.

475 See para. 12.3 above.

476 See *R (on the application of British Aggregates Association and others) v. C & E Commrs*, above.

477 Ager, *Taxation*, 15 January 2004, p. 357.

Ltd v. C & E Commrs,[478] the ECJ had to address a challenge to the imposition by the UK of a special high rate of insurance premium tax ('IPT') equal to the standard rate of VAT on travel insurance policies sold through travel agents and domestic appliance insurance sold through retailers of those goods. This had been introduced in order to prevent VAT avoidance through price shifting. The main challenge mounted related to compatibility with the Sixth VAT Directive.[479] A secondary challenge, however, was also mounted on the basis that it constituted unlawful state aid under Art. 87(1), European Treaty (ex 92(1)).[480] The ECJ's reasoning was as follows:[481]

> ... even on the assumption that the introduction of the higher rate of IPT involves an advantage for operators offering contracts subject to the standard rate, the application of the higher rate of IPT to a specific part of the insurance contracts previously subject to the standard rate must be regarded as justified by the nature and the general scheme of the national system of taxation of insurance. The IPT scheme cannot therefore be regarded as constituting an aid measure within the meaning of Art. 87(1) EC.

State aid issues remain pressing ones in relation to the design of environmental taxes, however. This applies, not only to the structure of exemptions and reliefs within the taxes, but also to the destination of the proceeds raised by them. The general issues relating to this point have already been explored in some detail.[482] The issue has become particularly pressing in relation to landfill tax, as is evidenced by governmental concern at the state aid restrictions upon the scope available to it in relation to the disposal of the anticipated 2005/06 landfill tax surplus.[483] As analysed,[484] what is required is a grant aid, or possibly an interest-free loan,[485] regime to fund the construction of recycling plants. A central problem appears to be, however, that, although Community state aid law permits assistance to small and medium-sized enterprises, such enterprises are unlikely to be of sufficient size to warrant the construction of their own discrete recycling capacity. A possible solution to this appears to be the making of grants through venture capital funds promoting such ventures.

---

[478] Case C–308/01, [2004] STC 961.
[479] See para. 1.2.1.4 above.
[480] See para. 12.2.7.1 above.
[481] C–308/01, para. 78.
[482] See paras 11.3 and 12.3.5 above.
[483] See *An Assessment of Options for Recycling Landfill Tax Revenue*, Final Report, prepared by Integrated Skills Ltd for HM Treasury (February 2004).
[484] Tax-subsidised research and development seeming to have reached its capacity and enhanced capital allowances being considered to have no role to play in this sector.
[485] Which, it seems to the authors, might go a long way towards solving the council tax capping problems being faced by county councils, if made available to local authorities. Funding issues can lie at the root of county councils' ability to construct recycling facilities (see para. 21.4.1 above). The report does not, however, address this possibility.

PART III, SECTION B
GREENING THE UK TAX SYSTEM

Division 1
Removing Subsidies and
Creating Incentives

Chapter 22

Excise Duties

22.1   Introduction

Until very recently, there had been comparatively little coordination of excise duties within the EU.[1] This was not for want of trying on the part of the European Commission, given that disparities between Member States can have a very distortive effect on local economies. There appears, furthermore, to be a reluctance to regard significant differences in the rates of tax as contrary to public policy.[2]

The general provisions of the European Treaty that are designed to promote trade between Member States tend to operate against direct tax anomalies (for example, the inability of a resident of one Member State working in the neighbouring state to obtain full reliefs in the latter).[3] A country, such as the UK, which has historically had (and, in order to balance its Budget, will seek to maintain) high levels of excise duty on spirits and tobacco, will find that this tends to promote, rather than to restrict, imports.[4]

In such circumstances, the general principles of the European Treaty leave the UK a much greater freedom of action in relation to excise duties than in relation to other areas of fiscal policy.[5] That said, the means by which HM Customs and Excise have sought to counter unofficial trading in goods subject to excise duties[6] have, however, been found to be inconsistent with the European Single Market Directive.[7]

22.2   Motor cars and their fuel

At least four different levies are imposed on the manufacture, sale and operation of motor cars. First, there is a car tax, which is levied via the manufacturer at the time of purchase and, having no environmental element, is not addressed further in this book. There is then vehicle excise duty ('VED').[8] Thirdly, there are excise duties on

---

[1] See para. 12.3.4 above.
[2] *Zurstrassen v. Administrations des Contributions Directes*, C–87/99, [2001] STC 1102.
[3] See, for example, *Finanzamt Koln-Altstadt v. Schumacher*, C–273/93, [1995] STC 306.
[4] Which is, of course, what the European Single Market is supposed to be meant to achieve!
[5] *Imperial Chemical Industries plc v. Colmer*, C–264/96, [1998] STC 874.
[6] Recently a public health aspect has been brought into the equation, with it being revealed in December 2004 that a lot of counterfeit cigarettes contained dangerous substances.
[7] 92/12/EEC, in *Hoverspeed Ltd and others v. C & E Commrs*, [2002] EWHC 1630 (Admin).
[8] Vehicle Excise and Registration Act 1994.

fuel[9] and, finally, VAT is payable, in addition to the aforementioned levies, both on the purchase of the car and upon the purchase of its fuel.[10]

22.2.1   Vehicle excise duty

Evasion of vehicle excise duty has for long been a concern to UK governments. Although it is only payable for cars kept on the road, it is now the case that, when the annual renewal comes up, the 'keeper' will find that he will not be able to avoid paying by claiming that the car is to be kept off the road.[11] The tax is administered by the Driver and Vehicle Licensing Agency (DVLA) on behalf of the DFT.[12]

This is an annual registration duty originally called the road fund licence. It applies to all vehicles except for emergency vehicles and those farm vehicles which need to make only minimal use of the public highway. The annual rate was, originally, intended to reflect the relative wear and tear. Thus motorcycles pay less than cars, trade vehicles more than private, and heavy lorries more than smaller ones.

With effect from 1 May 2003,[13] the annual[14] rates of VED payable were varied to the following, in order to give an environmental incentive:[15]

motorcycles, up to 150cc		£15.00
151–400cc		£30.00
410–600cc		£45.00
over 600cc		£60.00
private and light goods vehicles registered before 1 March 2001:		
up to 1549cc		£110.00
over 1549cc		£165.00
light goods vehicles registered 2001–2003:		£165.00
light goods vehicles registered after May 1, 2003:		£110.00
private vehicles registered thereafter are banded according to CO_2 emissions and the type of fuel used:		
up to 100g/km:	diesel	£75.00
	petrol	£65.00
	alternative	£55.00
101–120g/km:	diesel	£85.00
	petrol	£75.00
	alternative	£65.00

---

9   Hydrocarbon Oils and Duties Act 1979.
10   Value Added Tax Act 1994.
11   Vehicle Excise and Registration Act 1994, s.1(1C): substituted by Finance Act 2002, Sched. 5, para. 2.
12   See para. 4.2.1.2(1) above.
13   Vehicle Excise and Registration Act 1994, s.2(2)–(7): substituted by Finance Act 2002, Sched. 5, para. 3.
14   Six-monthly renewals are also permissible, albeit at different rates, for example, £60.50 as against £110.
15   References to specialist vehicles have been excluded from the summary below.

121–150g/km:	diesel	£115.00
	petrol	£105.00
	alternative	£95.00
151–165g/km:	diesel	£135.00
	petrol	£125.00
	alternative	£115.00
166–185g/km:	diesel	£155.00
	petrol	£145.00
	alternative	£135.00
over 185g/km:	diesel	£165.00
	petrol	£160.00
	alternative	£155.00
trade licences are available for vehicles		£165.00
and bicycles and tricycles for		£60.00

22.2.2   Excise duties on fuel

Under the previous Conservative government, it was decided to encourage motorists to switch to unleaded petrol, by subjecting this to a lower rate of excise duty than the traditional leaded. This was bolstered by the indexation of the rate of duty on petrol, known as the 'fuel duty escalator'.

By the end of the century, the general switch by the majority of motorists to unleaded petrol demonstrated that the Lawson initiative had achieved a large measure of success. However, the rapid rise in oil prices during the summer of 2000 led to a nationwide organised blockade of oil distribution depots and the government had to agree to discontinue the escalator.

The use of diesel is increasing in the UK, helping to reduce emissions.[16] Such engines are 20 per cent more fuel-efficient than petrol.[17] In Europe, sales of diesel cars have doubled over seven years and are expected to make up half the total by 2006.[18] The Commission is, however, preparing new emissions rules. Whilst making diesel engines cleaner, they will make them more expensive and less efficient.[19] The cost is to increase with the introduction of the Euro 4 rules in 2005. The subsequent rule change, which is scheduled for 2010, is likely to centre on the control of nitrous oxide, which will create a problem for diesel engines.[20]

Mention was made in the 2002 Pre-Budget Report of fiscal encouragement of bio-fuels. However, this merely stimulated the House of Commons' Environmental Audit Committee, in its Fourth Report for 2002–03 issued on 1 April 2003,[21] to urge that a more coherent strategy be adopted.[22]

---

16   Hunt, *Financial Times*, 16 October 2003 [special report Sustainable Business], p. 4.
17   *Ibid.*
18   *Ibid.*
19   *Ibid.*
20   Mackintosh, *Financial Times*, 15 October 2003, p. 14.
21   HC 167.
22   *Ibid.*, para. 16.

The government-funded Energy Saving Trust[23] had been distributing most of its Energy Budget grants to vehicles fuelled by LPG, the market in which had grown from 1,000 to 90,000 in five years. This body considered that a radical change in the duty differential would undermine this. Yet, in contrast to biofuels and hydrogen, this fuel offered no carbon dioxide advantage over diesel.[24] Nonetheless, Lord Whitty, then the junior farming Minister, subsequently told the House of Commons Environmental Audit Committee that the 2002 Budget cut of 20p had proved an insufficient incentive, in effect supporting the view that increased use of biodiesel and bioethanol should be seen as an effective means of contributing to the attainment of the Kyoto targets.[25] On 1 March 2004, Mr David Jamieson, MP, Under-Secretary of State, Department of Transport, revealed that the Energy Saving Trust's Powershift grants[26] were estimated to have assisted in the purchase or conversion of 13,516 vehicles to LPG in the context of LPG vehicle totals of 39,000 in 2000 rising to 107,000 in 2003.[27] On 2 March 2004, the same minister made it clear that, even though the rate of excise duty on LPG was to be increased gradually over the following three years, support for LPG still continued through the Powershift conversion grant scheme and reduced VED rates for gas powered vehicles.[28]

In the Budget on 9 April 2003, a package of measures had been announced:[29]

1   the creation of a lower VED band was created, for cars emitting less than 100 grams of carbon dioxide per kilometer;[30]
2   the setting of the duty on bioethanol at 20p per litre lower than for conventional low-sulphur fuel, leaving LPG unchanged;[31]

---

23   See para. 4.2.1.2(1) above.
24   Houlder, *Financial Times*, 23 April 2003, p. 6.
25   Mason, *Financial Times*, 16 September 2003, p. 5.
26   Which had been undersubscribed, but was anticipated to overspend by £8m if applications continued at the rate current at the end of 2003 (see HC Written PQs, 1 March 2004, col. 622W, WA 156200 and 156201).
27   Written PQs, cols 622–623W, WA 156245.
28   Written PQs, col. 763W, WA 157450.
29   Press Release PN 04 (9 April 2003).
30   On 13 May 2003, the Economic Secretary to the Treasury (Mr John Healey, MP) told a Committee of the Whole House on the Finance Bill that evaluation had begun of a carbon dioxide-based vehicle excise duty system, which was due for completion in the autumn: col. 225. But the House of Commons' Environmental Audit Committee, in its Tenth Report of Session 2003–04, *Budget 2004 and Energy*, HC 490, published on 11 August 2004, found it extraordinary that the Minister, in evidence given to the Committee, did not accept that one of the main objectives should be to influence buying decisions. They were concerned, furthermore, that the government's own evaluation of the VED scheme showed that current differentials were insufficient to prompt behavioural changes. A radical increase in those differentials was therefore recommended (*ibid.*, para. 51).
31   HM Treasury's Budget 2003 (9 April 2003), para. 7.30, indicated that the government also intended to exempt hydrogen for a limited period following a consultation with 'stakeholders'.

3   the introduction of a duty incentive of 0.5p per litre for sulphur-free fuels from 1 September 2004;[32] and

4   the further tightening in 2005/06 of the emissions criteria for company car scales, albeit less stringent than that previously announced for 2004/05.[33]

The Pre-Budget Report of 10 December 2003,[34] was a severe disappointment to manufacturers of cars powered by LPG because of the Chancellor's decision that the 41.7p per litre differential was no longer justifiable.[35] Instead, emphasis was given to an Alternative Fuels Framework,[36] committing the government to a rolling three-year period of differential certainty, within a general context of 'blunt' duty differentials being considered against other means of support, such as capital incentives, grants and regulatory solutions.

Budget 2004 contained the following provisions:[37]

1   The road fuel mixing charge was to be removed,[38] and the Energy Products Directive's[39] requirements[40] were to be implemented, with effect from Royal Assent to the Finance Bill.[41]

---

32   Press Release CE 30 (9 April 2003).
33   See para. 23.2 below. The success of this scheme was noted by the House of Commons' Environmental Audit Committee, in its Tenth Report of Session 2003–04, *Budget 2004 and Energy*, HC 490, published on 11 August 2004, para. 47, and the suggestion made that differentials be widened in order to increase the incentives for purchasing very low-emission vehicles.
34   Cm 6042.
35   Mason, *Financial Times*, 11 December 2003, p. 10.
36   Cm 6042, Box 7.1. The introduction of the Alternative Fuels Framework was welcomed by the House of Commons' Environmental Audit Committee, in its Tenth Report of Session 2003–04, HC 490, *Budget 2004 and Energy*, published on 11 August 2004, para. 40.
37   In July, it was announced that the general increases scheduled for 1 September 2004, were to be deferred in the light of the steep increases in the oil price on international markets. The government no doubt feared a repetition of the protests which took place in 2000. The generally accepted view is, however, that (unlike in 2000) the price increases should not be seen as short term, being generated very substantially by the very large increase in the size of the economy of the People's Republic of China during 2003–04. But the UK Government's long-term solution, the substitution of satellite-based road user charging, does seem to be beset with technical difficulty (see Wright, *Financial Times*, 24–25 July 2004, p. 11). Nonetheless, against the background of carbon emissions from transport still moving in the wrong direction and notwithstanding the fuel protests of 2000, the House of Commons' Environmental Audit Committee, in its Tenth Report of Session 2003–04, *Budget 2004 and Energy*, published on 11 August 2004, urged the government to implement the duty rises 'at the earliest opportunity' (*ibid.*, paras 44–5).
38   Press Release CE 28 (17 March 2004).
39   See para. 12.3.4 above.
40   Press Release CE 29 (17 March 2004).
41   22 July 2004.

2   Bioethanol used as a machinery fuel or additive was to be charged at 28.52 per litre from 1 January 2005, 20p[42] below the prevailing sulphur-free petrol rate.[43]

3   Sulphur-free petrol and diesel[44] was to be charged at 48.52p per litre from 1 September 2004, 0.5p less than that for ultra-low sulphur petrol and diesel.[45] For this the rate was to be 49.02p. Other unleaded petrol was to be 51.70p. LPG was to be 13.03p,[46] natural gas 11.10p[47] and biodiesel 28.52p.[48]

Following concern at the possibility of further protests following the increase in oil prices during the summer, the 2004 Pre-Budget Report contained a single change to these excise duties, a reduction of the duty on bioethanol from 28.52p per litre to 27.10p, from 1 January 2005.[49]

22.3   Heavy goods vehicles and their fuel

Vehicle excise duty rates for heavy goods vehicles are banded both by reference to axle weight and numbers[50] and by reference to whether they are built to standard or reduced pollution standards.[51] In relation to the latter, the differential in band G is £500 (that is, £1,350 as against £1,850).

Excise duty on vehicle fuel is far higher in the UK than, for instance, in France. Cross-border truckers tend therefore to plan their journeys on the basis that they will be able to refuel on the other side of the Channel.

In the Pre-Budget Report 2000, the government announced its commitment to a range of road fuels in order to achieve long-term reductions in the emission of both the six Kyoto GHGs[52] and the eight pollutants identified in *The Air Quality Strategy*

---

[42]   In Committee, the Chairman of the House of Commons Environment, Food and Rural Affairs Committee (The Rt Hon. Michael Jack, MP, a former Conservative Financial Secretary to the Treasury) challenged the Economic Secretary to the Treasury (Mr John Healey, MP) as to why the rebate was as low as 20p and the impression gained by some members of the Committee that Budgetary reasons lay at the root of this: the debate is reported in *Standing Committee A*, 6 May 2004, cols 41–59.

[43]   Press Release CE 27 (17 March 2004). This was linked with a climate change levy change: Press Release CE 20 (17 March 2004): see para. 14.3 above. The differential was to be maintained until 2007.

[44]   The Freight Transport Association (see para. 2.6 above) said that the increase in diesel duty would add £260m a year to transport costs (see Griffiths, *Financial Times*, 18 March 2004, pl.12).

[45]   Press Release CE 25 (17 March 2004).

[46]   There was to be a 1p differential until 2007.

[47]   For this, the differential was to be frozen until 2007.

[48]   Press Release CE 26 (17 March 2004). The DFT was to conduct a Consultation on biofuels. The results were made known on 2 December 2004.

[49]   Excise Duties (Surcharges or Rebates) (Bioethanol) Order 2004, S.I. 2004 No. 3162.

[50]   Vehicle Excise and Registration Act 1994, s.62(1).

[51]   *Ibid.*, Sched. 1, para. 9(1): inserted by Finance Act 2002, Sched. 2, para. 2.

[52]   See para. 8.3.1.4 above.

for England, Scotland, Wales and Northern Ireland published in January 2000, that is, benzene, 1,3 butadiene, carbon monoxide, ozone, particulates and sulphur dioxide.

In Finance Act 2001, s.3,[53] Customs were given power to reduce or rebate duty in relation to experimental fuels. In 2002, the Financial Secretary to the Treasury invited the submission of applications for such recognition, on a pilot basis, to be made to Customs by 31 July 2002. For this purpose, lists of fuels were published:

1   capable of inclusion were hydrogen,[54] ethanol and bioethanol (including blends of the latter), methanol, biogas, alternative methods of treating biogas and novel treatments of organic oils (rather than esterification); while
2   excluded were fuels not for transport use, conventional fossil based fuels, road fuel gases (other than those specified), esterified organic oils (that is, biodiesel), blends of biodiesel and fossil fuels, fuel additives and emulsions (fuels mixed with water).

Furthermore, in early May 2003, in *Modernising the taxation of the haulage industry – lorry road user charge: Progress Report Two*, the government revealed that it had achieved the necessary changes to the (then) draft Energy Products Directive,[55] which had been sent to the European Parliament for approval, in order to enable it to reduce the duty on diesel fuel after the introduction of the lorry road user charge in 2006.[56] This was in order to enable the UK haulage industry to compete with its EU counterparts. The Directive was scheduled to take effect from January 2004, and, under it, substantial increases in fuel duties would take effect across the European Union from 1 January 2010.[57]

22.4   The emissions problem

Customs launched a *Consultation on Duty Differentials for More Environmentally Friendly Rebated Oils* on 9 July 2003. This explained that there were three main pollutants associated with heavy oil combustion, that is, primary particulates, sulphur dioxide; and nitrogen oxides, for all of which emissions were expected to fall, but as to which consideration was being given to the role of fiscal incentives in meeting the targets[58] set by the Air Quality Framework Directive[59] and its first 'daughter'

---

[53]   Inserting Hydrocarbon Oil Duties Act 1979, s.20AB.
[54]   As to the problem areas becoming apparent in relation to hydrogen, see Roberts, *Financial Times*, 13 May 2003, p. 21. The voicing of these concerns has, however, led to a counter attack by the manufacturers of LPG cars (see Macintosh, *Financial Times*, 25 August 2003, p. 4).
[55]   See para. 12.3.4 above.
[56]   See para. 27.3 below.
[57]   *Ibid.*
[58]   Within the National Emissions Ceilings for Certain Atmospheric Pollutants Directive (see para. 12.2.6.2 above).
[59]   *Ibid.*

directive.[60] The latter laid down targets to be achieved for sulphur dioxide[61] and primary particulates by 1 January 2005, and for nitrogen dioxide by 1 January 2010. However, under the revised Large Combustion Plant Directive,[62] more stringent limits are imposed and, under the revised Fuel Quality Directive,[63] the maximum sulphur content of gas oil sold as diesel road fuel has to fall by 80 per cent by 1 January 2009.[64]

The purpose of the consultation referred to above was to establish whether the previously successful policy of successive UK governments in setting duty differentials[65] was likely to be repeatable, in circumstances in which rebated gas oil and fuel oil continued to contribute to local air quality problems.[66]

---

[60] *Ibid.*, n.
[61] See also para. 12.2.6.3(5) above.
[62] See para. 12.2.6.2 above.
[63] See para. 12.2.6.3(5) above (see also the Sulphur Content Directive, *ibid.*).
[64] Proposals were, furthermore, before the Commission to tighten the emissions standards for diesel engines still further.
[65] That is, in encouraging the switch between 1997 and 1999 from conventional gas oil to ultra-low sulphur diesel, and similar success in relation to ultra-low sulphur petrol.
[66] Consultation, para. 4.4. It is worthy of note, that, in September 2004, the US State of California took a different approach. New emissions reduction features were to be incorporated in all vehicles between 2009 and 2016, at an estimated cost of US$1,000 per car or lorry (see Simon, *Financial Times*, 25–26 September 2004, p. 8).

Chapter 23

Employee Taxes

23.1 Introduction

The general principles governing the income tax treatment of employees in the UK are that both cash remuneration[1] and benefits in kind[2] are brought into charge, but expenditure incurred by the employee is only relieved if 'wholly, exclusively and necessarily' incurred in the performance of his duties.[3] Suffice it to say that, as a general principle, it has never been regarded as necessary to eat or sleep in order to work, nor has it ever been so to travel from home to one's place of work.[4]

What follows should be read subject to the outline of income tax given in Chapter 1 above[5] and the reader is referred to one of the standard works on the subject[6] for income tax in general.

23.2 Income tax treatment of company cars

In the late 1960s and early 1970s, the fleet car business was seen as an essential means of supporting the British motor industry.[7] In times of very high personal taxation (the top rate of income tax for earnings after 1974 became 83 per cent), the provision of benefits in kind became the norm, and that of a 'company car' the lead item. The car benefit scales were, indeed, so low that professional partnerships took to forming service companies to provide cars to their equity partners.

The ground rules changed very suddenly, with the doubling of the scales in 1988. From 1991, this was supplemented with the imposition, on the employer, of Class 1A NICs.[8] The employee's liability was nonetheless calculated by reference to a

---

[1] Income and Corporation Taxes Act 1988, ss.19, 131; Income Tax (Earnings and Pensions) Act 2003, ss.10(2), 67(2).

[2] Income and Corporation Taxes Act 1988, ss.145 (residential), 154 (other); Income Tax (Earnings and Pensions) Act 2003, ss.97, 201 respectively.

[3] Income and Corporation Taxes Act 1988, s.198; Income Tax (Earnings and Pensions) Act 2003, s.328.

[4] *Ricketts v. Colquhoun*, (1924) 10 TC 118.

[5] See para. 1.4.3.1 above.

[6] See John Tiley, *Revenue Law*, 4th edn (Oxford: Hart Publishing, 2000); and Lesley Browning *et al.*, *Revenue Law: Principles and Practice*, 22nd edn (London: Lexis Nexis Tolley, 2004).

[7] At that time there was still a major British owned car maker and exporting was seen as the primary method of keeping the balance of payments under control. According to conventional wisdom at that time, sustaining a substantial home market was seen as a pre-requisite to export growth.

[8] Social Security Contributions and Benefits Act 1992, s.10.

466 *Environmental Taxation Law*

combination of the number of cars made available to him and his family, the amount of business mileage done by him and the cost, age and engine size of the car.

From 6 April 2002, this system has been changed to one based on carbon dioxide emissions, without any discount for the age of the car or high business mileage.[9] In Budget 2004,[10] it was announced that the minimum level of charge would be fixed as follows:

2004/05	145 grams per kilometre of carbon dioxide
2005/06	140 grams per kilometre of carbon dioxide

Fuel used for private motoring is subject to a separate charging regime from that for the provision of the car.[11] Suffice it to say that, for 2003/04, a combination of the increased taxable amount and employer's Class 1A national insurance contributions has made it economic for this to be foregone at a litreage of as little as 3,000.[12] The maximum charge for that year was £14,400 and this has been frozen for 2004/05.[13]

Following schemes to provide vans for employee use, the government instituted a consultation to encourage the use of more environmentally friendly vans.[14] When it appeared on 8 May 2003,[15] this did not have many firm proposals in it, apart from the extension of free fuel scale charging from cars to vans.[16] The underlying problem in extending the emissions-based provision of vehicle scale from cars was that, for vans, there was no published data on them. The alternatives appeared to be the replacement of a scale reduction for older vehicles by a supplement; having a different level of charging after the introduction of the Euro IV emission standard in 2006, with non-compliant vans being charged a supplement; and giving a reduction for vehicles running on alternative fuels, such as liquid petroleum gas[17] or compressed natural gas.

In Budget 2004, it was announced that the existing full private use charges of £500, reduced to £350 for vans four years old or over, would continue in a revised scheme to be introduced from 2005/06, when fuel scale charges would be added. From 2007/08, however, the full availability charge would be increased to £3,000, with no abatement for the van's age.[18] Fleet management companies indicated that these changes would mean that additional administrative (and perhaps security and tracking) arrangements would have to be put in place by employers.[19] In the

---

9   Income and Corporation Taxes Act 1988, Sched 6; Income Tax (Earnings and Pensions) Act 2003, ss.133–44.
10 Press Release REV BN 41 (17 March 2004).
11 Income and Corporation Taxes Act 1988, s.158(2); Income Tax (Earnings and Pensions) Act 2003, s.150.
12 Crowther, *Simon's Tax Briefing*, 14 October 2002, p. 8.
13 Press Release REV BN 41 (17 March 2004).
14 HM Treasury's Budget 2003 (9 April 2003), para. 7.37.
15 *Employer-provided Vans*.
16 *Ibid.*, proposal 4.
17 The Inland Revenue author does not seem to have been conscious of the doubts being expressed about LPG (see para. 22.2.2 above).
18 Press Release REV BN 42 (17 March 2004).
19 Martin, *Financial Times special report, Fleet Management*, 31 March 2004, p. 1.

Responses to the Consultation Document[20] both employers and professional advisers had emphasised that, for commercial reasons, vans tended to have to be taken home at night (and it was accepted in the RIA[21] that home-to-work travel should not be regarded as private mileage), that commercial considerations militated against the notion that the employee would have a choice of make or that a van (in the state it was likely to be after business use) would not be an idea vehicle for genuinely private motoring, with the result that private usage (even if permitted under the terms of the employee's contract)[22] would be likely to be insignificant. After consideration of such submissions, the RIA saw the environmental impact of the changes as likely to be as follows:

1   a reduction in the number of company vans used for private purposes;[23]
2   decreased attractiveness of a company van in place of a company car;[24]
3   a decrease in the number of company vans for which there is no restriction on private use;[25] and
4   a reduction in private mileage by those who have to pay for it;[26] but
5   although the abolition of the older van reduction would remove the incentive to retain such vehicles, no significant increase in the number of such vehicles being scrapped was to be anticipated.[27]

On 29 April 2004, the Inland Revenue published a *Report on the Evaluation of the Company Car Tax Reform*.[28] This suggested that between 0.15m and 0.2 m tonnes of carbon in road emissions (0.1 per cent of the total) had been saved in 2003, and would lead to reductions of between 0.5m and 1m in the longer term. This was as a result of reductions in business travel of between 300m and 400m miles after the removal of incentives had induced a high level of awareness in drivers and employers.

23.3   Environmentally-friendly transport

The zero-rating of passenger rail and bus transport for VAT[29] should provide a degree of incentive to commuters to use public transport, but the effect is very marginal. The completing services to the private car are seen, frequently as both relatively expensive and (more important) not sufficiently reliable.

---

20   Published in April 2004.
21   Signed 1 April 2004.
22   Proof of which the RIA indicated would be likely to be required by the Inland Revenue under the redesigned system.
23   Consultation, para. 23.
24   *Ibid.*, para. 24.
25   *Ibid.*, para. 25.
26   *Ibid.*
27   *Ibid.*, para. 26.
28   www.inlandrevenue.gov.uk.
29   Value Added Tax Act 1994, Sched. 8, Group 8, item 4(a).

In 1999, incentives were introduced for works' buses,[30] employer subsidisation of buses,[31] employer provision of cycling equipment[32] and the payment of approved passenger subsidies.[33] Further incentives were considered, but only minor palliatives were, in the event, brought into effect (in 2002), that is, up to six free meals per annum for employees cycling in to work[34] and the use of a works minibus for shopping trips (of up to ten miles) from work.[35]

It has, therefore, to be concluded that the government's scope for providing meaningful assistance in this field has run its course. This has, however, to be seen against the background of the Report of the Sustainable Development Commission issued in April 2004: *Shows Promise. But Must Try Harder*. This mentions a 'fundamental need ... for more sustainable transport and planning policies that reduce the need to travel where possible'.[36]

---

[30]   Income and Corporation Taxes Act 1988, s.197AA; Income Tax (Earnings and Pensions) Act 2003, s.242.

[31]   Income and Corporation Taxes Act 1988, s.197AB; Income Tax (Earnings and Pensions) Act 2003, s.243.

[32]   Income and Corporation Taxes Act 1988, s.197AC; Income Tax (Earnings and Pensions) Act 2003, s.244.

[33]   Income and Corporation Taxes Act 1988, s.197AE; Income Tax (Earnings and Pensions) Act 2003, s.233. For each of the above and the pre-1999 position, see J. Snape, 'Tax Law Aspects of Adapting and Operating Green Transport Plans' (1999), 1 ELR 92–124.

[34]   Income Tax (Exemption of Minor Benefits) Regulations 2002, S.I. 2002 No. 205, reg. 3. From 2003/04, the limit of six free breakfasts for cyclists on official cycle-to-work days was lifted: REV BN 04 (9 April 2003) , Income Tax (Exemption of Minor Benefits) (Amendment) Regulations 2003, S.I. 2003 No. 1434.

[35]   Income Tax (Exemption of Minor Benefits) Regulations 2002, S.I. 2002 No. 205, reg. 4.

[36]   *Ibid.*, para. 134. Paragraph 135 then goes on to describe the aviation situation as 'even more acute'. See, *ibid.*, paras 29, 74 and 132. See also para. 4.2.1.4 above.

Chapter 24

Business Taxes

24.1   Introduction

By contrast with the method of calculating the taxable emoluments of employees,[1] the profits of businesses are taxed on the basis of accounts drawn up under the best current practice.[2]

Any employer who incurred expenditure on the congestion charge or any workplace parking levies would therefore be able to deduct that expenditure in computing the profits of his trade. (Whether any employee whose vehicle was involved would be taxable on that expenditure is a different issue.)[3] Furthermore, where LATS are concerned,[4] corporation tax problems will not arise because local authorities are not treated as carrying on trades in relation to the performance of their statutory functions.[5]

Adjustments are made for certain matters specified by law, which fall within two general heads – those required in order to prevent the deduction of capital or personal expenditure,[6] and those required as a matter of public policy (for example, business entertainment or involvement in criminal activity).[7] The reader is referred to the outlines in Chapter 1 above and to the standard texts on the subject for the background to the provisions discussed below.[8]

24.2   Nature and role of capital allowances

For the purposes of computing trading profit, all capital expenditure has to be deleted.[9] Expenditure on the acquisition of, say, plant and machinery will only very rarely come within the scope of revenue deductions. When drawing up accounts, the appropriate deduction against revenue is obtained by estimating the useful life of the asset in question and amortising its cost over that period through the medium of depreciating its costs.

---

[1]   See para. 23.1 above.
[2]   Finance Act 1998, s.42(1).
[3]   See, however, para. 18.2 above in relation to congestion charge.
[4]   See para. 20.7 above.
[5]   Income and Corporation Taxes Act 1988, s.519(1). The same is intended to apply for VAT, although the position is rather complex: see Value Added Tax Act 1994, s.33.
[6]   See Income and Corporation Taxes Act 1988, s.74.
[7]   Income and Corporation Taxes Act 1988, ss.577, 577A.
[8]   See CCH Editions, *The CCH Tax Handbook 2003–2004* (Banbury: Croner CCH, 2003); John Tiley, *Revenue Law*, 4th edn (Oxford: Hart Publishing, 2000); and Lesley Browning et al., *Revenue Law: Principles and Practice*, 22nd edn (London: Lexis Nexis Tolley, 2004).
[9]   Income and Corporation Taxes Act 1988, s.74(1)(f).

Depreciation is not expenditure as such and, accordingly, had to be written back in order to accord with the concept of ascertaining a profit by deducting expenditure from receipts, as comprehended by the late nineteenth-century judiciary. Since the Second World War, both Labour and Conservative Governments have made provision for series of capital allowances (or, at one stage, investment grants) in order to redress the balance.

Capital allowances take two main forms: the possibility of an initial allowance to give accelerated relief for new expenditure and a writing down allowance for the balance, and the balance between the two changes regularly.

24.3   Motor cars

As indicated in the previous chapter,[10] the provision of company cars to virtually all and sundry started off as a means of propping up the British motor industry. The only restriction imposed on employers was the imposition of a cost limit, currently £12,000, above which the cost of cars was not included in the general 'pool' of expenditure by reference to which 25 per cent reducing balance allowances were generated, and is pooled separately subject to an annual cap of £3,000.[11]

From 2002, this cap does not apply[12] where a 100 per cent first year allowance is available because it is a new car that is either electrically-propelled or has low carbon dioxide emissions.[13]

24.4   Environmentally-friendly equipment

From 2001, 100 per cent allowances have been available for energy-saving plant or machinery that is of a type specified by the government.[14] In 2002 and 2004, further technologies were added to the list. In 2003, certain water technologies were awarded the same treatment.[15]

In August 2003, in *Economic Instruments to Improve Household Energy Efficiency: Consultation Document on Specific Measures*,[16] it seemed that the only realistic course of action might be to request European Commission consent to extend the lower (5 per cent) rate of VAT to the purchase of energy-efficient equipment. It was noted, however, that this would require the unanimous consent of the other Member States.

---

[10]   See para. 23.2 above.
[11]   Capital Allowances Act 2001, ss.74(2)(b), 75(1).
[12]   *Ibid.*, s.74(2)(c).
[13]   *Ibid.*, s.45D.
[14]   *Ibid.*, ss.45A–45C.
[15]   See para. 21.3.1 above (see also Capital Allowances Act 2001, ss.45H–J).
[16]   HM Treasury and Defra, following-on an early Consultation in July 2002.

24.5　Urban regeneration

The taxes considered previously in this chapter need to be distinguished from the fiscal changes made as a result of the report of Lord Rodgers' Urban Task Force, *Towards an Urban Renaissance*. The linkage between these changes, which were designed to cost the Treasury as little as possible, and the 100 per cent first year allowances for generating equipment designed to sit alongside the 80 per cent reduction in climate change levy for those with IPPC problems who entered into climate change agreements, is purely coincidental.[17]

The point has, furthermore, to be made that, in the course of the Marrakesh and Johannesburg Environmental Summits, the focus of the 'sustainable development', which the participant governments participating discussed, was shifted, subtly but discernibly, from the protection of the environment to the assistance of the poor.[18]

24.5.1　Clean-up expenditure by purchasers

Where contaminated land is purchased and the purchaser carries out remedial works, he can, in principle, obtain tax relief on a notional 150 per cent of expenditure incurred after 11 May 2001.[19] Relief is only available for acquisitions as trading stock or as a capital asset of a trade or a Schedule A business.[20] The limitations within those categories are, however, significant, and reduce very materially the scope of this concession:

1　Perhaps understandably, it is not available where the purchaser, or a person connected with him, created the contamination.[21]
2　Also understandably, the relief should not be available if the expenditure was grant-aided. But this particular disqualification also extends to situations in which the purchaser obtains an indemnity from the vendor.[22]
3　It is not available if the expenditure is deductible as a trading expense,[23] thus, it seems, eliminating developers from the scope of beneficiaries. The Inland Revenue have accepted, however, that land that is trading stock can be within the relief.[24]
4　The relief is not available if the expenditure qualifies for capital allowances.[25]

---

[17]　That is, in current British political parlance, it is pure 'spin'.
[18]　Which would, indeed, accord with the British position: see Chapter 5 above.
[19]　As to this relief, see Malcolm Gammie, QC, and Jeremy de Souza, *Land Taxation*, Release 53 (London: Sweet and Maxwell, looseleaf), para. C1.466B.
[20]　That is, acting as landlord.
[21]　Finance Act 2001, Sched. 22, paras 1(5), 31(3),(4).
[22]　*Ibid.*, Sched. 22, paras 2(6), 8.
[23]　*Ibid.*, Sched. 22, para. 1(4)(a).
[24]　*Ibid.*, Sched. 22, para. 12 (see *Taxation*, 7 March 2002, p. 546).
[25]　*Ibid.*, Sched. 22, para. 1(4)(b).

24.5.2   Flats above shops

From the same date,[26] 100 per cent capital allowances were made available against rental income for the conversion of space over shops into small self-contained flats, provided that the rental level was modest.[27] The 'small print' in the legislation is, however, such that its effect will be limited.[28]

24.5.3   Stamp duty exemption in disadvantaged areas

It was intended to supplement these measures with total exemption from stamp duty for purchases in disadvantaged areas.[29] However, when it came to be implemented, a value cap of £150,000 was imposed.[30]

The government subsequently proposed that European Commission consent should be sought with a view to this being lifted for non-residential properties, and implemented enabling legislation in 2002 to facilitate this in due course.[31] Consent was duly granted on 21 January 2003,[32] albeit on terms that the scheme should only last until 31 December 2006; that, recognition having been given to the limit nature of the subsidy, further aid was not to be combined with it; and, note having been taken of the serious environmental problems involved with the redevelopment of brown-field sites, a series of reports were to be submitted to the Commission on the effect of the subsidy on physical regeneration. The relief was brought into effect on 10 April 2003.[33]

24.6   Private residential landlords

From 6 April 2004 until 5 April 2009,[34] landlords who pay income tax[35] are able to deduct from rental income up to £1,500 per building containing one or more dwelling

---

[26]   That is, 11 May 2001.

[27]   Capital Allowances Act 2001, s.393E.

[28]   *Ibid.*, ss.393A–393V (as to this relief, see Gammie and de Souza, *op. cit.*, para. C1.466C).

[29]   Finance Act 2001, s.92.

[30]   Nonetheless, 35,000 houses came within the relief in the first year (see Economic Secretary to the Treasury (Mr John Healey, MP), HC written answer to PQ, 7 November 2002, vol. 392, No. 203, col. 759W). See de Souza and Snape [2002] BTR 57–64.

[31]   Finance Act 2001, ss.92A, 92B.

[32]   Up to that date 758 commercial transactions had been exempted: *Hansard*, HC, 26 February 2003, vol. 400, No. 54, col. 574W.

[33]   Stamp Duty (Disadvantaged Areas) (Application of Exemptions) Regulations 2003, S.I. 2003 No. 1056. Finance Act 2003, Sched. 6, extended this to stamp duty land tax.

[34]   A limitation that only became clear with the publication of the Finance Bill.

[35]   Press Release REV BN 31. Companies resident in, or trading through a permanent establishment in, the UK pay corporation tax and will not be able to participate in this relief. Individuals, trustees and non-resident companies not trading in the UK pay income tax on rental income.

houses[36] of capital expenditure on the installation of loft or cavity wall insulation in let dwelling houses.[37]

24.7 Emissions trading scheme corporation tax and VAT treatment

Whilst no official statement has been made on this issue, the tax treatment of the receipts and payments from UK ETS[38] should not present problems. Although, under Finance Act 1998, section 42(1), taxed profits have to be ascertained primarily by reference to best UK GAAP, there is an overrider where an issue of law arises. And it is generally considered that the critical distinction between capital and income has still to be determined under this overrider.

The generally accepted test for whether a deduction is on revenue account is that laid down by the House of Lords in *British Insulated and Helsby Cables Ltd v. Atherton*,[39] that is, whether an enduring advantage is being obtained from the expenditure. The test for whether a receipt is on revenue account is similar.

It follows from this that the premium paid to direct participants in the 2002 UK ETS[40] would have been a capital receipt, subject to capital gains tax, rather than to income tax. But, because the rights were not generated out of an existing asset, the only deductions available would have been the professional fees incurred in entering the process. The net amount would have been brought into charge at the recipient's marginal tax rate.

Other receipts and payments are, however, likely to be of a revenue nature because the rights sold or acquired would be confined to the current compliance year. It is clear that this is the case for trading by both direct participants and those involved in climate change agreements.[41] It also seems difficult to form any other conclusion in relation to dealings in Green Certificates, irrespective of whether by one-way conversion into UK ETS or by informal trading between power stations.[42] In principle, it seems that a revenue deduction should arise where, instead of trading Green Certificates, a power station opted to make the full alternative payment to

---

[36] See Energy Savings Items (Deductions for Expenditure, etc) Regulations 2004, S.I. 2004 No. 2664, regs 1(2) and 2, of which a draft was published during the Committee Stage of the Finance Bill. With domestic energy consumption one of the two areas in which the UK was regressing against its Kyoto targets, the House of Commons' Environmental Audit Committee, in its Tenth Report of Session 2003–04, *Budget 2004 and Energy*, HC 490, published on 11 August 2004, expressed disappointment that it had not been possible to provide a more significant package of measures than this (and two small VAT) concessions and, in the light of the savings to be expected from caving wall insulation, that the Energy Saving Trust had not been involved more fully (*ibid.*, paras 63, 66).

[37] Income and Corporation Taxes Act 1988, ss.31A, 31B.

[38] See Chapter 20 above.

[39] (1925) 10 TC 177.

[40] See para. 20.2.

[41] See para. 20.4.

[42] See para. 20.6.

Ofgem before the deadline.[43] Where the obligation had not been balanced by the deadline and voluntary payments to other suppliers were made by arrangement with Ofgem, the position is likely to be the same.[44] However, in the case of uncooperative defaulters, penalties under any enforcement action taken[45] might not qualify for a tax deduction.[46]

The position may be different for EU ETS because the plant's allocation of allowances appears to be permanent.[47] In this event, the allocation would be a non-wasting[48] asset for capital gains tax, albeit without any acquisition cost.[49] The possibility that, under international accounting practice, recipients from the initial allocation on 1 October 2004 may have to attribute an acquisition value – on the analogy that the allocation is a government subsidy[50] – would not mean that there could be a taxable event for this purpose. Most plant operators will be subject to corporation tax and, for them, there will be nothing to index and so each disposal will be taxable in full at their marginal rate.[51] Sole traders, trusts and partnerships of individuals will, however, be entitled to claim business asset taper relief once the allocation has been held for a year.[52] Clearly total disposals are only likely to take place on the sale or closure of a plant. More frequent will be partial realisations of the allocation following either downsizing or (hopefully more frequently) technical upgrading. These will constitute part disposals for CGT purposes[53] or partial realisations in the event of the corporation tax intangible fixed asset code applying.[54]

Where a 'permanent' allocation is acquired, this will clearly be capital expenditure under the *British Insulated and Helsby Cables* test and therefore disallowed as a trading expense for income tax computation purposes.[55] But where the acquirer is subject to corporation tax, if it is classed as an *intangible* asset for accounting

---

43   Ofgem's first *Renewables Obligation Annual Report*, published on 27 February 2004, reveals that, out of 66 companies, 13 opted for this course of action.

44   As happened in five cases. The accounting period in which relief would be obtained would, of course, be a matter for accounting practice by reference to SSAP 17.

45   It was being considered in two cases.

46   See *IRC v. Alexander von Glehn & Co Ltd*, (1919) 12 TC 232.

47   See Chapter 28 below.

48   Taxation of Chargeable Gains Act 1992, s.44.

49   There appear to be similarities to milk quota, which was originally only transferable with land. This did not, however, enable farmers to amalgamate it with that land for capital gains tax base value purposes (see *Cottle v. Coldicott*, (1995) SpC 40). It would, of course, be inconsistent with the concept of trading permits for there to be a land linkage for any form of emissions trading scheme.

50   See Perry, ACCA's Corporate Sector Review, April 2004, pp. 1–3, and Casamento, *Accounting & Business*, April 2004, pp. 36–9. The authors are grateful to Judith Dahlgreen, University of Leeds, for drawing this material to their attention.

51   Because there will have been no acquisition expenditure, it seems that this will also be the case if the allocation is regarded for accounting purposes as an *intangible asset*, subject to tax under Finance Act 2002, Sched. 29.

52   Taxation of Chargeable Gains Act 1992, Sched. A1, para. 5(1A).

53   *Ibid.*, s.42.

54   Finance Act 2002, Sched. 29, para. 19(2).

55   Income and Corporation Taxes Act 1988, s.74(1)(f).

purposes[56] its cost may either be written down on an accounting basis[57] or, by election,[58] on a 4 per cent reducing balance basis.[59]

However, if the allocation is traded on a 'current compliance period only' basis,[60] then both the receipt and the expenditure will be treated as of revenue.[61]

For value added tax purposes, a disposal of either UK ETS or EU ETS, or any derivative related to them, will be a supply of services[62] chargeable at the standard rate. Acquirers will be able to recover this in full by way of input tax unless they are partially exempt, when a proportion will be disallowed either under the residuary formula[63] or a special scheme.[64] Where both parties are in the UK, no output tax accountability problem arises because this is clearly the responsibility of the vendor. Where the counterparty is in another EU Member State, VAT will be chargeable in accordance with the law of the vendor's Member State.[65]

24.8   Direct tax treatment of environmental trust contributions

One of the reliefs available in relation to landfill tax is to make contributions to environmental trusts.[66] Under the scheme, it is of the essence that the reduction of landfill tax is limited to 90 per cent.

Logically, the question should, therefore, arise as to whether the 10 per cent additional liability falls to be deducted from the operator's trading profits under

---

[56] Finance Act 2002, Sched. 29, paras 2(1), 3(1).

[57] *Ibid.*, Sched. 29, para. 9.

[58] *Ibid.*, Sched. 29, para. 10.

[59] *Ibid.*, Sched. 29, para. 11.

[60] Assuming this to be permissible. It will not be easy to integrate EU ETS with UK ETS without such a facility.

[61] Subject, however, to the correct accounting treatment in the event of the 'totality' having been acquired, in which case Finance Act 2002, Sched. 29, para. 19(1)(b) may produce a different computation.

[62] Value Added Tax 1994, s.5(2)(b); Sixth VAT Directive, Arts 2(1), 5(1) (see para. 1.2.1.4 above).

[63] Value Added Tax Regulations 1995, S.I. 1995 No. 2518, reg. 101(2)(d); Sixth VAT Directive, Arts 17(5), 19(1) (see para. 1.2.1.4 above).

[64] Value Added Tax Regulations 1995, reg. 102. It should be noted that such arrangements may not be backdated.

[65] Value Added Tax Act 1994, s.7(10). A reverse charge under *ibid.*, s.8 would only arise if it could be said that the allocation was a *similar right* to a *copyright, patent, licence or trademark*: see *ibid.*, Sched. 5, para. 1; Sixth VAT Directive (see para. 1.2.1.4 above), Art. 9(1)(e). Initially, Customs' view did not accord with this analysis, but, from 25 October 2004, they have agreed to follow the practice of other Member States, which had been following that interpretation (see *Business Brief* 28/04 (25 October 2004)). That *Business Brief* also indicated that brokers' services needed to be evaluated differently. Where 'sufficient intermediation is performed', the supply of those will take place in the Member State where the 'customer' belongs. But where the broker's service is of 'mere introduction', the supply will take place where the 'introducer' belongs.

[66] See para. 21.3.2 above.

Income and Corporation Taxes Act 1988, s.74(1)(a), which provides that deduction is only permitted where the liability is incurred *wholly and exclusively for the purposes of the trade*.[67] However, this was not how the Inland Revenue saw matters, at least initially. Their view was that the problem was the deductibility of the 90 per cent.[68]

Nonetheless, operators seem to have been very keen to take up their full quota and the authors surmise that the problem may have been got round, in practice, by presenting the whole operation as an advertising exercise (rather like sports sponsorship). For example, the Annual Report and Accounts for the year to 27 March 1999 of Shanks and McEwan Group plc revealed that 100 'awards' had been made and £14m distributed under the scheme since the inception of the rebating arrangements. On p. 4 of the report, the following observation was made, that: 'the success of these initiatives brings considerable credit to the Group'.

If the operator is a close company, a further fiscal problem can arise unless charitable recipients are selected.[69] Because the intention is to confer a *gratuitous benefit*, exemption cannot be available under Inheritance Tax Act 1984, s.10(1). An immediate inheritance tax charge may therefore arise because a company cannot make a potentially exempt transfer.[70]

---

[67]   See Malcolm Gammie, QC, and Jeremy de Souza, *Land Taxation*, release 53 (London: Sweet and Maxwell, looseleaf), para. E3.008A.

[68]   *Tax Bulletin* No. 23 (June 1996).

[69]   Category (f) recipients – that is, the provision of financial, administrative and other services confined to other environmental trusts – are, in particular, unlikely to be in this category.

[70]   Inheritance Tax Act 1984, s.3A(6): see Gammie and de Souza, *op. cit.*, para. E3.008A.

Division 2
The Provisions in Operation

Chapter 25

Policies in Practice (2)

Most of the 'green' tax changes addressed in Chapters 22 to 24 have been of very short duration. Their success (or otherwise) has not therefore been easy to judge independently.[1]

Road transport has been the subject of a number of such initiatives. Following on the apparent success of the Lawson differentiation of fuel duty in favour of unleaded petrol, the present government has introduced more targeted measures both in relation to VED (from 2003)[2] and excise duty (from 2004),[3] but both have to be seen against the background of the abandonment of the 'escalator' as a result of the organised protest boycott in 2000. One major area of uncertainty which has still to be resolved is the position of LPG, in relation to which considerable investment appears to have been made by manufacturers, only to have its environmental properties questioned.[4] The relaxed capital allowances regime introduced for environmentally friendly cars in 2002 may have some effect in the longer term (once the viability of electric cars has become clearer),[5] but this may be difficult to measure because the Inland Revenue do not ask taxpayers for declarations by reference to specific allowances.[6]

Similarly the official investigation into the use of experimental fuels for commercial vehicles,[7] which seems likely to be overtaken by the introduction of a lorry road user charge,[8] may not bear fruit.

Turning, then, to the employee benefits aspects, the 2003 car scale reform (which has been declared a success by the Inland Revenue)[9] might have a major long-term effect.[10] The provision of income tax relief on employer assistance to come to work other than by car has, of necessity, been rather limited and appears to have run its course.[11]

Finally, it is necessary to address the incentives made available in relation to premises. Capital allowances for environmentally friendly equipment[12] may be having some effect, but the cost of these to the Exchequer cannot be ascertained because the Inland Revenue do not keep their statistics in a manner which permits

---

[1]   In which context, it should be noted that in only one case has there been an official declaration of success: see para. 23.2 above.
[2]   See para. 22.2.1 above.
[3]   See para. 22.2.2 above.
[4]   See para. 22.2.2 above. See also ENDS Report 350 (March 2004), pp. 28–30.
[5]   See para. 24.3 above.
[6]   See ENDS Report 349 (February 2004), pp. 29–31.
[7]   See para. 22.3 above.
[8]   See para. 22.4 above.
[9]   See para. 23.2 above.
[10]   But see ENDS Report 352 (May 2004), pp. 12–13.
[11]   See para. 23.3 above.
[12]   See para. 24.4 above.

them to measure it.[13] Industrial sources suggest, moreover, that the success which there has been in introducing, respectively more efficient boilers, driving motors and fluorescent lighting was not affected by the enhanced capital allowances available.[14] The 2002 first year allowance for the conversion of flats above shops[15] and the 2004 revenue deduction for residential landlords' expenditure on insulation[16] have been drawn in such a way as to make it difficult for, respectively, almost all and some potential users to qualify. The corporation tax relief supplement introduced in 2001 for purchasers who incur clean-up expenditure on contaminated land has been the subject of sufficient academic comment[17] to lead the authors to conclude that use is being made of it, albeit perhaps more in the field of asbestos removal than 'brown field' site restoration.[18]

With one exception, the jury must, therefore, be regarded as still out on the concessions discussed in the immediately preceding chapters.

The stamp duty and stamp duty land tax disadvantaged areas concessions[19] are extremely difficult to justify on environmental grounds, not least because the discrimination in favour of commercial properties is difficult to defend. The availability of relief is determined by the average income level of residents in the areas in question, rather than that of those who work there. This has led to commercial conveyancing transactions being exempted from stamp duty land tax in the London Docklands office complex known as Canary Wharf, the street in Leeds in which the up-market ladies clothes shop, Harvey Nicholls, has its branch, the site of one of the northeast's largest retailing complexes and the area where a number of leading private client solicitors' firms have their offices, Lincoln's Inn Fields. Suffice it to say that the justification for this end result can only be regarded as obscure.[20]

---

13   See ENDS Report 349 (February 2004), pp. 29–31.
14   *Ibid.*
15   See para. 24.5.2 above.
16   See para. 24.6 above. See also ENDS Report (August 2004), p. 32.
17   Judging by the number of articles in journals, for example, Sheridan, *Law Society's* Gazette, 11 May 2001, p. 34; O'Keefe, *Taxation*, 21 August 2001, p. 290, Davidson and Hallpike, *Property Law Journal*, 17 September 2001, p. 13, Plager, *Taxation*, 11 October 2001, pp. 32–8, Jamieson, *Taxation*, 21 February 2001, pp. 498–500, Woodward and Yaard, *The Tax Journal*, 18 August 2003, pp. 13–14, *Taxation* Query T16,279, 18 September 2003, p. 680, Beaumont, *Taxation*, 25 September 2003, pp. 694–95, and Oliver and Butlin, *Estates Gazette*, 22 November 2003, pp. 146–8.
18   In respect of which it is of the essence that the purchaser has not obtained a cost indemnity from the vendor. The uncertainties are such that a prudent professional would be unlikely to advise a purchaser against taking such an indemnity if one could be negotiated.
19   See para. 24.5.3 above.
20   But, to be fair, the same comment could not be made about the proposals, to be put to the European Commission, in *Capital Allowances: Renovation of Premises in Disadvantaged Areas*, although these have no environmental content.

PART IV
PROSPECTS AND
NEW DIRECTIONS

Chapter 26

Environmental Taxes and the Tax Base

26.1   Introduction

It now falls to us to consider both the future prospects for exploiting environmental policy as a source of government funding and the continuing viability of the historic UK tax base. In this chapter, consideration is given to the concerns which must be being addressed by the Chancellor of the Exchequer and his advisers. In Chapter 28, the EU ETS, against which all environmental developments will, in the future, need to be judged, is analysed at length. In Chapter 29, some tentative conclusions have been drawn as to the position at which we found ourselves at the end of December 2004.

Although environmental taxes have been on the UK statute book since 1996, as far as revenue raising capability is concerned, their potential has not yet been seen. However, following the publication of *Waste Not, Want Not*, this could be about to change.[1] The catalyst is New Labour's need to raise enormous sums in taxes to pay for improvements in public services, the seriousness of continuing funding problems for education,[2] health[3] and transport[4] having become apparent during the first half of 2003, and (with the assistance of having to pay for the forces in Iraq) acute by autumn 2004.

Despite the relative novelty of environmental taxes, with the advent of a combination of the European Single Market[5] and internet-based international e-commerce, the UK (and, indeed, other) governments may come (and perhaps even need) to rely on them

---

[1]   See para. 21.4.1 above.

[2]   Witness the university tuition fees issue, which nearly brought down the Blair Government.

[3]   Where the vast amount of money raised by the 1 per cent NIC surcharge (see para. 21.2 above) does not seem to have made a material effect on the published performance statistics for the National Health Service.

[4]   Where Railtrack (a listed company) has been replaced by Network Rail, a 'not for profit' company with only a nominal capital, and which is therefore reliant on loan finance. The bondholders of the latter subscribed on the basis of a government guarantee. Others dealing with the company are, presumably, only doing so on the assumption that, if things went wrong again, the UK Government would have to bail the company out.

[5]   To which must be coupled the facility with which the ECJ has found it possible to undermine Member States' domestic tax base by reference to the freedom of establishment principle in Art. 43 (ex 52) of the European Treaty. This has led to it being concluded by the retiring chairman of the Law Society of England and Wales's Tax Committee that the UK Government's August 2003 consultation on the future of corporation tax revealed concern at the freedom of business movement underlying the decisions of the ECJ (see Troup, *Financial Times*, 20 August 2003, p. 17). See also para. 26.4 below.

for an increasing proportion of their revenue.[6] In the *Economist*'s 'Globalisation and Tax' survey in 2000, there was made the telling comment:[7]

> And the harder it gets to tax mobile people and businesses, the bigger the burden that will have to be borne by the immobile. Land taxes, which used to be one of the most important revenue-earners, may regain their former pre-eminence. Consumption taxes on purchases of physical goods and services which the taxman can track may have to rise. In future, local environmental taxes – say pollution taxes or road prices – may well come to look attractive to tax starved governments.

This seems to be the pattern developing in the UK. Almost the last fling of Prime Minister John Major's Conservative Government was the introduction of landfill tax,[8] the saving to be reflected in a reduction in employers' NICs. Labour went further, first increasing and then reformulating stamp duty on property transactions, then introducing climate change levy and aggregates levy, having made available to local authorities the powers to levy road user and workplace parking charges.

The theme was continued in relation to non-environmental land taxes. Partial Regulatory Impact Assessment, *Modernising Stamp Duty*, published the same day as Finance Bill 2003,[9] said that stamp duty land tax, which was to take effect on 1 December 2003, was intended to be 'revenue raising through improved compliance and redefinition of the scope of the charge'[10] and suggested that 5 per cent of commercial property deals could be affected, with a slight reduction in transactions anticipated.[11]

The UK has not joined the euro, but if it were to do so, as its Prime Minister, the Rt Hon. Tony Blair, MP, still intends,[12] then the associated requirement to comply with the Stability Pact[13] is likely to put pressure on the UK Government to increase its yield from taxation rather than resort to increased borrowing.

---

[6]   There may be a repetition, where there is the option of additional environmental controls at some future date of the dilemma which arose in July 1968 when Customs' advised the Wilson Government a proposed tobacco advertising ban proposed would have jeopardised the government's intention to raise an additional £30m from tobacco duty that year, with subsequent consequences to the UK economy. This was revealed upon the release of public archives in July 2003 (see Milne, *Financial Times*, 25 July 2003, p. 3).

[7]   Bishop, *Economist*, 29 January 2000, p. 18.

[8]   See Finance Act 1996, ss.39–68, and Sched. 5. The last property-based measure was the tightening up of the VAT option to tax in Finance Act 1997, s.37.

[9]   16 April 2003.

[10] *Modernising Stamp Duty*, para. 28.

[11] *Ibid.*, para. 29.

[12] Although this project has had to be put on the back burner, not only as a result of unfavourable economic conditions on the continent of Europe, when compared with the UK, but also because the government has had to promise that the Constitutional Treaty will be submitted to a referendum which it might be very hard pressed to win.

[13] Of which the UK was alleged to be in breach in the autumn of 2004.

26.2   The consolidation of the tax base

The UK is not the only country to be increasing its reliance on property based taxation in the face of the internet challenge to compliance.

26.2.1   *Property transfer and planning permission taxes*

The introduction of stamp duty land tax with effect from 1 December 2003, allegedly to close loopholes in the old regime, could well presage an increase in the rates over and above those proposed for leases.[14] It needs to be remembered in this context that it was the Labour Government who introduced the concept of value banding shortly after taking office in 1997. With the introduction of stamp duty land tax, they took the opportunity to increase the tax on lease rents[15] and their new scheme for the taxation of partnership transactions[16] will also result in an increased tax take.[17]

Furthermore, the suggestions made during 2004 for residential planning consent and flood plain development levies[18] would, if implemented, be additional property-based taxes.

26.2.2   *The local government tax base*

Council tax and uniform business rates, which are the bedrock of local government finance in mainland Britain, are also property based. Indeed they have built into them fall-back collection provisions against, respectively, the owner and the (physical) occupier which have the effect of providing an element of guarantee in favour of collection as against the old community charge (or 'poll tax') for which default rates were high in some areas. It is worthy of note, in this context, that, in its Ninth Report of Session 2003–04, entitled *Local Government Revenue*, published on 16 July 2004,[19] the House of Commons ODPM: Housing, Planning, Local Government and the Regions Committee considered both congestion charges (which had only been implemented in two cities) and workplace parking levies (which had been implemented nowhere) in the context of local authorities' future income-raising potential.

---

[14]   Finance Act 2003, Sched. 5.
[15]   Albeit apparently ameliorated by the deduction of the threshold in all cases under changes announced on 20 October 2003. However, when the draft statutory instruments were published on 30 October 2003, some of the changes made appeared to bring agreements for lease made before 11 July 2003, and where possession had been taken between that date and 30 November 2003, within the charge to tax while still within their original contractual term if there was a variation in their terms after 30 November 2003. This problem appears to have been shut down under the Budget Resolutions which have been passed, with effect from 17 March 2004.
[16]   Draft clauses were published for consultation the same day as the revised rental duty scheme. An amended version of them was enacted by Finance Act 2004, Sched. 41.
[17]   See also Thomas and Keenay, *The Times*, 4 November 2003, Law 4.
[18]   See paras. 1.2 above and 27.2 below.
[19]   HC 402.

26.2.3   National environmental taxes

The national environmental taxes[20] are similarly based, and especially so following the introduction of secondary liability (that is, deemed guarantors) in relation to some of them.[21] An example of this approach, albeit in an unrelated context, is the criminal liability imposed in 2002 for the 'manager' of premises where unmarked tobacco products are sold.[22] It is, furthermore, significant that such draconian legislation was introduced in the light of the difficulties faced by Customs in preventing (excise duty) smuggling, following the introduction of the single market for VAT purposes.

Indeed, the same line of thinking may lie behind the apparent persistence of HM Customs and Excise in seeking the introduction of a pesticides levy,[23] the disastrous effect such an impost would have on British farming (as especially the apple growers) notwithstanding.[24] It may turn out to be very fortunate for crop farmers that the postulated fertiliser tax was seen off on environmental grounds very early in the day before bureaucratic time had been 'invested' in consultations and draft legislation.

Similar concerns in relation to small quarries and those outside the core south eastern region were insufficient to prevent the introduction of the aggregates levy, after Customs had 'invested' in a number of consultations, including two rounds of draft legislation.[25] There cannot be said to be any economic case for this tax, the export rebating provisions of which alone undermine the alleged justification of supporting 'environmentally friendly quarrying'.

It seems difficult to avoid the conclusion that environmental taxes, and the associated trading markets in allowances and (in effect) permits, will increasingly become a revenue producing milch cow. Indeed, the very existence of such markets will enable taxes to be imposed on them in the same way as proved possible with the financial futures markets through (for instance) stamp duty reserve tax. Furthermore, in the 2004 Pre-Budget Report,[26] it was revealed that, out of the future increases in landfill tax, local authorities would be insulated against the tax increases for 2005/06 (which was expected to be an election year),[27] and the allocation during the same year of £45m from the 'reform' of the Waste Minimisation and Recycling Challenge Fund.[28] There was also to be a series of subscriptions[29] to the Business Resource Efficiency and Waste ('BREW') programme to be launched in April 2005, with a funding allocation of £43m during its first year, £95m for 2006/07 and £146m for 2007/08.[30]

---

20   See Chapters 13–16 above.
21   See para. 16.16 above.
22   Tobacco Products Duty Act 1979, s.8H(8).
23   See para. 21.9.5 above.
24   See para. 21.5 above.
25   See para. 21.10 above.
26   Entitled *Opportunity for All: The Strength to Take the Long-term Decisions for Britain*, Cmn 6408, 2 December 2004, www.hm-treasury.gov.uk.
27   Para. 7.55. In relation to which, as they are the biggest users, one might well ask what effect the increase in the landfill tax rate was, in consequence, expected to have.
28   Para. 7.61.
29   Para. 7.55, which only quantified that for 2007/08.
30   Defra News Release 468/04, dated 22 November 2004 (www.defra.gov.uk), which headlined the total of £284m!

26.3 The future of non-property-based taxes in the internet world

26.3.1 The non-resident

The most notable item in Budget 2003 may well turn out to have been the tacit acceptance that bringing long-term residents who retained foreign domicile within the income and capital gains tax systems on a world wide receipts basis remained an improbability. Labour's attempt to do this was not, of course, the first, the Conservatives had also tried, but had, apparently, to capitulate to a coalition of Arab oil royalty recipients and Norwegian and Greek shippers.

26.3.2 The avoidance/evasion borderline

Although some ministers and Inland Revenue officials seem to regard the terms as interchangeable,[31] it must be borne in mind that tax *avoidance* is, in principle, legal at any rate where the taxpayer has put his cards on the table, but that *evasion* is always illegal and normally a serious criminal offence. The distinction is, of course, a critical one in the context of money laundering, because it is of the essence that a criminal offence has been committed.

On the domestic front, the government's campaign against single worker companies (known as 'IR 35'), the result in many cases of the non-availability of full time employment for professionally qualified and skilled people in the market place (including its extension to nannies in 2003),[32] as well as the incredibly wide scope of the Inland Revenue's interpretation of professional firms' prospective money laundering obligations,[33] are indicative of the fact that the *black economy* is considered to be getting out of control. Clearly the end product of such a development would be a serious reduction in the yield from both income tax and NICs (the latter being, as indicated above, the lynchpin of the proposed funding of increased expenditure in the public services).

Furthermore, while it remains to be seen how effective the e-commerce VAT arrangements introduced by the EU from 1 July 2003 will be, it is hardly a secret that enforcement is the major potential problem. The 'carousel fraud' based on the formation of the single market for VAT had reached such proportions that, in 2003, the UK Government sought legislative amendments which could have the effect of making those in the same trading chain potentially liable for frauds committed by others.[34] This drew severe criticism on ECHR grounds from, inter alia, the House

---

[31] This development appears to have arisen in parallel with the use of 'evasion' in the English translation of EU Directives to cover both concepts. The retention of the distinction has not been assisted by the adoption by the ECJ in cross-border VAT cases of the purposive construction concept known as *fraus legis* in the Netherlands and *abus de droit* in France (see *Diamantis v. Greece*, C–373/97, [2001] 3 CMLR 41, and *Emsland-Stärke v. Hauptzollamt Hamburg-Jonas*, C–110/99).

[32] Finance Act 2003, s.136.

[33] *Working Together 13* (June 2003): see www.inlandrevenue.gov.uk.

[34] Value Added Tax Act 1994, s.77A and the revised Sched. 11, para. 4(2).

of Lords' Economic Affairs' Sub-Committee on the Finance Bill.[35] However, a subsequent Court challenge on these grounds failed,[36] although it did succeed in having the issue of the validity of the provisions in question under Community Law referred to the ECJ for expedited determination, albeit subject to Customs receiving encouragement from a majority of the Court of Appeal to exercise their statutory powers in the meantime.[37]

26.3.3   The losing battle against tax avoidance

26.3.3.1   The historic position

The traditional judicial approach to the interpretation of tax legislation in the UK has been a literal one,[38] well illustrated by the following famous quotation from Rowlatt, J. in *Cape Brandy Syndicate v. IRC:*[39]

> One has to look merely at what is clearly said. There is no room for any intendment. There is no equity about a tax. There is no presumption as to a tax. Nothing is to be read in, nothing is to be implied. One can only look fairly at the language used.[40]

The high water mark of this approach was the well-known pre-War case in which the Duke of Westminster was enabled to pay his gardener by deed of covenant.[41] Since 1981, however, the UK courts have been taking a less open handed attitude to tax avoidance. The background to this can perhaps be summarised as follows:

1   Since 1960, UK tax law has become increasingly complex, as government after government enacted specific anti-avoidance provisions targeted against identified loopholes in the statutory code; leading to
2   taxation becoming an increasingly specialist subject,[42] in turn leading to
3   the creation of large tax departments in professional firms,
4   whose growth[43] had become dependent by the end of the twentieth century upon the initiation of work through the marketing of artificial schemes.

---

35   HL 109, ordered to be printed 10 June 2003: Chapter 5.
36   *R (on the application of Federation of Technological Industries and others) v. C & E Commrs and Attorney-General,* [2004] STC 1008.
37   *C & E Commrs and Attorney-General v. Federation of Technological Industries,* [2004] EWCA (Civ) 1020.
38   *Ayrshire Employers Mutual Insurance Association v. IRC,* (1945) 27 TC 331.
39   (1920) 12 TC 358.
40   *Ibid.,* at 366.
41   *IRC v. Duke of Westminster,* (1935) 19 TC 490.
42   In the mid-1960s, it was still quite often possible for the audit manager at one of the large accountancy firms to compute the statutory adjustments to the accounting profit. By the end of the century, even tax departments were sub-departmentalised, with the ability to take an overall view increasingly limited to very senior people and specialist Counsel.
43   A prerequisite for partners in the relevant department under the normal criteria for profit distribution in large professional firms in the 1990s.

The critical case relating to tax avoidance in the UK[44] was *W T Ramsay Ltd v. IRC*.[45] Under what is called the *Ramsay Principle*, the Courts may nullify a tax advantage sought to be obtained where there is a *pre-ordained series of transactions* or a *single composite transaction*, having one of the following characteristics:

1   circularity, such as (in *Ramsay* itself) the creation of two 'debts', one of which increased in value but was considered for technical reasons not to generate a profit for CGT purposes and the other creating a 'loss', the House of Lords holding that the latter was not a *real* one; or

2   having steps inserted which have no commercial purpose other than the avoidance of tax, such as (in the case of the sale of a company) arranging for it to be taken over by an offshore company through a share exchange, with the latter proceeding with the cash sale to the original purchaser.[46]

This doctrine is not a new one, merely a rule of statutory construction[47] evolved from previous case law. This limitation has, however, enabled a line to be drawn in a number of cases.[48]

As a rule of statutory construction, it became clear fairly early on that the Courts were likely to adopt a more broad brush approach where the statutory word was vague – such as *loss*[49] or *disposal*[50] rather than an ordinary English word such as *payment*. Indeed, when the House of Lords was confronted with the latter in 2001, in *Macniven v. Westmoreland Investments Ltd*,[51] Lord Hoffmann drew the distinction between wording for which it was appropriate to adopt a normal (alias *legal*) construction – when the *Ramsay* doctrine would be unlikely to be applied – from that for which a *commercial* meaning was more appropriate. The second category of case was clearly more vulnerable to the *Ramsay* approach.

In subsequent cases, the lower courts in the UK and those in other jurisdictions found this distinction hard to operate in practice, and the issue was therefore clarified

---

44   There had been no such reluctance in the US (see, *inter alia, Gregory v. Helvering*, 293 US 465 (1935), and *Laidlaw Transportation Inc v. Commissioner*, 75 TCM (CCH) 2598 (1998)).

45   [1981] STC 174; it was followed by *Burmah Oil Co Ltd v. IRC*, [1982] STC 30, *Furniss v. Dawson*, [1984] STC 153, *Ensign Tankers (Leasing) Ltd v. Stokes*, [1992] STC 226 and *IRC v. McGuckian*, [1997] STC 908.

46   As in *Furniss v. Dawson*, [1984] STC 153. Lord Brightman's formulation in this case also involved the deployment of the concept of secondary findings of fact, by way of inference from the primary findings. This part of his speech may enable some of the overseas case law in the year preceding the House of Lords' cases decided in November 2004 to be distinguished.

47   For which purpose, it is necessary to distinguish between intended reliefs and artificial avoidance (see *Commissioner of Inland Revenue v. Challenge Corporation Ltd*, [1986] STC 548).

48   For example, *Craven v. White*, [1988] STC 476; *IRC v. Fitzwilliam*, [1993] STC 502; *Macniven v. Westmoreland Investments Ltd*, [2001] STC 237.

49   In *Ramsay* itself, [1981] STC 174.

50   In *Furniss v. Dawson*, [1984] STC 153.

51   [2001] STC 237.

by the House of Lords on 25 November 2004, in *Barclays Mercantile Business Finance Ltd v. Mawson.*[52] Their Lordships said that the modern approach to statutory construction involved having regard to the purpose of a particular provision and, insofar as possible, interpreting its language in a way which best gave effect to that purpose.[53] The essence of this approach was to adopt a *purposive* construction,[54] and not just for tax legislation.[55] But it was necessary to *focus carefully* on the particular statutory provision and identify its requirements, in which context the legal/commercial distinction which had been formulated by Lord Hoffmann (who also sat in *Barclays Mercantile*) while 'not an unreasonable generalisation' had not been 'intended to provide a substitute for a close analysis of what the statute means'.[56] In *Barclays Mercantile*, the legislation under consideration related to a finance lessor's entitlement to capital allowances (in lieu of depreciation) for expenditure incurred. There was no doubt that such expenditure had been incurred by the lessor, and the fact that, as far as the lessee was concerned, all the money apart from the capital allowances went round in a circle was immaterial for the purpose of determining the lessor's entitlement to those allowances.[57]

Insofar as the authors have been able to discover, the only widespread avoidance which has occurred in relation to environmental taxes involved the environmental trust deduction for landfill tax, and led to the rules having to be changed (that is, resort to the traditional targeted anti-avoidance legislation).[58] The Courts have, however, on occasion resorted to a *purposive* approach, more in line with European trends[59] (where EU Directives tend to be construed by reference to their objectives, as ascertainable from their recitals and the preceding legislative history).[60] And there is some evidence of the Courts being willing to adopt such an approach in relation to the ordinary construction of landfill tax.[61] A tribunal has, however, declined to extend that approach to aggregates levy,[62] and, in view of the lack of symmetry between climate change levy and the electricity supply change,[63] this would seem to be inconceivable in relation to the latter tax.

---

[52]   [2004] UKHL 51, in an 'opinion' to which all five law lords contributed and which was (unusually) not formulated as a 'speech' attributed to one of them and assented to by the others. The same feature was to be found in a second *Ramsay* case decided on the same day, *IRC v. Scottish Provident Institution*, [2004] UKHL 52, in which it was held that the Courts were entitled to ignore 'a commercially irrelevant contingency' inserted by the parties in order to make it possible to say that the course of events had not been pre-ordained.

[53]   [2004] UKHL 51, para. 28.

[54]   *Ibid.*, para. 32.

[55]   *Ibid.*, para. 33.

[56]   *Ibid.*, para. 38.

[57]   *Ibid.*, paras. 39–42.

[58]   See para. 21.3.2 above.

[59]   And, only subsequently, the views of the House of Lords!

[60]   As to which, see, in a tax avoidance context, *IRC v. McGuckian*, [1997] STC 908.

[61]   *C & E Commrs v. Parkwood Landfill* Ltd, [2002] STC 1536 (see para. 15.3 above) and perhaps also *Ebbcliff Ltd v. C & E Commrs*, [2004] STC 1496 (see para. 15.4 above).

[62]   See *Pat Munro-(Alness) Ltd v. C & E Commrs*, (2004) A2, para. 13.2n above.

[63]   See para. 21.6 above.

As a means of discovering Parliament's *purpose*, the Courts may well decide to have more regard to extraneous evidence, such as the Explanatory Notes now published with Bills[64] or, in some circumstances, *Hansard*.[65]

26.3.3.2 The general anti-avoidance rule solution

In 1998 the UK Government decided that the traditional piecemeal approach, under which each loophole which was identified became the subject of a targeted antidote, was not working. It therefore went out to consultation on the possibility of enacting general anti-avoidance rules ('GAARs'), along the lines of those present in the Australian and New Zealand tax codes. These were not proceeded with and a system of notification instituted in 2004 instead. There appear to have been two problem areas, that is, the fact that, despite the relevant statutory wording, the Courts have sometimes rejected the attempts of the tax authorities to invoke such provisions, as well as the lack of qualified staff to provide pre-activation clearances on the scale likely to be required. The UK does not operate the US system of binding rulings.

26.3.3.3 The 'disclosure' solution

In Budget 2004, provoked by a carefully marketed and concealed income tax reduction scheme involving the use of gilt edged derivatives, the UK Government announced that, in future, the promoters or, in some cases, the users of schemes the main object of which is to reduce any form of tax may have to register them and, in the case of promoters, have to obtain a scheme number for their clients to quote in their tax returns.[66] In theory, these provisions could be extended (by statutory instrument) to any tax, irrespective of whichever part of HM Revenue and Customs (as the amalgamated body is to be named)[67] is responsible for collection or (indeed) to locally collected taxes (although the latter is unlikely). The exact disclosure requirements are to be promulgated in regulations from time to time.

In the case of VAT, there was to be an obligation for those with a turnover exceeding £600,000 to notify those on a list to be published by Customs and those with a turnover in excess of £10m to notify other deliberate tax reduction arrangements.[68]

---

[64] Which may be resorted to under *Westminster City Council v. National Asylum Support Service*, [2002] UKHL 38.

[65] To which reference may be made in order to ascertain whether the promoters gave clear guidance on the point in question where the text is ambiguous, unclear or manifestly unreasonable: *Pepper v. Hart*, [1992] STC 808, *R v. Secretary of State for the Environment, Transport and the Regions, ex p Spath Holme Ltd*, [2001] 1 All ER 195.

[66] Budget Day Press Release REV BN 28 (17 March 2004); Finance Act 2004, ss. 306–19.

[67] See para. 4.2.1.2(2) above.

[68] Budget Day Press Release BN 01/04 (17 March 2004); Value Added Tax Act 1994, Sched. 11A (inserted by Finance Act 2004, Sched. 2); draft regulations – to take effect on 1 August 2004 – were published on 28 May 2004. In the event those actually promulgated differed significantly from the original published draft.

Such arrangements had previously been pioneered in the US, where regulatory enforcement activity was in progress against some of the international accountancy practices, the promotional activities of whose tax departments had been the subject of major controversy since the collapse of Enron in late 2001.

26.3.4   Cross-border transfer pricing

Furthermore, at the end of April 2004, an agreement in principle was reached between the governments of Australia, Canada, UK and the US to set up an international tax avoidance task force in New York to coordinate a crack down on tax avoidance. It was mentioned, in this connection, that this was taking place at a time when 'the public finances of many developed countries [were] under pressure'.[69] This position was highlighted in a series of articles on transfer pricing and thin capitalisation in the *Financial Times* at the end of July 2004.[70]

26.3.5   The need to fund social spending

This problem has arisen at a time when a shortfall in tax collection would have a serious effect on the Blair Government's plans for the improvement of public services as set out in the *Spending Review 2004*.[71] And yet the researches undertaken by the House of Commons' Environmental Audit Committee for its Report, *Budget 2004 and Energy*,[72] revealed that the revenue from environmental taxes as a proportion of total taxes in 2003 stood at 9 per cent, the same figure as in 1993. Even this represented a recovery from 8.6 per cent in 2001, after a high of 9.8 per cent in 1999.[73] The Committee expressed concern, in this context, that the Economic Secretary to the Treasury[74] did not appear to be aware of these figures.[75]

26.4   International trade treaty problem for federations and economic areas

It is not just the world wide web which has been causing problems for government tax collections in the developed world, however. Free trade areas[76] share with

---

[69] Crooks and Balls, *Financial Times*, 24–25 April 2004, p. 1.
[70] Plender, 21 July 2004, p. 15; Plender and Simons, 2 July 2004, p. 15; second leader, 24–25 July 2004, p. 18; Plender, 26 July 2004, p. 20.
[71] Cm 6237 (July 2004).
[72] Tenth Report of Session 2003–04, HC 490, published 11 August 2004.
[73] *Ibid.*, para. 10.
[74] Mr John Healey, MP.
[75] HC 490, para. 12.
[76] While the EU is the best known, both NAFTA (Canada, Mexico and USA) and MERCOSUR (Argentina, Brazil, Paraguay and Uruguay) are now significant, even though each is dominated by one country.

'genuine' federations[77] one fundamental legal principle, the need to prevent state protectionism to the detriment of residents of other Member States.

Where, in a 'genuine' federation, it is the state element which has constitutional protection, it is necessary to ascertain, in each case, whether it is the centre or the states which hold the residuary powers. Except in the case of Canada, where the UK Parliament deliberately made the federal part predominant, in the aftermath of the American Civil War, and Belgium, where formal federalism was the product of constitutional evolution between 1945 and 1993, the states have tended to hold the residuary powers, usually because it was they who created the Federation, such as in the case of the US and Australia.

In reality, however, courts have tended to interpret the devolved powers widely, to the detriment of the holder of residuary power. This became particularly apparent during the economic crisis of the 1930s, when federal governments tried to impose what, in the US, was called New Deal legislation. The end result of that litigation was the opposite of what one would have expected. Whilst the federal and Commonwealth governments came out on top in the US and Australia,[78] in Canada much of the Dominion (that is, federal) legislation was struck down,[79] making the Dominion Government extremely weak politically as against the Provincial (that is, State) Governments.

It is, however, not necessary for a multinational entity to be on the road to union for the issue of 'interstate' discrimination to arise. The North American Free Trade Area ('NAFTA') came into being on 1 January 1994. The investment chapter of the treaty enabled the US-based Ethyl Corporation to obtain both an apology and compensation from the Canadian Government following the introduction by the latter, in 1996, of a ban on the importation, and inter-Provincial trade, of a petroleum additive, MMT. Ethyl Corporation was the sole producer and importer of MMT. The basis of the company's complaint was that there was insufficient evidence to introduce a trade ban on health and environmental impact grounds.[80]

Whilst some Member States may see the EU as developing into a federation, the current structure cannot be so described. Indeed, there is a strong movement in a number of Member States, especially the UK, to prevent any further development

---

[77] This qualification is needed to distinguish countries, such as India, where there is a federal structure, but the federal government has reserve powers which allows it to override the rights of state governments, such as (in that case) the ability to dismiss a state government and declare President's (that is, federal government) rule.

[78] Where the tax-raising powers are split between the Commonwealth and the States, but the former appears to have the power to nullify its exercise by the latter (see *Federal Commissioner of Taxation v. Farley*, (1940) 63 CLR 278, 314 *et seq.* (Dixon, J.). Tax-raising powers have been split in a more recent federal creation, Spain, which is within the EU.

[79] In *Attorney-General for British Columbia v. Attorney-General of Canada*, [1937] AC 377, *Attorney-General of Canada v. Attorney-General of Ontario*, [1937] AC 326 and *Attorney-General of Canada v. Attorney-General of Ontario*, [1937] AC 355.

[80] See Luke Eric Peterson, *UK Bilateral Investment Treaty Programme and Sustainable Development: Implications of bilateral negotiations on investment regulation at a time when multilateral talks are faltering* (RIIA: Sustainable Development Programme Briefing Paper No. 10, February 2004).

along those lines. Nonetheless, the *freedom of establishment* provision of the European Treaty[81] seems to be performing, in relation to corporation tax, a very similar role to that performed by the *interstate commerce clauses* in the New Deal litigation.[82] This development is especially significant in the EU because Member States have not been able to achieve the degree of unanimity necessary to provide for direct tax convergence through the route of Community legislation. But the ECJ, which is going along a route which seems likely to achieve this, a development which has implications[83] for the soundness of the tax base of individual Member States,[84] driving them further down the road towards greater reliance on property-based, including environmental, taxation.

Indeed it might well be thought, from a Decision of the Court in November 2003, that the Court had gone too far. In *Proceedings brought by Lindman*,[85] a Finnish resident who won on a ticket bought in a Swedish lottery was assessed to income tax in Finland. Under Finnish law, only winners in Finnish lotteries were entitled to exemption from that tax. The Court held, however, that such discrimination was contrary to Art. 49, European Treaty (ex 59), on the basis that it constituted a restriction or obstacle to freedom of services, in this case those of gaming!

The problem which is staring the UK's Chancellor of the Exchequer in the face in the autumn of 2004, at a time when the country's budget deficit is above the EU's 3 per cent norm, is the prospect of its future corporate tax take being undermined by the ECJ, perhaps to the extent of £30bn per annum.[86] And what goes for the UK must clearly be capable of applying, albeit to differing extents, to other Member States.

All this, which has its genesis in the *interstate commerce clause* concept, will accelerate the trend resulting from e-commerce, in making national governments more dependent upon property-based taxes, included in which must be many environmental taxes.

26.5   Employer's national insurance contributions

Until 2002, it was accepted that a by-product of the introduction of environmental taxes would be a reduction in a major business burden, the employer's NICs. But

---

[81]   Art. 43, European Treaty (ex 52). See the *Hoechst* case, C–410/98, [2001] STC 452.
[82]   Especially the Canadian version of it (see above).
[83]   On top of (and, in the public perception, overshadowed by) the internet revolution.
[84]   Potentially one of the most serious avenues of potential loss for the UK is apparent from the claim, upon which Park, J. refused to adjudicate outside the normal appeals procedure – although the Court of Appeal, [2004] STC 1054, reversed his decision – in *The Claimants under the Loss Relief Group Litigation Order v. IRC*, [2004] STC 594, that the European Treaty combined with the non-discrimination clause in the UK-US double taxation treaty enabled a UK parent company to claim relief in the UK for losses made by its US trading subsidiaries.
[85]   C–42/02 [2005] 873.
[86]   Attributed to unnamed Inland Revenue officials by Brown, *Financial Times*, 11–12 September 2004, p. 1.

with the imposition of a flat 1 per cent on both employer and employee, this objective has plainly gone into suspended animation.[87]

Nonetheless it was not until the Pre-Budget Report in November 2002 that it became apparent that the formal linkage would be broken (albeit then in a form which was not conceded as having this result). While landfill tax rates had been increased without any reduction in NICs, the tax had not been self-supporting when account was taken of the quantum of rebates made through the medium of environmental trusts.[88] The diversion of some of this money to 'publicly funded' projects is, in reality, an appropriation of environmental tax money into the general tax pool.

This could turn out to be as significant as the removal many decades ago, of the link between the road fund licence and the road building programme. Unfortunately the current social spending obligations of the UK Government provide an ideal background to the exploitation of environmental taxes as a general milch cow, rather than as a targeted incentive to improved environmental compliance.

In its report issued in April 2004,[89] the Commission on Sustainable Development[90] deplored the failure to make further progress on environmental taxation at a time when the depletion of natural resources and the built-up of long-term pollution and the burden of waste was not adequately priced in. The abandonment of the fuel duty escalator, the postponement of a pesticides tax and the confinement of climate change levy to sectors of industry were instanced as examples.[91]

There had been no significant proportionate increase in the UK take of environmental taxation over a five year period[92] and the government needed to counter the political pressures which make progress difficult with steps to raise public awareness to enable the burden of taxation to be shifted from the taxation of 'desirable goods'.[93]

Finally, it is, surely, worthy of note that, in relation to the consultation opened in June 2004 in relation to catchment-sensitive farming,[94] no pretence was made that the net yield on any levies would be used to reduce employers' NICs.

26.6   The advantage of environmental taxes

Environmental taxes are, generally speaking, indirect taxes.[95] Indirect taxes, however, are not necessarily also environmental taxes. Nonetheless both direct and indirect taxes may have a part to play in environmental fiscal incentivisation.

---

[87] See para. 21.2 above.
[88] See para. 21.3.2 above.
[89] *Shows Promise. But Must Try Harder*, paras. 42, 43.
[90] See para. 4.2.1.4 above.
[91] *Shows Promise. But Must Try Harder*, para. 61.
[92] *Ibid.*, para. 71.
[93] For example, employment, which, it is not mentioned, is actually taxed far higher on the continent of Europe, especially in the UK's major industrial competitors there, France and Germany (*ibid.*, para. 77).
[94] See para. 21.9.4 above.
[95] See Chapter 1 above, especially para. 1.2.1.4 above.

Where a tax is levied at a flat rate (or a flat rate with a ceiling and/or a cap), it may be difficult to build an environmental role into its structure. The obvious example is employers' NICs. These are a fixed levy on payroll costs collected through the PAYE system. It would therefore be difficult to build into such a business tax a differential based, for instance, on the role of particular employees. Such a differential can only really be achieved through taxing on a differential basis either the profit attributable particular activities, or turnover by reference to the nature of the items sold. An example of the former is the (very limited) income tax relief for clean-up costs incurred on the purchase of polluted land where the purchaser is not responsible for indemnification.[96] An example of the latter is the application of the lower rate (5 per cent) of VAT to the installation of grant-funded renewable source heating systems introduced from 1 June 2002.[97]

On the other hand, it would be possible to build an environmental variable into the taxation of employees, for example, by giving direct relief from income tax and employee's NICs for qualifying commuting costs;[98] or, alternatively, by encouragement in the form of differential congestion charging.[99] The last of these is, of course, *per se* an environmental tax, and, indeed, one which can have a more marked effect on business behaviour,[100] albeit one which was only espoused in principle in mid-2003 before it was clear that the necessary technology to give effect to it would be available. In the light of this development, the minor concessions already in the public domain in relation to commuting costs, would be unlikely to be added to, even were it not apparent that the Treasury were not minded to proceed further along this route.

26.7   European Union limitations

It will also be apparent that the environmental encouragement made available to businesses through the taxation of profits and taxing turnover is of a very marginal nature. In relation to the government's response to the Urban Task Force Report,[101] the suspicion must be that the extraordinarily detailed limitations on the special reliefs enacted subsequently was motivated by the Treasury's desire (and, indeed, need) to 'cap' the subsidy thereby made available out of public funds.

There have also, however, been other limitations caused by the UK's Membership of the EU.[102] One of those reliefs, that from stamp duty and stamp duty land tax in disadvantaged areas,[103] could not be implemented on other than a short-term basis (and only then because of its marginal effect) because of the rules against state aid.[104]

---

96   See paras. 21.3.3 and 24.5.1 above.
97   Value Added Tax Act 1994, Sched. 7A, Group 3, Notes 4A and 4B (see also para. 21.3.1 above).
98   See para. 23.3 above.
99   See para. 18.5 above.
100 See para. 18.4 above.
101 See paras. 21.3.3 and 24.5 above.
102 See Chapter 12 above.
103 See para. 24.5.3 above.
104 See paras. 12.2.7 and 12.3.5 above.

Where VAT is concerned, there is, furthermore, a need to justify the application of the lower rate of tax by reference to social criteria.[105] Not only has this limited the scope for subsidisation of environmentally-friendly goods and services, but, where new construction is concerned, the principal/ancillary supply rule has restricted its application.[106] Where a tax is a turnover tax, Art. 33 of the Sixth VAT Directive[107] may prevent its implementation, although it has not, hitherto, in similar types of case.[108]

A further limitation is the prohibition, under Art. 92, European Treaty (ex 98), of charges on imports and the rebating of charges on exports where 'direct' taxes are concerned.[109] One of the objections taken, albeit unsuccessfully, to aggregates levy was that it might be such a tax.[110] This is likely to confine such taxes[111] to ones such as the regionally-imposed levy[112] of one euro per day between 2002 and 2003 on holiday makers in the Balearic Islands.[113] This situation should be contrasted with that in the US, where a tax per ton was imposed on petroleum and chemical products under Hazardous Substance Response Revenue Act 1980.

It has been seen that landfill is a major problem area for the UK.[114] The underlying pressure here comes, however, from European Directives. Here again, the UK Government hopes that trading, this time between local authorities, will have a role to play in the outcome.[115]

Finally, all domestic environmental legislation, taxation included, has, henceforth, to operate against the background of the EU ETS. This is discussed in the next chapter.

---

[105] See the two infraction proceedings cases: *Commission of the European Communities v. United Kingdom*, C–353/85, [1988] STC 251, and C–416/85, [1988] STC 456.

[106] See para. 21.3.1n above.

[107] See para. 1.2.1.4 above.

[108] See, for example, the consideration of, respectively, national meat and dairy products marketing charges in *Fazenda Pública v. Fricarnes SA*, C–28/96, [1997] STC 1348, and *Fazenda Pública v. União das Cooperativas Abestacedoras de Leite de Lisboa*, C–347/95, [1997] STC 1337, stamp duty on construction contracts in *Fazenda Pública v. Solisnor-Estaleiros Navais SA*, C–130/96, [1998] STC 191, and a local tax upon the gross receipts from entertainment performances in *NV Giant v. Commune of Overijse*, C–109/90, [1993] STC 651. It should be noted that this objection was not, however, raised in relation to municipal parking charges in *Fazenda Pública v. Câmera Municipal do Porto (Ministério Público, third party)*, C–446/98, [2001] STC 560.

[109] See para. 12.3.3.1(6) above.

[110] See *R (on the application of British Aggregates Association and others) v. C & E Commrs*, [2002] EWHC 926 (Admin), [2002] 2 CMLR 51, paras. 68-78. See paras. 12.3.3.1(6) and 21.10 above.

[111] If, indeed, such can be classified as 'environmental' taxes, which the authors consider to be very debatable (see para. 1.2.1.5 above).

[112] Which is interesting because it is an example of a state-level tax within a federal structure (see para. 26.4n above).

[113] It seems that this levy was introduced at the instance of Green Party membership of a rainbow coalition, and abolished after that administration lost office after the next following election (see para. 12.3.2n above).

[114] See paras. 2.3 and 21.4.1 above.

[115] See para. 20.7 above.

Chapter 27

Government Proposals

27.1 Introduction

At the time of writing, there are a number of proposals for environmental taxes and other economic instruments at various stages of design and implementation. This chapter seeks to review what is so far known about them.

During the present government's final period in opposition, the shadow cabinet member responsible for environmental policy, Mr Michael Meacher, had undertaken a detailed study into the part taxation might play in achieving the next government's environmental objectives.[1] Following the government's return in the May 1997 General Election, Mr Meacher (a left winger) was not given a Cabinet seat, but was created a Privy Councillor and appointed to the position of Minister of State at the mammoth DETR,[2] with responsibility for environmental policy. The 'grey' discussion document, *Economic Instruments for Water Pollution*[3] was the result of Mr Meacher's researches while in Opposition. It concentrated on three possibilities: a tax on fertilisers, a tax on pesticides and transferable water pollution permits, with special reference to the Thames Estuary.

Much more recently, in a *Study into the Environmental Impacts of Increasing the Supply of Housing in the UK*, published by Defra in May 2004, it has been recommended that an assessment be made of the extent to which economic instruments could have an effect on the location of housing development,[4] and also managing its demand and addressing the environmental impacts of construction and occupation.[5]

Although each of the above proposals may, or may not, yet come to fruition, the government has now taken the decision to introduce a satellite-based national scheme for charging heavy lorries by reference to all road usage, with the possibility of extending it to other vehicles. Moreover, by May 2003, the DFT had indicated that it was considering the possibility of replacing VED on cars[6] with a 'main road' congestion charge, possibly even before 2010. In addition, the government has signalled its intention to work for the inclusion of aviation in the EU ETS[7] from 2008, something which it will have the opportunity to do on taking up the EU presidency in 2005.

With the creation of recycling facilities having an unacceptably small effect on the amount of waste consigned to landfill, the possibility of giving recycling a further

---

1 Landfill tax had already been introduced by the pre-1997 Conservative Government (see paras 11.4 and 15.1 above).
2 See para. 4.2.1.2(1) above.
3 Published by the DETR on 20 January 1998.
4 Para. 6.3.
5 Para. 6.3.1.
6 See para. 22.2.1 above.
7 See Chapter 28 below.

boost by the imposition of a tax on waste incineration has also been considered, but, at the moment, there appears to be no intention of taking this forward.[8]

27.2   Economic instruments and housing development

As mentioned in the Introduction to the present chapter, in a *Study into the Environmental Impacts of Increasing the Supply of Housing in the UK,*[9] published by Defra in May 2004, it was recommended that an assessment be made of the extent to which economic instruments could be used in connection with housing. The assessment could embrace the extent to which economic instruments could have an effect on the location of housing development,[10] and also managing its demand and addressing the environmental impacts of construction and occupation.[11] In the case of a tax, this would have to reflect the *external* cost.[12] The inability of the market to reflect the uncompensated effect on third parties was considered to be a form of *market failure,*[13] suggesting the possibility of government intervention via economic instruments to:

1   protect undeveloped land;
2   encourage the regeneration of brown field land;
3   reduce the environmental impact of construction; and
4   reduce the environmental impact of occupation.[14]

1 and 2 were currently addressed primarily through the planning system,[15] but some tax incentives had been introduced.[16] Forms of development or land-holding tax had also been proposed by the Urban Task Force[17] and by the Barker Report.[18] As to 3, the aggregates levy was already in existence.[19] As to 4, landfill tax addressed waste

---

8   See para. 27.7 below.
9   See above.
10   Para. 6.3.
11   Para. 6.3.1.
12   As to which more work needed to be done (*ibid.*, paras 6.3.2, 6.3.3). An external effect arises where the *welfare* of *another agent* is affected without the *producer* paying compensation (*ibid.*, para. 2.7.1n.1). See also para. 5.4 above.
13   *Ibid.*, para. 2.7.2.
14   *Ibid.*, para. 2.7.3.
15   *Ibid.*, para. 2.7.4.
16   *Ibid.*, para. 2.7.5 (see para. 24.5 above).
17   *Ibid.*, paras 2.7.6, 2.7.7.
18   To which the study was supplemental (*ibid.*, para. 1.1.12). Her *Final Report* was published on 17 March 2004, and suggested the replacement of the collection of social housing contributions by local authorities through the mechanism of agreements under Town and Country Planning Act 1990, s.106 (which was not being done on a consistent basis) by a charge payable by the landowner when detailed planning consent for housing was granted by the local authority. The government is currently considering this recommendation. It is understood from the Defra consultation, initiated in July 2004, *Making Space for Water*, para. 16.19, that a 'package' is likely to emerge by the end of 2005. It also seems possible, from para. 7.21, that developers in flood plains may have to fund remediation works.
19   Para. 2.7.8 (see para. 11.3.2 above, Chapter 13 above and (especially) para. 21.8 above).

generation,[20] climate change levy energy use,[21] and efficient appliances certain other problems.[22] Water costs were reflected in its price.[23]

27.3   Road pricing for heavy lorries

The government has, as mentioned at the beginning of the chapter, decided to introduce a satellite-based[24] national scheme for charging heavy lorries[25] by reference to all[26] road usage.[27] If such a scheme could be set up, it could also be extended to other vehicles.[28] Leeds was, originally, intended to be used for a pilot study.[29]

---

20   See Chapter 15 above.
21   See Chapter 14 above.
22   See para. 24.4 above.
23   *Study into the Environmental Impacts of Increasing the Supply of Housing in the UK*, para. 2.7.9.
24   It is hoped that this will dovetail in with an embryo European scheme. Switzerland, Germany and Austria operate on the basis of micro-based systems: HM Treasury, *Modernising the Taxation of the Haulage Industry – Lorry Road User Charge: Progress Report Two* (London: HM Treasury, 2003). The failure of the less sophisticated German scheme in early 2004 was, however, clearly likely to delay the introduction of the UK one: Benoit and Wright, *Financial Times*, 18 February 2004, p. 8. The system was, however, resuscitated on 1 March 2004, albeit to take effect in 2006 (see BBC *Business News*, 1 March 2004).
25   The definition of heavy lorry which has been adopted is that used by Switzerland (which introduced a scheme on 1 January 2001), Germany, whose scheme was supposed to start on 31 August 2003, and Austria, whose scheme starts on 1 January 2004: HM Treasury, *op. cit.*. In the event, although the German technology became operational on 31 August 2003, tolls did not become payable under it under 2 November 2003, and the German Government's support for domestic hauliers had become the subject of a discrimination enquiry by the European Commission: Williamson, *Financial Times*, 2 August 2003, p. 6. The problem was settled after the German Minister of Transport gave an undertaking not to introduce subsidies for German hauliers without the Commission's consent: (see Williamson, *Financial Times*, 27 August 2003, p. 10).
26   As in Switzerland, Germany and Austria confining their schemes to motorways: *Modernising the Taxation of the Haulage Industry – Lorry Road User Charge: Progress Report Two* (6 May 2003).
27   Under Finance Act 2002, s.137: this had been anticipated to be put in place in 2005 or 2006, with offsetting reductions to ensure that the UK haulage industry did not pay more as a result (see Written Answer by the Economic Secretary to the Treasury (Mr John Healey, MP), *Hansard*, HC, 15 January 2003, Vol. 397, No. 193, col. 664W). In the event, the charging system adopted would apply to lorries from other Member States and the UK industry would be able to benefit from reductions in diesel duty under the terms of the proposed Energy Products Directive (Council Directive 03/96/EC, (2003) OJ L283 51): HM Treasury, *op. cit.* For details of the Energy Products Directive, see para. 12.3.4 above.
28   See para. 19.5 above. A proposal for such a system, albeit not then based on satellite technology was put forward as early as 1966 by A. Roth, *A Self-Financing Road System* (London: Institute of Economic Affairs, 1966).
29   Guha, *Financial Times*, 23 July 2002, p. 2.

It seemed possible, however, that the original scheme (which would also be applied to lorries from other Member States of the EU)[30] might not be permissible under Community Law.[31] There seems, nonetheless, to be political reluctance to substitute a system of Community-wide road tolls for the *Eurovignette* licence for lorries to use the motorways of the seven Member States who do not levy such tolls.[32]

The government's intention is to introduce a lorry road-user charge based on distance travelled. The target date had been 2006,[33] with the technology is expected to be put out to tender at the end of 2003, and the legislation contained in Finance Bill 2004.[34] In March 2004, this had, however, been put back to 2008, both in the light of problems with the German scheme and in order to provide time for pilot testing of an arrangement based on an on-board unit. Truckers would be compensated with a fuel duty rebate, via a repayment scheme. There would, furthermore, be an electronic scheme for occasional users.[35] On 9 June 2004, however, Customs indicated in an *Information Guide*, that the charge would be introduced in 2007/08 for lorries of above 3.5 tonnes and there would be both: a Frequent User Scheme for regular users, based on on-board equipment; and an Occasional User Scheme for infrequent users, who would be able to opt for a low-use on-board unit.

The quantum of the charge may vary depending upon axle weight and whether environmentally-friendly emission control systems are installed.[36]

Progress Report 2 has already been referred to in connection with excise duties.[37] *Progress Report 3* was published on Budget Day 2004.[38] It revealed that the projected lorry road user charge would be administered by a LRUC Management Authority within Customs.[39] The procurement arrangements, which had to be run in accordance with EU advertising and evaluation procedures, would consist of three contractual packages – central services, roadside equipment and on-board

---

[30] This problem is to be covered, as in Switzerland and Germany, by enabling occasional users to pay for a ticket along a predetermined route before travelling. Austria requires such users to have the micro-device installed (see HM Treasury, *op. cit.*).

[31] *Andreas Hoves Internationaler Transport-Service SARL v. Finanzamt Borker*, C–115/00.

[32] Dombey, *Financial Times*, 7 February 2003, p.8. The UK's first toll motorway bypassed a particularly congested part of the M6, where traffic speeds average 17 mph, and was only 27 miles long (see www.m6toll.co.uk). Its standard lorry charge of £11, including VAT, was to be the most expensive in Europe: Jowit, *Financial Times*, 7 May 2003, p. 2. But haulier pressure led to it being cut from its initial level of £10 to £6 until the end of 2005 (see Guthrie, *Financial Times*, 23 July 2004, p. 5). The government's road programme, announced in July 2004, had envisaged an additional parallel paying extension of the M6. Whether this will turn out to be a project contractors are prepared to bid for remains to be seen.

[33] HM Treasury's Budget 2003 (9 April 2003), para. 7.38.

[34] HM Treasury, *op. cit.*

[35] Wright, *Financial Times*, 18 March 2004, p.12: *Modernising the Taxation of the Haulage Industry – Lorry Road User Charge: Progress Report 3* (17 March 2004).

[36] That is, as in Germany (see HM Treasury, *op. cit.*).

[37] See para. 22.3 above.

[38] 17 March 2004.

[39] On the same day, it was announced in the Budget that Customs would be merged with the Inland Revenue! (See para. 4.2.1.2(2) above.)

equipment. Finance Bill 2004[40] contained the initial legislation, but that relating to structure, collection and administration was deferred until 2005. Under the published timetable, it was envisaged that contracts would be awarded by the end of 2005. The roadside equipment would be installed, staff recruited and trained and the necessary regulations promulgated during 2006. Equipment would be installed in vehicles during 2007 and the system would go live in early 2008. A slippage in the timetable of two years had, therefore, to be accepted.

27.4   Nationwide satellite-based congestion charging

On 25 February 2003, the Commission for Integrated Transport[41] published two *Fact Sheets*, Nos 8 and 9, on its website.[42] The first[43] advocated the use of global positioning technology, using position-fixing satellites, by imposing flexible charges on 47 per cent of travelling time, and producing revenue either to lower fuel duty by 12p or to abolish vehicle excise duty and reduce fuel duty by 2p. This technology would:

1   detect vehicles entering areas for which charges are to be made (including 10 per cent of the motorway and trunk road network);
2   distinguishing not only the published periods when they are in force; but also
3   waiving the charge if the route was not in fact congested at the time; and also
4   identifying the type of vehicle, so that a reduced charge could be imposed for *green* vehicles (it being estimated that heavy goods vehicles would be paying 11 per cent of the total charges); with
5   users paying by card or monthly invoice.

It was estimated that congestion would be reduced by 44 per cent nationally, with savings in journey times of 34 per cent in Central London, 5 per cent in inner London, 35 per cent in the conurbations, 11–12 per cent in small urban areas and 11 per cent on rural roads.

The anticipated reductions in traffic would be:

1   5 per cent nationally;
2   20 per cent in Central London;
3   16 per cent in inner London;
4   11–13 per cent in outer London, and the inner areas of Birmingham, Manchester, Leeds and Liverpool;

---

40   Published 8 April 2004.
41   See para. 4.2.1.4 above.
42   See www.cfit.gov.uk.
43   The second gave examples of the effect on a variety of road users and estimated how the imposition of such a scheme might both affect their current routine and enable them to make savings in their transport costs.

5   8–9 per cent in large cities, such as Hull, Southampton, Leicester and
     Middlesborough;
6   2.6 per cent on motorways; and
7   2.4 per cent on rural roads.

By the end of April, however, it had become clear that the government's Ten-Year
Transport Plan[44] would not be capable of being modified to encompass so big a
change, indeed progress overall was likely to regress as against the original proposals
made under the stewardship of Mr John Prescott.[45] Nonetheless, at the end of May,
there were indications that Mr Alastair Darling, the Secretary of State for Transport,
was thinking in terms of replacing vehicle excise duty on cars[46] with a 'main road'
congestion charge, and possibly even before 2010.[47] This was confirmed when the
government's £7bn road investment programme was unveiled in July 2003,[48] albeit
subject to the proviso that, technologically, this involved a far bigger step forward
than the lorry scheme.

Nonetheless, in the White Paper published the following month, *The Future
of Transport*, it was clear that this was to be the cornerstone of the government's
forward planning.[49]

27.5   Pricing air passenger transport

Airliners are very heavy consumers of petroleum products and, although less so than
formerly, it goes without saying that they are also very noisy.[50] It might be thought,
therefore, that, carbon dioxide emissions from this source in the UK having increased
by 85 per cent between 1990 and 2000, and being estimated to rise by 30 per cent

---

[44]   *Transport 2010: the Ten Year Plan* (DETR, July 2000), which had been severely affected
       by the tragedy of the Hatfield rail crash of October 2001.
[45]   Jowit, *Financial Times*, 26–27 April 2003, p. 4.
[46]   See para. 21.2 above.
[47]   Newman, *Financial Times*, 24–25 May 2003, p. 1: this was supported by the remarks made
       by The Rt Hon. Alastair Darling, MP (Secretary of State for Transport) in an interview
       with Mr Jeremy Paxman on BBC2's *Newsnight* programme on 29 May 2003. These
       indications were followed, on 16 June 2003, by the publication of *Transport Pricing:
       Better for Travellers* by the Independent Transport Commission, a body set up in 1999
       under the auspices of the University of Southampton. In this document, the ITC (which
       had published a number of studies and reports on the wider implications of government
       policy as set out in *Transport 2010: The Ten Year Plan*), advocated nationwide satellite-
       based road user pricing in place of excise duties on fuel. The document itself was based on
       the more extensive Summary and Technical Reports, *Transport Pricing and Investment
       in England*, by Prof. Stephen Glaister and Dr Dan Graham of the Department of Civil
       Engineering, Imperial College, London, published by the ITC on 31 May 2003.
[48]   Guha, Bream and Blitz, *Financial Times*, 10 July 2003, p. 5.
[49]   Wright, *Financial Times*, 24–25 July 2004, p. 11.
[50]   See John Sheail, *An Environmental History of Twentieth-Century Britain* (Basingstoke:
       Palgrave, 2002), pp. 191–6.

during the current decade, aircraft emissions would be thought to be an appropriate subject for environmental taxation.[51] However, air transport falls outside the Kyoto Protocol.[52] Airline travel by passengers is zero-rated for VAT.[53] The Energy Products Directive[54] exempts airliners from excise duties on fuel consumption within the EU. Under the Chicago Convention the taxation of fuel for international flights is prohibited.[55] Additional restrictions are imposed under bilateral air service agreements to which the UK is a party.[56]

Although air passenger duty was introduced in 1994,[57] it was as a revenue-raising measure, and, despite the UK government classifying it as an environmental tax, there are no environmental aspects to it.[58] The current standard rates of duty are £10 per passenger per flight to destinations in the European Economic Area[59] and £40 to anywhere else. Although there appears to have been a strong lobby in favour of increasing air passenger duty at the time the Air Transport White Paper was coming up to completion,[60] ministers decided not to follow that route in the event.[61]

As a prelude to the preparation of an Air Transport White Paper, for publication at the end of 2003, the government issued a discussion paper[62] on how economic measures could be used to encourage the industry to take more account of its environmental impact.[63] This concentrated on the external costs relating to four aspects:

1   climate change – an issue because aircraft engines emit both carbon dioxide (a GHG in itself) and nitrogen oxides, which produce ozone, both being contributors to global warming, and therefore climate change;[64] estimates for 2000 suggested

51   Houlder, *Financial Times*, 16 December 2002, p. 2.
52   See para. 8.3.1.4 above.
53   Value Added Tax Act 1994, Sched. 8, Group 8, item 4(a),(c).
54   See para. 12.3.4 above. It is worthy of note, however, that two of the incoming Barroso Commissioners, Mr Jacques Barrot and Mr Stavros Dimas, advocated a jet fuel tax on inter-EU flights in their European Parliament confirmation hearings (see Minder, *Financial Times*, 30 September 2003, p.8).
55   See para. 8.5 above.
56   *Aviation and the Environment: Using Economic Instruments* (HM Treasury, Department for Transport, 14 March 2003), Box 2 (Legal Issues).
57   See para. 1.2.1.5(2) above.
58   Indeed, it seems clear that the 1996 rate increases hit tour operators rather than BAA plc or airlines (see *Investors Chronicle*, 29 November 1996, p.10).
59   And this appears to have been the result of pressure from the European Commission (see Skapinker, *Financial Times*, 11 November 1999, p. 11).
60   Newman, *Financial Times*, 13–14 December 2003, p. 1. Done, *Financial Times*, 17 December 2003, p. 4.
61   Newman, *Financial Times*, 15 December 2003, p. 9.
62   *Aviation and the Environment: Using Economic Instruments* (HM Treasury, Department for Transport, 14 March 2003). This was followed by the Fourth Report of the House of Commons' Environmental Audit Committee, issued on 1 April 2003, HC 167, para. 39 of which was not complimentary about the Department for Transport's attitude to this issue.
63   HMT 39/03 (14 March 2003).
64   *Aviation and the Environment: Using Economic Instruments* (HM Treasury, Department for Transport, 14 March 2003), paras 3.3–3.11, Annexes A–D.

that civil passenger aviation was responsible for 5 per cent of UK carbon dioxide emissions, with an increase of 2.3 times in prospect by 2030 giving rise to between 10 and 12 per cent of such emissions;[65]

2   noise;[66] and

3   local air quality, which is the product of take-offs, landings and associated ground vehicles, for which mandatory EU limits for the two principal gases are to be in place by 2010;[67] as well as touching on

4   congestion in the skies and around airports, albeit with *external* costs thought likely to be minimal, except in relation to surface traffic around airports.[68] The observation was made, however, that the lack of market mechanisms in the current slot allocation system resulted in an inefficient use of airport capacity.[69]

This approach did not go down well with the House of Commons Environmental Audit Committee, who, in *Budget 2003 and Aviation*, their Ninth Report of Session 2002–03, published on 29 July 2003.[70] considered that the passive acceptance of a 4 per cent annual increase in traffic for 30 years, coupled with a decrease in fares of 40 per cent gave rise to concern. In the Committee's view, government policy should aid to decouple growth in air travel from economic growth; when the increase in cost of aviation emissions would wipe out the economic case for runway expansion; with the impact of noise pollution underestimated;[71] and no account having been taken of catastrophic changes to the atmosphere.[72]

The Environmental Audit Committee could, furthermore, see no reason why aviation should be treated differently from motoring as far as fiscal policy was concerned[73] and recommended the replacement of air passenger duty with an emissions charge[74] levied on flights which was 'displayed clearly' on travel documentation. It also recommended that examination should be given to a dual-till system to ensure that airlines paid a greater share of infrastructure costs; and that the European Commission should institute regular auctions of slots, so that demand was reflected in the price.

---

[65]   *Ibid.*, paras 3.9, 3.11. The government's calculations indicated that the external climate change cost could rise from £1.4bn to £4.8bn over this 30 year-period (*ibid.*, para. 3.22).

[66]   *Ibid.*, paras 3.12–3.15, Annex E.

[67]   *Ibid.*, paras 3.16–3.18, Annex F.

[68]   *Ibid.*, paras 3.19–3.20.

[69]   In relation to which the government had been pressing the European Commission for reform (*ibid.*, para. 3.21).

[70]   HC 672.

[71]   *Ibid.*, paras 57–9.

[72]   It is worthy of note that the Committee contrasted the approach in this case with the initial approach to landfill tax and aggregates levy (*ibid.*, para. 53).

[73]   *Ibid.*, paras 71–3.

[74]   Because it is not possible to tax airline fuel (*ibid.*, para. 74), where reference is made to an emissions charge levied at Zurich Airport and Norway's carbon-emission based national aviation green tax. In evidence given to the House of Commons' Environmental Audit Committee on 12 May 2004, Q233, Mr John Healey, MP, the Economic Secretary to the Treasury, emphasised the legal framework within which any proposal to 'reform' air passenger duty into a tax which might operate as an environmental instrument.

Following this, the Commission for Integrated Transport called for, albeit without much expectation of the same being adopted,[75] an EU-wide carbon dioxide emission charge of £70 per tonne; and a charge based on aircraft noise, with both being funded by increased fares. In early December 2003, however, this issue became subsumed in the controversy over whether the UK Government had the ability, under the Air Quality Framework Directive,[76] to authorise the construction of a third runway at Heathrow (the airline industry's preference), as against a second runway at Stanstead.[77]

The White Paper itself, *The Future of Air Transport*,[78] published on 16 December 2003, sought to provide a 30-year framework for tackling the enormous growth in air traffic anticipated over that period.[79] The central issue was the provision of additional runway space in the south east, with the first choice being a second runway at Stanstead by 2011–12[80] and the second a third runway at Heathrow by 2015–20[81] if emissions control permitted. The alternative was a second runway at Gatwick, which it had been agreed as part of a planning agreement would not be built before 2019. The environmental problem relating to Heathrow was the EU restriction on the level of pollutants in the air, irrespective of source, coming into effect in 2005 for particulates and 2010 for nitrogen dioxide.[82] The proposed solution to the emissions problem was to work, up to and during the UK presidency of the EU in 2005, for the inclusion of aviation in the EU ETS from 2008.[83]

The White Paper was, however, savaged by the House of Commons' Environmental Audit Committee in its *Pre-Budget Report 2003: Aviation Follow-Up*, published on 10 March 2004.[84] The central criticism was that the White Paper 'actively' promoted a huge growth in air travel over the next 30 years and that there would be a 'massive' environmental impact, in terms of emissions and aviation's contribution to global warning. In the opinion of the Committee, the DFT had failed to give adequate consideration to the latter.[85] Furthermore, by restricting its economic appraisals to the provision of new runways, the DFT had failed to provide an overall appraisal of the economic impact from air travel. The Committee was also sceptical as to whether there was any real prospect of the UK Presidency of the EU securing the inclusion of aviation within EU ETS,[86] even assuming that that, or a similar, trading

---

75   Spiegel, *Financial Times*, 1 September 2003, p. 2.
76   See para. 12.2.6.2 above.
77   Newman, *Financial Times*, 29–30 November 2003, p. 1, Wright, *Financial Times*, 1 December 2003, p. 4, Done, *Financial Times*, 3 December 2003, p. 9, Done, *Financial Times*, 6–7 December 2003, p. 11.
78   Cm 6046.
79   *Ibid.*, Annex A.
80   Which, it was anticipated, would lead to low-cost airlines 'resisting' higher fees from BAA plc (see para. 2.6 above) to finance this (see Done, *Financial Times*, 18 December 2003, p. 3).
81   Cm 6046, para. 11.66.
82   *Ibid.*, para. 3.29.
83   *Ibid.*, Annex B.
84   HC 233.
85   Even in its subsequently-produced supporting paper, *Aviation and Global Warming*.
86   See Chapter 28 below. See also Houlder, *Financial Times*, 23 September 2004, p. 9.

system was capable of delivering carbon reductions of the size required on the basis of the envisaged expansion of civil aviation. Indeed, in the opinion of the Committee, an increase in aviation emissions of the size envisaged by the DFT would make the UK's 60 per cent carbon emissions reduction target set by the UK Government in 2003 'meaningless and unachievable'. The best the Committee felt it would be realistic to hope for would be 35 per cent.

The DFT published its response to the Environmental Audit Committee's report in June 2004[87] and, in the same month, the Committee issued its (third) Report,[88] to which the DFT responded in August and the Committee its (fourth) Report in September 2004.[89] Suffice it to say that the Committee thought it necessary to record that there remained 'fundamental and apparently irreconcilable differences' between it and the ministry.[90]

27.6   Litter taxes

The packaging problem generally has been discussed in Chapter 19 above. It was reported, however, in May 2002, that Mr Michael Meacher (the Environment Minister at that time) was considering the imposition of a tax on the 8m plastic bags used by retailers every year.[91] It was envisaged that this would be similar to that, imposed at 15c per bag, the previous March in Eire, which, it later transpired, raised EUR 3.5m in its first three months and reduced usage by 90 per cent.[92] In the UK only one in 200 was recycled (when shredded and compacted plastic bags can be used as bedding for roads).[93] Denmark also has such a tax. One leading supermarket operator, Safeway, let it be known, however, that it considered that the 1997 Regulations imposed a similar levy.[94] At the beginning of September 2002, the Co-op introduced a type of plastic bag which did biodegrade substantially between 18 months and three years after manufacture and it was understood that other UK supermarkets were likely to follow suit.

On 22 October 2002, however, the Economic Secretary to the Treasury announced, in a written answer, that there were no plans to introduce such a tax.[95] Nonetheless, shortly afterwards, Durham Council let it be known that they were considering such

---

[87]   Cm 60163.
[88]   HC 233.
[89]   HCC 1063: see also HC 263 [Seventh Report of 2003–04].
[90]   HCC 1063, para. 6.
[91]   Mason, *Financial Times*, 21 May 2002, p. 2.
[92]   Albeit offset by a significant increase in the sale of bin liners (see Tighe, *Financial Times*, 19 December 2002, p. 4). Success seems, however, to be infectious because, on 15 July 2003, the BBC *6 O'Clock News* reported that the Irish Government was planning to introduce a hypothecated litter levy on chewing gum, polystyrene wrappers and cash till receipts.
[93]   Peachey, *The Independent*, 21 August 2002, p. 3.
[94]   Mason, *Financial Times*, 21 May 2002, p.2,
[95]   Official Report HC, Vol. 391, No. 194, col. 212W (Mr John Healey, MP).

a levy on a local basis.[96] A one-month pilot project supported by three major retailers suggested, however, that shoppers could be persuaded to buy 'bags for life' in large enough numbers to make a tax incentive unnecessary.[97]

Levies of this nature can, however, have unexpected results. At the instance of its Green Party Environment Minister, Mr Trittin, the German Government sought to promote the substitution of refillable drinks containers for cans by the imposition, from 2002, of a deposit of 25 to 50 cents payable at the time of purchase and only refundable upon return at the point of sale. This made the purchase of canned drinks at petrol stations unattractive to the public. More materially, the decline in trade created by that generally resulted, by June 2003, in four major retailers deciding not to stock drinks in such a form. This undermined the incentive for the retail industry to set up a national collection infrastructure. The resulting 60–70 per cent decline in canned beer sales led to the German canning industry laying off workers, with the prospect of painful restructuring by five large brewers.[98] Modifications were introduced.

On the technical side, the effect on can manufacturers in other EU Member States attracted the attention of the European Commission, from the point of view of non-compliance with the free market parts of the Treaty.[99] The problem centres on the fact that German manufacturers use reusable glass bottles, whereas imports (mainly from France) are in disposable plastic bottles.[100]

27.7   Taxing incineration

An incineration tax was a possibility raised at the end of 2002 in *Waste Not Want Not*,[101] although it was not taken any further at that time. *Waste Not Want Not* did, however, note that there were planning problems. As a solution, incineration is not one which it is easy for a local authority to adopt under the current town and country planning system. On 23 April 2003, the House of Commons Environmental Audit Committee recommended a freeze on planning consents for new incinerators.[102]

The issue next returned to the public domain during the House of Commons Report Stage of the Waste Emissions and Trading Bill on 28 October 2003, when both main Opposition parties expressed concern at the effect of the tabled new clauses

---

[96] Tighe, *Financial Times*, 19 December 2002, p. 4.
[97] Tighe, *Financial Times*, 6 November 2003, p. 4. The debate over biodegradable bags continued, with concern at the possibility of their finding their way to recycling facilities and contaminating the plastic there (see Houlder, *Financial Times*, 28–29 February 2004, p. 5). The development of such bags was, furthermore, not being supported by the UK Government because of the release of methane in the breaking-down process (see Mr Elliot Morley, MP, Minister of State for the Environment, *Hansard*, 23 February 2004, WA 153745, col. 124W).
[98] Benoit and Williamson, *Financial Times*, 9 July 2003, p. 12.
[99] That is, the old 'inter-state commerce clause problem' which bedeviled the New Deal legislation in the 1930s (see para. 26.4 above).
[100] BBC 1/News 24, *The World Today*, 22 September 2003.
[101] Downing Street Strategy Unit: see paras. 6.3.2 and 21.4.1 above.
[102] In *Waste – An Audit*, Fifth Special Report of Session 2002–03, HC 99, para. 63.

on Ministerial Reporting to Parliament, which were to form Chapter 3 of the new Act. Their concern was that a regime which limited the use of landfill in the manner proposed would inevitably result in local authorities resorting to increased use of incineration,[103] and indeed, in the light of the stringency imposed by the Incineration Emissions Directive,[104] to the construction of new large[105] facilities on the basis of fixed term contracts,[106] which would create not only a chimney emissions problem but also a transport emissions one[107] and the dispatch of fly ash to landfill.[108]

There appears, however, to have been a difference of approach between the two main opposition parties to the use of incineration as an acceptable alternative to landfill. The Liberal Democrats were totally opposed, being concerned at the additional environmental damage which would be inflicted. The Conservatives, however, seemed to envisage some circumstances in which an incineration option would be acceptable, in particular those in which the incinerator could be the source of CHP generation.

The minister in charge said, during this debate, that although it had been suggested by various bodies that an incineration tax might be an option, the government had no plans for the introduction of such a tax.[109]

At the same time as the release of estimates showing that the external cost of incineration was £3–£4 per tonne higher than that for landfill,[110] it was said in the 2004 Pre-Budget Report[111] that the government was not convinced that there was a strong case for the introduction of a tax on incinerated waste.[112]

27.8   Conclusions

It is ironic that New Labour's first environmental instruments candidate in 1997 was water,[113] that nothing transpired in relation to that,[114] but that, come the summer of

---

103 See, especially, Mr Norman Baker, MP (Liberal Democrat), Official Report HC debates, col. 190.
104 2000/76/EC. As to this, see the remarks of the Minister for the Environment (Mr Elliot Morley, MP) at *ibid.*, col. 233.
105 The only amelioration considered possible from this was the possibility, mentioned by Mr Bill Wiggin, MP (Conservative) of the creation of CHP.
106 Mr Norman Baker, MP, Official Report HC debates, col. 227.
107 *Ibid.*, col. 223.
108 *Ibid.*, col. 227.
109 Mr Elliot Morley, MP, *ibid.*, col. 239.
110 *Combining the Government's Two Health and Environmental Studies to Calculate Estimates for the External Costs of Landfill and Incineration*: published 2 December 2004 on www.hmce.gov.uk.
111 Entitled *Opportunity for All: The Strength to Take the Long-term Decisions for Britain*, Cmn 6408, 2 December 2004, www.hm-treasury.gov.uk.
112 Para. 7.60.
113 See para. 21.9.1 above.
114 Primarily, it seems, because US experience related to discrete catchment areas and had not been a great success: see para. 20.3 above.

2004 and the pending advent of an EU Directive, the government's attention has had to be redirected to this possibility, albeit through a reconsideration of the possibility of taxing fertilisers.[115] But following the responses to the consultation[116] showing no enthusiasm for an economic instrument based solution to the catchment area problem, no further mention was made of this possibility in the written Parliamentary Statement made by the Minister for the Environment and Agri-Environment on the same day as the Pre-Budget Report.[117]

Related to this has, throughout, been the possibility of taxing pesticides, which is partly an issue of public safety and (in the current financial/political climate) partly of broadening the tax base.[118] But Defra-watchers' attention on the former became diverted by the GM crops issue, which has only recently (summer 2004) been defused.[119]

At the same time, interest has been aroused, and then dampened, in the waste disposal sphere by the possibility of plastic bag taxes[120] and an incineration tax.[121]

Instead the possibility of economic instruments has been considered in the fields of housing development,[122] road pricing for lorries,[123] nationwide satellite-based congestion charging,[124] and building the environmental cost into air passenger transport pricing.[125] It is worthy of note, in the context of transport, that the Consultation on the review of the UK Climate Change Programme[126] revealed that carbon dioxide emissions increased between 1999 and 2002 by 7 per cent from road transport and 35 per cent from domestic air transport.[127] The responses to the government's April 2004 consultation on a possible biofuels strategy[128] revealed some scepticism with (and strong opposition from the oil companies to) the idea of a renewables obligation.[129] Nonetheless the government announced that it would be undertaking a feasibility study into the possibility of a Renewable Transport Fuels Obligation.[130]

---

[115] See para. 21.9.4 above.
[116] Published on www.defra.gov.uk on 2 December 2004.
[117] Mr Elliot Morley, MP, HC, 2 December 2004, col. 51WS.
[118] See para. 21.9.5 above. In the 2004 Pre-Budget Report, Cmn 6408, paras 7.65–66, the government's options were held open in relation to what happens when the voluntary programme comes to an end in 2006.
[119] See para. 21.9.6 above.
[120] See para. 27.6 above.
[121] See para. 27.7 above.
[122] See para. 27.2 above.
[123] See para. 27.3 above.
[124] See para. 27.4 above.
[125] See para. 27.5 above.
[126] Issued 2 December 2004: www.defra.gov.uk.
[127] Para. 3.5: it should be noted that those from international air transport were also increasing but not quantified because they are not part of the Kyoto Protocol.
[128] See http://www.dft,gov.uk/stellent/groups/dft_roads/documents/page/dft_roads_033085. hcsp.
[129] Para. 2.
[130] See *Review of the UK Climate Change Programme*, para. 8.17, and http://www.hm-treasury.gov.uk/media/92C/1C/pbr04_ch07_323.pdf.

Chapter 28

The EU Emissions Trading Directive

28.1   Introduction

The European Community is a party to the 1992 United Nations Framework Convention on Climate Change[1] and to its famous Kyoto Protocol of 1997.[2] Moreover, the Community's Sixth Environmental Action Programme of 2002 nominated the tackling of climate change and, specifically, the ratification and implementation of the Kyoto Protocol as one of four priority areas for action.[3] As already discussed,[4] the significance of Kyoto resides to a considerable extent in the introduction of the three 'Kyoto' or 'flexible' mechanisms' for use by industrialised countries in meeting their emissions reduction targets. In addition to Joint Implementation ('JI') and the Clean Development Mechanism ('CDM'),[5] these flexible mechanisms include the development of emissions trading schemes.

The EU Emissions Trading Directive ('the EU ETS Directive')[6] 'establishes a scheme for GHG allowance trading within the Community ... in order to promote reductions of greenhouse gas emissions in a cost-effective and economically efficient manner'.[7] Addressed to the Member States, the Directive entered into force on 25 October 2003, being the date of its publication in the *Official Journal*.[8] In what follows,[9] the scheme is referred to as the EU ETS, to distinguish it from the UK ETS, which was discussed in the previous chapter. Although in conception, the EU ETS has points of similarity both with the UK ETS and with similar schemes in the United States,[10] there are also, as will become apparent from the present chapter, some very significant differences between the EU ETS and existing national schemes. Indeed, at the time of writing, an intractable issue is the working out of a relationship between the UK ETS and the EU ETS. Some of the highly complex issues raised by the relationship are discussed in para. 28.4 below. Meantime, however, it is necessary to examine the background to the EU ETS Directive and to outline its principal provisions.

---

[1]   Approved by Council Decision 94/69/EC, (1994) OJ L33 11.
[2]   See paras 8.3.1.3 and 8.3.1.4 above.
[3]   See para. 12.2.3 above.
[4]   See para. 8.3.1.4 above.
[5]   *Ibid.*
[6]   European Parliament and Council Directive 03/87/EC Establishing a Scheme for Greenhouse Gas Emission Allowance Trading within the Community and Amending Council Directive 96/61/EC, (2003) OJ L275 32. (For Directive 96/61/EC (that is, the IPPC Directive), see para. 12.2.2 above.)
[7]   Directive 03/87/EC, Art. 1.
[8]   That is, of the EU (see Directive 03/87/EC, Art. 32).
[9]   That is, mainly in para. 28.3 below.
[10] As to which, see Joanne Scott, *EC Environmental Law* (London: Longman, 1998), p. 55n.

Whatever the strengths and weaknesses of the EU ETS, its main importance will remain the fact that it is the first such multinational scheme to make the transition from dream to reality.[11] Its realisation stands in marked contrast to the Commission's 1992 proposal,[12] subsequently withdrawn, for a harmonised tax on carbon dioxide emissions and energy content.[13] Moreover, it is crucially important for the EU ETS to be seen to be successful, since, without a credible programme for reducing GHG emissions, it may yet be extremely difficult for the Community to persuade industrialised third countries to ratify the Kyoto Protocol.

28.2   Background to the EU Emissions Trading Directive

28.2.1   General

The Treaty base of the Directive is Art. 175(1), European Treaty (ex 130s),[14] which requires the Council,[15] acting in accordance with the co-decision procedure,[16] and after consulting ECOSOC[17] and the Committee of the Regions,[18] to decide what action is to be taken by the Community to achieve the objectives in Art. 174, European Treaty (ex 130r).[19] The co-decision procedure, set out in Art. 251, European Treaty (ex 189b), is designed to prevent the adoption of a measure without the approval both of the Council and of the Parliament.[20] The emphasis in Art. 251 on the need for a jointly approved legislative text[21] is vividly illustrated by the fact that, from the Commission's first transmitting the text of the proposed Directive to the Council and the Parliament,[22]

---

[11]   See Philippe Sands, *Principles of International Environmental Law*, 2nd edn (Cambridge: Cambridge University Press, 2003), p. 161.

[12]   Proposal for a Council Directive Introducing a Tax on Carbon Dioxide Emissions and Energy, COM (92) 226 final, (1992) OJ C196 1.

[13]   The 2003 Energy Products Directive (Council Directive 03/96/EC: see para. 12.3.4 above) is the emasculated successor, not only to the 1992 proposal, with its harmonised carbon/energy tax proposal (amended in 1995 (see COM (95) 172 final, available on PreLex, at www.europa.eu.int/prelex) but also to a proposal of 1997 for a Council Directive restructuring the Community framework for the taxation of energy products, (1997) OJ C139 14. For the story of the 1992 proposal, see Scott, *op. cit.*, pp. 45–50 and, for early (economic) views on a Community-wide carbon/energy tax, see, for example, Carlo Carraro and Domenico Siniscalco (eds), *The European Carbon Tax: An Economic Assessment* (Dordrecht: Kluwer, 1993).

[14]   See para. 12.2.1 above.

[15]   See para. 4.3.1 above.

[16]   See para. 4.3.3 above.

[17]   See para. 4.3.4 above.

[18]   See para. 4.3.6 above.

[19]   See para. 12.2.1 above.

[20]   See para. 4.3.3 above.

[21]   See Paul Craig and Gráinne de Búrca, *EU Law: Text, Cases and Materials*, 3rd edn (Oxford: Oxford University Press, 2003), p. 144.

[22]   23 October 2001. (All details of the procedure in this chapter are from PreLex, at www. europa.eu.int/prelex.)

to the EU ETS Directive's final signature by the two institutions, was a period of almost two years.[23] Throughout, the Environment DG of the Commission, under the Environment Commissioner, Mrs Margot Wallström,[24] has had primary responsibility for the Directive.

28.2.2   The Commission's Green Paper on emissions trading

In March 2000, against the backdrop of the ECCP,[25] the Commission[26] adopted a Green Paper on GHG emissions trading within the EU.[27] Referring to Art. 17 of the Kyoto Protocol,[28] the Green Paper identified emissions trading schemes as one of the three Kyoto mechanisms, which together would facilitate the Protocol's cost-effective implementation.[29] Having accepted the differentiated reduction targets in Annex B to the Protocol,[30] each of the parties thereto was, emphasised the Paper, entitled to participate in international emissions trading under the Protocol. This being so, the Paper invited all governmental and non-governmental stakeholders to give their opinions on how the EU should 'strike the right balance' in the use of emissions trading.[31] To this end, the Green Paper asked respondents 10 specific questions about the design of such a scheme.[32]

According to the Green Paper, the main economic rationale for emissions trading[33] was the use of market mechanisms to make sure that 'emissions reductions required to achieve a pre-determined environmental outcome take place where the cost of reduction is the lowest'.[34] Although there had hitherto been no major application of tradable allowances in the Community's environmental policy, the Green Paper emphasised that Community law did contain a number of similar concepts, for

---

[23]   That is, 23 October 2001 to 13 October 2003.

[24]   Mrs Margot Wallström was confirmed as Environment Commissioner in September 1999 and expressed her belief in emissions trading at her confirmation hearing (see 296 ENDS Report (1999)). Commissioner Wallström has had a long domestic political career, both as a Member of the Swedish Parliament and in the Swedish Government, having been Minister for Consumer Affairs, Women and Youth (1988–1991), Minister for Culture (1994–1996) and Minister for Social Affairs (1996–1998). She is the only member of the 'Prodi Commission' to be reappointed to the post-2004 'Barroso Commission'.

[25]   See para. 12.2.6.2 above.

[26]   See para. 4.3.2 above.

[27]   COM (00) 87 final, available on EurLex, at www.europa.eu.int (referred to below as the 'Green Paper').

[28]   See para. 8.3.1.4 above.

[29]   Green Paper, para. 1.

[30]   See para. 8.3.1.4 above.

[31]   Green Paper, para. 2.

[32]   Green Paper, paras 6.3, 7.4, 8.5 and 9.4.

[33]   Emissions trading was defined in the Green Paper as 'a scheme whereby companies are allocated allowances for their emissions of greenhouse gases according to the overall environmental ambitions of their government, which they can trade subsequently with each other' (Green Paper, para. 3).

[34]   Green Paper, para. 3.

example: quotas for ozone-depleting chemicals under the 1987 Montreal Protocol;[35] fish catch quotas under the common fisheries policy;[36] and milk quotas under the CAP.[37] Alluding to the burden-sharing agreement between Member States,[38] the Green Paper stressed that the agreement did not constrain the use of the Kyoto mechanisms,[39] but, equally, that any EU-wide emissions trading scheme would not necessarily be identical to an international emissions trading scheme as envisaged by Art. 17 of the Kyoto Protocol.[40] In particular, said the Green Paper, the fact that Art. 17 did not refer to the involvement of 'entities' in emissions trading schemes did not disturb the Commission's view that the involvement of companies in emissions trading 'represent[ed] a unique opportunity for a cost-effective implementation of the Kyoto commitments'.[41] The prudent approach to the development of emissions trading at Community level, according to the Paper, was for the Community to confine itself initially to 'large fixed point sources of carbon dioxide', since this would facilitate the monitoring and supervision of the scheme.[42] This was a significant limitation on the scope of emissions trading, since, for the present at least, it would be excluded from the scope of any scheme of the other five GHGs. A Community-wide trading scheme would lead to a single price for allowances traded by scheme companies.[43] Although the scheme would need to be carefully designed to protect the internal market and so as not to infringe international treaty obligations,[44] it would be possible to organise the scheme at one of at least three different levels: least ambitiously, a series of national schemes in the Member States, overseen by the Commission; more ambitiously, a coordinated Community scheme that nonetheless allowed Member States a degree of discretion; or, most ambitiously, a harmonised Community-wide scheme which would limit the discretion of Member States and in which the design of all of the essential elements of the scheme would be agreed at the level of the Community.[45]

On the question of which industrial sectors should be included within the scheme, the Green Paper identified six sectors which together were responsible for something over 45.1 per cent of carbon dioxide emissions within the (then) 15 Member States, that is:

---

[35]   See European Parliament and Council Regulation EC/2037/00. See also para. 8.3.1.5 above.

[36]   See Council Regulation EC/2371/02, (2002) OJ L358 59.

[37]   See Council Regulation EC/856/84, (1984) OJ L90 10; Commission Regulation EC/1392/01, (2001) OJ L187 19; and Council Regulation EC/1788/03, (2003) OJ L270 123.

[38]   See para. 8.2.2n above.

[39]   Green Paper, para. 4.1.

[40]   *Ibid.*, para. 4.2.

[41]   *Ibid.*, para. 4.2.

[42]   *Ibid.*, para. 4.3.

[43]   *Ibid.*, para. 5.1.

[44]   *Ibid.*, para. 5.2.

[45]   *Ibid.*, para. 5.3.

1   electricity and heat generation;
2   the iron and steel industries;
3   oil refining;
4   the chemical industry;
5   the glass, pottery and building materials (including cement) industries; and
6   the paper and printing industries (including paper pulping).[46]

The selection of these six sectors, according to one early commentator on the Green Paper, represented not only a significant limitation on the proposed scheme, but also showed the potential difficulties of relating it to the UK ETS, then at a relatively early planning stage. In identifying the six sectors, the Commission had bowed to the political realities preventing the creation of an upstream scheme which, in embracing all energy sources, might therefore have covered all economic sectors, including not only industrial sectors such as transport, but also non-industrial sectors, such as households. The same commentator also predicted that it was unlikely that electricity or heat generators would be eligible to participate in the UK scheme, given that the UK ETS was designed to operate within the structural constraints imposed by the climate change levy which was, of course, a downstream tax on the final use of energy by industrial/commercial consumers.[47]

The Green Paper then turned to the question of the speed at which any scheme should be introduced and the stages in which the relevant sectors would be covered. It put forward three possibilities:[48]

1   a common Community scheme, which would apply from the outset to the same sectors in all the Member States;
2   a coordinated 'opt-in' scheme, agreed on at the outset by Member States, which Member States might 'opt-in' to; and
3   a coordinated 'opt-out' scheme, again agreed on at the outset by the Member States, under which they would agree as to which sectors should in principle be covered by the scheme but which would leave them free to 'opt-out', either completely or in respect of certain sectors.

The Green Paper emphasised, however, that possibilities 2 and 3 above would be subject to the condition that sectors not covered by the scheme should be regulated by other measures 'that represent at least a similar economic effort in terms of emissions abatement'.[49] Although not developed in detail in the Green Paper, it was expressly recognised that possibilities 2 and 3 raise complex competition concerns.[50]

Potentially more contentious even than issues of sector coverage and coordination, however, were questions of emissions allocation. In para. 7.2 of the Green Paper, the

---

[46]   Green Paper, para. 6.1, Table 1.
[47]   See 302 ENDS Report (2000). In the event, this indeed proved to be the case, electricity and heat generation being excluded from the UK ETS (see para. 20.4 above).
[48]   Green Paper, para. 6.2.
[49]   *Ibid.*, para. 6.2.2.
[50]   *Ibid.*

Commission carefully avoided a categorical statement of the role of the Community in the allocation of emissions allowances to Member States, preferring instead to refer generally to the Commission's role in intervening to safeguard fair competition and freedom of establishment within the internal market:

> However, the need for and nature of such intervention will depend very much upon the choices that are made. If the Community were to agree on the quantity of the emissions of the trading sectors in each Member State, possible distortive allocations to individual sectors or companies would be significantly limited. Hence, existing guidelines for state aid in the field of the environment would be sufficient to check whether allowances allocated to companies would respect EC competition law.[51]

Nonetheless, '... lack of agreement on what quantity of emissions should be allocated to the trading sectors in each Member State will require detailed and tight guidelines on how allocations are made to individual sectors and companies, and close scrutiny of every single case'.[52]

In addition to the contentious and no doubt highly political issues just discussed, there were also issues of a more technical nature and, in this regard, the Green Paper emphasised that it was still for clarification as to how technical regulation (for example, the Large Combustion Plant Directive[53] and the IPPC Directive),[54] taxation (for example, the Energy Products Directive)[55] and environmental agreements (for example, those with ACEA, JAMA and KAMA)[56] were respectively substitutes for, or complementary to, an emissions trading scheme.[57] Moreover, there were issues as to compliance and enforcement, both in relation to individual companies and to Member States.[58]

On the basis of all of the foregoing, the Green Paper was able to ask the ten questions referred to above on the design of the scheme. These are worth setting out in full, since they show the highly purposeful and 'directed' nature of the consultation, as well as highlighting certain issues that tend to recur in discussions of the EU ETS Directive:

1   as to which sectors should be covered by the scheme (that is, whether it should be tied into the IPPC regime);[59]
2   as to whether there should be a common Community scheme;[60]

---

[51]   Green Paper, para. 7.2.1.
[52]   *Ibid.*, para. 7.2.1.
[53]   See para. 12.2.6.2 above.
[54]   See para. 12.2.2 above.
[55]   See para. 12.3.4 above.
[56]   See para. 12.2.6.4(2) above.
[57]   Green Paper, para. 8.
[58]   *Ibid.*, para. 9.
[59]   *Ibid.*, Question 1, para. 6.3.
[60]   *Ibid.*, Question 2, para. 6.3.

3   as to whether there should be a coordinated 'opt-in' or 'opt-out' scheme, with regard to the requirements of the internal market and the potential for complexity in either of these possibilities;[61]

4   as to the scope, if any, for Member States to include more sectors in their domestic trading scheme than might be covered by a Community scheme;[62]

5   as to whether the overall amount of allowances allocated should be subject to agreement at Community level;[63]

6   as to whether the way in which allowances are allocated should be subject to agreement at Community level or whether detailed guidelines based on existing competition law would be sufficient;[64]

7   as to the balance to be struck between sectors involved in the scheme and other sectors;[65]

8   as to how the mix of emissions trading, energy taxes and environmental agreements might be geared towards environmental effectiveness and transparency;[66]

9   as to whether, given that the scheme would require a robust enforcement regime in order to be successful, pre-existing Community enforcement mechanisms would be adequate or whether it would be necessary to develop additional ones;[67] and

10 finally, as to whether compliance and enforcement should be harmonised or coordinated at Community level.[68]

28.2.3 Developments subsequent to the Green Paper

To the ten specific questions raised in the Green Paper, the Commission subsequently reported that there had been approximately 100 responses, of which the majority were 'overwhelmingly in favour' of emissions trading.[69] Whilst this may well have been the case, such a statement did not however reveal anything about the shades of opinion on the ten specific questions thereby raised. That the creation of an EU-wide GHG emissions trading scheme was seen by the Commission and the Council at least as an imperative was however abundantly clear from the conclusions of its 2334th meeting in Brussels, on 8 March 2001,[70] when the Council singled out the Commission's work on the Green Paper as being, together with the ECCP, of the greatest importance.[71]

---

61   *Ibid.*, Question 3, para. 6.3.
62   *Ibid.*, Question 4, para. 6.3.
63   *Ibid.*, Question 5, para. 7.4.
64   *Ibid.*, Question 6, para. 7.4.
65   *Ibid.*, Question 7, para. 8.5.
66   *Ibid.*, Question 8, para. 8.5.
67   *Ibid.*, Question 9, para. 9.4.
68   *Ibid.*, Question 10, para. 9.4.
69   See the Explanatory Memorandum to the Proposal for the EU ETS, discussed below, para. 1.1.
70   At which the UK was represented by The Rt Hon. Michael Meacher, MP, (then) Minister of State for Environment at the (then) DETR (see para. 4.2.1.2(1) above), and Mr Sam Galbraith, MSP, then Minister for the Environment, Sport and Culture, in the Scottish Executive (see para. 4.2.2 above).
71   See PRES/01/93, 8 March 2001, available from www.europa.eu.int. See also para. 12.2.6.2 above.

The Commission adopted its Proposal for the EU ETS Directive[72] on 23 October 2001,[73] sending it both to the Council and to Parliament, in accordance with the Art. 175 procedure.[74] According to the Explanatory Memorandum that formed part of the Proposal, the latter relied on two central concepts, that is, a GHG 'permit', which would be required by all installations falling within the scope of the proposed scheme, and GHG 'allowances', denominated in metric tonnes of carbon dioxide equivalent and which would entitle the holder to emit a corresponding quantity of GHGs.[75] The six sectors listed in the Green Paper became the nine types of installation which were eventually to be listed in the EU ETS Directive itself.[76]

Certain features of the Proposal were perhaps rather surprising. For instance, the indications in the Green Paper[77] that the Commission was prepared to consider a transitional introduction of the EU ETS, with trading not initially being compulsory for all six sectors were not, however, reflected in the Proposal. The Proposal did, however, embody a concession in relation to the method of allocating allowances. By Art. 10 of the Proposal, it was envisaged that, between 1 January 2005, and 1 January 2008, Member States should allocate allowances free of charge, with a harmonised scheme to apply for the five-year period beginning on 1 January 2008. This was in response to the criticism made by UNICE and others[78] that, to leave allocation methods in the hands of Member States (albeit subject to the state aid rules)[79] would confer unfair competitive advantages on industries in certain States. Allocating allowances free of charge, however, would involve grandfathering, that is, basing allowances for 2005–2007 on past emissions. In order to avoid the possibility that grandfathering would therefore allow firms to derive windfall gains from selling surplus credits thus generated, the criteria for national allocation plans contained in Annex III to the Proposal included the requirement that '[q]uantities of allowances to be allocated shall be consistent with the technological potential of installations to reduce emissions'. Such national allocation plans were, in any event, subject to the Commission's veto.[80] The scheme thus envisaged disclosed an important point of difference with the UK ETS. Given the electronic auction of allowances under the UK ETS,[81] the grandfathering rights in the EU ETS did not arise in relation to the UK ETS, '... especially as regards so-called 'direct participants' outside the sectoral climate change agreements. They presume a depth of knowledge about, and official

---

[72]   COM (01) 581 final, (2002) OJ C75E 33.

[73]   All details of the institutional progress of the Directive are taken from the PreLex database on www.europa.eu.int/prelex.

[74]   What follows in this para. traces only the main features of the legislative process. The process, including, for example, the contributions of ECOSOC and the Committee of the Regions, can be traced in full at www.europa.eu.int/prelex.

[75]   See Proposal, para. 1.2.

[76]   See para. 28.3 below.

[77]   See Green Paper, para. 6.2.

[78]   See 321 ENDS Report (2001).

[79]   As envisaged by Green Paper, para. 7.2 (see above). As to state aids, see para. 12.2.7 above.

[80]   COM (01) 581 final, Art. 9.

[81]   See para. 20.2 above.

second-guessing of, greenhouse gas abatement opportunities across industry which would erode the attractiveness of trading to both business and governments'.[82]

Although, following the objections to the Green Paper just discussed, the Proposal contained a concession on the question of the method of allocation of allowances, it confirmed the view in para. 4.3 of the Green Paper that the scheme should initially cover carbon dioxide only, given the perceived difficulties of measuring the other five GHGs. On this basis, Annex I to the Proposal confined the gases to be subject to the scheme to carbon dioxide, whilst giving the Commission the opportunity to make a further proposal by 31 December 2004, for the inclusion of other GHGs.[83]

Despite the differences with the UK ETS referred to above, in at least one other area, the Proposal appeared to be consistent with other Community and domestic regulation, that is, in relation to the criteria for national allowance allocation plans contained in Annex III to the EU ETS. Criterion 4 in Annex III required national allocation plans to be consistent with other EC legislative and policy instruments:

> In particular, no allowances should be allocated to cover emissions which would be reduced or eliminated as a consequence of Community legislation on renewable energy in electricity production, and account should be taken of unavoidable increases in emissions resulting from new legislative requirements.[84]

This part of the Proposal was consistent with the national indicative targets, already discussed in this study, for renewables output in 2010, and which are contained both in the Renewables Directive[85] and in domestic UK law.[86]

With the Proposal indicating considerable scope for difficulty, it may have seemed that the omens for an amicable discussion of the Proposal at the 2457th meeting of the Council in Luxembourg, on 17 October 2002,[87] were hardly auspicious. Denmark, then holding the Presidency of the Council,[88] had tabled a compromise proposal on the Proposal but, given that some Member States said that they were not yet ready to discuss the details, the discussion became a general debate, with political agreement on the Proposal being held over until the Council Meeting scheduled for 19 December 2002.[89] Following the October 2002 Council Meeting, a number of issues remained, therefore, including: whether participation in the proposed scheme should

---

82 321 ENDS Report (2001).

83 COM (01) 581 final, Art. 26(1).

84 *Ibid.*, Annex III, para. 4.

85 See paras 12.2.6.3(1) and 12.2.6.3(2) above.

86 See paras 6.4 and 12.2.6.3(2) above. See also paras 6.4.3.1(2) and 21.5 above, for Renewable Obligation (or 'Green') Certificates ('ROCs').

87 See PRES/2002/320, available on PreLex, at www.europa.eu.int/prelex. The UK was there represented by The Rt Hon. Mrs Margaret Beckett, MP, Secretary of State for Environment, Food and Rural Affairs and The Rt Hon. Michael Meacher, MP (then) Minister of State for Environment at Defra (see para. 4.2.1.2(1) above).

88 See Art. 203, European Treaty (ex 146). Since the 2457th Council Meeting concerned the environment, the President was Mr Hans Christian Schmidt, the Danish Environment Minister.

89 The account of the proceedings here is taken from 333 ENDS Report (2002).

be compulsory or voluntary; whether the proposed scheme should be restricted to the nine types of installation contained in the Proposal; whether allowances should be grandfathered or auctioned; and whether Member States which were struggling to meet their targets under the burden-sharing agreement[90] should be able to buy-in credits from emissions reduction schemes carried out in third countries under JI and CDM.[91] The European Parliament had already registered its view on each of these matters on 10 October 2002, when, at the first reading of the Proposal, it had voted for 73 amendments to the Proposal, including:

1   allowing Member States to apply to the Commission for the temporary exclusion of certain installations from the proposed scheme, provided that the installations thus excluded had satisfied the Commission that their emissions would be reduced as much as if they had participated in the scheme;[92]
2   enabling Member States to extend the scheme to additional sectors, activities and installations;[93]
3   requiring the auctioning of 15 per cent of allowances for both 2005–2007 and 2008–2012;[94] and
4   excluding credits from CDM/JI projects in third countries under the Kyoto Protocol until 2008.

Only on 4, prior to the October 2002 Council meeting, were Commission and Parliament agreed. As to the others, Commissioner Wallström expressed the views, in a speech to a Eurelectric Workshop on 15 October 2002, that voluntary participation would produce competitive distortions; that the extension of the scope of the scheme to GHGs other than carbon dioxide would overburden the scheme; and that any amount of auctioning, however small, would make it even harder for businesses to accept emissions trading.[95] These Commission objections were accordingly reflected in the amended Proposal adopted by the Commission on 27 November 2002.[96] The amended Proposal did not include any provision for voluntary participation, the Commission insisting on a mandatory scheme as of 2005.[97] Neither was there

---

90   See para. 8.2.2 above.
91   The UK was one of a group of Member States who wanted participation to be voluntary until the beginning of 2008 (see 333 ENDS Report (2002)). This position no doubt owed something to the originally voluntary nature of the UK ETS.
92   See (2003) OJ C279E 20 and the *Report of the Committee on the Environment, Public Health and Consumer Policy on the Proposal for a European Parliament and Council Directive establishing a Scheme for Greenhouse Gas Emission Allowance Trading within the Community and amending Council Directive 96/61/EC*, A5-0303/2002, 13 September 2002, Part 1, p. 32 (the Portuguese MEP, Mr Jorge Moreira da Silva, Rapporteur), amendment 50 (inserting a new Art. 23a into the Proposal), available on PreLex, at www.europa.eu.int/prelex.
93   See A5–0303/2002, Part 1, p.15, amendments 16 and 17 (amending Art. 2 of the Proposal).
94   See 333 ENDS Report (2002).
95   Speech 02/481, 15 October 2002, available from www.europa.eu.int.
96   COM (02) 680 final.
97   *Ibid.*, p. 11.

any provision for the voluntary and unilateral 'opt-in' of additional activities, the Commission's argument being that this would distort competition and undermine the scheme's environmental integrity.[98] Finally, since the Commission was opposed to any auctioning at all in the period 2005–2008, the allocation of allowances was to be on a 'free of charge' basis.[99]

Accordingly, on 9 December 2002, the amended Proposal having been transmitted to the Council and Parliament on 27 November the two institutions reached political agreement on a common position on the amended Proposal. This was that Parliament and the Council would provide the Member States with the option of applying to the Commission for the temporary exclusion of certain installations and activities until 31 December 2007; that Member States would be at liberty to allocate allowances free of charge in the period 1 January 2005 to 31 December 2007, 90 per cent being allocated free of charge in the period 1 January 2008 to 31 December 2012; that Member States would be allowed to grant operators that carry on the same activity to form a pool, with a trustee having responsibility for managing the allowances on their behalf;[100] and, finally, that penalties of 40 euros would be fixed for the period 1 January 2005 to 31 December 2007 and of 100 euros in the period 1 January 2008 to 31 December 2012, for each tonne of carbon dioxide emitted that exceeded the allowances granted. Speaking on the same day, Commissioner Wallström was reported as saying:

> This is a landmark decision for the EU's strategy to fight climate change … It proves that the EU is taking action on climate change and gets emissions down, and that we do so in a way that minimises the cost to the economy. The world's eyes have been upon us to see whether we will succeed in creating the biggest emissions trading scheme world-wide so far. We *have* succeeded. It will help all member States, as well as the EU as a whole, to reach their Kyoto targets while cutting costs at the same time.[101]

The result was satisfying to the UK, since it had seen most of its demands met. In particular, the opt-out would allow it some flexibility in managing its transition from the UK ETS to the EU ETS.[102] How the opt-out criteria were to be met was a different matter, however, given that the electricity generation and heat generation sectors were excluded from the UK ETS. However, it was also reported that Parliament's Rapporteur for the Directive, the Portuguese MEP, Mr Jorge Moreira da Silva, was predicting that Parliament was unlikely to agree to the amended proposal at its second reading. This, of course, gave rise to the possibility that the adoption of the EU ETS Directive could be delayed as it worked its way through the conciliation procedure.[103] A defeat for the UK, however, was the fact that the Commission

---

98   COM (02) 680 final, p. 7.
99   *Ibid.*, p. 9.
100 This had been included to accommodate Germany's concerns as to the preservation of its sectoral climate change agreements (see 335 ENDS Report (2002)).
101 Press Release IP/02/1832, available from www.europa.eu.int.
102 See 335 ENDS Report (2002). See further para. 28.4 below.
103 See 335 ENDS Report (2002). For the conciliation procedure, see Craig and de Búrca, *op. cit.*, pp. 80–81 and 148.

retained its right of veto over national allocation plans, for the deletion of which veto the UK Minister of State for Environment, The Rt Hon. Michael Meacher, MP, had previously announced he would be pressing strongly.[104]

This was how matters stood at the time of the 2494th meeting of the Council in Brussels, on 17–18 March 2003.[105] There, the Council formally adopted a common position[106] on the amended Proposal and, in accordance with the co-decision procedure, forwarded the common position to Parliament for its second reading. In its statement of reasons annexed to the common position, the Council stated:

> Of the 73 amendments proposed by the European Parliament in first reading, the common position incorporates 23 (totally, in part or in principle, by means of identical or similar wording, or in spirit). The Council considers that the common position does not alter the approach and aims of the original proposal from the Commission and notes that the Commission also supports the common position as it stands.[107]

In a Communication of 25 March 2003, the Commission, pursuant to Art. 251(2), European Treaty (ex 189b), notified Parliament fully of its position.[108] Parliament sought to amend the common position further at the second reading of the Proposal,[109] adopting 17 amendments to the Council's common position, all of which the Commission subsequently approved, considering 'that the principal aims of the EC emissions trading scheme are safeguarded by the compromise package that these amendments constitute'.[110] Accordingly, the Commission amended the Proposal pursuant to Art. 250(2), European Treaty (ex 189a(2)), the Council finally adopting the EU ETS Directive at its 2524th meeting in Brussels on 22 July 2003.[111]

The EU ETS Directive was ultimately signed off by both Parliament and the Council on 13 October 2003.

---

104 See 335 ENDS Report (2002).
105 See PRES/2003/76, available on PreLex, at www.europa.eu.int/prelex. The UK was there represented by Lord Whitty, Parliamentary Under-Secretary of State, Defra, and Mr Ross Finnie, MSP, Minister for Environment and Rural Development in the Scottish Executive (see para. 4.2.2 above).
106 Common Position (EC) No 28/2003 adopted by the Council on 18 March 2003, (2003) OJ C 125 E 72.
107 *Ibid.*, 89.
108 SEC (03) 364 final.
109 See *Recommendation for Second Reading on the Council Common Position for adopting a European Parliament and Council directive establishing a scheme for greenhouse gas emission allowance trading within the Community and amending Council Directive 96/61/EC*, A5–0207/2003, 12 June 2003 (Mr Jorge Moreira da Silva, Rapporteur).
110 COM (03) 463 final, available from PreLex, at www.europa.eu.int/prelex.
111 See PRES/2003/215, available on PreLex, at www.europa.eu.int/prelex. On this occasion, the UK was represented by Mrs Margaret Beckett, above, and by Mr Ben Bradshaw, MP, Parliamentary Secretary, Privy Council Office.

28.3   Provisions of the EU ETS Directive

The installations and the single GHG to which the Directive currently applies are set out in Annex I thereto. Annex I draws within the scope of the EU ETS Directive emissions of carbon dioxide only from each of the following:[112]

Energy activities
Combustion installations with a rated thermal input exceeding 20 MW (except hazardous or municipal waste installations)
Mineral oil refineries
Coke ovens

Production and processing of ferrous metals
Metal ore (including sulphide ore) roasting or sintering installations
Installations for the production of pig iron or steel (primary or secondary fusion) including continuous casting, with a capacity exceeding 2.5 tonnes per hour

Mineral industry
Installations for the production of cement clinker in rotary kilns with a production capacity exceeding 500 tonnes per day or lime in rotary kilns with a production capacity exceeding 50 tonnes per day or in other furnaces with a production capacity exceeding 50 tonnes per day
Installations for the manufacture of glass including glass fibre with a melting capacity exceeding 20 tonnes per day
Installations for the manufacture of ceramic products by firing, in particular roofing tiles, bricks, refractory bricks, tiles, stoneware or porcelain, with a production capacity exceeding 75 tonnes per day, and/or with a kiln capacity exceeding 4m³ and with a setting density per kiln exceeding 300 kg/m³

Other activities
Industrial plants for the production of
(a) pulp from timber or other fibrous materials
(b) paper and board with a production capacity exceeding 20 tonnes per day

The list in Annex I is worth setting out in full in this way, since it is identical to that set out in Annex I to the original Proposal for the EU ETS Directive of October 2001,[113] and that has so far been alluded to only briefly in this discussion. Furthermore, although Annex II contains the full list of GHGs contained in Annex A to the Kyoto Protocol,[114] there is provision for the Commission to make a proposal to Parliament and the Council by 31 December 2004, to amend Annex I to include other activities and emissions of other GHGs listed in Annex II.[115] Although, as mentioned above, Parliament had wanted to include a provision for Member States

---

[112]   See Henry van Geen, 'Emission Allowance Trading in the EU', [2003] 1 IELTR 299–305.
[113]   See para. 28.2.3 above.
[114]   See para. 8.3.1.4 above.
[115]   Directive 03/87/EC, Art. 30(1).

to be entitled to extend the scheme, its temporary confinement to carbon dioxide reflects the Commission's fears about 'overburdening' it. The amendment to the Commission's proposal agreed on 9 December 2002, which allows Member States to apply for the temporary exclusion of certain installations from the EU ETS until 31 December 2007, appears as Art. 27.

Article 4 of the Directive requires Member States to ensure that, from 1 January 2005, in any case where the operator of an installation carries on any of the activities mentioned above and the activity results in the emissions specified, then that operator must hold a permit duly issued by a competent authority.[116] By Art. 6(2) of the Directive, in addition to certain administrative details, such GHG emissions permits must contain:

> an obligation to surrender allowances equal to the total emissions of the installation in each calendar year, as verified in accordance with Article 15,[117] within four months following the end of that year.[118]

For the three-year period beginning on 1 January 2005 ('the first phase'), as well as the five-year period beginning on 1 January 2008 and each five-year period thereafter ('the second and subsequent phases'), each Member State must develop a national plan, setting out the total quantity of allowances that it intends to allocate for that phase and how it proposes to allocate them.[119] By Art. 9(1) of the Directive, the criteria for the allocation, on which the plan is based, must include those set out in Annex III to the Directive and, by Art. 9(3), the Commission (who must be notified of the plan) may (with reasons) accept or reject the plan.

For the first phase, each Member State must decide on the total quantity of allowances to be allocated for that phase and the allocation of those allowances to the operator of each installation.[120] For the second and subsequent phases, each Member State must decide on the total quantity of allowances it will allocate for that phase and begin the process for the allocation of those allowances to the operator of each installation.[121] Each of these allocation rounds must take place within certain time periods, must be based on the national allocation plan of the Member State in question and must take due account of comments from the public.[122] Allocation decisions must also take account of the provisions of the European Treaty, especially

---

[116]   Applications, conditions for and contents of, GHG emissions permits are covered in Directive 03/87/EC, Arts 5 and 6. Article 8, Directive 03/87/EC, requires Member States to take the measures necessary to ensure that the conditions for the issue of a GHG emissions permit are coordinated with those for the issue of a permit under the IPPC Directive; the requirements for the issue of a GHG emissions permit under Directive 03/87/EC may be integrated with those applicable under the IPPC Directive (see para. 12.2.2 above).

[117]   See below in this para.

[118]   Directive 03/87/EC, Art. 6(2)(e).

[119]   *Ibid.*, Art. 9(1).

[120]   *Ibid.*, Art. 11(1).

[121]   *Ibid.*, Art. 11(2).

[122]   *Ibid.*, Arts 11(1) and 11(2).

[123]   See para. 12.2.7.1 above.

those on state aids,[123] as well as of the need to provide access to allowances for new entrants.[124] For the first phase, at least 95 per cent of the allowances must be allocated free of charge and; in the subsequent phase, at least 90 per cent of the allowances must also be allocated free of charge.[125] The scope thus embodied in the Directive for at least some auctioning of allowances in the period 2005–2007 was part of the final compromise struck between Parliament and the Commission in July 2003.[126]

Article 12 of the Directive deals with the transfer, surrender and cancellation of allowances. Four distinct obligations are placed on Member States. First, Member States must ensure that allowances can be transferred, not only between persons within the Community, but also, subject to the provisions of Art. 25,[127] between persons within the Community and persons in third countries.[128] Secondly, Member States must ensure that, no later than 30 April each year:

> ... the operator of each installation surrenders a number of allowances equal to the total emissions from that installation during the preceding calendar year as verified in accordance with Article 15, and that these are subsequently cancelled.[129]

Thirdly, Member States must make sure that, for the purposes of meeting the operator's obligation to surrender allowances, allowances issued by a competent authority of another Member State are duly recognised.[130] Finally, Member States must take the necessary steps to ensure that allowances will be cancelled at any time at the request of the person holding them.[131]

Other Articles provide for the adoption of guidelines for the monitoring and reporting of emissions;[132] the adoption of rules for the imposition of penalties for breach of the national provisions adopted pursuant to the Directive;[133] access to decisions relating to matters such as the allocation of allowances;[134] the designation of the competent authority and registry for accounting for the issue, holding, transfer and cancellation of allowances;[135] the designation of a central administrator;[136] and for reporting by Member States.[137] Rules on the imposition of penalties are contained in Art. 16,

---

[124] *Ibid.*, Art. 11(3).
[125] *Ibid.*, Art. 10.
[126] COM 903, 463 final. See para. 28.2.3 above.
[127] See below.
[128] Directive 03/87/EC, Art. 12(1).
[129] *Ibid.*, Art. 12(3).
[130] *Ibid.*, Art. 12(2).
[131] *Ibid.*, Art. 12(4).
[132] *Ibid.*, Art. 14. Such guidelines are to be based on the principles for monitoring and reporting set out in Annex IV to the Directive.
[133] *Ibid.*, Art. 16.
[134] *Ibid.*, Art. 17.
[135] *Ibid.*, Arts 18 and 19.
[136] *Ibid.*, Art. 20.
[137] *Ibid.*, Art. 21.

which provides both for the imposition of 'effective, proportionate and dissuasive' penalties[138] and for the naming and shaming of operators who are in breach of their obligation to surrender sufficient allowances under Art. 12.[139] In particular, by Art. 16(3):

> Member States shall ensure that any operator who does not surrender sufficient allowances by 30 April of each year to cover its emissions during the preceding year shall be held liable for the payment of an excess emissions penalty. The excess emissions penalty shall be EUR 100 for each tonne of carbon dioxide equivalent emitted by that installation for which the operator has not surrendered allowances. Payment of the excess emissions penalty shall not release the operator from the obligation to surrender an amount of allowances equal to those excess emissions when surrendering allowances in relation to the following calendar year.[140]

The amendments to the IPPC Directive necessitated by the EU ETS Directive are contained in Art. 26. These consist of the addition to Art. 9(3) of the IPPC Directive[141] of four further sub-paragraphs. Article 9(3) relates to the inclusion in permits of emission limit values based on the BATs for certain pollutants. The main addition to Art. 9(3), provided for by Art. 26, is that, where emissions of a GHG from an installation are specified in Annex I to the Directive,[142] in relation to an activity carried out in that installation, 'the permit shall not include an emission limit value for direct emissions of that gas unless it is necessary to ensure that no significant local pollution is caused'.[143] The amendment to the IPPC Directive contained in Art. 26 has been included in anticipation of gases other than carbon dioxide being included within the scope of the EU ETS Directive,[144] in which event IPPC permits will not generally contain a limit value for any such additional gas or gases.

Article 15 of the Directive provides that, when emissions are reported, they are to be verified in accordance with the criteria specified in Annex V to the Directive. Furthermore, in the absence of such verification by 31 March in each year, for emissions during the preceding year, the operator in question may not make any further transfers of allowances.

The pooling of installations, as required by Germany especially,[145] is provided for by Art. 28 of the Directive.

Finally, Art. 25 of the Directive provides for the linking of the EU ETS to other GHG emissions trading schemes. Article 25(1) states that agreements should be concluded with third countries[146] to provide for the mutual recognition of

---

138　*Ibid.*, Art. 16(1).
139　See above in this para.
140　But note that a lower penalty of EUR 40 will be applicable in the initial period referred to above (that is, the three years beginning with 1 January 2005).
141　See para. 12.2.2 above.
142　See above in this para.
143　Directive 03/87/EC, Art. 26.
144　*Ibid.*, Art. 30(1).
145　See para. 28.2.3n above.
146　That is, as listed in Annex B to the Kyoto Protocol, such third countries having ratified the Protocol (see para. 8.3.1.4 above).

allowances under other GHG emissions trading schemes. At the time of writing, some progress has been made with regard to such linking, a Proposal for a Directive of the European Parliament and of the Council, amending the EU ETS, in respect of the Kyoto Protocol's flexible mechanisms, having been adopted by the Commission in July 2003 under Art. 175, European Treaty (ex 130s).[147] This has reached the stage of being discussed by the Council at its 2556th meeting in Brussels on 22 December 2003.[148] Whilst the Commission has argued that the Proposal for the linking Directive will have economic benefits, environmental groups contend that it will have the effect of flooding the market with cheap credits from schemes having little or no environmental integrity.[149]

28.4   Transposition of the EU ETS Directive into UK law

Article 31 of the Directive requires Member States to bring into force the measures necessary to comply with the Directive no later than 31 December 2003.[150] Characteristically, the UK has sought to comply with its Treaty obligations with an exhaustive 67-page piece of secondary legislation,[151] the Greenhouse Gas Emissions Trading Scheme Regulations 2003,[152] which came into force on 31 December 2003.[153]

The basis for Regulations is the European Communities Act 1972, s.2(2), which allows any designated Minister or Department[154] to make provision for the purpose of implementing any Community obligation of the UK.[155] This is without, at this stage, an RIA having been conducted and less than three months after the publication of the EU ETS Directive in the *Official Journal*.

The purpose of the Regulations is, of course, to transcribe the EU ETS Directive into UK law. Although, the Regulations apply to the UK as a whole, provision is made in reg. 2 for the Scottish Ministers, the Department of the Environment in Northern Ireland and the National Assembly for Wales[156] to act as the 'appropriate authority' for their respective Principalities. As regards England and any of its offshore

---

[147] COM (03) 403 final, available on PreLex, at www.europa.eu.int/prelex.

[148] See PRES/2003/376, available from www.europa.eu.int.

[149] See 343 ENDS Report (2003).

[150] The issues surrounding implementation of the Directive are discussed in detail in Fiona Mullins and Jacqueline Karas, *EU Emissions Trading: Challenges and Implications of National Implementation* (London: Royal Institute of International Affairs, November 2003).

[151] A bulk which illustrative of a characteristically British activity, known as 'gold-plating' (see Moules, *Financial Times*, 3 January 2004, p. 3).

[152] S.I. 2003 No. 3311.

[153] *Ibid.*, reg. 1.

[154] In this case, The Rt Hon. Elliot Morley, MP, Minister of State for Environment, being a minister designated for the purpose of s.2(2), European Communities Act 1972, in relation to air pollution.

[155] The Regulations were made on 18 December 2003.

[156] See para. 4.2.2 above.

installations, the appropriate authority is the Secretary of State for Environment, Food and Rural Affairs.[157]

Schedule 1 to the Regulations specifies the installations[158] and activities covered by the Regulations by transcribing, with only slight adjustments, the table in Annex I to the EU ETS Directive.[159] Excluded from the scope of the Regulations, as also from the Directive itself, are installations used for research, development and testing of new products and processes.[160] Even so, in the UK, the total number of installations expected to be subject to the scheme has been estimated at between 1,500 and 2,000.[161] Regulation 11 contains the provision allowing operators of installations to apply to the responsible authority[162] for temporary exclusion of an installation from the scheme, as provided for by Art. 27(2) of the Directive. The announcement in the 2003 Pre-Budget Report that firms within sectoral climate change agreements (that is, umbrella agreements)[163] will retain their climate change levy discount if they opt into the EU ETS should minimise the number of opt-outs under Regulation 11.[164] Nevertheless, one of the big differences between the UK ETS and the EU ETS is the inclusion in the latter of the heat generation and electricity generation sectors.

The Regulations themselves are divided into three substantive parts, dealing respectively with the granting, transfer and revocation of GHG Emissions Permits;[165] the allocation and issue of allowances;[166] and the monitoring and enforcement of compliance.[167] There are then six other Parts, dealing with matters such as: interpretation and commencement of the Regulations; appeals; information; offences and civil penalties; the powers of the appropriate authorities in respect to regulators, etc.; and consequential amendments to other legislation.[168] The regulator of the EU ETS in relation to installations located in England and Wales will be the Environment Agency.[169]

---

[157]	See para. 4.2.1.2(1) above.
[158]	See S.I. 2003 No. 3311, Sched. 1 and reg. 2(1).
[159]	See para. 28.3 above, where this is set out in full.
[160]	Directive 03/87/EC, Annex I, para. 1; S.I. 2003 No. 3311, Sched. 1, Part 2, para. 2.
[161]	See 347 ENDS Report (2003). The same source gives a figure of 17,000 for the number of installations to be in subject to the Directive in both pre-2004 and post-2004 Member States.
[162]	Which is not necessarily the same as the 'appropriate authority' relevant to other regs (see S.I. 2003 No. 3311, reg. 11(7)).
[163]	See para. 14.6 above.
[164]	See HM Treasury PN3, *Protecting and Improving the Environment*, 10 December 2003. See also 347 ENDS Report (2003).
[165]	S.I. 2003 No. 3311, Part 2.
[166]	*Ibid.*, Part 3.
[167]	*Ibid.*, Part 4.
[168]	*Ibid.*, Parts 1 and 5–9.
[169]	*Ibid.*, reg. 2(1). See para. 4.2.1.3 above. In relation to installations located in Scotland, the regulator is SEPA and in relation to those located in Northern Ireland, it is the chief inspector duly constituted under the Pollution Prevention and Control Regulations (Northern Ireland) 2003, S.R. (N.I.) 2003 No. 46.

Part 2 of the Regulations deals with the matters covered by Arts 4–8 and 12 of the EU ETS Directive,[170] that is, applications for GHG Emissions Permits,[171] the conditions which must be included therein,[172] their variation,[173] their transfer[174] and their revocation.[175] There is a slight difference of terminology between the Directive and the Regulations, however, in that, whilst Art. 12 of the former refers to the transfer of 'allowances' between persons in the Community, etc., reg. 14 of the latter refers to the transfer of GHG Emission Permits between operators of installations. By reg. 10 of the Regulations, GHG Emissions Permits must contain, among other matters, such conditions as the regulator considers appropriate '... to ensure that the operator surrenders allowances equal to the annual reportable emissions from the installation within four months of the end of the scheme year[176] during which those emissions arose'.[177]

The allocation and issue of allowances is, as mentioned above, covered by Part 3 of the Regulations. Regulation 18 provides for the Secretary of State to develop a national allocation plan, in accordance with Arts 9(1) and 10 of, and Annex III to, the Directive, for the first phase (2005–2007) and for the second (2008–2012) and subsequent phases.[178] The Secretary of State is empowered to decide on three matters, that is, '(1) the total quantity of allowances to be allocated for the phase in question; (2) the allocation of allowances to each installation', including the number of those allowances to be issued in each scheme year in that phase; and (3), where there is more than one GHG Emissions Permit relating to an installation, the division of the allowances allocated to that installation under (2).[179] The final allocation decision for the first phase must be made by 1 October 2004.[180] The power to make this final allocation decision may, in relation to installations situated within the Principalities, only be reached by agreement with the devolved administrations.[181] However, there is a power for the Secretary of State to act in default of agreement, where this is necessary to ensure that the UK complies with its obligations under the Directive.[182] The Secretary of State published her draft national allocation plan on 19 January 2004, based on a reduction in carbon dioxide emissions of 20 per cent on 1990 levels by 2010, and requesting responses both to the draft plan and to other important aspects of the EU ETS, by 12 March 2004.[183] The reason for

---

[170]   See para. 28.3 above.
[171]   S.I. 2003 No. 3311, reg. 8.
[172]   *Ibid.*, reg. 10.
[173]   *Ibid.*, reg. 13.
[174]   *Ibid.*, reg. 14.
[175]   *Ibid.*, reg. 16.
[176]   That is, a year beginning with 1 January in the first, second or subsequent phases of the scheme (see S.I. 2003 No. 3311, reg. 2(1)).
[177]   See Directive 03/87/EC, Art. 12(3).
[178]   See para. 28.3 above.
[179]   S.I. 2003 No. 3311, reg. 19.
[180]   *Ibid.*, reg. 19(3)(a).
[181]   *Ibid.*, reg. 19(15)(a).
[182]   *Ibid.*, reg. 19(16) and 19(17).
[183]   The consultation documentation is available from www.defra.gov.uk.

the truncated consultation period is that, according to Art. 9(1) of the Directive, the national allocation plan must be 'published and notified to the Commission and to the other Member States by 31 March 2004 at the latest'. Such a period is, of course, far too short for a measure of such complexity as the allocation of allowances under the Directive,[184] even allowing for Defra's two-stage consultancy project on allocation methodologies conducted in summer 2003.[185]

The registry required to be set up by Art. 19 of the Directive[186] is provided for by reg. 20 of the Regulations and reg. 21 allows one or more operators of installations to make a joint application to the appropriate authority to form a pool[187] for either or both of the first or second phases of the scheme.

Regulations 22 to 25 make up the final substantive Part of the Regulations, Part 4, on the monitoring and enforcement of compliance. The basic rule, which is contained in reg. 22, is that, where a GHG Emissions Permit is in force, the regulator has the duty of taking such action under the Regulations 'as may be necessary to ensure that the monitoring and reporting conditions are complied with'. In addition to powers to determine the reportable emissions[188] from an installation[189] and to serve enforcement notices in respect of likely or actual contravention of monitoring and reporting conditions,[190] the Secretary of State has power to authorise suitable persons in writing to enter offshore installations[191] to check for compliance.[192]

A number of issues not discussed above, such as the tax treatment of the EU ETS mechanisms for corporation tax and VAT purposes, remain under discussion.[193]

28.5   Developments since transposition of the EU ETS Directive

By the spring of 2004, however, progress across the EU was somewhat behind schedule. Only five Member States met the 31 March deadline for notifying their allocation plans to the Commission.[194] Surprisingly, perhaps, the UK was not among

---

[184]   Cold comfort may, however, have been afforded to interested parties by the holding of a Defra-DTI-Devolved Administrations seminar on the consultation on 28 January 2004, at the National Exhibition Centre, Birmingham! (See Defra News Release, *UK Announces Consultation on Draft National Allocation Plan for the EU Emissions Trading Scheme*, 19 January 2003 [*sic*].)

[185]   See 341 ENDS Report (2003).

[186]   See para. 28.3 above.

[187]   *Ibid.*

[188]   Defined in S.I. 2003 No. 3311, reg. 2(1).

[189]   *Ibid.*, reg. 24.

[190]   *Ibid.*, reg. 23.

[191]   *Ibid.*, reg. 2(1).

[192]   The powers are framed by reference to the Offshore Combustion Installations (Prevention and Control of Pollution) Regulations 2001, S.I. 2001 No. 1091.

[193]   Mullins and Karas, *op. cit.*, p. 27. See para. 24.7 above.

[194]   Luxembourg and the Netherlands had filed in April 2004. By contrast, Latvia and Slovenia, two of the ten new Member States who joined the EU on 1 May 2004 (who had to file by 30 April), had already done so (see Minder, *Financial Times*, 20 April 2004, p. 10).

them, the DTI having estimated that emissions allocations would result in electricity price rises of 3 per cent for domestic consumers and 6 per cent for industrial ones.[195] Spain had also failed to file, being on the verge of an unexpected change of Government following the tragedy of the Madrid railway bombings. Spain was at that time, in any event, over target in emissions by 17 per cent and there was a deep division between generators as to the pros and cons of adopting either historical or current generating mixes as the basis of allocation.[196] Although Germany had filed in time, after a major political row within the governing coalition, its plan appeared to permit the older lignite-based plants to continue operating.[197] Indeed, there was some disquiet among commentators and governments at the plans which had been filed by the deadline. As one commentator said:

> Those that have published allocation plans have, mostly, erred on the side of leniency – so lenient that it is by no means impossible that the Commission will reject them out of hand.[198]

This cautionary note was echoed in a Joint Statement issued by Mrs Margaret Beckett, the Secretary of State for the Environment, Food and Rural Affairs,[199] and Sir Digby Jones, Director-General of the CBI,[200] on 9 June 2004, following the UK's own late filing of its plan:

> At stake is the credibility of emissions trading as a mechanism for engaging other key countries, including the United States. The UK national plan, recently submitted to Brussels, involves real effort by business that will help ensure that we not only meet our Kyoto target, but go beyond it.
> Many other EU countries face a stiffer challenge simply to meet their Kyoto targets. The Commission must see to it that their plans at the very least deliver on these commitments. British business sees it as vital that they are operating on level ground.
> The European Commission has an important job to do. The British Government and British business will be watching to make sure it does the job properly.[201]

The UK's final filing, in October 2004, revealed an increase of 7.6 per cent in the number of installations covered, but less than 3 per cent in the number of allowances.[202]

---

[195] The UK's late filing was, however, accepted by the Commission on 7 July 2004: IP/04/862. Those from Austria, Denmark, Eire, Germany, the Netherlands, Slovenia and Sweden were accepted on the same date. In the case of three countries, including the UK, approval was conditional upon the making of technical changes. In the case of Austria and Germany, this included the abandonment of an automatic reallocation mechanism and, in the case of the UK, the provision of adequate information on how new entrants would be able to participate.

[196] Lex, *Financial Times*, 19 April 2004, p. 20.

[197] *Ibid.*

[198] *Ibid.*

[199] See para. 4.2.1.2(1) above.

[200] See para. 2.2n above.

[201] See www.defra.gov.uk.

[202] Mrs Margaret Beckett, HC, 27 October 2004, col. 49WS.

But this did not deal with the allocation of that total between individual installations, a task the DTI was to complete in early 2005. This was the cause of some concern to British industry.[203]

On 7 December 2004, Mrs Beckett indicated that the government's 2010 emissions target was unlikely to be met.[204] The following day, Defra launched *Consultation on the Review of the UK Climate Change Programme*,[205] which revealed that the progress which had been made since 1990 had been mainly due to energy supply industry restructuring.[206]

28.6   Concluding observations

The EU ETS will surely prove to be of decisive importance for the future of the flexible mechanisms for the mitigation of climate change in the Kyoto Protocol. If it is seen to be successful, it may yet be of crucial importance in persuading those countries who have not yet done so to ratify the Kyoto Protocol. The vigour with which the EU ETS project has been pursued at the European level is no doubt greatly attributable to the vision and drive of the Environment Commissioner at that time, Mrs Margot Wallström, but, given the dramatically foreshortened timetable for the introduction of the EU ETS by 1 January 2005, it seems unlikely that such qualities, even if replicated in the relevant government ministers of the Member States, will be enough to ensure the smooth introduction of the EU ETS across Europe.[207]

The authors feel that it is appropriate to end this chapter by putting on record that, in early December 2004, it was understood that some Member States had still not yet filed their allocations with the Commission!

---

[203]   Harvey and Eaglesham, *Financial Times*, 27 October 2004, p. 2; Harvey, *Financial Times*, 29 October 2004, p. 5.

[204]   Eaglesham and Harvey, *Financial Times*, 8 December 2004, p. 4.

[205]   www.defra.gov.uk.

[206]   Para. 3.2. See paras. 2.4.1 and 6.4.3.1 above.

[207]   Trading in such permits through investment banks had, however, been taking place for some months: Harvey, *Financial Times*, 11 October 2004, p. 5.

PART V
PROVISIONAL ASSESSMENT

Chapter 29

The Current State of Play

29.1   The problem of measuring success

It will have been seen that the system of environmental taxation in the UK is to a large extent still in its exploratory stages. And major parts of it may need to be reformulated in the context of experience with the EU ETS. But this development should not obscure the fact that environmental taxation has had one spectacular success during the 1990s, namely as a carrot to both the private and business motorist to switch to more environmentally friendly fuels.[1] It is therefore appropriate to choose HM Customs and Excise as the source of what environmental taxation is all about:

> Environmental taxation must, however, meet the general tests of good taxation. It must be well designed, to meet objectives without undesirable side-effects; it must keep compliance costs down to a minimum; distributional impact, including regional impact and effects on social inclusion, must be acceptable; and care must be had to the implications for the international competitiveness of UK business.[2]

29.2   How do the UK taxes measure up to this standard?

It is therefore appropriate, not least in the context of the government need to find additional sources of revenue,[3] to ask whether the other environmental taxes under Customs' care measure up to this standard:

1   One cannot other than start by saying that it is difficult to envisage taxes which are more anti-UK business than aggregates levy or levies on fertilisers and pesticides.

  • One of the objectives of aggregates levy is to preserve virgin sources, and yet the tax is rebated if extracted aggregate is exported.[4]

---

[1]   See Chapter 22 above. But note the doubts expressed as to the likely effectiveness of the proposed degree of relief for bioethanol during the Committee Stage of Finance Bill 2004 (see para. 22.3n above).

[2]   *Hydrocarbon Oil Duty: Consultation on Duty Friendly Differentials for More Environmentally Friendly Rebated Oils* (9 July 2003), para. 4. 5.

[3]   Giles and Houlder, *Financial Times*, 29 October 2003, p. 3.

[4]   See paras 8.4.5.1 and 21.10 above.

- It has been admitted that a levy on nitrogenous fertiliser would harm British farmers unless measures taken by other Member States restored their competitive price position.[5]
- It is also clear from official research that the imposition of a pesticides levy would undermine English apple growers.[6]

2   It would be difficult to design a tax which is more complicated, or difficult to reconcile with environmental objectives, than climate change levy.[7] In *The State of the Nation 2004*,[8] the Institution of Civil Engineers described the background against which this tax has to operate:

- the reduction in coal generation from 35 per cent in 2004 to 17 per cent in 2010 being central to the attainment of the UK Government's target of a 20 per cent reduction in CO_2 emissions by the latter date; and
- with nuclear reducing from 22 per cent to 17 per cent over the same period; and
- with no realistic savings having been made in commercial and domestic heating and transport emissions; and
- a prospective reliance on gas for 60 per cent by 2010; when
- North Sea gas reserves will be reaching exhaustion, necessitating the expansion of imported gas storage facilities;[9] with
- renewables, energy efficiency and CHP intended to fill the gap against the background of increasing prices; but
- wind farm development hindered by planning delays and Defence Ministry objections; coupled with
- a need to extend the National Grid to link in sources of renewable energy.

Furthermore, the House of Commons' Environmental Audit Committee, in its Tenth Report of Session 2003–04, *Budget 2004 and Energy*, published on 11 August 2004,[10] drew attention to the fact that domestic carbon emissions was one of the two sectors (the other being transport) in which movement had been in the wrong direction.

---

[5]   See para. 21.9 above.
[6]   *Ibid.*
[7]   The reason is, of course, the policy constraint indicated in para. 21.4.2 above. In this context, it has to be said that the debates on the House of Lords' amendments to the Energy Bill in the House of Commons' Standing Committee B on 20 May and 8 June 2004 can only be seen as indicative of a reluctance on the part of the UK Government machine to be tied down to a realistic Parliamentary Kyoto progress control mechanism which would satisfy not only the main Opposition parties but also its own backbenchers.
[8]   Published on 15 June 2004: available from www.ice.org.uk.
[9]   This issue was gone into by the House of Lords European Union Committee's 17th Report of Session 2003–04, *Gas: Liberalised Markets and Security of Supply*, HL 105, published 24 June 2004. Interestingly, the Committee's most serious concern was the possibility of winter peak demand not being met in the shorter term (*ibid.*, para. 112).
[10] HC 490.

- A 'carbon gap' between the savings expected from the government's Climate Change Programme and the target to reduce carbon emissions by 20 per cent by 2010 had been identified by the Carbon Trust,[11] with:

  ○ difficulty in addressing energy efficiency against baseline forecasts which turn out to be inaccurate;[12] and
  ○ the Action Plan which did not address how to achieve this in relation to building procurement.[13]

- In the narrower field of domestic energy efficiency – where because of the 5 per cent VAT rate and exemption from climate change levy, the only means of fiscal incentivisation available has to take the form of tax concessions on the necessary capital expenditure:

  ○ the Committee were not convinced that the income tax[14] and VAT[15] concessions in the 2004 Budget would have any significant impact and expressed disappointment that a consultation which had taken place over two years had failed to come up with something more significant;[16] and
  ○ noted,[17] and (in the context of the Energy Efficiency Commitment) shared,[18] the concern of the Energy Saving Trust over the degree of commitment to achieve the 70 per cent savings expected by the EU from cavity wall insulation, requiring 4.5m installations.

3   With the exception of Lloyd George's Constitution-shattering 1910 land tax[19] and the various attempts to tax development value,[20] it would be difficult to name a tax which had been more ineffective in achieving its objective (recycling) than landfill tax.[21]

Between 1996/97 and 2002/03, municipal waste in England grew by 19.2 per cent. Although, within the context of that difference:

- the proportion being disposed of as landfill fell from 84 per cent to 75 per cent;

---

[11] *Ibid.*, para. 54.
[12] *Ibid.*, para. 56.
[13] *Ibid.*, para. 57.
[14] See para. 24.6 above.
[15] See para. 21.3.1 above.
[16] HC 490, para. 63.
[17] *Ibid.*, para. 65.
[18] *Ibid.*, para. 66.
[19] Finance (1909–10) Act 1910, ss.1–42, which included a mineral rights duty (*ibid.*, ss.20–24).
[20] That is, its nationalisation under Town and Country Planning Act 1947, Land Commission Act 1967, Finance Act 1974 (development gains tax) and Development Land Tax Act 1976. See also para. 27.2 above as to possible variants under current consideration.
[21] See, especially, para. 21.3.2 above.

- the proportion being incinerated with the production of electricity from waste rose from 6 per cent to 9 per cent; and
- disposal by recycling or composting rose from 7 per cent to 16 per cent;

there was still an increase of 6.49 per cent in the amount sent to landfill.

In *The State of the Nation 2004*,[22] the Institution of Civil Engineers highlighted the following:
- the growing waste generation problem 'is not going to go away';
- nobody accepting responsibility for coordinating the management of hazardous waste disposal, for which restrictions were one month off coming into force; and with
- non-domestic waste exceeding domestic;
- a serious risk of increased fly-tipping;
- the hostile attitude of the public towards planning applications for the necessary new facilities, at a time when commissioning takes five years and the first EU target for landfill reduction only six years away; and
- between 1,500 and 2,300 new facilities required by 2020.

4   Air passenger duty, although held out as an environmental tax, not only is not,[23] but seems to have had no effect on the increase in civil aviation, where international treaties prevent effective taxation being imposed.[24] Furthermore, it is instructive to note, in this context, that:

- the DFT's rejection[25] of the House of Commons' Environmental Audit Committee's Report[26] on *The Future of Air Transport* White Paper was itself criticised by the Committee[27] as 'poor and not of the standard we would normally expect',[28] leading to the conclusion that the government's aviation policy 'remains the most glaring example of the failure of government to put sustainable development at the heart of policy making';[29] and
- in *The State of the Nation 2004*,[30] the Institution of Civil Engineers commented that, in the long term, uncontrolled growth of air travel could not be sustained without the installation of additional surface transport facilities.

5   The House of Commons' Environmental Audit Committee, in its Tenth Report of Session 2003–04, *Budget 2004 and Energy*, published on 11 August 2004,

---

22   Published on 15 June 2004: www.ice.org.uk.
23   See para. 1.2.1.5(2) above.
24   See paras. 8.5 and 27.5 above.
25   That is, in May 2004.
26   See para. 27.5 above.
27   On 7 June 2004.
28   HC 623, *Conclusions and Recommendations*, paras. 1, 7.
29   *Ibid.*, para. 30.
30   Published on 15 June 2004: www.ice.org.uk.

highlighted the fact that carbon emissions from transport were moving in the wrong direction.[31] It represented one of the most serious problems being faced.[32] The following concerns were expressed by the Committee:

- that the voluntary agreement with European car-makers might not deliver the anticipated emissions reductions;[33]
- the attempt to set additional targets for low carbon vehicles in 2010 and 2020 had not been particularly helpful; and
- there had been insufficient government support for, or coordination between, the various bodies involved in dispensing capital grants and investment subsidies;[34]
- while the introduction of the Alternative Fuels Framework[35] was welcomed, major choices needed to be made and long-term investment in alternative fuels would only come from a longer-term strategy than that provided by this;[36]
- the fuel duty rises (which, for political reasons,[37] had been put on hold following the summer 2004 surge in international oil prices) should be implemented 'at the earliest opportunity';[38]
- disappointment was expressed at the failure of the DFT's White Paper, *Future of Transport*, to tackle the emissions issue, since the introduction of national road charging would take between ten and 15 years;[39] and
- furthermore, such an instrument would be far more 'blunt' than the current system under which it was possible to promote a shift[40] to low carbon vehicles through the use of differentials in the fuel duty and VED systems;[41] although
- according to the government's own evaluation, the current VED differentials were insufficient to promote behavioural change.[42]

The most serious criticisms of the Environmental Audit Committee were, however:

---

[31]   *Ibid.*, para. 25.
[32]   *Ibid.*, para. 45.
[33]   *Ibid.*, para. 29.
[34]   *Ibid.*, para. 33.
[35]   See para. 22.2 above.
[36]   www.ice.org.uk, para. 40.
[37]   *Ibid.*, para. 44.
[38]   *Ibid.*, para. 45.
[39]   *Ibid.*, para. 46.
[40]   Surprise was expressed, however, that the responsible minister did not appear to accept that one of the main objectives of the VED changes had been to influence buying decisions (*ibid.*, para. 51).
[41]   *Ibid.*, para. 46. It was also noted that the company car tax scheme afforded additional potential in this respect (*ibid.*, para. 47).
[42]   *Ibid.*, para. 51.

Environmental Taxation Law

- the absence of reliable data making assessment of the impact of the various measures and initiatives taken difficult to assess;[43] and
- concern that the reviews to be undertaken by HM Treasury might fail to exploit opportunities for more imaginative policy initiatives which might deliver:

  - step changes; rather than
  - steady incremental progress.[44]

29.3   Success stories

However, on the positive side, there is evidence that two minor tax changes are having a material environmental impact, that is,[45] against most forecasts,[46] the Central London congestion charge;[47] and, according to initial official research,[48] the revised company car (employee benefits) code.[49]

29.4   Comparing 1997 with 2004

One closing observation has, however, to be made.

Despite the *Statement of Intent on Environmental Taxation* published as a Annex to a Press Release on the occasion of its first Budget in July 1997, that the new government would 'over time ... reform the tax system to increase incentives to reduce environmental damage', New Labour's approach to environmental taxation has been distinctly strange.

While there are clearly some fiscal objectives, not least in the original objective of relieving general business costs through reductions in employers' national insurance contributions, an environmental objective was said to be behind each of the taxes introduced or amended under the Blair government's stewardship.

---

43   *Ibid.*, para. 78.
44   *Ibid.*, para. 79.
45   Perhaps fortuitously, these are in a sphere, transport, in which GHG emissions rose by 47 per cent between 1990 and 2002, as against a reduction of 15 per cent in other industries (see Houlder, *Financial Times*, 23 July 2004, p. 4).
46   See also Wolf, *Financial Times*, 6 August 2004, p. 17.
47   See para. 18.2 above. The DFT hopes to build on this both in the short term through offering financial inducements to other cities to adopt congestion charging and subsequently through the adoption of a nationwide satellite-based road charging system (as to which see para. 27.4 above) in place of current taxes (see *The Future of Transport*, Cm 6234, a White Paper published on 22 July 2004, and the criticism of it by Wright, *Financial Times*, 24–25 July 2004, p. 11.
48   Noted by the House of Commons' Environmental Audit Committee, in its Tenth Report of Session 2003–04, *Budget 2004 and Energy*, HC 490, published on 11 August 2004, para. 47.
49   See para. 23.2 above.

The question has to be asked as to why it was decided to proceed with the introduction of climate change levy before the publication of the 2003 Energy White Paper?[50]

A similar issue arises as to why the new threshold rate of £35 per tonne for landfill tax was announced not only before the public consideration of 10 Downing Street's *Waste Not, Want Not*[51] but also, apparently, the commissioning of the necessary research on comparative costings eventually published 15 months later.[52]

It is also, to put it at its lowest, very difficult to see how the introduction of aggregates levy had any economic or environmental justification,[53] not least in the context of the Government's road programme being the largest user of the end product.

Finally, the authors think it is relevant to draw attention to the fact that, shortly before this book went to press, the principal Minister responsible for climate change policy, Mrs Margaret Beckett, conceded that events were 'not presently on track' for the attainment of the government's main climate change target, cutting carbon dioxide emissions by 20 per cent by 2010.[54]

---

[50] See para. 21.6 above.
[51] See paras. 6.3.2 and 21.4.1 above.
[52] See para. 21.4.1 above.
[53] See para. 21.8 above.
[54] Eaglesham and Harvey, *Financial Times*, 8 December 2004, p. 4.

Postscript

We mention elsewhere that, over the period of writing the book, new governmental material has appeared almost weekly. In the main text, we have tried to reflect developments, legal and otherwise, as at December 2004, the day on which that year's Pre-Budget report was published.

The pace of change in the subsequent 10-month period, over which the manuscript has been reviewed and prepared for publication, has been somewhat less feverish. Nonetheless, there have been a number of important developments, some of which we think it is worth recording here.

The question of whether any or all of the levies discussed in the book satisfy the technical definition of a 'tax' has arisen in two, rather different, contexts.[1] One of these is the proposal for a plastic bag levy in Scotland. The projected levy, which is being promoted by the Liberal Democrat, Mike Pringle, MSP, seeks to 'place a levy on all plastic bags made wholly or in part of plastic ... [including] paper bags that have a laminate coating ... [the levy being] set at 10p, to be paid by the customer receiving the bag'.[2] The detail of the proposal is contained in the Environmental Levy on Plastic Bags (Scotland) Bill,[3] which was introduced in the Scottish Parliament on 17 June 2005. It is understood that, rather than relying on some non-existent tax-raising power, the Bill seeks to makes use of the Scottish Parliament's power to legislate in relation to 'local taxes to fund local authority expenditure'.[4] It remains to be seen what the Bill's fate will be.[5]

The other context in which the question of the definition of a 'tax' may become acute is in relation to the recent and somewhat humorously publicised dispute over whether foreign diplomats in London should be liable to pay the congestion charge.[6] If the 'charge' is a tax, rather than a charge or fee, properly so called, then it looks more like a direct tax than an indirect one and, as such, there must at least be a technical question as to whether there is a liability on diplomatic staff to pay it.[7] This

---

[1]   See paras 1.2.1.1, 7.2 and 7.3 above.

[2]   See *Environmental Levy on Plastic Bags (Scotland) Bill: Policy Memorandum* (Edinburgh: Scottish Parliamentary Corporate Body, 2005), para. 57 (available from www.scottish. parliament.uk).

[3]   Also available from www.scottish.parliament.uk.

[4]   See para. 4.2.2 above. The authors would like to thank David Cullum of the Non-Executive Bills Unit, Scottish Parliament, for clarifying this point.

[5]   For developments thus far, see 353 ENDS Report (2004) 53; 358 ENDS Report (2004) 33–34; 366 ENDS Report (2005) 37; and 368 ENDS Report (2005) 29.

[6]   See paras 1.4.2.4 and 18.2 above.

[7]   See Sherwood, *Financial Times*, 19 October 2005, p. 3. The report suggests that the US mission (and presumably also the German one, which is also refusing to pay) is seeking to rely on the 1961 Vienna Convention on Diplomatic Relations (presumably Art. 34 thereof,

question aside, however, there is a much more interesting issue as to whether, given the almost universal problem of traffic congestion in developed economies,[8] foreign diplomats nonetheless *ought* to pay it.[9]

The main institutional development, already envisaged in the text itself,[10] has been the amalgamation of the Inland Revenue and Customs and Excise in a new Department, called HM Revenue and Customs ('HMRC').[11] The creation of the new Department promises a greater accountability in environmental tax administration, as in tax administration generally. In this sense, the amalgamation of the two old Departments in a new statutory framework[12] is a very significant development. However, it does not affect the discussion in the main text; references to 'the Revenue' and to 'Customs' can simply be read as references to HMRC.[13]

February 2005, as is well known, saw the long-awaited entry into force of the Kyoto Protocol.[14] However, no sooner had one treaty became a reality than the likelihood of another taking effect dramatically receded: the European Constitutional Treaty was, of course, rejected in the Dutch and French referenda. Nonetheless, the significance of the rejection for Community environmental law and policy could be overstated. Bell and McGillivray conclude that: '[t]he environmental provisions of Article 2 EC[15] remain essentially unchanged, and although the environmental integration principle in Article 6 EC[16] has been moved to a position further back in the draft Constitution its legal status is probably unaffected'.[17]

A succession of both EU and domestic policy documents continues to appear. At the UK national level, there is, however, evidence of the effectiveness of the various levies and other economic instruments being subject to ever-closer scrutiny. Some salient features of these developments are as follows.

---

as to which see, for example, Luke T. Lee, *Vienna Convention on Consular Relations* (Durham N.C.: Rule of Law Press, 1966), ch. 18; Grant V. McClanahan, *Diplomatic Immunity: Principles, Practices, Problems* (London: Hurst, 1989), pp. 56–67). The way in which Art. 34 is drafted may also make it necessary, at some stage, to consider the concept of 'dues' in the present context.

[8] Traffic congestion, rather than emissions, being the chief justification for such levies.

[9] As, it would appear, are diplomats from the Russian, Japanese and Spanish missions.

[10] See para. 4.2.1.2(2) above.

[11] See www.hmrc.gov.uk.

[12] Commissioners of Revenue and Customs Act 2005, s. 4.

[13] See the useful discussion in the new edition of John Tiley's incomparable book (see John Tiley, *Revenue Law*, 5th edn (Oxford: Hart Publishing, 2005), pp. 62–65).

[14] See paras 8.3.1.4 and 20.2 above. See also the criticisms of Kyoto's target-based approach in House of Lords Economic Affairs Committee, *2nd Report. The Economics of Climate Change* (House of Lords Papers, Session 2005–2006, HL Paper 12–I) (see para. 4.2.1.1(3)n above).

[15] See para. 12.2.1 above.

[16] *Ibid.*

[17] See Stuart Bell and Donald McGillivray, *Environmental Law*, 6th edn (Oxford: Oxford University Press, 2005), p. 198.

January 2005 saw the launch of the European Commission's work programme[18] for 2005.[19] Although, by placing emphasis on the three 'key priorities' of 'prosperity', 'solidarity and security' and 'enhanced external responsibility', the programme has raised concerns about the relative weight being given to environmental questions,[20] a document containing seven environmental strategic themes is expected later in the autumn. Meantime, there have been developments in relation to a number of matters of detail, among them the adoption by the Council of Ministers of a common position on modernising the Waste Shipment Regulation[21] and agreement on the new directive which will amend European Parliament and Council Directive 1999/62/ EC on lorry charging (the 'Eurovignette' Directive).[22] Whilst it obviously does not obligate Member States to introduce lorry-charging schemes, the proposed amending Directive does insist on a common model and, as such, has had rather more success than the UK's plans for a national scheme for charging heavy lorries,[23] which were abandoned in July.[24] For the reasons discussed in the text,[25] this represents a rare, possibly unique, case of the scrapping of an environmental levy actually angering the industry sector affected by it (i.e. truckers).[26]

On the emissions front, the implementation of the EU ETS seems also to have lost much of the momentum that it once had. In April 2005, the government applied to the European Commission for the exclusion from the first phase of the EU ETS of parties to climate change agreements, together representing nearly a third of the installations in the UK NAP.[27] This followed an application to the CFI for review of the Commission's refusal to allow the UK to increase its emissions allocation.[28] Against this background, Defra's own refusal, in its *Consultation Paper on the EU Emissions Trading Scheme Phase II (2008–2012)*[29] unilaterally to extend the UK's implementation of the EU ETS so as to cover GHGs other than carbon dioxide seems a realistic one. It should be noted in this regard that the original 2003 regulations

---

[18] See para 4.3.2 above.
[19] See COM (2005) 15 final. The proposals contained in the work programme are subject to integrated impact assessment (see *Communication from the Commission on Impact Assessment*, COM (2002) 276 final). We are indebted to Helen Toner for drawing the Communication to our attention.
[20] See 361 ENDS Report (2005) 52–54; 365 ENDS Report (2005) 22–24.
[21] Council Regulation 259/93/EEC, (1993) OJ L30 1 (see para 12.2.5.1(4) above); see 367 ENDS Report (2005) 45–46.
[22] 364 ENDS Report (2005) 44 (see para. 12.2.6.4(2) above).
[23] See para 27.3 above.
[24] See 366 ENDS Report (2005) 44, which comments that, at the point at which the plans were dropped, £39 million had been spent on its development.
[25] See para 2.6n above.
[26] See 366 ENDS Report (2005) 44.
[27] See Defra, *UK Application for Approval of Temporary Exclusion from the EU Emissions Trading Scheme for Climate Change Agreements Participants*, 23 March 2005 (available from www.defra.gov.uk).
[28] See 363 ENDS Report (2005) 43–44.
[29] (London: Defra, 2005) (available from www.defra.gov.uk).

implementing the EU ETS in the UK, as well as the 2004 amendment regulations, have already been consolidated in the Greenhouse Gas Emissions Trading Scheme Regulations 2005.[30]

In the light of all this, it would be surprising if Parliamentary Select Committees were not beginning to ask ever-more searching questions about policy choices and their effectiveness. Although Budget 2005 did not make changes to the existing 'main' environmental taxes, namely, aggregates levy, climate change levy and landfill tax, it was the occasion of the Chancellor of the Exchequer's claim that, not only was the government reaching its environmental policy objectives, it was actually exceeding them.[31] In April 2005, the Environmental Audit Committee,[32] in a substantial report,[33] criticised the lack of new environmental taxation measures in Budget 2005.[34] Indeed, the Committee opined:

> ... It is reasonable to ask whether the Treasury has an environmental tax strategy at all. In previous reports, we have welcomed the Statement of Intent[35] as a strategic aim but criticised the Treasury for failing to put in place an adequate strategy to implement it – an omission which the Treasury's 2002 document, *Tax and the Environment: Using Economic Instruments*,[36] signally failed to rectify.[37]

The Committee also pressed the government to consider again proposals for a 'Green Tax Commission'.[38]

Since it has been an election year, 2005 has also seen, not one, but two, Finance Acts. Consistent with Budget 2005, however, neither of these contains any relevant legislative developments in relation to the three 'big' environmental taxes. Instead, there has been a certain amount of 'tinkering', via secondary legislation, with climate change levy and landfill tax exemptions. The Climate Change Levy (Miscellaneous Amendments) Regulations 2005[39] bring in changes, as from 22 July 2005, in the way in which exemptions and reliefs from the levy are administered.[40] Likewise,

---

30   S.I. 2005 No. 925 (see para 28.4 above).
31   See HM Treasury, *Budget 2005 – Investing for Our Future: Fairness and Opportunity for Britain's Hard-Working Families* (House of Commons Papers, 16 March 2005, HC 372), p. 151. The Budget Report refers to a generally favourable Cambridge Econometrics evaluation of climate change levy (*ibid.*, p. 156).
32   See para 4.2.1.1(3) above.
33   See House of Commons Environmental Audit Committee. *7th Report. Pre-Budget 2004 and Budget 2005 – Tax, Appraisal, and the Environment* (House of Commons Papers, Session 2004–2005, HC 261).
34   *Ibid.*, para. 10.
35   See para. 5.3 above.
36   Discussed in Chapter 5 above.
37   House of Commons Environmental Audit Committee, *op. cit.*, para. 31.
38   *Ibid.*, para. 36.
39   S.I. 2005 No. 1716.
40   The regulations amend the Climate Change Levy (General) Regulations 2001, S.I. 2001 No. 838, and 'elucidate' part of the Climate Change Levy (Registration and Miscellaneous Provisions) Regulations 2001, S.I. 2001 No. 7.

the Finance Act 2003, Sections 189 and 190, (Appointed Day) Order 2005,[41] make certain provisions in relation to supplies of taxable commodities, as regards basing the exemption for CHP source electricity on efficiency percentages[42] and supplies of taxable commodities when circumstances change.[43] Rules relating to CHP-source electricity are now consolidated in the Climate Change Levy (Combined Heat and Power Stations) Regulations 2005.[44] Various amendments to landfill tax exemptions have been made in the Landfill Tax (Site Restoration, Quarries and Pet Cemeteries) Order 2005[45] and the Landfill Tax (Amendment) Regulations 2005,[46] to take account of developments on the IPPC front.

Finally, amendments to the Capital Allowances (Environmentally Beneficial Plant and Machinery) Order[47] have been made, such as to replace existing definitions and augment the technologies that qualify for special direct tax treatment.[48] A similar order has been made in relation to energy-saving plant and machinery.[49]

For reasons of space, it has not been possible to do more than to highlight certain salient developments in this Postscript. The interested reader is referred to the various websites detailed in the footnotes for further information. (The Defra website is particularly detailed.)

J.S.,
J. de S.,
25 October 2005.

---

[41] S.I. 2005 No. 1713.

[42] Finance Act 2003, s. 189.

[43] *Ibid.*, s. 190.

[44] S.I. 2005 No. 1714.

[45] S.I. 2005 No. 725 (see *Business Brief* 9/05 (4 April 2005)).

[46] S.I. 2005 No. 759.

[47] S.I. 2003 No. 2076 (see para. 24.2 above).

[48] See the Capital Allowances (Environmentally Beneficial Plant and Machinery) (Amendment) Order 2005, S.I. 2005, No. 2423.

[49] See the Capital Allowances (Energy-saving Plant and Machinery) (Amendment) Order 2005, S.I. 2005, No. 2424 (amending the Capital Allowances (Energy-saving Plant and Machinery) (Amendment) Order 2001, S.I. 2001, No. 2541.

Select Bibliography

For reasons of space, the list below does not include every publication to which reference is made in the footnotes. Readers are referred to the relevant paras of the text for fuller citations.

Andersen, Mikael Skou, *Governance by green taxes: Making Pollution Prevention Pay* (Manchester: Manchester University Press, 1994)

Arnold, Brian and McIntyre, Michael, *International Tax Primer*, 2nd edn (The Hague: Kluwer, 1995)

Aust, Anthony, *Modern Treaty Law and Practice* (Cambridge: Cambridge University Press, 2000)

Baldwin, Robert and Cave, Martin, *Understanding Regulation – Theory, Strategy and Practice* (Oxford: Oxford University Press, 1999)

Baldwin, Robert, Scott, Colin and Hood, Christopher (eds), *A Reader in Regulation* (Oxford: Oxford University Press, 1998)

Barlow, Ian and Milne, David (eds), *CCH Tax Statutes and Statutory Instruments 2003–04 with Concessions and Statements of Practice* (Banbury: Croner CCH, 2003)

Barnett, Joel, *Inside the Treasury* (London: André Deutsch, 1982)

Bateman, Ian J. and Willis, Ken G., *Valuing Environmental Preferences – Theory and Practice of the Contingent Valuation in the US, EU and Developing Countries* (Oxford: Clarendon Press, 1999)

Beck, Peter, *Prospects and Strategies for Nuclear Power* (London: Earthscan 1994)

Beck, Peter and Grimston, Malcolm, *Double or Quits? The Global Future of Civil Nuclear Energy* (London: RIIA Briefing Paper, April 2002)

Beetham, David and Lord, Christopher, *Legitimacy and the European Union* (Harlow: Longmans, 1998)

Bell, Stuart and McGillivray, Donald, *Ball and Bell on Environmental Law: The Law and Policy relating to the Protection of the Environment*, 5th edn (London: Blackstone Press, 2000)

Berlin, D., *Droit Fiscal Communautaire* (Paris: Presse Universitaire Francaise, 1988)

Birnie, Patricia and Boyle, Alan, *International Law and the Environment*, 2nd edn (Oxford: Oxford University Press, 2002)

Blackstone, Sir William, *Commentaries on the Laws of England*, 16th edn, ed. C.J.T. Coleridge (London: Butterworths, 1825)

Brack, Duncan, Falkner, Robert and Goll, Judith, *The Next Trade War? GM Products, the Cartagena Protocol and the WTO* (London: RIIA Briefing Paper No. 8, September 2003)

———, *Trade and Environment in the WTO: after Cancun* (London: RIIA Sustainable Development Programme Briefing Paper No. 9, February 2004)

Browning, Lesley, *et al.*, *Revenue Law: Principles and Practice*, 22nd edn (London: Lexis Nexis Tolley, 2004)

Brownlie, Ian, *Principles of Public International Law*, 5th edn (Oxford: Oxford University Press, 1999)

Burnett-Hall, Richard, *Environmental Law* (London: Sweet and Maxwell, 1995)

Cameron, Peter, *Competition in Energy Markets* (Oxford: Oxford University Press, 2002)

Choi, Won-Mog, *Like Products in International Trade Law: Towards a Consistent GATT/WTO Jurisprudence* (Oxford: Oxford University Press, 2003)

Christiansen, Peter Munk (ed.), *Governing the Environment: Politics, Policy and the Organization in the Nordic Countries* (Copenhagen: Nordic Council of Ministers, 1996)

Confederation of British Industry Business Environment Brief, *Green Taxes: Rhetoric and Reality* (London: April 2002)

Craig, Paul and Búrca, Gráinne de, *EU Law: Text, Cases and Materials* 3rd edn (Oxford: Oxford University Press, 2003)

Daintith, Terence and Willoughby, Geoffrey (eds), *United Kingdom Oil and Gas Law* (London: Sweet and Maxwell, 1984)

Dam, Kenneth W., *The GATT: Law and International Economic Organization* (London: University of Chicago Press, 1970)

Davies: Principles of Tax Law, ed. Geoffrey Morse and David Williams, 4th edn (London: Sweet and Maxwell, 2000)

Dell, Edmund, *The Chancellors* (London: HarperCollins, 1996)

Easson, Alexander, *Taxation in the European Community* (London: Athlone Press, 1993)

Ellerman, A.D., Jacoby, H.D. and Decaux, A., *The Effects on Developing Countries of the Kyoto Protocol and Carbon Dioxide Emissions Trading* (Washington DC: World Bank Policy Research Paper 2019, 2000)

Esty, Daniel, *Greening the GATT: Trade, Environment and the Future* (Washington DC: Institute for International Economics, 1994)

European Environment Agency, *Environmental Taxes – Information and Environmental Effectiveness* (Copenhagen: European Environment Agency, 1999)

Evans, Andrew, *European Community Law of State Aid* (Oxford: Clarendon Press, 1997)

Farmer, Paul and Lyal, Richard, *EC Tax Law* (Oxford: Clarendon Press, 1994)

Fauchald, Ole Kristian, *Environmental Taxes and Trade Discrimination* (London: Kluwer Law International, 1998)

Friends of the Earth Briefing, *Time for a Sustainable Economy?* (London: Friends of the Earth, November 2003)

Gammie QC, Malcolm and de Souza, Jeremy, *Land Taxation* (Sweet and Maxwell looseleaf: up to Release 53)

Giddens, Anthony, *The Third Way: Renewal of Social Democracy* (Oxford: Polity Press, 1998)

————, *The Third Way and its Critics* (Cambridge: Polity Press, 2000)

Glaister, Stephen and Graham, Dan, *Transport Pricing and Investment in England* (London: The Independent Transport Commission, 2003)

Goulder, Lawrence, *Environmental Taxation and the 'Double Dividend': A Reader's Guide* (Cambridge, MA: National Bureau of Economic Research, 1994)

Greaves, Rosa, *EC Transport Policy* (Harlow: Pearsons Education, 2000)

Griffiths and Ryle on Parliament: Functions, Practice and Procedures, ed. by Blackburn, Robert and Kennan, Andrew, 2nd edn (London: Sweet and Maxwell, 2003)

Grubb, Michael, Brewer, Tom, Müller, Benito, Drexhage, John, Hamilton, Kistry, Sugiyama, Taishi and Aiba, Takao, *A Strategic Asssessment of the Kyoto-Marrakesh System: Synthesis Report* (RIIA Sustainable Development Programme Briefing Paper No. 6, June 2003)

Halsbury's Laws of England (London: Lexis Nexis)

Hamilton, Penny (ed.), *Environmental Taxes Handbook 2001–02* (London: Tolley LexisNexis, 2001)

Hancher, A. Leigh, Ottervanger, Tom and Slot, Piet Jan, *EC State Aids*, 2nd edn (London: Sweet and Maxwell, 1999)

Harlow, C. and Rawlings, R., *Law and Administration*, 2nd edn (London: Butterworths, 1997)

Hart, H.L.A, *The Concept of Law* (Oxford: Clarendon Press, 1961)

Heady C.J. *et al.*, *Study on the Relationship Between Environmental/Energy Taxation and Employment Creation* (Bath: University of Bath, 2000)

Healey, Denis, *The Time of My Life* (London: Michael Joseph, 1989)

Helm, Dieter, *Energy, The State, and the Market British Energy Policy since 1979*, revised edn (Oxford: Oxford University Press, 2004)

Herrera, Pedro M., 'Legal Limits on the Competence of Governments in Spain', in *Critical Issues in Environmental Taxation: International and Comparative Perspectives: Volume I*, ed. by Janet Milne, Kurt Deketelaere *et al.* (Richmond: Richmond Law and Tax, 2003), pp. 111–23

Hertzog, Robert, 'Environmental Fiscal Policy in France', in *Taxation for Environmental Protection. A Multinational Legal Study*, ed. by Sanford E. Gaines and Richard A. Westin (London: Quorum Books, 1991)

Hilson, Chris, *Regulating Pollution – A UK and EC Perspective* (Oxford: Hart Publishing, 2000)

Hood Phillips, O. and Jackson, Paul, *Constitutional and Administrative Law*, ed. by Paul Jackson and Patricia Leopold, 8th edn (London: Sweet and Maxwell, 2001)

Hughes, David *et al.*, *Environmental Law*, 4th edn (London: Butterworths, 2002)

Ison, Stephen, *Road User Charging: Issues and Policies* (Aldershot: Ashgate, 2004).

Jackson, John H., *World Trade Law and the Law of GATT* (Indianapolis: Bobbs-Merrill, 1969)

————, *The World Trading System*, 2nd edn (London: MIT Press, 1997)

Jans, Jan H., *European Environmental Law* (The Hague: Kluwer, 1995)

Karas, Jacqueline, *Russia and the Kyoto Protocol: Political Challenges* (London: RIIA, Sustainable Development Programme, 2004)

Krämer, Ludwig, *EC Treaty and Environmental Law*, 2nd edn (London: Sweet and Maxwell, 1995)

———— (ed.) *European Environmental Law* (Aldershot: Ashgate/Dartmouth, 2003)

Kratena, Kurt, *Environmental Tax Reform and the Labour Market: The Double Dividend in Different Labour Market Regimes* (Cheltenham: Edward Elgar, 2002)

Kynaston, David, *The Secretary of State* (Lavenham: Terence Dalton, 1978)

Lawrence, Duncan, *Waste Regulation Law* (London: Butterworths, 1999)

Limon, Donald, McKay, W.B. *et al.*, *Sir Thomas Erskine May's Treatise on the Law, Privileges, Proceedings and Usages of Parliament*, 22nd edn (London: Butterworths, 1997)

Lodge, Juliet, *The 1999 Elections to the European Parliament* (Basingstoke: Palgrave, 2001)

Lomborg, Björn, *The Skeptical Environmentalist: Measuring the Real State of the World* (Cambridge: Cambridge University Press, 2001)

Loveland, Ian, *Constitutional Law – A Critical Introduction*, 2nd edn (London: Butterworths, 2000)

Lyons, Timothy, *EC Customs Law* (Oxford: OUP, 2001)

Määttä, Kalle, *Environmental Taxes – From an Economic Idea to a Legal Institution* (Helsinki: Finnish Lawyers' Publishing, 1997)

McEldowney, J.F. and S., *Environmental Law and Regulation* (Oxford: Blackstone Press 2001)

McGoldrick, Dominic, *International Relations Law of the European Union* (London: Longman, 1997)

Mander, J. and Goldsmith, E., *The Case Against the Global Economy and for a Turn Toward the Local* (San Francisco: Sierra Club Books, 1996)

Mill, John Stuart, *Principles of Political Economy with Some of their Applications to Social Philosophy* (London: Longmans, Green, 1892)

Milne, Janet (ed.), *Critical Issues in Environmental Taxation* (Richmond: Richmond Law and Tax, c.2003)

Moe, Arild and Tangen, Kristen, *The Kyoto Mechanisms and Russian Climate Politics* (London: RIIA 2000)

Moore, Michael, *A World Without Walls – Freedom, Development, Free Trade and Global Governance* (Cambridge: Cambridge University Press, 2003)

Mourato, Susana and Pearce, David, *A Review of the London Economics Report* (DETR: London, 1998)

Mullins, Fiona and Karas, Jacqueline, *EU Emissions Trading: Challenges and Implications of National Implementation* (London: RIIA, 2003)

Neunraither, K. and Weiner, A., *European Integration after Amsterdam* (Oxford: Oxford University Press, 2000)

Noll, R. (ed.), *Regulatory Policy and the Social Sciences* (Berkeley, CA: University of California Press, 1985)

Nugent, Neill, *The Government and Politics of the European Union*, 5th edn (London: Macmillan, 2003)

————, *The European Commission* (Basingstoke: Palgrave, 2001)

Ocana, Carlos *et al.*, *Competition in Energy Markets* (OECD/IEA, 2001)

Ogus, Anthony: *Regulation – Legal Form and Economic Theory* (Oxford: Clarendon Press, 1994)

———— (ed.), *Regulation, Economics and the Law* (Cheltenham: Edward Elgar, 2001)

Opschoor, J.B. and Vos, H.B., *Economic Instruments for Environmental Protection* (Paris: OECD, 1989)

Organisation for Economic Co-operation and Development, *Environmental Policy: How to Apply Economic Instruments* (Paris: OECD, 1991)

————, *Environmental Taxes and Green Tax Reform* (Paris: OECD, 1997)

————, *Environmentally-Related Taxes in OECD Countries: Issues and Strategies* (Paris: OECD, 2001)

O'Riordan, Timothy (ed.), *Ecotaxation* (London: Earthscan, 1997)

Park, Patricia D., *Energy Law and the Environment* (London: Taylor and Francis, 2002)

Patterson, Walt, *Power from Plants* (London: Earthscan 1994)

————, *Transforming Electricity* (London: Earthscan 1999)

————, *The Electricity Challenge: Keeping the Lights On* Working Paper No. 1 (London: RIIA Sustainable Development Programme, March 2003)

————, *Generating Change: Keeping the Lights On* Working Paper No. 2 (London: RIIA Sustainable Development Programme, September 2003)

————, *Networking Change: Keeping the Lights On* Working Paper No. 3 (London: RIIA Sustainable Development Programme, June 2004)

Peterson, Luke Eric, *UK Bilateral Investment Treaty Programme and Sustainable Development: Implications of Bilateral Negotiations on Investment Regulation at a Time when Multilateral Talks are Faltering* (London: RIIA, Sustainable Development Programme Briefing Paper No. 10, February 2004)

Pigou, Arthur Cecil, *A Study in Public Finance*, 3rd edn (London: Macmillan, 1947)

————, *The Economics of Welfare* 4th edn (London: Macmillan, 1952)

Ricardo, David, *The Works and Correspondence of David Ricardo: Volume IV*, ed. Piero Sraffa with M.H. Dobbs (Cambridge: Cambridge University Press, 1951)

RICS Education Trust, *Can the waste planning system deliver?* (London: Royal Institute of Chartered Surveyors, 2004)

Roggenkamp, Martha, *et al.* (eds), *Energy Law in Europe: National, EU and International Law and Institutions* (Oxford: Oxford University Press, 2001)

Roth, Andrew, *A Self-Financing Road System* (London: IEA 1966)

Sands, Philippe, *Principles of International Environment Law*, 2nd edn (Cambridge: Cambridge University Press, 2003)

Sands, Philippe and Klein, Pierre, *Bowett's Law of International Institutions*, 5th edn (London: Sweet and Maxwell, 2001)

Schelling, Thomas C., *Incentives for Environmental Protection* (London: MIT Press, 1983)

Scott, Joanne, *EC Environmental Law* (London: Longmans, 1998)

Seneviratne, Mary, *Ombudsmen: Public Services and Administrative Justice* (London: Butterworths, 2002)

Shaw, Malcolm, *International Law*, 5th edn (Cambridge: Cambridge University Press, 2003)

Sheail, John, *An Environmental History of Twentieth-Century Britain* (Basingstoke: Palgrave, 2002).

Smith, Adam, *The Wealth of Nations*, ed. D.D. Raphael (London: David Campbell, 1991)

Smith, Stephen, *'Green' Taxes and Charges: Policy and Practice in Britain and Germany* (London: Institute for Fiscal Studies, 1995)

Smith, Bailey and Gunn on the Modern English Legal System, ed. S.H. Bailey, M.J. Gunn and D.C. Ormerod, 4th edn (London: Sweet and Maxwell, 2002)

de Smith, Woolf and Jewell, *Judicial Review of Administrative Action*, 5th edn (London: Sweet and Maxwell, 1995)

Stern, Jonathan, *Security of European Natural Gas Supplies* (London: RIIA, 2002)

Stevens, A. (with Stevens, H.), *Brussels Bureaucrats? The Administration of the European Union* (Basingstoke: Palgrave, 2001)

Sunkin, Maurice, Ong, David and Wight, Robert, *Sourcebook on Environmental Law*, 2nd edn (London: Cavendish, 2002)

Terra, B. and Wattel, P., *European Tax Law* (Amsterdam: Kluwer, 1993)

Thain, Colin and Wright, Maurice, *The Treasury and Whitehall* (Oxford: Oxford University Press, 1995)

Tietenberg, Tom, *Environmental and Natural Resource Economics*, 3rd edn (New York: HarperCollins, 1992)

Tiley, John, *Revenue Law*, 5th edn (Oxford: Hart Publishing, 2005)

Trebilcock, Michael J. and Howse, Robert, *The Regulation of International Trade*, 2nd edn (London: Routledge, 1999)

Tromans, Stephen and Fuller, Carl, *Environmental Impact Assessment: Law and Practice* (London: LexisNexis Butterworths, 2003)

Tudway, Robert H. *et al.*, *Energy Law and Regulation in the European Union* (London: Sweet and Maxwell, 1999)

Turner, Adair, *Just Capital: The Liberal Economy* (London: Macmillan, 2002)

Upton, David, *Waves of Fortune: the Past, Present and Future of the United Kingdom Offshore Oil and Gas Industries* (Chichester: John Wiley, 1996)

Vogel, Klaus on *Double Taxation Conventions* (The Hague: Kluwer Law, 1997)

Wade, E.C.S. and Bradley, A.W., *Constitutional and Administrative Law*, ed. by A.W. Bradley and K.D. Ewing, 11th edn (London: Longman, 1993)

Weatherill, Stephen and Beaumont, Paul, *EU Law*, 3rd edn (London: Penguin, 1999)

Weiss, Friedl, Denters, Erik and de Waart, Paul (eds), *International Economic Law with a Human Face* (The Hague: Kluwer Law International, 1998)

Werksman, Jacob (ed.), *Greening International Institutions* (London: Earthscan, 1996)

Williams, David, *EC Tax Law* (London: Longman, 1998)

Willis, J.R.M. and Hardwick, P.J.W., *Tax Expenditures in the United Kingdom* (London: Heinemann, 1978)

Wolf, Susan and Stanley, Neil, *Principles of Environmental Law*, 4th edn (London: Cavendish, 2003)

World Commission on Environment and Development, *Our Common Future* (Oxford: Oxford University Press, 1987)

World Trade Organization, *The Legal Texts: The Results of the Uruguay Round of Multilateral Trade Negotiations* (New York: Cambridge University Press, 1999)

Index

9 780367 604196